ALFRED BENZON SYMPOSIUM V

Transport Mechanisms
in Epithelia

TRANSPORT MECHANISMS IN EPITHELIA

Proceedings of the Alfred Benzon Symposium V
held at the premises of the Royal Danish Academy
of Sciences and Letters, Copenhagen, 10–14 September 1972

EDITED BY

H. H. USSING

N. A. THORN

Published by

Munksgaard, Copenhagen

In North and South America:

Academic Press, New York

SCANDINAVIAN UNIVERSITY BOOKS
DENMARK MUNKSGAARD *Copenhagen*
NORWAY UNIVERSITETSFORLAGET *Oslo, Bergen, Tromsø*
SWEDEN ESSELTE STUDIUM *Stockholm, Gothenburg, Lund*

Published simultaneously in the U.S.A. by Academic Press, Inc.
LIBRARY OF CONGRESS CATALOG CARD NO. 73–2087
ISBN 0-12-709550-0

Printed in Denmark by Andelsbogtrykkeriet, Odense
ISBN 87 16 01122 8

CONTENTS

LIST OF PARTICIPANTS

ANDERSEN, Bernhard
Institute of Biological Chemistry A
University of Copenhagen
13 Universitetsparken
2100 Copenhagen Ø
DENMARK

BERLINER, Robert W.
Department of Health,
Education & Welfare
National Institutes of Health
9000 Rockville Pike
Bethesda, Maryland 20014
USA

CRABBÉ, Jean
Université de Louvain
Departement de Physiologie
Unité d'Endocrinologie
6 Dekenstraat
3000 Louvain
BELGIUM

CURRAN, Peter F.
Yale University
Department of Physiology
New Haven, Connecticut 06510
USA

EDELMAN, Isidore S.
Cardiovascular Research Institute
University of California
San Francisco Medical Center
San Francisco, California 94122
USA

FRÖMTER, Eberhard
Max-Planck-Institut für Biophysik
6 Frankfurt a. Main – Süd 10
70 Kennedyallee
GERMANY

GIEBISCH, Gerhard
Yale University School of Medicine
Department of Physiology
333 Cedar Street
New Haven, Connecticut 06510
USA

GOTTSCHALK, Carl W.
University of North Carolina
Department of Medicine
Chapel Hill, North Carolina 27514
USA

HARVEY, William R.
Temple University
Department of Biology
Philadelphia, Pennsylvania 19122
USA

HESS THAYSEN, Jørn
Rigshospitalet
Department of Medicine P
9 Blegdamsvej
2100 Copenhagen Ø
DENMARK

HOGBEN, C. Adrian M.
The University of Iowa
College of Medicine
Department of Physiology & Biophysics
Iowa City, Iowa 52240
USA

HOSHIKO, Tomuo
Department of Physiology
Case Western Reserve University
Medical School
Cleveland, Ohio 44106
USA

HVIID LARSEN, Erik
Zoophysiological Laboratory A
University of Copenhagen
13 Universitetsparken
2100 Copenhagen Ø
DENMARK

KEYNES, Richard D.
Agriculture Research Council
Institute of Animal Physiology
Babraham, Cambridge
ENGLAND

KIRSCHNER, Leonard B.
Department of Zoology
Washington State University
Pullman, Washington 99163
USA

KOEFOED-JOHNSEN, Valborg
Institute of Biological Chemistry A
University of Copenhagen
13 Universitetsparken
2100 Copenhagen Ø
DENMARK

KRISTENSEN, Poul
Institute of Biological Chemistry A
University of Copenhagen
13 Universitetsparken
2100 Copenhagen Ø
DENMARK

LEAF, Alexander
Harvard Medical School
Massachusetts General Hospital
Boston, Massachusets 02114
USA

LEYSSAC, Paul P.
Institute of Experimental Medicine
University of Copenhagen
71 Nørre Allé
2100 Copenhagen Ø
DENMARK

LINDEMANN, Bernd
II. Physiologisches Institut
der Universität des Saarlandes
Medical Faculty
665 Homburg (Saar)
GERMANY

LOEWENSTEIN, Werner R.
University of Miami School of Medicine
Department of Physiology & Biophysics
P. O. Box 875, Biscayne Annex
Miami, Florida 33152
USA

MAETZ, Jean
Groupe de Biologie Marine
Département de Biologie du C. E. A.
Station Zoologique de Villefranche-
sur Mer (06)
FRANCE

MEARES, Patrick
University of Aberdeen
Department of Chemistry
Meston Walk, Old Aberdeen AB9 2UE
SCOTLAND

MOREL, François
Collège de France
Laboratory of Cellular Physiology
11, Place Marcelin Berthelot
Paris Ve
FRANCE

NEDERGAARD, Signe
Institute of Biological Chemistry A
University of Copenhagen
13 Universitetsparken
2100 Copenhagen Ø
DENMARK

NIELSEN, Robert
Institute of Biological Chemistry A
University of Copenhagen
13 Universitetsparken
2100 Copenhagen Ø
DENMARK

ORLOFF, Jack
Department of Health,
Education & Welfare
National Institutes of Health
Laboratory of Kidney & Electrolyte
Metabolism
National Heart & Lung Institute
9000 Rockville Pike
Bethesda, Maryland 20014
USA

OSCHMAN, James L.
Northwestern University
Department of Biological Sciences
Evanston, Illinois 60201
USA

PETERSEN, Ole Holger
Institute of Medical Physiology C
University of Copenhagen
71 Rådmandsgade
2200 Copenhagen N
DENMARK

SACHS, George
The University of Alabama Medical
Center
Department of Medicine
Division of Gastroenterology
1919 7th Avenue South
Birmingham, Alabama 35233
USA

SCHMIDT-NIELSEN, Bodil
The Mount Desert Island
Biological Laboratory
Salisbury Cove,
Maine 04672
USA

SCHOFFENIELS, E.
University de Liège
Laboratoire de Biochemie
17, Place Delcour
4000 Liège
BELGIUM

SCHULTZ, Stanley G.
Department of Physiology
School of Medicine
University of Pittsburgh
Pittsburgh, Pennsylvania 15213
USA

SKADHAUGE, Erik
Institute of Medical Physiology A
University of Copenhagen
28 Juliane Maries Vej
2100 Copenhagen Ø
DENMARK

STEN-KNUDSEN, Ove
Department of Biophysics
University of Copenhagen
28 Juliane Mariesvej
2100 Copenhagen Ø
DENMARK

STEVENS, Charles E.
New York State Veterinary College
Department of Physiology,
Biochemistry & Pharmacology
Cornell University
Ithaca, N. Y. 14850
USA

THORN, Niels A.
Institute of Medical Physiology C
University of Copenhagen
71 Rådmandsgade
2200 Copenhagen N
DENMARK

TOSTESON, Dan C.
Duke University Medical Center
Departments of Physiology &
Pharmacology
Durham,
North Carolina 27710
USA

ULLRICH, Karl J.
Max-Planck-Institut für Biophysik
6 Frankfurt a. Main 70,
70 Kennedyallee
GERMANY

Ussing, Hans H.
Institute of Biological Chemistry A
University of Copenhagen
13 Universitetsparken
2100 Copenhagen Ø
DENMARK

Wall, Betty
Northwestern University
Department of Biological Sciences
Evanstone, Illinois 60201
USA

Van Bruggen, John T.
University of Oregon Medical School
Department of Biochemistry
3181 S. W. Sam Jackson Park Road
Portland, Oregon 97201
USA

Voûte, C. L.
Medizinische Universitetsklinik
Laboratorium für Experimentelle
Nephrologie
Bürgerspital Basel
4000 Basel
SWITZERLAND

Whittembury, Guillermo
Centro de Biofisica & Bioquimica
Instituto Venezolano de Investigaciones
Cientificas,
IVIC, P. O. Box 1827
Caracas
VENEZUELA

Windhager, Erich E.
Cornell University
Department of Physiology
Medical College
1300 York Avenue
New York, N. Y. 10021
USA

Zadunaisky, José A.
Yale University School of Medicine
Department of Ophthalmology and
Visual Science and Department of
Physiology
New Haven, Connecticut 06510
USA

Zerahn, Karl
Institute of Biological Chemistry A
University of Copenhagen
13 Universitetsparken
2100 Copenhagen Ø
DENMARK

I. Welcome
II. The Effect of Antidiuretic Hormone on the Transport Paths through the Isolated Toad Skin

Hans H. Ussing

Mr. Chairman, Ladies and Gentlemen:

My first duty at the opening of this symposium is to express my thanks to the participants who have come from all over the world to discuss epithelial transport with us here in Copenhagen. Furthermore, my sincere thanks are due to the Alfred Benzon Foundation for making this meeting possible. I know how well the Foundation has arranged previous meetings and I am sure that the officials of the Foundation will again surpass themselves to make this meeting a success. The rest is up to us.

I find it appropriate at the outset to say a few words about the choice of participants. In a rapidly developing field like ours it would be impossible to gather all of those whom the organizing committee would have liked to see. We might have tried then to bring together people who have never met before, people with different backgrounds and different orientations. Such a gathering could certainly have been stimulating, but much time would have been wasted on semantics. Instead, we have chosen to gather people who speak the same scientific language, who know each other and each others work. This means that we can skip preliminaries and immediately start discussing experiments and interpretations. For many of us it means continuing discussions which started 1, 5, 10 or maybe 20 years ago.

Epithelia may be considered as membranes of the second order, that is, membrane – covered cells arranged in membranelike sheets. Indeed, in the past, quite satisfactory descriptions of epithelial function have been based on a model which is essentially a "black box membrane" provided with certain permeability and active transport properties. The most spectacular advances in recent years have been concerned with attempts to peep into the black box. There is hope that we can, to some extent, resolve the events

»Transport Mechanisms in Epithelia«, Munksgaard, Copenhagen.

in the epithelium, in cell membrane processes, permeation processes in leaky seals between cells and through cell junction, and finally, the complications arising from flow and diffusion in the interspace system.

The philosophy behind this meeting is the assumption that the mechanisms underlying transport of substances across different epithelia are basically the same, the epithelia differing only with respect to the combinations and intensities of the basic mechanisms. Thus we may hope that by studying many types of epithelia, we may find some where one or the other mechanism is either dominating or absent. This is the reason why we have brought together scientists who work with such diverse tissues as frog skin, insect intestine, and kidney tubule.

If you watch the program closely, you will notice that the first three speakers are not epithelial-transport workers of the usual kind. They come from friendly neighbouring territories. Professor Loewenstein will tell us about recent developments in the field of coupling adjacent cells, whereas Professors Meares and Tosteson will act as our teachers and consultants with respect to transport through "simple" model systems.

The program is arranged in such a way that we begin with systems which are mainly of value as model systems (frog skin, toad bladder, etc.) and end with discussion of transport in kidney tubules. This sequence is didatic, but it is also an expression of an ambivalent attitude towards the scientist's choice of object. On the one hand we feel the right to pursue any problem which is intellectually challenging, but on the other hand, we feel a unique satisfaction if what we do can be of practical use, for instance in medical science. Thus, it is our hope that some of the results of our work will not only satisfy our curiosity, but that they will be handed on to the clinicians.

Effect of ADH on Transport Paths in Toad Skin

For some time we have been trying to clarify the transport paths through amphibian skin. During the last decade it has become increasingly clear that passive fluxes through epithelia can be both cellular and extracellular. We therefore felt the need for a fresh study of the sites of increase in permeability, when a toad skin is treated with antidiuretic hormone (ADH). In all experiments the dose was 80 I. U. per liter inside solution. As a tracer for extracellular path, we chose sucrose. The experiments were performed with skin halves, so that influx and outflux could be measured simultaneously under practically identical conditions. The concentration of sucrose was in all experiments 10 mmoles per liter on both sides and C^{14} labelled sucrose in tracer amounts was added to one side.

Table I. shows two such experiments with NaCl Ringer's solution as the medium on both sides of the skins. A total of 6 were performed all with similar results. There is hardly any effect of the hormone. In experiments,

Table I. Effect of ADH on sucrose fluxes in toad skin. NaCl-Ringer both sides.

date		P. D. max. mV	P_{in} cm $\times$ sec^{-1} $\times$ 10^{-8}	flux ratio (P_{in}/P_{out}) (from paired skins)
		Effect of ADH on sucrose fluxes in toad skin NaCl-Ringer both sides		
9–5 72	Control	66	5.8	3.4
	"	66	9.6	5.1
	ADH	90	9.1	2.1
	"	82	11.0	2.4
10–5 72	Control	70	13.6	5.7
	"	62	12.9	3.5
	ADH	83	11.6	2.2
	"	84	9.0	1.4

»Transport Mechanisms in Epithelia«, Munksgaard, Copenhagen.

not shown here, the outflux increased slightly, whereas the influx did not change. It is evident from the column showing the flux ratios that the influxes are larger than the outfluxes. Influxes significantly larger than the outfluxes were found in all of a total of 12 experiments. (In 6 of these experiments ADH was not added).

Each experiment comprised 4 experimental periods each one lasting one hour. Of course it is conceivable that these flux ratios indicate active transport of sucrose. It is, however, much more likely that the phenomenon is related to the non-specific transport of substances in the gall bladder, which according to Diamond (1962) is a consequence of the intercellular flow maintained by the NaCl gradient in the interspaces.

Table II. Effect of ADH on sucrose fluxes in toad skin. NaCl-Ringer inside, 1/10 NaCl-Ringer outside.

date		P. D. max. mV	P_{in} cm × sec^{-1} × 10^{-8}	flux ratio (P_{in}/P_{out}) (from paired skins)
		\multicolumn{2}{c}{Effect of ADH on sucrose fluxes in toad skin}	
1–5 72	Control	73	18.2	16.5
	"	73	17.3	12.4
	ADH	83	17.6	13.5
	"	92	17.7	8.4
2–5 72	Control	97	10.2	17.0
	"	95	13.1	10.9
	ADH	98	16.5	18.3
	"	105	15.9	11.4

Table II shows examples from a series of 7 experiments where the conditions were exactly like the ones mentioned above, with the exception that the outside medium was 1/10 Ringer's solution, so that there was an osmotic gradient across the skin. Again there was no significant effect of ADH on the sucrose fluxes, although the skins do respond to the hormone as evidenced by the increase in potential difference. The flux ratios are even higher than in the experiments with Ringer's solution on both sides. The very high flux ratios were found in all 7 experiments except one. The exception was an extremely tight skin where the flux ratios were around 2. The high flux ratios in these series most likely are due to solvent drag in the interspaces, brought about by the osmotic transport of water.

If we now consider a smaller test molecule like glycerol, the results are quite different. Table III shows examples from a series of experiments with NaCl Ringer's solution on both sides. Like before, the test substance is present in 10 millimolar solution on both sides. The total number of experiments is 9. In every case (that is 9 influx and 9 outflux experiments), ADH gave rise to a large increase in glycerol flux. The flux ratios were mostly in the vicinity of 1.

Table III. Effect of ADH on glycerol fluxes in toad skin. NaCl-Ringer both sides.

date		P. D. max. mV	P_{in} cm $\times$ sec^{-1} $\times$ 10^{-8}	flux ratio (P_{in}/P_{out}) (from paired skins)
	\multicolumn{4}{l}{Effect of ADH on glycerol fluxes in toad skin NaCl-Ringer both sides}			
17–4 72	Control	60	16.0	1.1
	"	50	20.6	1.2
	ADH	69	58.3	0.6
	"	60	227.8	1.1
20–4 72	Control	60	26.9	1.1
	"	50	30.0	1.1
	ADH	67	40.3	0.8
	"	64	56.9	0.8

Four double experiments, with glycerol as the test substance and 1/10 Ringer's solution as the outside medium, gave similar results (Table IV). Here ADH also increases the glycerol permeability, but despite the osmotic inward flow of water, the flux ratios remain low compared to those found in the corresponding experiments with sucrose (Table II). It is tempting to conclude from the foregoing that sucrose and glycerol follow different pathways. Sucrose passes mainly or exclusively through one pathway where there is a pronounced solvent drag and one which is not acted upon by ADH. Glycerol follows a path where diffusion is strongly augmented by ADH. If we maintain the view that sucrose follows an extracellular pathway, the obvious conclusion is that ADH somehow opens a cellular pathway for glycerol.

The large difference in flux ratios for sucrose and glycerol in experiments with 1/10 Ringer's solution outside and Ringer's solution inside may also speak in favour of the hypothesis that sucrose follows an intercellular path

Table IV. Effect of ADH on glycerol fluxes in toad skin. NaCl-Ringer inside, 1/10 NaCl-Ringer outside.

date		Effect of ADH on glycerol fluxes in toad skin NaCl-Ringer inside, 1/10 NaCl-Ringer outside		
		P. D. max. mV	P_{in} $cm \times sec^{-1} \times 10^{-8}$	flux ratio (P_{in}/P_{out}) (from paired skins)
5–4 72	Control	70	33.6	1.7
	"	68	35.6	1.6
	ADH	76	45.1	1.2
	"	80	68.1	1.4
6–4 72	Control	74	16.7	1.6
	"	72	20.9	1.6
	ADH	85	33.9	1.0
	"	90	82.1	1.0

where the solvent drag is exerted over a considerable length of passageway, whereas glycerol may mainly pass through the cell membrane so that any solvent drag can only be exerted over a short distance.

The flux ratio equation for a case like the one under consideration can be expressed as follows (Koefoed-Johnsen & Ussing 1953, Andersen & Ussing 1957):

$$(1) \quad \ln (M_{in}/M_{out}) = \frac{\triangle w}{D} \int_0^{x_0} \frac{1}{A} \, dx \qquad , \text{where } M_{in} \text{ and } M_{out} \text{ are the}$$

inward and outward fluxes, $\triangle w$ the volume flow of water per unit area, D the free diffusion constant of the substance, A is the fraction of the area which is available to flow for a given value of x (the distance in the membrane from one surface) and x_0 the total thickness of the membrane. It is quite apparent that the length of the pore has a decisive effect on the flux ratio. If we compare two substances with different diffusion coefficients passing through the same pore, we obtain

$$(2) \quad \ln(M'_{in}/M'_{out}) = (D''/D') (\ln(M''_{in}/M''_{out}) \qquad , \text{or}$$

$$(3) \quad M'_{in}/M'_{out} = (M''_{in}/M''_{out}) \, D''/D'$$

Now the free diffusion coefficient for glycerol is roughly twice that for sucrose. Thus the flux ratio for sucrose should be that for glycerol to the second power. Our materiel is hardly good enough for a safe comparison, but it does seem as if the difference in flux ratios is considerably larger than would be expected if the two substances followed the same path.

As we have shown several years ago (Andersen & Ussing 1957) ADH increases the permeability for acetamide and thiourea in the toad skin and we have now seen that ADH increases the permeability to glycerol. Thus, in this preparation the permeability increase is not as restricted as it is in the toad bladder (Leaf 1960), where the permeabilities for thiourea, glycerol, and chloride remain unaffected by the hormone. Even the outflux of sodium remained the same after ADH treatment in the toad bladder, and the same was found in the case of the frog skin by Morel & Bastide (1965) and Bastide & Jard (1968). From this it was concluded that the influx and outflux paths for sodium are not the same.

Table V shows that the situation is different in the toad skin. This table shows typical experiments from a series of experiments where sodium outflux, urea outflux, and short-circuit current were measured simultaneously on the

Table V. Effect of ADH on fluxes in toad skin. NaCl-Ringer on both sides.

date			Effect of ADH on fluxes in toad skin NaCl-Ringer on both sides	P_{out} (cm $\times$ sec^{-1} $\times$ 10^{-8})	
		μAmps $\times$ cm^{-2}		Na	Urea
	Control	13.0		9.3	26.5
	"	11.4		10.0	24.4
26–6	ADH	48.9		15.4	131.5
69	"	32.9		13.4	48.9
	Control	9.4		18.4	34.4
24–7	"	8.6		19.9	34.4
69	ADH	49.6		60.0	108.0
	"	41.4		57.3	59.0

same skin. It can be seen that all 3 quantities are increased markedly by ADH. A total of 11 experiments were performed. A distinct increase in Na-outflux was seen in 8 of these and an increase in urea flux in 9.

We have just completed a series of experiments to test the effect of ADH on the chloride flux in toad skin. A total of 7 experiments were performed (Conditions: short-circuit, NaCl Ringer's solution on both sides, influx and outflux determined in parallel experiments on skin halves). In 4 of the experiments there was no effect of the ADH, but in 3 experiments there was a distinct increase in chloride flux. In this context it may be worth mentioning that Kristensen (1970), in experiments with frog skin, failed to obtain any increase in chloride flux with ADH, but did obtain a strong response with theophyllin. Thus, whether or not one obtains a permeability increase for a given species in a given epithelium may be a quantitative rather than a qualitative problem.

The mechanism of the ADH effect is still an open question. One may speculate that the activation of the cyclic AMP system leads to the phosphorylation of a membrane protein. In itself such a phosphorylation might make the membrane more hydrophilic and thus increase its permeability to water solutable substances. Or, the phosphorylation might give rise to conformational changes.

Whatever primary effect the hormone has, there are still a number of possible secondary effects which need further study. We know that ADH induces an increase in the volume of the outermost living cell layer in the frog skin and that, at the same time, the interspaces are expanded (Voûte & Ussing 1968). The two changes may in turn affect the permeability of the intracellular and the extracellular pathway, respectively. Experiments are now in progress to study these secondary effects of the ADH action.

REFERENCES

Andersen, B., & Ussing, H. H. (1957) Solvent drag on non-electrolytes during osmotic flow through isolated toad skin and its response to antidiuretic hormone. *Acta physiol. scand. 39*, 228–239.

Bastide, F., & Jard, S. (1968) Actions de la noradrénaline et l'ocytocine sur le transport actif de sodium et la perméabilité à l'eau de la peau de grenouille. Rôle du 3,5'-AMP cyclique. *Biochim. biophys. Acta (Amst.) 150*, 113–123.

Diamond, J. M. (1962) The mechanism of the solute transport by the gall bladder. *J. Physiol. (Lond.) 161*, 474–502.

Koefoed-Johnsen, V., & Ussing, H. H. (1953) The contribution of diffusion and flow to the passage of D_2O through living membranes. Effect of neurophypophyseal hormones on isolated anuran skin. *Acta physiol. scand. 28*, 60–76.

Kristensen, P. (1970) The action of theophylline on the isolated skin of the frog (Rana temporaria). *Biochim. biophys. Acta (Amst.) 203*, 579–582.

Leaf, A. (1960) Some actions of neurohypophyseal hormones on a living membrane. *J. gen. Physiol. 43*, 175–189.

Morel, F., & Bastide, F. (1965) Action de l'ocytocine sur la composante active du transport de sodium par la peau de grenouille. *Biochim. biophys. Acta (Amst.) 94*, 609–611.

Voûte, C. L., & Ussing, H. H. (1968) Some morphological aspects of active sodium transport. The epithelium of the frog skin. *J. Cell Biol. 36*, 625–638.

DISCUSSION

KEYNES: Can you say anything about fixed charges in the extracellular pathway. It has sometimes seemed to me that one might want to regard that extracellular pathway as a specific channel for the anions to get through, in which case it would presumably have positive charges.

USSING: Yes, that is quite possible, but we have been unable to demonstrate any electroosmosis in the skin, so we do not know the size of the charges.

THORN: I would like to ask whether there is really any evidence that the hormone affects other membranes than the lumen-facing one. I know that there is some electronmicroscopial evidence for increased size or enlargement of the spaces. But there has been quite some criticism.

LEAF: The evidence that has been accumulated on toad bladder as far as the action of ADH goes would seem to be accounted for entirely by an action on the outward-facing or mucosal membrane of the bladder epithelium. The evidence from the micropuncture work of Dr. Civan & Frazier (1968) indicated that the resistance across this membrane is diminished. The evidence from the cell volume measurements made recently (Macknight *et al.* 1970) would indicate that there is an increase in sodium in this layer of cells as a result of the action of ADH. And if one couples that finding together with the increase in sodium movement across this layer of cells it seems that the logical explanation is that more sodium can enter the cell in the presence of ADH. The change in the intercellular spaces, however, that are observed, Civan & Di Bona felt could be accounted for by an action of ADH on the smooth muscle layer below the epithelium and that with relaxation of the smooth muscle layer the spaces between cells will open up so that this was an effect not necessarily associated with movement of water across the epithelium, but simply an effect of tension or the amount of stretching there was in the mucosal layer of cells. They were able to demonstrate these actions of ADH even in the absence of an increase in solute or water movement across the epithelium, although physiologically they would occur with transport. So

---

Civan, M. M. & Frazier, H. S. (1968) The site of the stimulatory action of vasopressin on sodium transport in toad bladder. *J. gen. Physiol. 51,* 589–605.

Macknight, A. C., Leaf, A., & Civan, M. M. (1970) Effects of vasopressin on the water and ionic composition of toad bladder epithelial cells. *J. Membrane Biol. 6,* 127–137.

the physiological effects we have seen in relation to increased permeability are best accounted for by change in permeability of the mucosal or outward-facing membrane of the bladder epithelium.

ORLOFF: Dr. Leaf's statement that an action of vasopressin on smooth muscle accounts for distension of the intercellular spaces is contrary to our conclusions. Dr. Ganote in our laboratory (Ganote *et al.* 1968) investigated this problem in the isolated collecting tubule of the rabbit. This is a pure epithelial preparation devoid of muscle. It consists of a single layer of cells surrounded by a basement membrane. ADH, as in toad bladder, increases water flow along an osmotic gradient. Under these circumstances the cells swell and marked dilatation of the intercellular spaces occurs. If the solutions bathing the luminal and serosal surfaces are isotonic, no net water movement occurs when ADH is added to the serosal bathing medium nor is there dilatation of the intercellular channels. ADH, however, is exerting an effect on the permeability of the luminal membrane as evidenced by an increase in the diffusional permeability to water. We have concluded that dilatation of the intercellular spaces occurs in consequence of osmotic flow of water from the cell into the intercellular channels. Grantham *et al.* (1971) in a recent study reexamined the phenomenon in the toad bladder and has reached conclusions similar to ours.

EDELMAN: With respect to Dr. Thorn's question, the concept of a simple two-barrier system with independent properties may not be adequate to describe the control mechanisms. In the steady-state the fluxes across each barrier are equal. In the transient state there will be discrepancies in these fluxes. The mechanism of coupling between the two barriers may be somewhat more complicated than we would like to believe. Similar results to those quoted by Dr. Leaf were obtained by Handler *et al.* (1972). Lipton and I (1971) did a similar study in a slightly different subspecies of toad, the Columbian toad. We found no significant change in intracellular sodium concentration after vasopressin. We concluded that there was a bipolar effect, involving a mechanism of coupling between the apical and basal-lateral boundary which hasn't been clearly defined. In any case I think that the issue

Ganote, C. E., Grantham, J. J., Moses, H. L., Burg, M. B., & Orloff, J. (1968) Ultrastructural studies of vasopressin effect on isolated perfused renal collecting tubules of the rabbit. *J. Cell Biol. 36*, 355–367.

is still complex. The conductance changes may be needed for the increase in active sodium transport to occur but additional effects at the basal-lateral boundary may also be determinants of the final steady-state response.

Grantham, J., Cuppage, F. E., & Fanestil, D. (1971) Direct observation of toad bladder response to vasopressin. *J. Cell Biol. 48*, 695–699.

Handler, J. S., Preston, A. S. & Orloff, J. (1972) Effect of aldosterone on the sodium content and energy metabolism of epithelial cells of the toad urinary bladder. *J. Steroid Biochem. 3*, 137–141.

Lipton, P. & Edelman, I. S. (1971) Effects of aldosterone and vasopressin on electrolytes of toad bladder epithelial cells. *Amer. J. Physiol. 221*, 733–741.

Cell Coupling

Werner R. Loewenstein

COUPLING INTERCELLULAR PASSAGEWAYS

The surface membranes at the junctions of epithelial cells are so organized as to form molecular passageways between cells (Loewenstein 1966). We know as yet little about the structure of the passageways, but their existence is well assured by electrical measurements with intracellular probes and by measurements with tracer molecules injected into the cells. These measurements define each passageway unit as a matched pair of membrane regions of high permeability *(junctional membranes)*, one region on either side of the cell junction, circumscribed by a seal *(junctional seal)* insulating the interior of the connected cell system from the exterior (Loewenstein 1966) (Fig. 1, top). As a result, molecules can flow from one cell interior to another without appreciable leakage to the exterior. Many such independent, insulated passageway units make up a junction; in some epithelia at least, there appears to be enough extracellular space between the units to permit an extracellular flow of inorganic ions through the junction perpendicular to and independent of the flow between the cell interiors (Frömter & Diamond 1972). The passageways have just been traced electronmicroscopically in *Chironomus* salivary glands with the aid of peroxidase injected into the cells by W. Larsen (Fig. 2, see figure insert opposite p. 24).

Junctional passageways form rapidly where the cells, that can make them, come into contact. Fully coupling passageway systems arise within 1–4 min. of bringing mammalian epithelial liver and lens cells into contact in tissue culture, and within 20–40 sec. of micromanipulating together embryonic newt cells. A large part, perhaps all, of the cell surface membrane in these cells is capable of passageway formation: When a junction is broken by

Department of Physiology and Biophysics, University of Miami School of Medicine, Miami, Florida 33152.

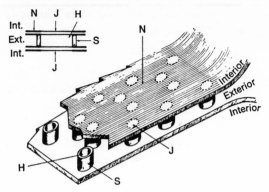

Fig. 1. Intercellular passageways. *Top.* Scheme of a unit passageway as defined by measurements of conductance and by fluorescent tracer diffusion (Loewenstein 1966). *J,* junctional membrane, highly ion permeable. *N,* a portion of nonjunctional membrane, relatively ion impermeable. *S,* junctional seal, a diffusion barrier of unknown nature, circumscribing aqueous channel (*H*). A J-element overlies each end of a channel. *Bottom.* Diagram representing a possible spatial relationship in a cell junction with extracellular space (open to exterior) between junctional units. Such a configuration allows 2 independent flows through the region of the intercellular junction: one from cell to cell through the insulated passageway units and, perpendicular to this, another entirely extracellular (not dealt with in this paper). Among several possibilities, a tubular passageway configuration is represented. (For possible electronmicroscopical correlates, see, e. g., Wiener *et al.* 1964, Revel & Karnovsky 1967, Bullivant & Loewenstein 1968, McNutt & Weinstein 1970, Gilula *et al.* 1970, Rose 1971.)

trypsin and chelator treatment or by micromanipulation, the junctional membranes seal (in the presence of sufficient (Ca^{++} or Mg^{++} in the external medium). New passageways are readily formed by manipulating the cells into contact at other arbitrarily chosen spots (Loewenstein 1967a, Ito & Loewenstein 1969).

The two apposing membranes need not be entirely equal for passageway formation. In culture, viable junctions are made between cells of different type and genus. For instance, a lens cell from rabbit makes passageways with a liver cell from rat or with fibroblasts from hamster or man (Michalke & Loewenstein 1971). Although the cells in the experiments had undergone some dedifferentiation in culture, they were genetically different; they had different morphological features, made different enzymes, and their membranes, in at least some cases, had different immunological properties. Thus, the capacity of forming communicative passageways seems to be a basic and general characteristic of cell membranes, at least in mammalian cells.

During passageway formation, the cell membrane must undergo a pronounced structural change; the membrane region at a passageway becomes several orders of magnitude more permeable than it was before. Membrane-bound Ca appears to play an important role in this transformation. The junctional membrane permeability is inversely related to the Ca^{++} activity in the medium in contact with the membrane. In an established junction where the junctional membrane, insulated from the exterior by the junctional seal, faces the low Ca^{++} activity ($<10^{-6}$ M) of the intracellular compartment on either side, the permeability is high. It falls when the Ca^{++} activity rises on one side. This is shown for the cytoplasmic side by injecting Ca^{++} into a cell with a micropipette (Loewenstein *et al.* 1967) or by exposing the junctional membrane to test media of varying Ca^{++} concentrations via a hole in the cell's (nonjunctional) membrane (Fig. 3). Junctional membrane permeability falls markedly when the Ca^{++} activity is thus raised above $4 - 8 \times 10^{-5}$ m in *Chironomus* salivary gland cells (Oliveira-Castro & Loewenstein 1971). At concentrations of $10^{-3} - 10^{-2}$M, such as those prevailing in the normal extracellular medium, the junctional membranes are no longer distinguishable in their permeability from the nonjunctional ones. Access to test media to the other side of the junctional membrane may be gained by breaking the junctional seal with chemical treatment. Again, junctional permeability falls with rising Ca^{++} activity in the medium under these conditions (Loewenstein *et al.* 1967, Loewenstein 1967b).

The change in membrane structure caused by Ca^{++} is fast. The fall in permeability is fast enough to limit transjunctional flux of Ca^{++}; only the junctional membranes to which Ca^{++} has direct access through the hole are affected in an experiment of the above type. The structural change produced by Ca^{++} is also fully reversible in low Ca^{++} medium under such experimental conditions, provided that the electrical potential across the (nonjunctional) cell membrane is sufficiently high (Oliveira-Castro & Loewenstein 1971).

Junctional membrane structure is also affected by a number of other experimental procedures. Among these are substitution of extracellular Li for Na (Rose & Loewenstein 1971), cooling of the cells, or their treatment with oligomycin, cyanide, dinitrophenol and other inhibitors of ATP synthesis or utilization. All these procedures produce depression of junctional membrane permeability (ouabain, a specific inhibitor of Na^+- and K^+-activated ATPase, does not depress junctional permeability) (Politoff *et al.* 1969). Injection of ATP into cells can prevent or reverse the permeability depression by

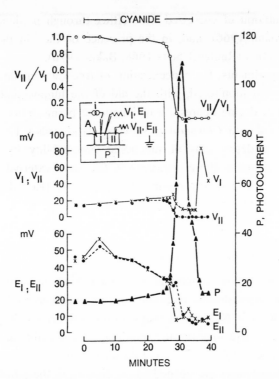

Fig. 4. Rise in cytoplasmic Ca^{++} during junctional uncoupling. *Chironomus* salivary gland was treated with cyanide (5 mM); cytoplasmic Ca^{++} concentration and coupling were measured simultaneously. The Ca-sensitive protein aequorin was injected into cell *I (inset);* aequorin emits light in proportion to the cytoplasmic Ca^{++} concentration (Shimamura & Johnson 1969); light output is measured by photomultiplier *P.* For measurement of coupling, an electric current (i = 4 × 10^{-8} amp) was pulsed with a microelectrode, between interior of cell *I* and the grounded exterior, and the potentials at zero current *(E)* and the steady changes in this potential *(V)* produced by the current were measured with two other microelectrodes, in cells *I* and *II*. The plot gives the values of the photocurrent (▲ arbitrary units) of E, V (recorded simultaneously on a chart recorder) and of the coupling coefficient V_{II}/V_I. The points before time zero are the corresponding values before cyanide application, constant over 15 minutes of observation. Note the close correlation between uncoupling and the rise in cytoplasmic Ca^{++} concentration. The gland is in Ca, Mg-free medium throughout the experiment. Fluctuations in photocurrent, ± 5 units at base line and ± 10 units at peak (Rose-Loewenstein & Loewenstein, unpublished).

dinitrophenol (Politoff *et al.* 1969). All of these procedures are known to cause a rise in cytoplasmic Ca^{++} in various kinds of cells. This rise is due to Ca^{++} release from mitochondria and other intracellular stores (cf A. Lehninger 1970), inhibition of energized Ca^{++} efflux through nonjunctional membrane (Schatzmann 1966), and, in the presence of Ca^{++} in the exterior, to excessive Ca^{++} influx (Reuter & Seitz 1968, Baker 1970).

In recent experiments, the concentration of free Ca^{++} in the cytoplasm of epithelial cells was monitored, with the aid of the luminescent Ca-sensitive protein aequorin, during depression of junctional permeability. (The protein was microinjected into *Chironomus* salivary gland cells.) The experiments showed that the depressions of junctional permeability by cyanide and dinitrophenol (including those occurring in Ca^{++}-free external medium) are associated with a rise in cytoplasmic Ca^{++} (Fig. 4) (Rose-Loewenstein & Loewenstein, unpublished).

SOME POSSIBLE ROLES OF COUPLING

One of the most obvious functional consequences of junctional coupling in epithelia is the establishment of a common intracellular pool for the small inorganic ions K^{+}, Cl-, Na^{+}, etc., throughout the connected cell system. Hence the transports of these ions through nonjunctional membrane draw on this pool and may conceivably be concerted throughout the epithelium.

Still wider possibilities are opened by the finding of the junctional passage of larger molecules. All coupling cell types heretofore examined show junctional passage of molecules of the order 500 w., and some up to the order of 10,000 (Loewenstein & Kanno 1964, Kanno & Loewenstein 1966, Furshpan & Potter 1968).

Most metabolites, many hormones and other cellular substances fall within the size range of molecules permeating the junctional membrane. There is thus ample room for cell-to-cell flow of molecules exerting regulatory action on cellular activities. A particularly exciting possibility is that the size range also includes molecules regulating gene activity, that is, molecules controlling cellular growth and differentiation. We have been exploring this possibility by genetic analysis of cells with defective coupling and growth. This approach is based on the idea that if the junction is indeed a pathway for growth-controlling molecules, interruption of this pathway by genetic defect should produce uncontrolled (cancerous) growth (Loewenstein 1968).

We have isolated so far 4 strains of epithelial cells in culture, which unlike their normal counterparts, are incapable of making coupling passageways. All 4 strains are also cancerous (Borek *et al.* 1969, Azarnia & Loewenstein

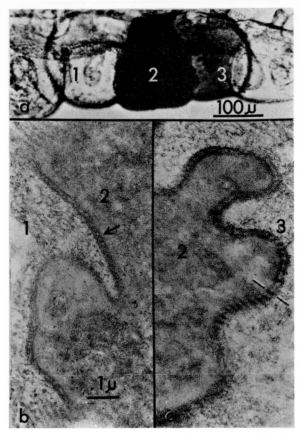

Fig. 2. Intercellular passageways as traced in electronmicrograph of a septate junction. Horse-radish peroxidase (Sigma, type VI) containing proteins with peroxidative activity ranging from <12,000 mol wt to 40,000 was injected into a cell (*2*) of *Chironomus* salivary gland; 30 min. after injection, the gland was fixed in glutaraldehyde and incubated in diamino benzidine and H_2O_2 to visualize the electronopaque reaction product of the peroxidase. Before injection, contiguous cell *1* was injured in medium containing 10^{-3}M Ca to uncouple cell junction *2/1;* fluorescein, injected as a probe of junctional coupling into cell *2* together with the peroxidase, was observed to pass from cell *2* to *3* but not from cell *2* to *1* (not photographed). *a,* Light micrograph of a whole mount of the gland showing the peroxidase reaction product in cell *2* and in the coupled cell *3,* but not in *1. b, c,* Electronmicrographs of the septate junctions *2/1* and *2/3* (same section). The dark peroxidase reaction product is seen (osmication, uranyl acetate) in cell *2,* with less density in cell *3,* and with highest density in the septa of cell junction *2/3.* The peroxidase did not penetrate detectably into the uncoupled junction *2/1.* × 122,000. Electronmicrographs by W. Larsen.

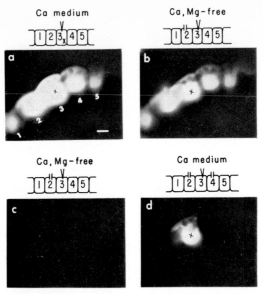

Fig. 3. Junctional permeability probed with fluorescent molecule. Cell-to-cell communication in normal epithelial cell system (isolated *Chironomus* salivary gland) (*a*). Cells are surrounded by medium containing 12×10^{-3}M Ca. Fluorescein (330 mol wt) (not detectably taken up by the cells from the exterior), is injected into cell *3* with a micropipette and the fluorescence photographed 6 min. thereafter. Fluorescein has spread through interior of cell chain. 20 min. later, the cell's fluorescence having diminished below photographic resolution, a hole (about 10 μ effective diameter) is drilled into the (nonjunctional) membrane of cell *2* while the cells are in Ca-free medium, and fluorescein is reinjected into cell *3* (*b*). Cell-to-cell flow of fluorescein continues unimpaired, but fluorescein now leaks out visibly through the hole in cell *2*. The Ca-free medium is replaced by Ca-containing medium (12×10^{-3}M), a hole is also made in cell *4,* and cell *3* is injected again (*d*). Fluorescein stays now within the confines of injected cell. Calibration 50 μ (Oliveira-Castro & Loewenstein 1971).

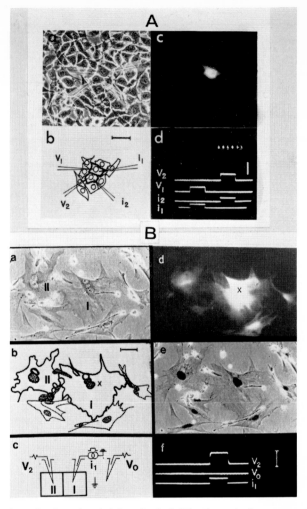

Fig. 5. Correction of a junctional defect by hybridization. *A*, the noncoupling parental epithelial cell (culture). *d*, pulses of current (i = 2 × 10⁻⁸ amp; 100 msec duration) passed between cell *1* and grounded exterior produce no detecable voltage (*V*) in adjacent cell 2. The lack of coupling is further demonstrated in *c* by the failure of cell-to-cell passage of fluorescein. Fluorescein is injected into cell 1 simultaneously with the electrical measurement (this anion carries the current in micropipette i_1). All cells are from one clone. *a*, phase contrast micrograph of the cell culture. *b*, tracing *c*, darkfield micrograph of fluorescence after fluorescein injection; calibration 50 μ. *d*, oscilloscope record; voltage calibration 500 mV. (Azarnia & Loewenstein 1971). *B*, the coupling hybrid, the primary product of fusion between the noncoupling epithelial cell and the coupling fibroblast. Hybrid *I* contains 2 nuclei from coupling cells and 1 nucleus from noncoupling cell. Hybrid *II* contains one nucleus from each parental cell. Coupling is probed electrically and with fluorescein 24 hr. after cell fusion. *a*, phase contrast photomicrograph. *e*, radioautograph; nuclei from the coupling paren- tal cells are ³H-thymidine labelled. *b*, tracing of the micrographs. *d*, darkfield micro- graph showing the results of fluorescein injection into hybrid *I*. Fluorescein is seen to have spread from hybrid *I* to *II* and to 3 parental fibroblasts in contact. Calibration, 50 μ. *f*, oscilloscope record from an electrical measurement between 2 hybrids: *i*, cur- rent pulse (2 × 10⁻⁸ amp; 100 msec. duration) passed between interior of hybrid *I* and grounded exterior (*c*). V₂, V₀, the resulting voltage changes in hybrid *II* and in the exterior. Voltage calibration 100 mV. (Azarnia & Loewenstein 1972).

1971). These cells were hybridized with coupling, normally growing cells (by cell fusion) with the aid of Sendai virus. The hybrid cells turned out to be normal in coupling and in growth; the genetic defects of coupling and growth were, in all cases, corrected together (Azarnia & Loewenstein 1972) (Fig. 5). This is very encouraging. The analysis of the segregants of these hybrids, now under way, should show whether the two defects are genetically correlated. Thus, we may reasonably hope to have soon the answer to the question of whether the junction is instrumental in the dissemination of gene-controlling molecules.

REFERENCES

Azarnia, R. & Loewenstein, W. R. (1971) Intercellular communication and tissue growth. V. A cancer cell strain that fails to make permeable membrane junctions with normal cells. *J. Membr. Biol. 6,* 368–385.

Azarnia, R. & Loewenstein, W. R. (1972) Parallel correction of cancerous growth and of a genetic defect of cell-to-cell communication. *Nature (Lond.)* In press.

Baker, P. F. (1970) Sodium-calcium exchange across the nerve cell membrane. In *Calcium and Cellular Function*, ed. Cuthbert, A. W., p. 96. St. Martin's Press, New York.

Borek, C., Higashino, S. & Loewenstein, W. R. (1969) Intercellular communication and tissue growth. IV. Conductance of membrane junctions of normal and cancerous cells in culture. *J. Membr. Biol. 1,* 274–293.

Brightman, M. V. & Reese, T. J. (1969) Junctions between intimately apposed cell membranes in the vertebrate brain. *J. Cell Biol. 40,* 468–492.

Bullivant, S. & Loewenstein, W. R. (1968) Structure of coupled and uncoupled cell junctions. *J. Cell Biol. 37,* 621–632.

Frömter, E. & Diamond, J. (1972) Route of passive ion permeation in epithelia. *Nature (Lond.) 235*:9.

Furshpan, E. J. & Potter, D. D. (1968) Low resistance junctions between cells in embryos and tissue culture. In *Current Topics in Developmental Biology*, Vol. 3, ed. Moscona, A. A., pp. 95–127. Associated Press, New York.

Gilula, N. B., Branton, D. & Satir, P. (1970) The septate junction – a structural basis for intercellular coupling. *Proc. nat. Acad. Sci. (Wash.) 67,* 213–218.

Ito, S. & Loewenstein, W. R. (1969) Ionic communication between early embryonic cells. *Devel. Biol. 19,* 228–243.

Kanno, Y. & Loewenstein, W. R. (1966) Cell-to-cell passage of large molecules. *Nature (Lond.) 212,* 629–630.

Lehninger, A. (1970) Mitochondria and calcium transport. *Biochem. J. 119,* 129.

Loewenstein, W. R. (1966) Permeability of membrane junctions. *Ann. N. Y. Acad. Sci. 137,* 441–472.

Loewenstein, W. R. (1967a) On the genesis of cellular communication. *Devel. Biol. 15,* 503–520.

Loewenstein, W. R. (1967b) Cell surface membranes in close contact. Role of calcium and magnesium ions. *J. Colloid Interface Sci. 25,* 34–46.

Loewenstein, W. R. (1968) Some reflections on growth and differentation. *Perspect. Biol. Med. 11*, 260–272.

Loewenstein, W. R. & Kanno, Y. (1964) Studies on an epithelial (gland) cell junction. I. Modifications of surface membrane permeability. *J. Cell. Biol. 22*, 565–586.

Loewenstein, W. R., Nakas, M. & Socolar, S. J. (1967) Junctional membrane uncoupling. Permeability transformation at a cell membrane. *J. gen. Physiol. 50*, 1865–1891.

Michalke, W. & Loewenstein, W. R. (1971) Communication between cells of different types. *Nature (Lond.) 232*, 121–122.

McNutt, J. & Weinstein, R. S. (1970) The ultrastructure of the nexus. A correlated thin section and freeze-cleave study. *J. Cell. Biol. 47*, 666–684.

Oliveira-Castro, G. M. & Loewenstein, W. R. (1971) Junctional membrane permeability. Effects of divalent cations. *J. Membr. Biol. 5*, 51–77.

Oschman, J. L. & Wall, B. J. (1972) Calcium binding to internal membranes. *J. Cell Biol. 55*, 58–73.

Politoff, A. L., Socolar, S. J. & Loewenstein, W. R. (1969) Permeability of a cell membrane junction. Dependence on energy metabolism. *J. gen. Physiol. 53*, 498–515.

Reuter, H. & Seitz, N. (1968) The dependence of Ca efflux from cardiac muscle on temperature and external ion composition. *J. Physiol. (Lond.) 195*, 541–560·

Revel, J. P. & Karnovsky, M. J. (1967) Hexagonal array of subunits in intercellular junctions of the mouse heart and liver. *J. Cell Biol. 33*, 37.

Rose, B. (1971) Intercellular communication and some structural aspects of membrane junctions in a simple cell system. *J. Membr. Biol. 5*, 1–19.

Rose, B. & Loewenstein, W. R. (1971) Junctional membrane permeability. Depression by substitution of Li for extracellular Na, and by long-term lack of Ca and Mg; restoration by cell repolarization. *J. Membr. Biol. 5*, 20–50.

Schatzmann, J. J. (1966) ATP-dependent Ca-extrusion from human red cells. *Experientia 22*, 364.

Shimamura, O. & Johnson, F. (1969) Properties of the bioluminescent protein aequorin. *Biochemistry 8*, 3991–3997.

Wiener, J., Spiro, D. & Loewenstein, W. R. (1964) Studies on an epithelial (gland) cell junction. II. Surface structure. *J. Cell Biol. 22*, 587–598.

DISCUSSION

ULLRICH: How is the cell fusion that you have shown here influenced by substances which change the surface coat of the cells like concanavalin or neuraminidase. And what is the influence of substances which change the surface charge as for instance acid glyco-proteins or polylysine?

LOEWENSTEIN: We haven't tried this yet. We have only just begun experimenting along these lines with concanavalin.

LEAF: Is this very rapid spread of the fluorescein dye to the salivary cells consistent with what would appear to be the very small area of the lateral cell surface, that must be involved in these communicating junctions?

LOEWENSTEIN: All our data fit with simple free diffusion; and this not only for fluorescein itself, but also for the larger fluorescein-tagged molecules which, of course, go much more slowly.

LEAF: Do you think components of the cytoplasm will spread like the dye?

LOEWENSTEIN: So long as they are small enough, yes. However, for macromolecules, say, nucleic acid informational molecules, the cell behaves like a unit. But things of up to the order of 1000 and, in some cases, even 100,000 mol. wt. seem to move quite freely from cell to cell.

SCHULTZ: I wonder if you would comment on the extent to which the different junctional complexes may be involved in transepithelial shunts as opposed to communication between adjacent cells.

LOEWENSTEIN: The structural correlates of what I call a communicative junction, a purely functional term, remain to be determined. There are a few candidates for it – one, the "gap junction" – is a very good candidate, but the evidence so far is only good circumstantial evidence. Another candidate is the septate junction. The experiments in our laboratory by Larsen, combining electron microscopy with intracellular injection of peroxidase implicate the septate junction as a cell to cell passageway. The passageway was blocked upon uncoupling. It may turn out that both junctional structures mediate coupling, perhaps couplings of different kinds with different molecular sieving properties. What you are asking me also is where does the transverse shunt flux go? If you remember my first slide (Figure 1), this flux may be entirely extracellular and independent of the cell-to-cell flux; it may go so to speak around the junctional seals.

Ion Transport across Thin Lipid Bilayer Membranes

D. C. Tosteson

In this paper, we introduce certain concepts and facts about the transport of ions across thin lipid bilayer membranes. We emphasize the formidable difficulties involved in defining transport mechanisms even when the object of study is as relatively simple as a bilayer. In such systems, the molecular composition area and thickness of the single membrane constituting a resistance to transport are relatively well known. Furthermore, the driving forces for transport (eg. differences in chemical composition, electrical potential, etc. of the solutions bathing the membrane surfaces) as well as the magnitude of the fluxes can all be controlled or measured by the investigator. In the light of these considerations, it is not surprising that the problem of transport across epithelia, which consists of at least two membranes arranged in series and connected in parallel by various types of junctions and separated by intracellular fluid in which chemical composition, electrical potential, etc. are difficult to measure and control, is so complicated and intractable.

We consider two aspects of ion transport across bilayers. First, we explore some of the implications of the fact that the relatively unstirred layers of external solutions bathing the membrane surfaces may, under certain circumstances, provide the major resistance to transport. Second, we describe some of the factors which influence the rates and selectivity of transport of monovalent cations across bilayers containing valinomycin or one of its analogues, peptide PV.

ROLE OF UNSTIRRED LAYERS

Bilayers, or indeed, any single biological membrane, may be thought of as comprising 3 regions arranged in series in the transport path. First, there

Department of Physiology and Pharmacology, Duke University Medical Center.

»Transport Mechanisms in Epithelia«, Munksgaard, Copenhagen.

are unstirred layers of external, aqueous solution; second, there are interfaces between these aqueous layers, and the hydrocarbon interior of the membrane which forms the third region. All 3 of these regions must be considered in the analysis of any transport process. Cases are known in which each may offer the major resistance to transport. For example, K transport across bilayers containing valinomycin is controlled by events occurring in the latter two regions (Stark *et al.* 1971), while diffusion of water and other rapidly penetrating small solutes is also influenced by the first region (Cass & Finkelstein, 1967).

The overall resistance of the membrane system to transport of component j may be described by a series equation of the following form:

$$1° \qquad (^tP_j)^{-1} = (^uP_j)^{-1} + (^{ms}P_j)^{-1} + (^{mi}P_j)^{-1}$$

where ^tP, ^uP, ^{ms}P, and ^{mi}P are the permeabilities of the total system, unstirred layers, membrane surfaces or interfaces, and membrane interior, respectively. The magnitude of ^uP_j is about 10^{-3}cm sec^{-1} since $^uP_j = {}^uD_j/\triangle$, where $\triangle$ is the effective thickness of the unstirred layer (about 10^{-2}cm for bilayers), and ^uD_j is the diffusion coefficient of j in water (about 10^{-5}cm^{2}sec^{-1} for small molecules and ions). Clearly, when the permeability of the overall membrane system, ^tP, is 10^{-3}cm sec^{-1} or more, ^uP becomes rate limiting. Since the membrane itself is only about 10^{-4} the thickness of the unstirred layer, the argument leads to the conclusion that ^uP is rate limiting for all transported substances for which the product of the intramembrane diffusion coefficient and the partition coefficient (membrane/external solution) is equal to or greater than 10^{-4} the value of its diffusion coefficient in the aqueous solutions bathing the membrane.

For example, consider the membrane transport of a gas which has a partition coefficient of 1, an intramembrane diffusion coefficient of 10^{-5} cm^2 sec^{-1}, and thus a ^{mi}P of 10 cm sec^{-1}. Assuming that $^{ms}P \geq 10$, the permeability of the entire system, ^tP, will clearly equal 10^{-3} cm sec^{-1}, the value of ^uP. Thus, for substances with membrane permeabilities in the range from 10^{-3} to 10 cm sec^{-1}, measurements of ^tP give little or no information about ^mP $[(^mP_j)^{-1} = (^{mi}P_j)^{-1} + (^{ms}P_j)^{-1}]$.

However, we will now show that the occurrence of chemical reactions between the transported component j and other substances, not necessarily transported across the membrane but present in excess in the unstirred layer, may lower the resistance of this region to transport of j and thus permit measurement of ^mP_j.

We first encountered the phenomenon of "facilitated" diffusion in the unstirred layer in the course of a study of bromide transport across bilayers (Gutknecht *et al.* 1972). The flux of Br across bilayers formed from sheep red cell lipids was measured with the use of tracer ^{82}Br. The Br flux in such a system with 0.1 M NaBr in the bathing solutions was found to be several orders of magnitude higher than would be predicted from the electrical conductance of the membrane. Furthermore, the Br flux was inhibited by reducing agents such as $S_2O_4^=$ and increased by addition of Br_2.

From these observations, we concluded that the tracer traversed the membrane mainly in the form of Br_2. However, when $^tP_{Br2}$ was computed from the measured magnitude of the Br flux and the known concentration of Br_2 in the bathing solutions, it was found to be 8 cm sec^{-1}, far higher than the computed value of 10^{-3} cm sec^{-1} for the permeability of the unstirred layer. The explanation for this paradox is that ^{82}Br exists in the system in two forms, Br_2 and Br^-, which equilibrate rapidly by the exchange reaction $^{82}Br^- + Br_2 \rightleftharpoons Br^- + {}^{82}Br_2$.

Assuming that ^{82}Br diffuses through the unstirred layer in both forms but moves through the membrane only as Br_2, we showed (Gutknecht *et al.* 1972) that:

2° $$(M_{Br})^{-1} = ({}^uP_{Br}\text{-}C_{Br}\text{-})^{-1} + ({}^mP_{Br_2}C_{Br_2})^{-1} + (\gamma\triangle{}^uP_{Br_2}C_{Br_2})^{-1}$$

where M_{Br} is the Br flux, C_{Br-} and C_{Br2} are the concentrations of Br^- and Br_2 respectively in the identical solutions bathing the two sides of the membrane, $^uP_{Br-}$, $^uP_{Br2}$, and $^mP_{Br2}$ are the corresponding permeability coefficients in the unstirred layers and membrane, $\triangle$ is the thickness of the unstirred layer, and γ^{-1} is the relaxation length of the isotopic exchange reaction. This expression described the data reasonably well except that it failed to account for the magnitude of the slope of the relation between bromine concentration and Br flux. The best values for $^mP_{Br2}$, $^uP_{Br2}$ and $^uP_{Br-}$ are about 10, 10^{-3} and 10^{-3} cm sec^{-1}, respectively. $^mP_{Br-}$ is assumed to be zero.

When the concentration of Br^- in the unstirred layer is much greater than the concentration of Br_2, ^{82}Br moves through this region almost entirely in the form of Br^-. The exchange reaction maintains the specific activity of Br_2 equal to that of Br^- except in the region within 1 relaxation length (not greater than 10^{-4} cm) from the membrane. Thus, the presence of an excess of Br^- maintains the specific activity of Br_2 close to the values in the bulk solution and thus effectively reduces the resistance of the unstirred layers to the diffusion of $^{82}Br_2$. The same explanation accounts for the anomalously high permeabilities of bilayers to I_2 observed by Läuger et al. (1967).

This effect of chemical reactions on diffusion through unstirred layers is general. It occurs whenever a substance, which can move through the membrane rapidly and is present in relatively low concentration in the unstirred layers, can react rapidly to form a compound present in much higher concentration in the unstirred layers, but for which the membrane has low or zero permeability. The occurrence of the rapid chemical reaction increases the effective concentration in the unstirred layer of the species which penetrates the membrane rapidly.

The effect is analogous to that which would be produced if the substance had a large partition coefficient in the unstirred layer to the bulk solution. However, in the case of the chemical reaction system, the penetrating species diffuses across the unstirred layer in a different chemical form, eg. Br^- rather than Br_2. Therefore, the extent of the "facilitation" of transport depends on both the diffusion coefficient of this form (eg. $^uP_{Br^-}$ in equation 2°) and the rate of the reaction (eg. the relaxation length, γ^{-1}, in equation 2°).

An example of the physiological significance of this process is the "facilitation" of O_2 diffusion in the cytoplasm of red cells which results from its reaction with hemoglobin to form oxyhemoglobin (Kreuzer & Hoofd 1972). Another example of potential importance in biology is that of un-ionized forms of weak acids and bases (eg. CO_2) which penetrate membranes rapidly and the corresponding ionized form (eg. HCO_3^-) to which membranes are generally much less permeable.

We have recently performed experiments to evaluate the role of this process in the diffusion of weak acids across thin lipid bilayer membranes (Gutknecht & Tosteson 1972). We measured the flux of ^{14}C labelled salicylic acid across bilayers prepared from egg lecithin. The flux of salicylate was measured as a function of the concentration of undissociated salicylic acid. All of these experiments were carried out with a concentration of ionized salicylate of 0.1 M, but with the different values of pH required to produce the desired concentration of salicylic acid. A progressive increase of flux with increasing concentration of salicylic acid was observed and is consistent with the conclusion that the membrane is much more permeable to salicylic acid than to anionic salicylate (Fig. 1).

However, Fig. 2 shows that increasing the concentration of salicylate

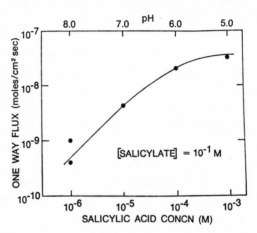

Fig. 1. Salicylate flux across a lecithin bilayer is plotted on the ordinate as a function of salicylic acid concentration on the abscissa. Salicylate concentration was maintained constant at 0.1 M and pH varied to produce different concentrations of salicylic acid. Flux was measured with ^{14}C labelled salicylate.

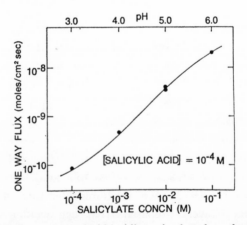

Fig. 2. Salicylate flux across a lecithin bilayer is plotted on the ordinate as a function of salicylate concentration on the abscissa. Salicylic acid concentration was maintained constant and pH varied to produce different concentrations of salicylate. Flux was measured with ^{14}C labelled salicylate.

at a constant salicylic acid concentration also increases the flux. This effect can be explained by "facilitation" of diffusion of ^{14}C salicylic acid across the unstirred layers by salicylate. It will be interesting to study in the future the role of this phenomenon in the many examples of transport of weak acids and bases such as CO_2 and NH_3 across biological membranes.

MECHANISMS OF CATION TRANSPORT ACROSS BILAYERS

During the past few years a large volume of literature has developed describing the selective increase in cation transport across bilayers and biological membranes produced by valinomycin (Läuger 1972). Under most circumstances, this compound induces a highly selective increase in K^+ permeability. Valinomycin (val) forms a complex with K^+ at the membrane surface, the complex moves through the hydrocarbon interior and releases the K^+ to the opposite side. In the process of complex formation, valinomycin undergoes a change in conformation to form a bracelet shaped cage surrounding the cation. In this structure, inwardly directed carbonyl oxygens in ester linkage in val replace completely the water oxygens in the hydration shell of the metal. The relatively non-polar methyl and isopropyl side chains are directed outwardly either axially or equatorially. The bracelet conformation of the cation complex is stabilized by intra-molecular hydrogen bonds involving all six peptide carbonyls and amide protons in the molecule.

In this paper, we will discuss two factors which influence the properties of cation carriers like valinomycin. First, we treat the role of the primary structure of the carrier. Second, we consider the effect of certain lipophilic anions in the cation selectivity of valinomycin.

Primary Structure of Carriers

As noted above, in complexes of valinomycin with monovalent cations, 6 carbonyl oxygens all in ester linkages interact with the metal ion. We report here experiments designed to investigate the effect of substituting carbonyl oxygens in amide linkage for those in ester linkage on the interactions with metal ions. In the design of the compound, we sought to preserve the alternating hydrogen bonding (amide) and non-hydrogen bonding (ester) character of the primary sequence in the parent molecule. This character is important in promoting formation of the 6 10-membered rings which stabilize the "bracelet" conformation of the metal complexes of valinomycin. Accordingly, we chose to substitute proline, which cannot participate in hydrogen-bonding when in amide linkage, for the hydroxy acids in valino-mycin. Explicitly, we have recently reported the synthesis of a cyclic dodecapeptide (PV) with the primary structure (DVal – LPro – LVal – DPro) which may be compared with valinomycin (DVal – LLac – LVal – DHyv) (Gisin et al. 1972).

That this compound forms complexes with monovalent cations is shown clearly by the proton NMR spectra of PV dissolved in $CDCl_3$ in the absence and presence of the K^+ salt of trinitrocresolate (TNC) (Gisin et al. 1972). The proton NMR spectrum of free PV is complicated and consistent with the presence of at least three conformers under these conditions. By comparison, the proton NMR spectrum of valinomycin in $CDCl_3$ is considerably simpler and suggestive of only one conformer. The K^+ complex of PV shows a considerably simpler proton NMR spectrum in which assignments for all resonances can be made by comparison with the proton NMR spectrum of valinomycin. The similarity in the proton NMR spectra of the K^+ complexes of PV and valinomycin strongly suggests that both are in the "bracelet" conformation stabilized by 6 10-membered rings.

In order to estimate the ion complexation properties of PV, we have performed two phase titrations in which PV dissolved in an organic solvent (CH_2Cl_2 or $CHCl_3$) was permitted to come to equilibrium with an aqueous

Table I. Two phase equilibrium.

Water, pH 7.0 XTNC 25°C		CHCl₃ Peptide, 10^{-4} M X-VAL-TNC	
	$K = \dfrac{(X\text{-PV-TNC})_o}{(X)_w(TNC)_w(PV)_o}$		
Peptide	X^+	K 10^{10} M^{-2}	$\dfrac{K_K}{K_{Na}}$
PV	K	12	6
PV	Na	2	
VAL	K	12×10^{-4}	2×10^3
VAL	Na	7×10^{-7}	

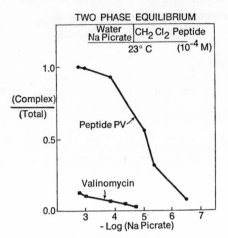

Fig. 3. Two phase titration of peptide PV (upper curve) and valinomycin (lower curve) in CH₂Cl₂ with Na TNC added to water.

solution containing a known concentration of one of the alkali metal salts of picrate or trinitrocresolate (TNC). The equilibrium concentration of picrate or TNC in the organic phase was measured spectrophotometrically and the apparent association constant (K) computed (Table I).

Fig. 3 shows a typical titration curve in which the ratio of complexed to total PV concentration in the organic phase is plotted as a function of the log of the Na picrate concentration in the aqueous phase. Note that PV has a much greater apparent affinity for Na^+ than does valinomycin. Values for K shown in Table I indicate that PV has a much greater apparent affinity but much lower selectivity for alkali metal cations than does valinomycin in this $CHCl_3 : H_2O$ equilibrium system.

Fig. 4 shows curves of the relation between current and voltage at zero frequency across thin bilayer membranes formed from sheep red cell lipids dissolved in decane. Measurements were made when the concentration of PV was 10^{-6} M or 10^{-5} M in both bathing solutions and when the back (B) and front (F) solutions contained 10^{-6} M and 10^{-5} M respectively. The bathing solutions contained up to 1% (v/v) ethanol which, by itself, did not affect the bilayer conductance. In all cases, the curves are of the "saturating" type in which the slope decreases with increasing potential difference. The conductance of the membrane increased with increasing concentration of PV. When the concentrations of PV in the back and front chambers

3*

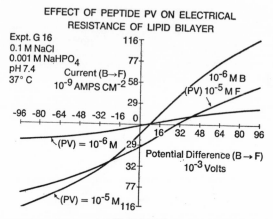

Fig. 4. Current across a sheep red cell lipid bilayer is plotted on the ordinate as a function of electrical potential difference plotted on the abscissa in the presence of the indicated concentrations of peptide PV added to the aqueous phases.

were different, the I-V curve displayed rectification and the potential difference at zero current was about 25 mV (with the chamber containing the lower concentration of PV positive). The development of this zero-current potential difference was observed when the bathing solutions contained KCl rather than NaCl as was the case in the experiment shown in Fig. 4.

Thus, a concentration difference for the uncharged peptide PV produces an electrical potential difference across a membrane in the absence of concentration differences for ions. This effect probably arises from the establishment of a concentration difference for the cation complexes of PV, eg. K^+PV, Na^+PV. If such is the case, the permeability of the bilayer must be low to the free, uncharged form of PV, since, with equal concentrations of KCl, the concentration of K^+PV can be higher on one side of the membrane than on the other only in the free PV concentration are also different.

Electrical potential differences at zero current have also been observed across bilayers separating identical salt solutions but different concentrations of valinomycin, but only when the salt is NaCl rather than KCl (Andreoli *et al.* 1967). However, the sense of the potential difference is opposite in the two cases. In contrast to the situation with PV, the side of the membrane exposed to the lower concentration of valinomycin is negative.

An important feature of the data shown in Fig. 4 is the relatively low value of the membrane conductance in bilayers bathed with 10^{-6} M PV. The slope conductance at zero-current and zero-potential is about $10^{-6} ohm^{-1} cm^{-2}$, about 1000 times *less* than the conductance of similar bilayers exposed to 10^{-6} M valinomycin, but about 100 times *greater* than unmodified bilayers. We have observed a similar disparity in the action of valinomycin and PV in red cell membranes. 10^{-6} M PV does not alter cation permeability in sheep red cells while 10^{-6} M valinomycin produces a marked selective increase in K^+ permeability. Thus, despite the greater affinity of PV for cations in two phase extraction systems (eg. Table I), this compound is much less potent than valinomycin in promoting cation movement across membranes (Table II).

Table II. Potency of ion carriers. KCl (0.1 M) 23° C

Carrier	K^+ Conductance	
10^{-6} **M**	$ohm^{-1}cm^{-2}$ Red Cells	Bilayers
Peptide PV	10^{-9}	10^{-6}
Valinomycin	10^{-6}	10^{-3}

We have estimated the ionic selectivity of bilayer membranes exposed to PV in 3 ways. First, we have measured the membrane conductance at zero current, potential difference and frequency when the membrane separates identical solutions containing the test cation. Second, we have measured the electrical potential difference across a bilayer separating a 0.1 M from a 0.01 M solution of the chloride salt of the test cation (V_m^{10}). Third, we have measured the so called bi-ionic potential (V_m^{Bi}), that is, the electrical potential difference at zero current across membranes separating equimolar solutions of two different cation chloride salts. All measurements were made at pH 7.4 (0.001 M $XHPO_4$) where X is a monovalent cation and at 23° C. Tables III and IV present the results of such measurements in systems containing 0.1 M XCl where X is the test cation.

Each figure in the table is the mean of all observations on at least two different bilayers. By methods 1 and 3 the sequence of selectivity is $K^+ > Rb^+ > Cs^+ > NH_4^+ > Na^+ > Li^+$. However, the observed values of V_m^{10} show that the interpretation of the measurements of G_m and V_m^{Bi} in terms of the quantitative selectivity for cations is difficult. The failure of V_m^{10}

Table III. Cation selectivity of peptide PV sheep red cell lipid bilayers.

PV (10^{-5} M)	XCl (0.1 M)	XHPO$_4$ (0.001 M)	pH 7.4 23° C
X$^+$	10^{-6}ohm^{-1}cm^{-2} G_m	V_m(X:X/8.6) V_m (K:X)	mV mV
Li	0.057	−32	−106
Na	0.34	−6	−76
NH$_4$	0.86	−47	−39
Cs	1.3	−21	−22
Rb	3.5	−11	+2
K	4.4	−33	0

to equal the equilibrium potential for X$^+$ can be interpreted in at least 2 different ways. First, PV could induce permeability of the membrane to chloride. Second, low membrane permeability leading to concentration differences of free PV could produce a deviation of the concentration ratio of PV-cation complexes expected for a 10-fold salt concentration ratio. The selectivity sequence for PV is similar to that for valinomycin. However, the magnitude of the selectivity for K$^+$ over Na$^+$ in bilayers is much less with PV (10^1–10^2) than it is with valinomycin (10^3–10^4).

The positions of NH$_4^+$ and Tl$^+$ in the sequence of selectivity are of interest because of the suggestion of Eisenman & Krasne (1972) that these ions can be used to test the number and type, repectively, of ligands involved in the interaction between cations and carriers. Table IV shows that PV, like valinomycin, prefers K$^+$ and Rb$^+$ to both NH$_4^+$ and Tl$^+$ in bilayers. Thus,

Table IV. Cation selectivity of peptide PV sheep red cell lipid bilayers.

PV 1 (10^{-5} M)	XCl (0.01 M)	pH 7.4 23° C	XHPO$_4$ (0.001 M)
X$^+$	G_m 10^{-6}ohm^{-1}cm^{-2}	V_m(X:X/4.2) mV	V_m (K:X) mV
Na	0.023	–	–
NH$_4$	0.22	−30	−66
Tl	0.38	−30	−29
Cs	0.84	−28	−5
Rb	2.1	−24	+26
K	0.76	−35	0

conversion of ligands from ester carbonyls (valinomycin) to amide carbonyls (PV) does not alter the K^+–Tl^+ selectivity.

In summary, we have shown that a new cyclic dodecapeptide of proline and valine (PV) displays a greater affinity for monovalent cations in two-phase extraction systems, a lower potency in increasing membrane conductance, and a lower selectivity for K^+ over Na^+ than does its analogue valinomycin. The similarities in the behavior of the two compounds are not surprising since both assume a similar "bracelet" conformation when they form complexes with cations. However, the explanation of the impressive differences is less obvious. The relative roles of such factors as solubility and conformational states in water and hydrocarbons of both free and complexed forms, rates of conformational transitions, ring size and the nature of ligand involved in complexation must all be assessed in future experiments.

EFFECT OF ANION ON SELECTIVITY OF CATION CARRIERS

When chloride is the major anion in the aqueous phases, valinomycin produces strikingly selective increases in K^+ permeability of sheep red cell membranes (Tosteson et al. 1967, also Fig. 5) and bilayers formed from lipids extracted from sheep red cell membranes (Andreoli et al. 1967), even when Na^+ is the major cation present in the system.

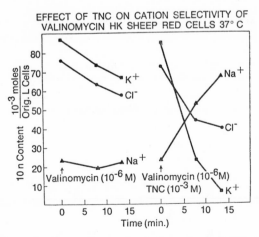

Fig. 5. The effect of valinomycin on the ionic content of HK sheep red cells in the absence and presence of 1 mM TNC.

However, when trinitrocresolate (TNC) is present in the medium, this selectivity is lost (Tosteson 1970, Tosteson 1971, also Fig. 5). Thus, in the experiment shown in Fig. 5, the presence of 1 mM TNC, which by itself has no effect on the cation permeability of sheep red cells (Gunn & Tosteson, 1971), alters the response to addition of valinomycin from loss of K^+ with no change in Na^+ content to simultaneous loss of K^+ and accumulation of Na^+.

The action of TNC in this system is analogous to the action of acetyl choline at the neuro-muscular junction where it converts the cation permeability of the muscle membrane in the motor endplate from its resting state of high K^+ selectivity to a state of approximately equal permeability to both K^+ and Na^+. What is the explanation for this remarkable change in the selectivity of the cation carrier valinomycin?

We believe that the answer to this question can be found in the dependence of the selectivity of valinomycin for cations on the solvent in which the interactions occur. Measurements of the two phase association constant of valinomycin for K^+ and Na^+ in water: decane and water: $CHCl_3$ systems reveal that selectivity for K^+ is greater in the latter than in the former system. This effect may be due to the greater extent of ion pair formation between Na^+val and TNC than occurs between K^+val and TNC.

In any case, the observations shown in Table V lead us to the following hypothesis to account for the effect of TNC on the selectivity of valinomycin

Table V. Effect of solvent on selectivity of valinomycin.

Water, pH 7.0 XTNC 25° C		Organic Solvent 10^{-4} M VAL X-VAL-TNC
	$K = \dfrac{(X)_w(TNC)_w(VAL)_o}{(X\text{-}VAL\text{-}TNC)_o}$	
Solvent	X^+	K 10^6 M^{-2}	$\dfrac{K_K}{K_{Na}}$
Decane	K	2.0×10^{-2}	10
Decane	Na	2.0×10^{-3}	
$CHCl_3$	K	12	2×10^3
$CHCl_3$	Na	6.6×10^{-3}	

as a cation carrier in red cell and bilayer membranes. Addition of TNC increases the concentrations of K and Na present as KTNC and NaTNC in the hydrocarbon interior of the membrane. In this region, the association constants of val for these cations are high and about equal. The permeabilities of the membrane to K^+ and Na^+ are proportional to the concentrations of K^+val and Na^+val in the membrane interior and are thus also about equal.

In contrast, in the absence of TNC, the concentrations of K and Na in the membrane interior are extremely low. Under these conditions, complex formation in the membrane interior is negligible but does occur at the membrane surfaces. In these regions, the affinity of val for K^+ far exceeds that for Na^+ and only K^+val complexes are formed and pass through the membrane.

Experiments designed to test this hypothesis are in progress in our laboratories. It is formulated here because it suggests a general mechanism for modification of selectivity of carriers for ions. In so far as these selectivities depend on the solvent in which the interaction between carrier and ion occurs, agents which affect the distribution of ions and carriers between the relatively polar and relatively non-polar regions of membranes may alter selectivity.

In summary, we have reviewed some observations and concepts arising from researches with lipid bilayers which may be of use in thinking about transport processes in epithelia. First, we emphasized the importance of the relatively unstirred layers on the surfaces of membranes as resistances to transport and, particularly, the significance of chemical reactions which the transported substance might undergo in the unstirred layers. Second, we described experiments which illuminate the roles of primary structure of carrier and aromatic anions in determining the selectivity of carrier mediated cation transport across bilayers.

REFERENCES

Andreoli, T. E., Tieffenberg, M. & Tosteson, D. C. (1967) The effect of valinomycin on thin lipid membranes. *J. gen. Physiol. 50*, 2527.

Cass, A. & Finkelstein, A. (1967) Water permeability of thin lipid membranes. *J. gen. Physiol. 50*, 1765.

Eisenman, G. & Krasne, S. (1972) Further considerations on ion selectivity of carrier molecules and membranes. *Proc. IV Intern. Biophys. Cong., Moscow.* In press.

Gisin, B. F., Davis, D. G., Kimura, J., Tosteson, M. T. & Tosteson, D. C. (1972) The interactions between a cyclic dodecapeptide and monovalent cations. *Proc. IV Intern. Biophys. Cong., Moscow.* In press.

Gisin, B. F. & Merrifield, R. B. (1972) Synthesis of a hydrophobic, potassium-binding protein. *J. Amer. Chem. Soc. 94,* 6165.

Gunn, R. B. & Tosteson, D. C. (1971) The effect of 2,4,6 trinitro-m-cresolate on cation and anion transport in sheep red cells. *J. gen. Physiol. 57,* 593.

Gutknecht, J., Bruner, L. J. & Tosteson, D. C. (1972) The permeability of thin lipid membranes to bromide and bromine. *J. gen. Physiol. 59,* 486.

Gutknecht, J. & Tosteson, D. C. (1972) Unpublished observations.

Kreuzer, F. & Hoofd, L. J. C. (1972) Diffusion and chemical reaction of oxygen with hemoglobin and myoglobin and facilitated diffusion in the presence of these pigments. Alfred Benzon Symposium IV, Oxygen Affinity of Hemoglobin and Red Cell Acid-Base Status. Astrup, P. & Rørth, M., ed. p. 451. Munksgaard, Copenhagen.

Läuger, P. (1972) Carrier mediated ion transport through artificial lipid membranes. *Science 178,* 24.

Läuger, P., Richter, J. & Lesslauer, W. (1967) Electrochemistry of bimolecular phospholipid membranes. *Ber. Bunsenges Phys. Chem. 71,* 906.

Stark, G., Ketterer, B., Benz, R. & Läuger, P. (1971) Kinetic analysis of carrier-mediated ion transport through artifical lipid membranes. *Biophys. J. 11,* 981.

Tosteson, D. C. (1970) Macrocyclic compounds as ion carriers in thin and thick lipid membranes and in biological membranes. In *Recent Advances in Microbiology,* ed. Perez-Miravete, A. & Pelaez, D, p. 191. Proc. XXth Internat. Cong. Microbiol., Mexico City.

Tosteson, D. C. (1971) Some characteristics of ion transport across thin lipid bilayer membranes containing macrocyclic compounds. *Proc. Symposium on Molecular Mechanisms of Antibiotic Action on Protein Biosynthesis and Membranes,* ed. Granada, D. Vazquez. In press.

Tosteson, D. C., Cook, P., Andreoli, T. E. & Tieffenberg, M. (1967) The effect of valinomycin on K^+ and Na^+ permeability of HK and LK sheep red cells. *J. gen. Physiol. 50,* 2513.

DISCUSSION

KEYNES: Can you say how much valinomycin there is in your bilayers and hence give a figure for the conductance of a single valinomycin molecule? People who work on the antibiotics like gramicidine which make tunnels through membranes come up with a figure for the conductance of a single channel which is of the order of 10^{-10} ohm^{-1}, which is in fact just about what we calculate for a single sodium channel in a nerve membrane. It is highly pertinent to ask what figure one gets for ionophores like valinomycin which shuttle ions across the membrane apparently in a very different manner.

TOSTESON: Estimates of the conductance per valinomycin molecule in bilayers range from 0.5×10^{-1} ohm^{-1} (Tosteson et al. 1968) to 7×10^{-14} (Läuger 1972), depending on the value taken for the partition coefficient of valinomycin between water and the bilayer. These values are about three orders of magnitude less than the single channel conductances in bilayers exposed to EIM (Bean et al. 1969) or gramicidin (Hladky & Haydon 1970). The conductance of single Na channels in axons from different nerves is 10^{-10} ohm^{-1} or more (Hille 1968), also much higher than the estimates of the conductance per valinomycin molecule.

The relatively low conductances per valinomycin molecule appear to be due both to the kinetics of the reactions of formation and dissociation of the ion-valinomycin complexes at the membrane surface *and* to the rates of translocation of both the complex and free valinomycin between the membrane surfaces (Läuger 1972). Recent observations by Melnik & Shkrob (1972) indicate that the relative magnitudes of the reaction rates and transport rates are different in different analogues of valinomycin. In this

Tosteson, D. C., Andreoli, T. E., Tieffenberg, M. and Cook, P. 1968. The effects of macrocyclic compounds on cation transport in sheep red cells and thin and thick lipid membranes. *J. Gen. Physiol., 51,* 3735–3845.
Läuger, P. 1972. Carrier-mediated ion transport. *Science, 178,* 24–30.
Bean, R. C., Shepherd, W. C., Chan, H. and Eichner, J. C. 1969. Discrete conductance fluctuations in lipid bilayer protein membranes. *J. Gen. Physiol., 53,* 741–757.
Hladky, S. B. and Haydon, D. A. 1970. Discreteness of conductance change in bimolecular lipid membranes in the presence of certain antibiotics. *Nature, 225,* 451–453.
Hille, B. 1968. Pharmacological modifications of the sodium channels of frog nerve. *J. Gen. Physiol., 51,* 199, 219.

frame of reference, the most unusual property of PV is the low rate of trans-location of the free carrier. The chemical basis of these differences in kinetic properties of different carriers is an important subject for future study.

SCHULTZ: Your data suggest that the specific structural features of the molecule that are responsible for the "caging" of amide and oxygen groups is very important. On the other hand, some recent data collected by Drs. Eisenman & Krasne (1972) suggest that the same ion selectivity can be ob-served using the individual ligands (e. g. formamide and its derivatives, and methyl ethers) without having to invoke the specific structural features of these ionophores. Could the difference arise from the fact that Krasne and Eisenman were examining equilibrium selectivity patterns as opposed to non-equilibrium or kinetic selectivity patterns?

TOSTESON: One reason that we synthesized PV was to try to get some insight into the problem which you pose. Eisenman's analysis of the basis of cation selectivity of neutral ion carriers (Eisenman et al. 1972) predicts that PV should have a considerably different selectivity from valinomycin, particularly with respect to the position of thallium ions in the sequence. This follows from the argument that the polarizability of the carbonyl oxygens in amide linkage (as in PV) is different from the polarizability of carbonyl oxygens in ester linkage (as in valinomycin), and that this property of the oxygen ligands which bind the metal ion is the most important determinant of selectivity. In fact, the position of thallium in the sequence is the same for PV and for valinomycin. This result could be interpreted to mean either that the polari-zability of the carbonyl oxygens in amide linkage with proline nitrogens are not different from the polarizability of the carbonyl oxygens in ester linkage or that polarizability of the oxygens which interact with the metal is not the most important determinant of selectivity in these particular compounds. In my opinion, at the present time we do not know the relative importance of

Melnik, E. I. & Shkrob, A. M. (1972) The conductivity of bilayer membranes in the presence of cyclodepsipeptides of valinomycin group. Abstr. of the *IV International Biophysics Congr.*, Moscow, p. 171.

Läuger, P. (1972) Carrier-mediated ion transport. *Science 178*, 24–30.

Eisenman, G., Laprade, R., Ciani, S. & Krasne, S. (1972) Experimentally observed effects of carriers on the electrical properties of bilayer membranes – equilibrium domain. Chapter 3 in *Membranes – A Series of Advances*, (Eisenman, G. ed.), Vol. 2, Marcel Dekker, New York.

several factors which can affect the relative magnitudes of the equilibrium constants of the reactions between neutral ion carriers and different cations. These factors include not only the number and bonding characteristics of the ligands which actually interact with the metal (e. g. ester carbonyl, amide carbonyl or ether oxygens) and how these bonding characteristics compare with those of the solvent molecules which compete for the metal ion. They also include the energetics of the conformational transitions required to bring the bonding ligands in the carrier molecule into proper relation with the cation. The magnitude of these terms depends not only on the primary structure of the carrier but also on its interactions with solvent molecules. In addition to these complexities which occur when carrier molecules are in solution in homogenous solvents, it is possible that additional factors may arise when complexation between carrier and ion occurs on the membrane surface. For example, Läuger (1972) and his colleagues have presented evidence which suggests that the association constant of valinomycin for K at the surface of a bilayer is 10^4 times less than the value observed in ethanol.

SACHS: You showed some change in optical spectra as you went through solvents of high dielectrical strength. You also had an NMR-spectrum on the effect of potassium on your synthetic peptide. In the presence of trinitrocresol do you have any NMR-spectrum conformational changes in the peptide as a function of the concentration?

TOSTESON: Dr. D. G. Davis in our laboratories has shown that the proton NMR spectrum of the K complex of valinomycin is not influenced by either solvent or anion. By contrast, the proton NMR spectrum of the Na complex is very dependent on the characteristics of the solvent but is also relatively independent of the nature of the anion. The proton NMR spectrum of the Na complex is most different from that of the K complex in relatively polar solvents and becomes progressively more like that of the K complex as the polarity of the solvent is decreased. He has also shown that the proton NMR spectrum of TNC is different when it is present with the K complex as compared with the Na complex of valinomycin in hydrocarbon solvents.

SACHS: Just to continue with that. You have the effect of potassium on the

Gunn, R. B. & Tosteson, D. C. (1971) The effect of 2, 4, 6-trinitro-h-cresol on cation and anion transport in sheep red blood cells. *J. Gen. Physiol. 57*, 593–609.

NMR spectrum in the absence of trinitrocresol. Sodium presumably has very little effect.

CURRAN: Have you tried any anions – other than trinitrocresol in these experiments. Is there anything special about it?

TOSTESON: Only picrate, which is very similar. We have not studied ions which are less soluble in hydrocarbon solvents and, therefore, do not know how unusual the properties of TNC might be in these systems. In more polar solvents, TNC and picrate do not seem to influence the interactions between valinomycin and cations. As you know, TNC has quite unusual properties in its effect on biological membranes (Gunn & Tosteson 1971).

MEARES: In the remarks you made about the unstirred layer of course the layer thicknesses to which you were referring apply to *in vitro* experiments, and to the permeabilities of complexed forms of these ions which involve not only the thickness of the layers or the mobilities or diffusion coefficients but also a concentration term. So that, for example, in an experiment where you have not got any valinomycin or other carrier in the bilayer, you are unlikely in studying either an ion transport or a whole salt transport to be influenced by the presence of these films. I wonder if you could comment on how far the unstirred layers are likely to create couplings in *in vivo* experiments where the distances between the surfaces are probably less than 10^{-2} cm anyway.

TOSTESON (blackboard): I was amazed in participating in a symposium on pulmonary function last spring to be reminded that the distance between the gas phase in the alveoli and the centre of a red cell in the adjacent capillary can be as low as 0.2 μ. Clearly the reduction in the thickness of unstirred layers is an adaptive characteristic of great functional importance in the lung and other biological systems. For example, the unstirred layer in whole blood can't possibly be more than 1 μ or so, because the average cell–cell distance is too short. On the other hand, in epithelia like the gut or the kidney, the cytoplasm may present an unstirred region about 10 μ thick. All of these are far less than the value of 100 μ which obtains with bilayers. Unstirred layers in biological systems probably do not constitute significant resistances to the transport of substances which traverse membranes relatively slowly, e. g. actively transported ions. By contrast, they may be very significant resistances to the transport of substances like gases, undissociated acids or bases and lipid soluble molecules which penetrate membranes rapidly.

KEYNES: Another way of putting these arguments is – again for the sodium channel in the nerve membranes – that if you work out the rate at which ions have to diffuse up to the membrane, to the mouth of the channel, in order to go through, it turns out that they go through the channel at a rate disconcertingly close to the maximum one at which they could possibly diffuse there, that is within about 1 order of magnitude. And if diffusion limitation doesn't even operate there, it is very unlikely to operate in sodium transport through epithelia, where the turnover time is much lower still. In your figure for valinomycin the turnover time was still quite high compared with what you get for one sodium-potassium ATPase.

TOSTESON: I agree.

GENERAL DISCUSSION

MEARES: I would like Dr. Ussing to comment on something on his 3rd slide which referred to glycerol transport. The initial and final potentials before and after the addition of hormone were both 60 mV, and the initial and final flux ratios were 1.1. The initial influx was 16 and the final influx was about 220. Thus there was a change of approximately 14-fold in both in- and outfluxes and no change in potential. It would appear that this could not be due to water drag because the water would affect the flux ratio.

USSING: The situation is rather involved in that the potential itself does not tell all about the hormone effects. It indicates that something happens, but even when the potential returns to the initial value it doesn't mean that the reaction is over. The skin is probably changed for several hours, as long as the hormone is present. The effect of the hormone means, for instance, probably a slight swelling of the cell. It probably means a swelling of the interspace system as well – at least we have clear evidence that it usually happens – which means that the diffusion paths may well change. So there is no way of saying whether the change should be up or down. All we really can say with certainty is that the glycerol permeability has been increased, because so many factors go into this quantity. I don't know whether this answers your question.

MEARES: Yes, in part, thank you.

ORLOFF: With respect to the effects of unstirred layer on diffusional permeability and its relevance to the pore theory, have you reinvestigated this problem?

USSING: In the first place we repeated the experiments of Dr. Hays on the frog skin and toad skin with different test molecules and we found that except for water where the effect was known already there was no effect whatsoever. We changed the rate of stirring from 1 revolution to 1000 revolutions per minute and the fluxes we found were the same. So our conclusion is that the toad bladder may respond to violent vibrations, whereas the frog skin doesn't, and that is why we do not find any change in the fluxes.

ORLOFF: Hays has stated that solvent drag is also eliminated by vigorous stirring. Do you still observe solvent drag at high rates of stirring?

USSING: Oh yes, that is another thing. We do.

VAN BRUGGEN: Dr. Ussing, in these toad skins with ADH – did you measure the net water influx?

USSING: Not in this series.

VAN BRUGGEN: Is the net water influx consistent with a solvent movement of the sucrose?

USSING: That we cannot say. These very small sucrose fluxes which may well be intercellular are so small compared to the total water movement of the system. We assume that the main osmotic flow in ADH reaction goes through the cell membrane. Thus, if there is a drag effect in the interspace, it may have any value relative to the total water flow, becauce quantitatively it is too small to be picked up.

VAN BRUGGEN: In frog skin we find about 4 μl/cm^2/h. Is toad skin iess than that?

USSING: It is about the same. But of course we don't know the areas through which this flow is going.

TOSTESON: May I ask Dr. Loewenstein a question. With contact-inhibited fibroblasts, for example chicken embryofibroblasts, is there any coupling?

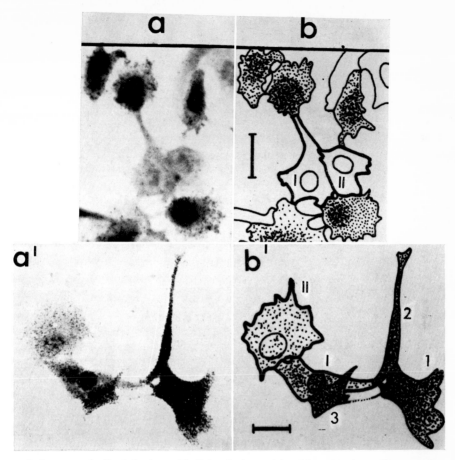

a, b, four genetically defective epithelial cells incapable of junctional coupling growing in contact with two mutant cells (I & II) fail to transfer ³H-labelled endogenous nucleotide. a', b' the control, showing nucleotide transfer from normally coupling cells (1, 2, 3) to mutant cells I & II. Left, radioautographs, Right tracings. From Azarnia *et al.* 1972.

LOEWENSTEIN: Yes. All normal fibroblasts and epithelial cells in culture so far examined are coupled.

HARVEY: Is there coupling between cells isolated from epithelia of different species? Do you ever obtain coupling between protozoon cells, which normally would never couple, and epithelial cells?

LOEWENSTEIN: I mentioned some experiments Dr. Michalke and I did with epithelial cells from different species. In these experiments we joined, in culture, liver cells from rat with lens cells from rabbit or fibroblasts from man, etc. We made eight such crosses, and by now we have made five or six more. All turned out to be nicely coupled. The cells may have undergone some dedifferentiation in culture; but they were clearly different in many respects from each other and their membranes had different immunological properties. Yet, in spite of this membrane inequality, the membranes can be matched to form communicative junctions. All these experiments were done with mammalian cells in culture. If one goes further afield, a different picture emerges. I did some experiments 5 years or so ago with sponge cells. I took a cell of the sponge Microciona, and tried to pair it with cells from Haliclona. The cells had convenient color markers. Microciona cells are red; Haliclona cells are yellow. The Microciona cells coupled with each other and so did the Haliclona cells with each other, but there was no cross-species coupling. I was at first a bit surprised by the two classes of results, but then I learned that these two sponge species are much further apart on the phylogenetic scale than mouse and man. So, somewhere there seems to be a limit for coupling between membranes of different kind. It would be interesting to make a systematic study of this aspect of membrane coupling.

HARVEY: The genetically defective cell hybridized with a genetically normal cell, suggesting that the normal cell contributes all that is necessary. Would a normal cell contribute all that is necessary to complex with, say, a red blood cell?

LOEWENSTEIN: I don't know. This hasn't been done. You are right, it seems that a normal cell can contribute the necessary information for correcting the genetic defect. All our four hybrids follow this pattern. But I don't know, whether one can do this with a red cell. For that matter, no one, to my knowledge, has shown whether red cells are capable of coupling by themselves. The cells are a bit small to do the experiments. I have just come from another membrane gathering where people were dealing with antigenic properties of membranes. One member there said that agglutinated lymphocytes also talk to each other.

SCHULTZ: Dr. Loewenstein, when you hybridize different epithelial cells, do they transfer specific cellular characteristics? For example, does the lens cell now begin to make collagen?

LOEWENSTEIN: This is a very interesting question. Azarnia, Michalke and I have recently addressed ourselves to this question whether endogeneous molecules involved in nucleic acid metabolism are transferred via junctions. We used for this purpose a mutant cell line which lacks inosinic pyrophosphorylase and hence cannot incorporate hypoxanthine into its pool of nucleic acid. Subak-Sharpe, Pitts and their colleagues at Glasgow had shown that when these mutant cells are grown in contact with wild-type cells, they incorporate hypoxanthine. A nucleotide or a nucleotide derivative appears to be transferred from the wild-type cell to the mutant cell bypassing the enzyme block in the latter. Dr. Subak-Sharpe has kindly given us the mutant cell and we paired it with our noncoupling cell strains. We asked, thus, the question whether the nucleotide derivative is transferred via junctions. The experiments are technically very simple: one loads the noncoupling cells with radioactive hypoxanthine (these cells have inosinic pyrophosphorylase), cultures them in contact with a mutant cell and counts the radioactive label in the radioautographs of the mutant cells. The result was that, in contrast to coupling cells, the noncoupling one did not transfer the nucleotide derivative to the mutant cell (Azarnia et al. 1972) (Figure 1, see figure insert, opposite p. 48). So, to answer your question, here we have an example of an exchange of endogenous molecules involved in nucleic acid metabolism.

Azarnia, R., Michalke, W. & Loewenstein, W. R. (1972) Intercellular communication and tissue growth. IV. Failure of exchange of endogenous molecules between cancer cells with defective junctions and noncancerous cells. *J. Membrane Biol. 10*, 247–258.

The Permeability of Charged Membranes

P. Meares

INTRODUCTION

Recent and precise measurements on the permeability and selectivity of synthetic charged membranes, which consist of crosslinked polyelectrolyte gels, have provided a reasonably clear picture of the functioning of such membranes in a variety of electrodialytic separation processes. This paper shows why such information is relevant also in a symposium devoted to transport in epithelia.

Charges on membranes may arise from strong physical adsorption of ions, particularly organic ions, but the most important charges originate from the dissociation of ionogenic groups chemically bound to the membrane material. Although there is evidence for the presence of such groups in the phospholipids and proteins which constitute cell membranes, their role is not clear because the amount and distribution of water in such membranes is unknown. The great importance of the water distribution is that it controls the state of ionization of the ionogenic groups in a medium of low average dielectric constant. In epithelia one is concerned not only with cell membranes but with a multi-cellular composite structure in which there is cytoplasm. There is no doubt about the presence of ionized groups in the cytoplasm and also in the material filling the interspaces around the cells.

Even in the case of an epithelium it is difficult to assess the role played by the bound charges in controlling permeation because the complexity of the structure complicates the identification of the important transport pathways through the system and the assignment to them of relative importances. The theory of simple transport processes in heterogeneous media is still relatively undeveloped despite considerable attention over a long period. A more immediate objective for a physical chemist interested in membrane

Biophysical Chemistry Unit, University of Aberdeen.

»Transport Mechanisms in Epithelia«, Munksgaard, Copenhagen.

4*

electro-chemistry and in its applications in biophysics is the explanation and prediction of passive transport phenomena in a wide range of model membranes and under a variety of ambient conditions. Until such passive processes are understood the unravelling of the mechanisms of transport coupled to metabolic processes will continue to present daunting difficulties.

AN IDEAL HOMOGENEOUS CHARGED MEMBRANE

The simplest system for theoretical analysis is a membrane whose properties are intrinsically uniform throughout. Such a membrane when brought into equilibrium with an electrolyte solution takes up a uniform electric potential different from that of the solution. This potential is a function of the nature and concentration of the solution and also of temperature and pressure which we shall assume constant throughout this paper.

This idealized membrane forms the logical basis for the Donnan equation when it is applied to describe the equilibrium distribution of ions between the solution and the membrane. The Donnan equation may be written in the form

$$\bar{a}_i = a_i \exp(-z_i F \Psi_D / RT) \tag{1}$$

where z_i is the valency (including charge sign) of the ionic species, a_i its activity in solution and $\bar{a}_i$ in the membrane. F, R and T have their usual meanings and Ψ_D is the Donnan potential i.e. the potential of the membrane interior relative to the external solution.

The number of ionic charges required to establish the potential difference Ψ_D is, in a reasonably thick membrane, a negligible fraction of the total ions present. Both phases may therefore be regarded as electrically neutral overall. When the membrane carries a set of ionized groups at concentration M equiv. kg $^{-1}$, the molalities $\bar{m}_i$ of all other ions in the membrane are related by

$$\Sigma z_i \bar{m}_i \pm M = O \tag{2}$$

where the positive sign applies when the fixed charges are positive and vice versa.

Combination of (1) and (2) gives the Gibbs-Donnan distribution. For a single salt which dissociates into v_g gegenions (i.e. ions of charge opposite from that of the fixed charges) and v_c co-ions the Donnan distribution can be written (Glueckauf 1962)

$$\bar{m}^{v_c} (\bar{m} + M/v_g)^{v_g} = (\alpha m)^v \tag{3}$$

Here $v = (v_c + v_g)$, m is the molality of the salt in solution and $v_c \bar{m}$ the molality of co-ions in the membrane. α is a ratio of activity coefficients which should be relatively insensitive to m and which, in a well-hydrated membrane, will be close to unity.

The essential consequences of eqn. (3) are that the concentration of co-ions is less and of gegenions greater than their concentration in the ambient solution and, further, that the concentration of co-ions increases with a power of m higher than the first (e.g. to the power v/v_c in dilute solutions). Both these predictions are qualitatively borne out in practice but there are important systematic quantitative deviations from eqn. (3) which are discussed below.

When a membrane which separates two different solutions is considered it is usual to assume that the distribution of ions is at equilibrium across each membrane/solution interface. The fluxes across the membrane are then controlled by a set of forces acting on the ions and molecules within the membrane itself.

It is clear that under a given set of forces per mole the concentrations of the ions in the membrane are as important as their mobilities in determining the fluxes. Factors which cause these concentrations to differ from the values predicted by eqn. (3) exert an important influence on the permeability. The experiments on ion distribution described below were designed to study these factors in a well-characterised system.

A SYNTHETIC »HOMOGENEOUS« CHARGED MEMBRANE

All the results in this paper have been obtained with a synthetic gel membrane bearing bound anions and which is therefore selectively permeable to cations. The membrane was prepared so as to be as nearly homogeneous as could be achieved by a random polymerization reaction while ensuring that the membrane charges were spatially fixed and not just constrained within the membrane phase, as they are in a liquid ion-exchange membrane.

Some natural membranes may have greater regularity in the spatial distribution of their charges but this does not necessarily lead to a greater uniformity of electrical potential on the local or micro-scale because the ions and molecules whose diffusion fluxes are important are mostly smaller than the structural elements of membranes. It is believed therefore that results obtained with this synthetic membrane give a useful guide as to how closely any membrane reasonably permeable to water, ions and other solutes may approach ideality in its behaviour.

The synthetic membrane, known as Permutit Zeo-Karb 315, was prepared by reacting together phenol and formaldehyde in the presence of sodium metabisulphite and in an aqueous alkaline medium. The bisulphite formed an adduct with some formaldehyde which led to the attachment of $-CH_2.SO_3Na$ on to the aromatic rings. The product was mixed with more phenol, paraformaldehyde and water and was reacted to form a non-crosslinked polycondensate in which about one-quarter of the phenolic residues carried $-CH_2.SO_3^-$ groups. These groups are probably distributed almost at random with perhaps a slight tendency to regular spacing because of the repulsion between the anionic charges.

The liquid polymer was cast into the form of the membrane and crosslinked by heating. No shrinkage occurred and the result was a randomly crosslinked molecular network encompassing the water and other substances present. On eluting these substances in water or a dilute salt solution the membrane swelled and its volume increased by 15–20 %. During this swelling, volume elements more loosely crosslinked than the average extended more than tightly crosslinked ones. Inevitably this led to some microscopic dispersity in the distribution of fixed charges in the finished membrane.

Some important characteristics of the Zeo-Karb 315 membranes are given in Table I. More details may be found in the original references cited throughout this paper.

Table I Properties of Zeo-Karb 315 in 0.1 mol dm^{-3} NaBr at 25°C.

$\bar{c}/\overline{Na}$	0.58 mole dm^{-3}
$\bar{c}/\overline{Br}$	0.031 mole dm^{-3}
water content	0.74 kg dm^{-3}
specific conductance	1.23 Ω^{-1} m^{-1}
electro-osmosis	7.5 mm^3 C^{-1}
transport number of Br$^-$	44.5 × 10^{-3}

CO-ION UPTAKE AND THE DONNAN DISTRIBUTION

The uptake of electrolytes from dilute solutions by ion exchange resins has been studied by several workers as a function of the solution concentrations. It has always been found that the uptake from dilute solutions is considerably larger than would be expected from the simple Donnan law. At higher concentrations the law is increasingly well obeyed. This behaviour has been confirmed with many electrolytes of different valence types in Zeo-Karb 315. Fig. 1 and 2, based on the results of Mackie and Meares (1955a) show data on NaCl and $MgCl_2$ as typical examples. In 0.01 m solutions the observed uptake was more than twice the ideal prediction but agreement was close above 0.5 m.

Various experimental artefacts, such as cracks and voids in the resin (Davies and Yeoman 1953), inability to separate completely the resin and solution for analysis, and variations among individual particles in a sample of bead resin, (Freeman 1960) have been suggested to explain these obser-

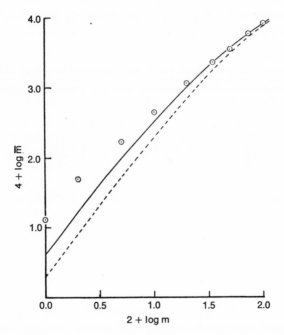

Fig. 1. Sorption of NaCl by Zeo-Karb 315. m is the molality in solution and $\bar{m}$ in the membrane. The points are experimental, the dotted curve was calculated from the ideal Donnan eqn. (3) with $\alpha=1$ and the full line from the theory of Katchalsky (1954) and Mackie & Meares (1955a).

vations. While all of these effects may contribute they are insufficient to explain all the deviations and, in particular, their universality.

The only measurable parameter of the membrane which appears in the Donnan equation is the macroscopic average concentration of fixed charges. Local variations in the potential, due to the discrete nature of the ionic charges, on a scale larger than the thickness of the ionic atmosphere in the membrane must lead to an ion uptake greater than the ideal prediction. Such local variations are inevitable where the mean distances between the fixed charges exceed 1nm i.e. where the concentration of fixed charges is less than about 10 molar.

Various attempts have been made to treat this problem theoretically. Three structural features may be described which could lead to local variations in the potential. The bound charges may be separated from one another along the macromolecules by distances large in the present context. The macromolecules in the swollen gel membrane may be separated by large distances. There may be irregularities in the microstructure of the macromolecules or in their distribution in the gel which produce local fluctuations in the fixed charge concentration on the scale 10–100 nm as well as on the molecular scale.

One group of theories concentrates on the second effect. The uniform charge of the ideal model is replaced by charges smeared along the macro-

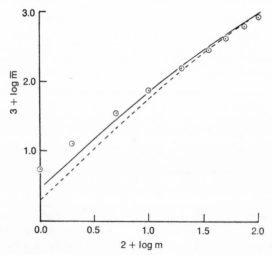

Fig. 2. As Fig. 1 but for $MgCl_2$.

molecules which are treated either as parallel plates (Lazare *et al.* 1956) or as rods (Manning 1969) or else more simply, as two regions of different uniform potential (Tye 1961). Such theories predict an effect in the direction observed but they do not take into account the discrete nature of the fixed ionic charges.

In the case of Zeo-Karb 315 the mean separation along the polymer chains of the fixed charges is about 2nm. This is similar to the mean separation between neighbouring polymer chains in the membrane and is several times the thickness of the ionic atmospheres. A smeared charge model is inappropriate here and the theory of solutions of flexible polyelectrolytes as extended to crosslinked gels (Katchalsky 1954) is more suitable. This theory takes into account the additional electrostatic and elastic contributions to the free energy which result from the interconnection of the fixed charges along the polymer chains and through the crosslinks.

Mackie and Meares (1955a) applied Katchalsky's theory to Zeo-Karb 315 and were able to predict co-ion uptakes on the basis of membrane structural parameters which fitted the observed data well down to 0.1m external concentration (see Fig. 1 and 2) and which deviated from observation far less than did the Donnan equation at lower concentration. Clearly the discreteness of ionic charge and the polymeric nature of the real system lead to substantial deviations from the Donnan equation.

At the lowest concentrations the uptake was still 50–60 % larger than could be accounted for. This is believed to reflect local irregularities in the distribution of fixed charges along the chains and of the crosslinks between them. The belief has been reinforced by observations of ionic mobilities and friction coefficients in the membrane (McHardy *et al.* 1969b).

Glueckauf (1962) showed how such ion uptake data could be analysed in terms of the Donnan equation to give information on micro-heterogeneity in the distribution of the fixed charges. Later research has shown that the empirical distribution function which described Glueckauf's membrane is not suitable in all cases. A more general distribution function would make possible the determination of important parameters of the membrane structure.

A plot of $1/v$ times the logarithm of the left side of eqn. (3) versus log m should be a straight line of slope unity and intercept $\beta \log \alpha$ when m = 1. Fig. 3 shows such plots for NaCl and MgCl$_2$ in Zeo-Karb 315. They are linear but the slopes, although almost equal for the two salts, are less than

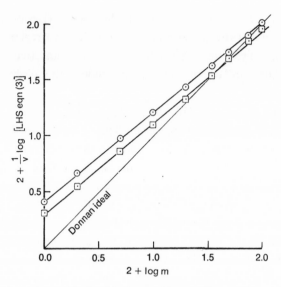

Fig. 3. Comparison of eqn. (3), i.e. ideal $\alpha=1$ Donnan line, with eqn. (4) for NaCl and MgCl$_2$. For NaCl $\alpha=0.991$ and $\beta=0.83$; for MgCl$_2$ $\alpha=0.807$ and $\beta=0.84$. $\odot$ NaCl: $\square$ MgCl$_2$.

unity while log α turns out, as hoped, to be almost zero in each case. Thus for the Zeo-Karb 315 membranes and perhaps for other essentially homogeneous membranes the ideal Donnan eqn. (3) can be replaced in practice by

$$\bar{m}^{v_c}\,(\bar{m} + M/v_g)^{v_g} = (\alpha m)^{\beta v} \qquad\qquad 4$$

Here α is a characteristic of the electrolyte but does not differ greatly from unity while β may be a valuable parameter characteristic of the internal dispersity of the charge distribution, the greater $(1-\beta)$ the greater the dispersity.

CONTRIBUTION OF DONNAN POTENTIALS TO MEMBRANE POTENTIALS

The sorbed co-ions (valency z_c) are not bound by the fixed charges and it is likely that the Donnan potential Ψ_D can be satisfactorily represented by

$$\Psi_D = RT \log(\alpha m/\bar{m})/z_c F \qquad\qquad (5)$$

provided that m is obtained from eqn. (4) with the appropriate values of α and β determined experimentally. This conclusion is important when attention is directed to a membrane which separates two solutions. The difference between the Donnan potentials at its opposite faces appears as a part of the membrane potential in the Teorell-Meyer-Sievers (TMS) theory (see e.g. Teorell 1953). It is clear from the sorption results that the Donnan potential at the dilute face of the membrane falls further below the ideal value than that at the concentrated face. This may lead to deviations greater than 20mV from the calculated TMS potentials.

SIMPLE FLUX RELATIONS

The TMS theory of fixed-charge membrane potentials originally used the Henderson equation to represent the internal diffusion contribution to the membrane potential. Later this was replaced by an appropriate solution of the Nernst-Planck equation. This had the advantage of giving also expressions for the ion fluxes.

According to the Nernst-Planck equation the ion fluxes $\emptyset_i$ are driven only by the gradients of concentration and potential in the membrane. They can be expressed by

$$\emptyset_i = - \bar{u}_i \bar{c}_i \, (RTd\ln \bar{c}_i + z_i \, Fd\Psi)/dx \qquad (6)$$

where $\bar{c}_i$ is the local concentration per unit volume in the membrane and x is distance through the membrane. $\bar{u}_i$ is the absolute mobility of the ions and, in a homogeneous membrane, is intrinsically independent of x but may be dependent on $\bar{c}_i$ and on the other substances present. Integration of eqn. (6) across the membrane has usually been based on assuming $\bar{u}_i$ to be constant and may, as in the Goldman (1943) equation assume $d\Psi/dx$ constant also.

Many particular cases have been considered and expressions obtained for ion fluxes and potentials (Schlögl 1964). These equations, despite the idealisations required in their derivation, have been widely used by biologists to interpret fluxes and potentials.

NON-EQUILIBRIUM THERMODYNAMICS AND FLUXES

More general flux equations have been developed from non-equilibrium thermodynamics (see e.g. Katchalsky & Curran 1965). These offer the possibility of including formally all forces, fluxes and the interactions between them. They suffer from two major disadvantages; their quantitative use requires the determination of a large number of empirical parameters for each system of membrane and ambient solutions and, in the linear form, they are applicable only to systems which are close to equilibrium. Usually the forces chosen are differences between intensive variables characteristic of the solutions and the coefficients which connect these forces with the fluxes represent average values across the membrane. In this "discontinuous" form equations with constant coefficients hold over such narrow ranges that they are likely to be invalid in many biological situations.

For these reasons non-equilibrium thermodynamics applied to biological transport processes is likely to prove of greater value in conceptual arguments than in the quantitative correlation of fluxes and forces. Biologists must continue to rely heavily on the simpler flux relations referred to earlier in order to examine the connections between passive fluxes and forces.

The application of non-equilibrium thermodynamics to synthetic membranes although difficult has now been achieved in detail with Zeo-Karb 315 (Meares *et al.* 1972, Foley & Meares 1971). The results of this work permit the assumptions used in the Nernst-Planck equation and its integrations to be tested.

Restricting attention to aqueous solutions of a single salt and using the subscripts 1 for cations (gegenions in Zeo-Karb 315) 2 for anions and 3 for water, the linear equations for the molar flux densities $\varnothing_i$ may be written

$$\varnothing_i = \sum_{k=1}^{3} L_{ik} \, \triangle \, \mu_k \tag{7}$$

The fluxes are positive in the direction opposite from that chosen for expressing increases in chemical potentials. $\triangle \mu_1$ and $\triangle \mu_2$ are differences in the electrochemical potentials of cations and anions between the external solutions and $\triangle \mu_3$ is the difference in the chemical potential of water.

In the case of a homogeneous membrane means can be adopted which permit the differential fluxes to be evaluated under conditions of vanishingly small forces (Krämer & Meares 1969). There is then no difficulty regarding linearity and the distance from equilibrium. The phenomenological conductance coefficients L_{ik} can be replaced by a set of differential conductances $\mathscr{L}_{ik}$ defined by

$$\mathscr{L}_{ik} = \underset{\text{all } \triangle \mu_i \to 0}{\ell im} \quad (\partial \varnothing_i / \partial \mu_k)_{\mu_{j \neq k}} \tag{8}$$

and where $\mathscr{L}_{ik} = \mathscr{L}_{ki}$. These conductances may be evaluated from experiments as functions of known membrane composition.

An inverse set of relations equivalent to eqn. (7) is also well known. It is

$$\triangle\mu_i = \sum_{k=1}^{3} R_{ik} \, \emptyset_k \tag{9}$$

from which differential resistance coefficients $\mathscr{R}_{ik}$ may be defined by

$$\mathscr{R}_{ik} = \lim_{\text{all } \emptyset_i \to O} (\partial\mu_i/\partial\emptyset_k)_{\emptyset_{j\neq k}} \tag{10}$$

COMPARISON WITH THE NERNST-PLANCK FORMULATION

The Nernst-Planck equation (6) is a local differential equation and in order to compare it with the non-equilibrium thermodynamic equation (7) this must also be written in differential form as follows

$$\emptyset_i = \ell[\mathscr{L}_{11}(\partial\mu_1/\partial x) + \mathscr{L}_{12}(\partial\mu_2/\partial x) + \mathscr{L}_{13}(\partial\mu_3/\partial x)] \tag{11}$$
$$(i = 1 \text{ or } 2)$$

where ℓ is the membrane thickness.

These equations are consistent with eqn. (6) only if all $\mathscr{L}_{ik}$ (i $\neq$ k) are zero. Eqn. (8) shows that $\mathscr{L}_{ik}$ connects the flux of i with the force on, and consequently the flux of, k. Use of the Nernst-Planck equation is suspect in circumstances where the fluxes of different species are coupled. This restriction is likely to be important for ionic fluxes through a membrane in which there is an osmotic flux of solvent. The generation of an electro-osmotic flux by the passage of electric current through a membrane is direct evidence of coupling between water and ion fluxes.

SOLVENT-DRAG AND THE NERNST-PLANCK EQUATIONS

The $\mathscr{L}_{ik}$ relate the flows to gross thermodynamic forces. A non-zero $\mathscr{L}_{ik}$ implies that a force on k affects the flux of i, it does not indicate direct interaction between i and k. The meaning of the $\mathscr{R}_{ik}$ is clearer. Eqn. (9) shows that R_{ik} represents the drag force on i produced by unit flux of k, all other fluxes, including $\emptyset_i$, being zero. Thus R_{ik} measures the direct interaction between i and k. A non-zero R_{ik} implies that if k is a mobile species a flux of k will directly influence the flux of i. R_{ii} represents the force required to produce unit flux of i all other fluxes being held at zero.

Interactions are not taken into account in the Nernst-Planck equation and it is useful to know when they are likely to be important. The degree of coupling q_{ik} introduced by Caplan (1965, 1966) is important here.

$$q_{ik} = - \mathscr{R}_{ik}/\sqrt{\mathscr{R}_{ii}\,\mathscr{R}_{kk}} \tag{12}$$

q_{ik} represents the degree to which a flow of i is created by a flow of k when there is no external force on i and all flows other than i and k are zero. q_{ik} may vary between zero in the absence of coupling and unity when, in the absence of other influences, the particles of i and k move with the same mean velocity.

The values of q_{13} and q_{23} for $NaBr_2$ in Zeo-Karb 315 are listed in Table II. The cation flux is more tightly coupled to the water flow than is the anion flux, a consequence probably of the greater hydration of cations, and in both cases the coupling is considerable.

Table II Degree of coupling of ion and water fluxes in Zeo-Karb 315.

soln. conc. (mol dm^{-3}) NaBr	q_{13}	q_{23}
0.995	0.65	0.46
0.495	0.60	0.38
0.290	0.56	0.26
0.085	0.50	0.13
0.020	0.56	−0.21
SrBr$_2$		
0.499	0.65	0.35
0.250	0.47	0.15
0.104	0.42	0.10
0.051	0.39	0.07
0.015	0.34	0.00

Ion/water flux coupling has been accommodated by adding a further term to the Nernst–Planck equation to give (Schlögl & Schödel 1955, Mackie & Meares 1955b)

$$\emptyset_1 = - \bar{u}_1\bar{c}_1 RT d\ell n\bar{c}_1 + z_1 Fd\Psi)/dx + \gamma\emptyset_3\bar{c}_1/\bar{c}_3 \tag{13}$$

here γ is a factor determined by the average degree of coupling of the ion and water fluxes. In the references cited γ was implicitly taken as unity i.e. complete coupling, and it was found that eqn. (13) over-corrected for solvent drag.

McHardy et al. (1969) determined γ, which includes also the effect of ion-ion drag, and found values for Na^+ and Sr^{2+} between 0.4 and 0.6, in good accord with q_{13} in table II. For Br^- γ was variable and was apparently strongly influenced by ion-ion interaction. It is likely that in many homogeneous membranes eqn. (13) would be an improvement on eqn. (6) if γ were set at 0.5.

IONIC MOBILITIES AND THE NERNST-PLANCK EQUATION

Integration of the Nernst-Planck equation is usually carried out under the assumption that the ionic mobilities u_i inside the membrane are independent of the concentrations outside. The nonequilibrium thermodynamic data enable this assumption to be tested.

Comparison of eqn. (6) and (11) shows that they can be made to agree irrespective of the values of the $\mathscr{L}_{ik}$ cross-coefficients provided all forces are zero except those on the species whose flux is being considered. Then

$$\mathscr{L}_{ii}\ell = \bar{u}_i\bar{c}_i \qquad (14)$$

provided the activity coefficient gradient of i may be neglected. Thus the constancy of the mobilities u_i may be judged from the dependence of $\mathscr{L}_{ii}\ell/\bar{c}_i$ on external solution concentration.

Plots of $\mathscr{L}_{11}\ell/\bar{c}_1$ versus the concentration of the external solution for Na^+ and Sr^{2+} and of $\mathscr{L}_{22}\ell/\bar{c}_2$ for the Br^- co-ions are shown in Fig. 4. It can be seen that for all the ions studied here there is a steady increase in the mobility with increasing external solution concentration.

It may be objected that in a membrane where flux coupling occurs $\mathscr{L}_{ii}\ell/\bar{c}_i$ measures the mobility of i against a moving background because fluxes of species other than i will be generated by the force on i. The reciprocal quantity $\ell/\bar{c}_i\mathscr{R}_{ii}$ measures the mobility of i relative to all other species held stationary. $\ell/\bar{c}_i\mathscr{R}_{ii}$ is plotted in Fig. 5 for Na^+, Sr^{2+} and Br^- in the Na^+ form of the membrane. It can be seen that although there is an upward trend in these mobilities also as the external concentration is increased

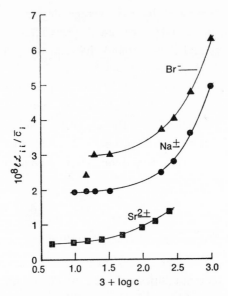

Fig. 4. Intrinsic conductance coefficients of ions in Zeo-Karb 315. The measurements on Br⁻ were made in the Na⁺ form of the membrane. c mol dm⁻³ is the solution concentration and $\mathscr{L}_{ii}/c_i$ is in m² mol J⁻¹ s⁻¹ and is scaled for a membrane of thickness lm.

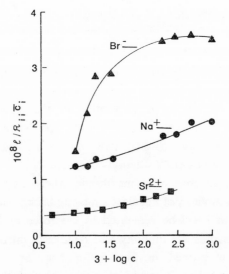

Fig. 5. Intrinsic reciprocal resistance coefficients of ions in Zeo-Karb 315. Other details as in Fig. 4. $l/\mathscr{R}_{11}\bar{c}_i$ is in m² mol J⁻¹ s⁻¹ for a membrane of thickness lm.

the effect is less marked than in Fig. 4. Probably the assumption of constant $\bar{u}_i$ in the Nernst-Planck integrations will not lead to serious errors if an appropriate mean mobility, which is a function of the solution concentrations, is used.

An increase in the gegenion mobilities with increasing solution concentration has been inferred on several occasions from isotopic tracer flux measurements but the behaviour of the co-ions is less uniform (see e.g. Meares 1968). Tracer diffusion coefficients in Zeo-Karb 315 are consistent with the data on $\mathscr{L}_{ii}\ell/\bar{c}_i$ and $\ell/\bar{c}_i\mathscr{R}_{ii}$ (McHardy et al. 1969b). There are quantitative differences between tracer mobilities and those given here which result from interactions between tracer and non-tracer ions of the same chemical species. These interactions are often ignored but measurements show they can be very significant. They will be dealt with in a later publication.

SUMMARY

The simplest and most popular formulations of the fluxes and potentials across charged membranes which can be used to interpret data on biological systems are still those based on the Teorell-Meyer-Sievers fixed charge model. Usually ionic distributions described by the ideal form of the Donnan equation are assumed to hold at the membrane faces and uncoupled ionic fluxes in the membrane are represented by the Nernst-Planck equation. In this paper these assumptions are tested on an almost homogenous synthetic polyelectrolyte gel membrane selectively permeable to cations.

It is found that:--

(a) the ideal Donnan equation is well obeyed at high concentrations but at low concentrations the co-ion uptake by the membrane is several times larger than the ideal value. This behaviour is a result of the polymeric character of the membrane matrix and of the fluctuation in local charge density on the micro-scale.

(b) Ion and osmotic water fluxes are strongly coupled, except for the co-ions in very dilute solutions. The omission of this factor from the Nernst-Planck equation may lead to considerable errors when in the presence of large water fluxes.

(c) Ionic mobilities evaluated through non-equilibrium thermodynamics approximately double as the solution concentration is increased from 0.01 to 1.00m. Provided an appropriate average value is used this variation may not greatly upset the integration of the Nernst-Planck equation but the average mobility has then to be regarded as a function of the concentration of both solutions separated by the membrane.

REFERENCES

Caplan, S. R. (1965) The degree of coupling and efficiency of fuel cells and membrane desalination processes. *J. phys. Chem. 69,* 3801–3804.

Caplan, S. R. (1966) The degree of coupling and its relation to efficiency of energy conversion in multiple flow systems. *J. Theor. Biol. 10,* 209–235, 346–347.

Davies, C. W. & Yeoman, G. D. (1953) Swelling equilibria with some cation exchange resins. *Trans. Faraday Soc. 49,* 968–974.

Foley, T. & Meares P. (1971) Linear transport coefficients in a cation-exchange membrane. Biological aspects of electrochemistry. *Experientia (Basel) Suppl. 18,* 313–319.

Freeman, D. H. (1960) Electrolyte uptake by ion-exchange resins. *J. Phys. Chem. 64,* 1048–1051.

Glueckauf, E. (1962) A new approach to ion exchange polymers. *Proc. roy. Soc. A 268,* 350–370.

Goldman, D. E. (1943) Potential, impedance and rectification in membranes. *J. gen. Physiol. 27,* 37–60.

Katchalsky, A. (1954) Polyelectrolyte gels. *Progress in Biophysics,* ed. Butler, J. A. V. and Randall, J. T. *4, 1–59.* Pergamon Press, London.

Katchalsky, A. & Curran, P. F. (1965) *Non-equilibrium thermodynamics in biophysics.* Harvard University Press, Cambridge, Mass.

Krämer, H. & Meares, P. (1969) Correlation of electrical and permeability properties of ion-selective membranes. *Biophys. J. 9,* 1006–1028.

Lazare, L., Sundheim, B. R. & Gregor, H. P. (1956) A model for crosslinked poly-electrolytes. *J. phys. Chem. 60,* 641–648.

McHardy, W. J., Meares, P., Sutton, A. H. & Thain, J. F. (1969a) Electrical transport phenomena in a cation-exchange membrane II. *J. Colloid and Interf. Sci. 29,* 116–128.

McHardy, W. J., Meares, P. & Thain, J. F. (1969b) Diffusion of radio-tracer ions in a cation-exchange membrane. *J. Electrochem. Soc. 116,* 920–928.

Mackie, J. S. & Meares, P. (1955a) The sorption of electrolytes by a cation exchange resin membrane. *Proc. roy. Soc. A 232,* 485–498.

Mackie, J. S. & Meares, P. (1955b) The diffusion of electrolytes in a cation-exchange resin membrane I. *Proc. roy. Soc. A 232,* 498–509.

Manning, G. S. (1969) Limiting laws and counterion condensation in polyelectrolyte solutions. I. *J. Chem. Physics 51,* 924–933.

Meares, P. (1968) Transport in ion-exchange polymers. *Diffusion in Polymers.* ed. Crank, J. & Park, G. S., pp. 373–428. Academic Press, London.

Meares, P., Thain, J. F. & Dawson, D.G. (1972) Transport across ion-exchange resin membranes. The frictional model of transport. *Membranes Vol. I.* (ed.) Eisenman, G., pp. 55–124. Marcel Dekker, New York.

Schlögl, R. (1964) *Stofftransport durch Membranen.* Steinkopf Verlag. Darmstadt, Germany.

Schlögl, R. & Schödel, U. (1955) Über das Verhalten geladener Porenmembranen bei Stromdurchgang. *Z. physik. Chem. N.F. 5,* 372–397.

Teorell, T. (1953) Transport processes and electrical phenomena in ionic membranes. In *Progress in Biophysics,* ed. Butler, T. A. V. and Randall, T. T. *3,* 305–369. Pergamon Press, London.

Tye, F. L. (1961) Absorption of electrolytes by ion exchange materials. *J. Chem. Soc.* 1961: 4784–4789.

68

DISCUSSION

KEYNES: I wonder whether the kind of consideration you have been presenting could conceivably explain one of the things about water movement in nerve membranes which has not been explained. If you measure the amount by which a nerve swells during a single impulse, it turns out to be much too big to fit with the net movement of ions. This measurement was made by D. K. Hill about 20 years ago (Hill 1950), and he suggested that it could be explained by supposing that some chloride went in with sodium, and correspondingly slightly less potassium came out to balance the charge movements. In fact the chloride movement has since been measured and it isn't big enough. But supposing there were water movements coupled to the sodium and potassium, then even if the fluxes of sodium and potassium were exactly equal, the movements of the water that accompanied them might not be equal. They are going through different channels, you see. And the fixed charges in the two channels might be quite different.

MEARES: Yes, I think that one would expect the coupling between sodium and water would be stronger than the coupling between potassium and water. The reason why I say this is because sodium tends to be more strongly hydrated and you find bigger coupling coefficients with strongly hydrated ions than with poorly hydrated ions.

KEYNES: Yes, that is the way I want it to be. In that case, if you made measurements with lithium instead of sodium you might find a bigger effect.

MEARES: Yes, and do you find such an effect?

KEYNES: No one has ever done that experiment.

CURRAN: I wonder if you have any ideas about why, in this really quite a loose membrane (75 percent water), the coupling coefficients between the ions and the solvent are only about 0.5. I would guess that they would be higher than that.

MEARES: Yes, let us talk about the counter-ions first, because I think this is essentially the difference between the counter-ions and the co-ions. The counter-ions of course are attracted by the ionic matrix, and consequently,

---

Hill, D. K. (1950) The volume change resulting from stimulation of a giant nerve fibre. *J. Physiol. (Lond.) 111*, 304–327.

although when you make a measurement on the amount of water present and the amount of ions present you know the average macroscopic concentrations, this tells you nothing about the local distribution within the system. The fact is that strontium, for example, has a weaker coupling than sodium. I admit that I said to Dr. Keynes, the more strongly hydrated the ion the greater is the coupling. He could say of course that strontium is more highly hydrated than sodium and yet it is weakly coupled. The reason in this case is that the strontium ions being doubly charged spend their time closer to the ionic matrix and indeed have a tendency to go from one fixed charge to the next and consequently are in less contact with the water stream.

At the same time, the water is interacting with the matrix itself so that the mean water velocity does not tell you what the local water velocity is. The way one carries out any thermodynamic analysis is of course as a statistical analysis. So the velocity of water which is implicit in this is a mean velocity of water but that is not the local velocity that the ions are seeing. If the local velocity in the environment of the particle being considered is less than the mean velocity, again you will get an apparently weaker coupling.

As far as the bromide co-ions are concerned, the first thing I have to say is that it is very difficult to get all these $R_{ik's}$ and $L_{ik's}$ with very high accuracy. When we get down to dilute solutions, when there are very few bromide ions present, I wouldn't like to argue about the precise value that we get. But at the high concentrations one finds the bromide about equally coupled with the solvent as are the sodium ions. I think the reason for this is that the bromide ions are not uniformly distributed. There is a structural dispersity in this resin. The factor B in the uptake equation is really a measure of dispersity, that is to say of irregularity – if you like – in the structure. The bromide ions when they are taken up tend to congregate in regions of low local Donnan potential. In those regions the water movement is probably slower becauce the osmotic activity is less than the average. So I rather think that both types of ions tend to find themselves in an environment where the water velocity is somewhat lower than the average.

SCHULTZ: Is the use of the Donnan distribution at the interface appropriate in view of the fact that it really does not discriminate between the distribution of different monovalent cations or different monovalent anions. Does your membrane discriminate between sodium and potassium?

MEARES: That is what the α is doing there, if you like. Of course, what I am trying to say is exactly what you are saying – that the ideal Donnan is not adequate. There is nothing wrong with the Donnan principle, which is a thermodynamical principle derived straight from Gibbs' principle that the chemical potential must be the same everywhere in a system in equilibrium, to which the particles have access. The question is: how are you going to express these chemical potentials in a meaningful way in a thermodynamic framework? The usual trick is to cover ignorance by introducing an activity coefficient. Provided you realize that this is what you are doing you won't leave the activity coefficient out, but what people have often done, and what was done, of course, in the original TMS theory, was to leave these activity coefficients out.

CURRAN: But what is α exactly. Is it the ratio of the activity coefficients?

MEARES: It is the ratio of the mean ionic activity coefficients of the salt, that is to say NaBr or whatever it is, inside and outside. The thermodynamic meaning is quite clear, but what it means in terms of the way the particles are behaving requires more thought. This is why for example α for magnesium chloride is lower than for sodium chloride: about 0.8 against 0.991. The reason for this is that there is a much more specific interaction between the ions in magnesium chloride when you have some ion pairing, even with the highly hydrated magnesium ions, than in the case of sodium chloride.

ZADUNAISKY: I wonder if these polyelectrolytes tend to swell according to the ionic composition of the solution. Does this change in the thickness of the membrane itself have any importance?

MEARES: I can answer this exactly because we have made many many measurements on this. When you make the membrane you have a liquid resin which you finally cure between glass plates in an oven. When you take this out and put it into an aqueous medium, a dilute solution or water, it swells in volume. It swells in volume by about 15 percent, $i.\ e.$ the thickness of the resin increases by $\sim$ 5 percent. Now this is an unfortunate fact of life and we haven't been able to improve on this. What this causes is: any regions which statistically are less strongly crosslinked do most of the swelling. This upsets the overall homogeneity of the structure. Once having got through this stage, if you transfer the resin from water to an IN solution of any of the salts I have been talking about to-day, the reduction in volume is less than 3 percent, $i.\ e.$ the reduction in thickness is less than 1 percent. The difference between the volume per electrochemical equivalent of fixed charges in the magnesium form and in the sodium form is about 2 percent. So the effect to which you allude is there. In working out our coefficients we introduce the membrane thickness, and so we take this swelling into account. But we are talking now in terms of a fraction of 1 per cent and I don't think it has any very large effect on the quantitative description.

ZADUNAISKY: I am glad you say that. It doesn't seem to have so much relevance to some of the biological membranes we work with. However, some other tissues that contain these polyelectrolytes tend to swell 2,000 times.

LOEWENSTEIN: How does your polyelectrolyte system behave with calcium in the medium. I am particularly interested to hear what happens in terms of volume. Does it shrink?

MEARES: I can speak about magnesium and about strontium and about many univalent ions, but I can't speak about calcium because I haven't done calcium work with this particular material. I have done calcium work with many other ion exchange materials. Calcium stands in the selectivity series between magnesium, strontium while barium goes on beyond it, so calcium stands in exactly the order we expect it to do. The volume of our membrane in the magnesium or strontium form, per equivalent of resin, is within 2 percent of the volume in the sodium or the hydrogen form. If you take a more highly crosslinked resin, some of the organic exchangers that people typically use in the lab, the volume change is rather bigger than that, but I think it would rarely go beyond 10 percent.

LOEWENSTEIN: There is one system which shows an important change with calcium. It is a cross linked gel of polymethacrylic acid, a system studied by A. Katchalsky and Zwick. I believe these workers have also a number of other polycarboxylic acid gels that undergo dimensional changes with divalent cations. I have one further question: If one puts hydrostatic pressure on your system, does one get appreciable charge transfers?

MEARES: Can I say something to you first about the volume change. You will get a big shrinkage if you use a weak acid resin. If you use a carboxyl type material then you get a big volume contraction with calcium, as opposed to sodium. Now, the question of pressure. The experiments we have done are rather like reverse osmosis. You put a pressure on one side of a membrane and allow the solution to come out of the other side. In this case one of the things we would like to be able to measure is the streaming potential because this actually would give us a means of checking on one of the Onsager reciprocal relations. The fact is that the streaming potentials for pressures around 10 atmospheres are rather less than 1 mV per atmosphere. It is quite difficult to measure this. We tried to make reversible electrodes, but they don't like pressure changes for some reason. Thus it is rather difficult for me to answer your question, except to say that the streaming potentials we observed were well within a few mV of those we expect from thermodynamics. There seems to be no very specific effect of pressure.

Kinetics of Sodium and Lithium Accumulation in Isolated Frog Skin Epithelium

François Morel and Gérard Leblanc

As first reported by Ussing (1963) and later described by Rawlins *et al.* (1970), the epithelial cell layers of the frog skin may be split from the underlying connective tissue by applying hydrostatic pressure *in vitro* to the inner face of the structure. But this treatment generally resulted in reduced D. C. resistance of the preparation and in morphological alterations of the inner border of the *stratum germinativum*. Such damage can be prevented when lower hydrostatic pressure, combined with collagenase treatment is used (Aceves & Erlij 1971 and Carasso *et al.* 1971).

Electron microscope examination of epithelia prepared by this combined treatment showed that the cells of the germinative layer were normal in appearance, although the basement membrane was no longer present (Carasso *et al.* 1971). In addition, isolated epithelia prepared by this procedure retained the electrical and transport properties of the whole skin (Aceves & Erlij 1971 and Rajerison *et al.* 1972a). Thus, as shown in Table I, short circuit current, transepithelial P. D., and D. C. resistance were found to be

Table I. *In vitro comparison of isolated epithelium and whole skin.*

	N	S.C.C. $\mu A/cm^2$ (± S.D.)	P.D. mV (± S.D.)	R KOhms × cm^2 (± S.D.)
Isolated epithelium	87	14.4 (± 1)	30.3 (± 15.9)	2.77 (± 1.75)
Whole skin	69	14.2 (±6.6)	33.3 (± 13.3)	2.46 (± 0.63)

Both sets of measurements were obtained from the same batches of *Rana esculenta*. The preparations were bathed with normal Ringer solution on both sides, and short circuited.
N : number of determinations; S.C.C. : short circuit current;
P.D. : transepithelial potential difference; R : d.c. resistance.
(Rajerison *et al.* 1972a)

Laboratoire de Physiologie Cellulaire Collège de France, 75231 Cedex 05, Paris, France and Département de Biologie C. E. N. de Saclay B. P. No. 2, 91190 Gif sur Yvette, France.

»Transport Mechanisms in Epithelia«, Munksgaard, Copenhagen.

equal both in isolated epithelia and pieces of whole skin prepared from the same batches of *Rana esculenta.*

It was also observed in isolated epithelia that both active sodium transport and osmotic permeability to water increased in response to oxytocin, cyclic AMP, theophylline (Rajerison *et al.* 1972a) and catecholamines (Rajerison *et al.* 1972b). Finally, tracer experiments on isolated epithelia placed in short circuit conditions made it possible to verify that the net sodium influx was equal to the short circuit current (Aceves & Erlij 1971), whereas chloride influx and outflux did not differ significantly from each other (Rajerison *et al.* 1972b).

It thus appears that preparations of isolated epithelia retain all the main properties of the whole skin *in vitro.* They are therefore very suitable for analysing ion movements and electrolyte cell composition as a function of the electrochemical gradients prevailing at the outer and inner borders of the preparation. A few observations recently made in our laboratory relating to this context will be reported.

When isolated pieces of epithelium were incubated with normal Ringer solution on both sides, the sodium contained within the cells amounted to a mean value of 111 nEq/cm^2 surface area in short circuit conditions, and to 80 nEq/cm^2 in open circuit conditions. Expressed in terms of concentration per liter of cell water, the corresponding values are 32 mM and 23 mM respectively. These measurements were made after an equilibration period of at least an hour, and both C^{14} and tritiated inulins were used simultaneously in order to introduce appropriate corrections for mucosal and serosal extracellular fluids. As a mean value for the cell water, we found 2.5 μl/cm^2; it should be mentioned that the thickness of the structure varied over a wide range from animal to animal. These values for the water and Na cell content are in close agreement with those measured in the same conditions in the isolated epithelium of *Rana pipiens* (Aceves & Erlij 1971).

It has been clearly established (Biber & Curran 1970) that lithium and sodium ions may compete for entering the outer barrier of the skin; in addition, it was observed (Hansen & Zerahn 1964) that lithium added to the outside medium accumulates in the structure, which suggests that lithium is poorly transported by the Na pump located in the intercellular and basal membranes of the epithelial cells.

Experiments were therefore performed using lithium ions in place of sodium ions in the solution bathing the outer border of isolated epithelia, in order to analyse further the role of the outer barrier in the establishment of the steady state ionic cell content (Leblanc, 1972). Normal sodium Ringer solution was used as the inner bathing medium, choline Ringer as the outer bathing medium. After a 40 min. equilibration period, LiCl was added outside up to 1, 2.5., 10 or 25 mM/1. The amount of lithium which entered the cells was then determined after different incubation times. Similar experiments were also performed in the presence of 10^{-4}M amiloride in the outside solution or 10^{-3}M cyanide in the inside solution.

Although some transepithelial net flux of lithium could be measured after an incubation of one hour, it clearly appeared that a large part of the lithium ions which entered the outer barrier accumulated in the structure. Lithium concentration was calculated assuming uniform distribution throughout the cell water of the entire structure. The accumulation tended to saturate as a function of the incubation time; saturation was also observed as a function of the lithium concentration used. After one hour's incubation, steady state Li concentrations ranging from 10 to 30 mM/1 were obtained, as shown in Fig. 1.

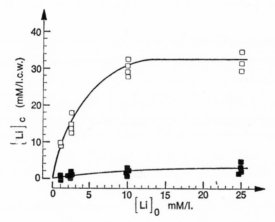

Fig. 1. Pieces of isolated skin epithelium *(Rana esculenta)* were incubated in short circuit conditions; inside bathing medium: normal Ringer solution; outside bathing medium: choline Ringer solution, to which various lithium chloride concentrations were added (as plotted in abscissa). After 1 h. exposure to lithium, the tissue was rapidly punched out and the lithium concentration in the cell water measured as indicated in the text.

Open symbols: lithium concentrations measured in cell water in control experiments.

Filled symbols: lithium concentrations measured when the incubation was performed in the presence of 10^{-4} amiloride in the outside medium.

(Leblanc, 1972).

Thus, lithium accumulation always took place against the concentration gradient; with the lowest external lithium concentration used (1 mM), chemical gradients of over 10 were obtained, both in short circuit and open circuit conditions. Lithium penetration was almost completely eliminated in the presence of amiloride (Fig. 1). It was also observed that the rate of accumulation increased when oxytocin was added to the serosal bathing medium.

All these observations can be explained by assuming a) that lithium ions penetrate the outer barrier of the structure through the same pathway and by the same mechanism as the sodium ions, and b) that the lithium ions which enter the outermost cell layer are only poorly extruded into the intercellular space by the sodium pump. Therefore, they may diffuse into deeper cells via intercellular junctions and accumulate in cell water.

In addition, the establishment of a chemical gradient for lithium ions larger than that expected from P. D. measurements made by several groups of authors (Cereijido & Curran 1965 and Rawlins et al. 1970) raises the possibility that lithium and sodium cell entry at the outer border could result from an active transport process. Such a possibility has already been discussed (Morel & Bastide 1965, Zerahn 1969, Biber et al. 1966, and Biber & Curran 1970).

The kinetics of sodium and lithium penetration and accumulation in isolated epithelia were analysed by using a different approach which we would now like to discuss. The following conditions were selected: a potassium chloride Ringer solution containing $10^{-4}M$ ouabain, oxytocin and only 10 mM sodium chloride was used as inside medium; choline chloride Ringer was used as outside solution; the preparation was short circuited and the S.C.C. curve recorded. In preliminary experiments using sodium Ringer as inside medium and choline Ringer as outside medium, steady state Na^+ concentrations of the order of 10 to 15 mM per liter cell water were measured. In the presence of ouabain, the sodium concentration in cell water increased up to $85 - 90$ mM/l, whereas potassium concentration dropped to $40 - 50$ mM/l. Potassium Ringer solution was therefore used inside, in order to maintain the Na cell concentration within the normal low range.

In addition, it was observed (Rawlins et al. 1970) that large inside K^+ concentrations completely depolarized the inner membranes of isolated epithelia and reduced the inner electrical resistance to nearly zero. It may thus be assumed that in the conditions we selected, most of the electrical resistance related to cation diffusion was located at the outer barrier of the preparation.

The low steady state negative S.C.C. observed (-4 to -8 $\mu A/cm^2$, Figs. 2 and 3) probably indicates that net outward K and Na diffusion was larger than inward choline diffusion. When either lithium or sodium ions were added to the outside medium, typical transient changes in S.C.C. were observed. Two such current curves, recorded successively with the same preparation, are shown in Fig. 2. As soon as sodium (or lithium) was added outside (up to a final concentration of 7.2 mM/1), the current abruptly switched to positive values, and then gradually dropped to zero and again to negative values; a new steady state level, not very different from that measured before adding the cation, was finally reached within about 15 to 20 min; if the outside medium was then washed out and replaced by pure

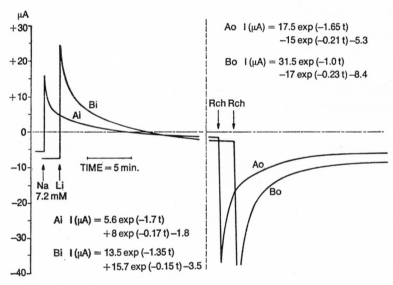

Fig. 2. The short circuit current was recorded as a function of time in a piece of isolated epithelium (*Rana esculenta* ventral skin; surface area: 1.5 cm^2) placed in the following conditions: inside bathing solution: KCl Ringer, containing 10 mEq/l Na, oxytocin (25 mU/ml) and ouabain (10^{-4}M); outside bathing medium: choline chloride Ringer solution. The curves show the transient change in S.C.C. which appeared when Na ions were added to the outer medium (curve A_i) and later washed out (curve A_o); B curves correspond to the subsequent addition of lithium ions to the same preparation (B_i) and to their washing out (B_o). These curves may be described as the sum of two exponential terms.

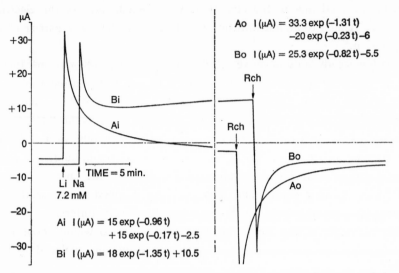

Fig. 3. Conditions similar to those described in Fig. 2, except that no ouabain was present in the inside bathing medium.

A curves: S.C.C. responses to the addition (A_i) of Li ions to the outside medium and to the return to pure choline Ringer (R_{ch}) solution (A_o). The same preparation was later similarly treated with Na ions. Note that in the absence of ouabain, the lithium curves are similar to those obtained in its presence (Fig. 2), whereas in the Na loading curve (B_i), the second exponential term is replaced by a constant, positive transport current.

Ringer choline solution, a similar transient change in S.C.C. again occurred, but in the reverse direction (right side Fig. 2).

Such responses are quite reproducible, and can be induced several times in the same preparation. We suggest the following interpretation: When Na or Li is added to the outside medium, a concentration gradient is created for this ion species at the outer cell border of the structure; the corresponding diffusion potential is shunted by the low resistances of both the inner membranes and the external short circuit, so that Na (or Li) ions diffuse into the epithelial cells through the outer border, whereas potassium ions diffuse out of the cells through the inner border of the preparation. It is thus assumed that the current measured (or, more precisely, the change in the current) is carried by sodium or lithium ions at the outer membrane, and by potassium ions at the inner membranes. As sodium or lithium ions accumulate in the cells, the diffusion gradient at the outer border progressively disappears until a new steady state is obtained.

The hypothesis that most of the lithium or sodium ions which penetrated the preparation remained trapped in the cell compartment and did not reach the serosal medium is supported by the observation that the transient current is reversed on returning to the initial conditions. We thus suggest that the curve of the current corresponds to the time course of sodium or lithium penetration and accumulation within the structure.

In this context, it is of interest to mention that these current curves can be described accurately as the sum of two exponential terms indicating that the structure contains two sodium or lithium pools with different degrees of accessibility from outside. The fast pool fills with a half time of 0.5 to 0.7 min; the slow pool (which contains about 5 – 10 times more sodium than the former) fills with a half time ranging from 4 – 7 min.; from the surface under the curves it may be calculated that about 30 – 40 nEq Na ions accumulated per cm^2 isolated epithelium when 7.2 mM was used as outside Na concentration. This figure would correspond to an increase of the Na concentration in cell water of about 10 mM per liter.

The following additional observations demonstrate that the two cellular pools are placed in series rather than in parallel with respect to the outer medium and that the fast pool might correspond to sodium located within the outermost active cell layer *(stratum granulosum)*, whereas the slow pool might correspond to sodium located within the deeper cell layers.

As exemplified in Fig. 3, experiments were also performed without blocking the serosal Na transport step, that is, by omitting ouabain from the potassium Ringer solution. The curves obtained by adding (or removing) the lithium ions outside were similar to those obtained in the presence of ouabain; when, on the contrary, sodium ions were used, the S.C.C. curves obtained were different. The transient change in current was of shorter duration, and may be described as a single, fast exponential term; the second exponential term was absent and was replaced by a positive plateau. On washing out the outside medium, the current curves of the sodium-loaded preparations again contained only a single, fast exponential term, whereas the curves of the lithium-loaded preparations contained two exponential terms.

Thus, in the absence of ouabain, the results indicate that the sodium ions entering the structure no longer accumulate in the slow pool; they are continuously extruded by the pump into the intercellular and serosal fluids, which accounts for the steady state net transport current observed. Lithium ions, on the contrary, still accumulate in two pools, which confirms that they are only poorly transported by the sodium pump.

Finally, the effects of amiloride ($5.10^{-5}M/l$ in the outside solution) were analysed. When added before lithium or sodium ions, amiloride completely prevented the changes in S.C.C. produced by these ions, both in the presence and absence of ouabain. This inhibition is reversible, since normal responses were again elicited after washing out amiloride. As shown in Fig. 4, when the amiloride was added during the negative current phase induced by washing out the outside sodium in a preloaded preparation (Curve e), the negative current was immediately blocked.

Also in Fig. 4, curve c corresponds to a control response obtained later in the same preparation, but without amiloride addition. The striking effects of amiloride confirm that the current passing through the outer border of the epithelium was definitely carried by sodium ions, and that no significant fraction of that current passed via non specific intercellular shunts. These results also indicate that amiloride not only blocks Na entry into the

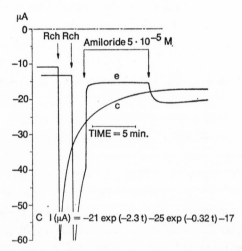

Fig. 4. Conditions similar to those described in Fig. 2. The preparation was successively loaded twice by adding Na (7.2 mM/l) to the outside medium; the figure shows the current curves obtained on washing out the outer Na by rinsing the outside chamber with pure choline Ringer solution (R_{ch}).

Curve e : amiloride was added outside during the response; the negative current phase was almost completely blocked by amiloride. This effect was reversible, since after washing out the amiloride and reloading the preparation with Na, a normal washing out curve was again obtained (curve c).

cells but blocks its release as well. Thus, the Na ions crossing the outer cell membrane from either side pass through amiloride-sensitive channels.

In conclusion, the results presented may be summarized in the way depicted in Fig. 5. Experimental conditions were used in which lithium and sodium ions accumulated into isolated frog skin epithelia from outside, or were released from the cells into the outer medium. The data confirm the great difference in cationic selectivity between the outermost functional barrier and the other cell membranes with respect to sodium and potassium ions, as first established by Koefoed-Johnsen and Ussing (1958).

Kinetic analysis of the results suggests that the sodium and lithium ions which entered the *stratum granulosum* may diffuse into the underlying cell layers; but this diffusion process is relatively slow, probably as a result of the small size and/or the low permeability of the intercellular junctions.

Consequently under normal conditions (that is, when the transport ATPase is operative), most of the sodium ions entering the *stratum granulosum* from outside are probably extruded into the intercellular spaces without being able to diffuse into deeper cell layers. That is why the so-called transport pool seems restricted to the superficial fraction of the sodium contained in

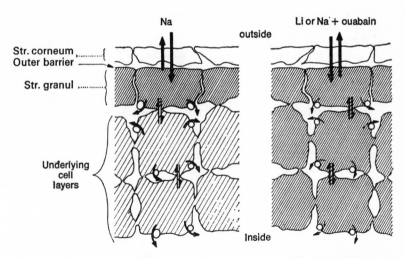

Fig. 5. Simplified diagram of the sodium transport pool of the isolated frog skin epithelium. Under normal conditions (left side) the Na ions which entered the *stratum granulosum* from outside are continuously pumped out into the intercellular space system, so that they only have little access to the deeper cell layers. When lithium ions are used, or when the pump is blocked by ouabain (right side) the lithium or sodium ions may accumulate in the deep cell layers.

the structure. This interpretation agrees with the general description of the active-sodium-transport path proposed by Ussing and Windhager (1964) on the basis of a very different experimental approach.

REFERENCES

Aceves, J. & Erlij, D. (1971) Sodium transport across the isolated epithelium of the frog skin. *J. Physiol. (Lond.) 212,* 195–210.

Biber, T., Chez, R. A. & Curran, P. F. (1966) Na transport across frog skin at low external Na concentrations. *J. gen. Physiol. 49,* 1161–1169.

Biber, T. & Curran, P. F. (1970) Direct measurement of uptake of sodium at the outer surface of the frog skin. *J. gen. Physiol. 56,* 83–99.

Carasso, N., Favard, P., Jard, S. & Rajerison, R. (1971) The isolated frog skin epithelium: I. Preparation and general structure in different physiological states. *J. Microscopy 10,* 315–330.

Cereijido, M. & Curran, P. F. (1965) Intracellular electric potentials in frog skin. *J. gen. Physiol. 48,* 543–557.

Hansen, H. & Zerahn, K. (1964) Concentration of lithium, sodium and potassium in epithelial cells of the isolated frog skin during active transport of lithium. *Acta physiol. scand. 60,* 189–196.

Koefoed-Johnsen, V. & Ussing, H. (1958) The nature of the frog skin potential. *Acta physiol. scand. 42,* 298–308.

Leblanc, G. (1972) The mechanism of lithium accumulation in the isolated frog skin epithelium. *Pflügers Arch. ges. Physiol., 337,* 1–18.

Morel, F. & Bastide, F. (1965) Action de l'ocytocine sur la composante active du transport de sodium par la peau de grenouille. *Biochim. biophys. Acta (Amst.) 94,* 609–611.

Rajerison, R. M., Montegut, M., Jard, S. & Morel, F. (1972a) The isolated frog skin epithelium: permeability characteristics and responsiveness to oxytocin, cyclic AMP and theophylline. *Pflügers Arch. ges. Physiol. 332,* 302–312.

Rajerison, R. M., Montegut, M., Jard, S. & Morel, F. (1972b) The isolated frog skin epithelium: presence of α and β adrenergic receptors regulating active sodium transport and water permeability. *Pflügers Arch. ges. Physiol. 332,* 313–331.

Rawlins, F., Mateu, L., Fragachan, F. & Whittembury, G. (1970) Isolated toad skin epithelium: transport characteristics. *Pflügers Arch. ges. Physiol. 316,* 64–80.

Ussing, H. H. (1963) Transport of electrolytes and water across epithelia. *Harvey Lect. 59,* 1.

Ussing, H. H. & Windhager, E. E. (1964) Nature of shunt path and active sodium transport path through frog skin epithelium. *Acta physiol. scand. 61,* 484–504.

Zerahn, K. (1969) Nature and localization of the sodium pool during active transport in the isolated frog skin. *Acta physiol. scand. 77,* 272–281.

DISCUSSION

KEYNES: One of the things about the physiology of lithium that strikes me is that there is no good estimate anywhere of the actual rate at which lithium is transported by the sodium pump. Dr. Swan and I made some measurements on frog muscle some time ago (Keynes & Swan 1959) from which we simply said that we thought that the sodium pump transported lithium of the order of 10 times more slowly than sodium but we didn't get at all a precise estimate. You say you think that lithium is poorly transported. You can't substitute a number for your 'poorly', can you?

MOREL: I cannot give an exact figure from the data I showed. I would say that Li is transported at least ten times slower than sodium.

USSING: If Dr. Morel's figure is right, then of course lithium has much larger areas through which to be actively transported. Therefore even though the pump maybe works slower in the individual cells, more cells are involved. Thus, the net results may be almost the same as for sodium. Isn't that your interpretation?

MOREL: I think the data we have obtained in the experimental conditions I have described, but omitting ouabain from the serosal bathing solution, can help to answer your question. When either Li^+ or Na^+ is added in the same low concentration to the outside Ringer-choline solution, the steady state increase in SCC obtained after the transient phase which corresponds to cellular accumulation is always definitely lower with Li^+ than with Na^+. This increase in current probably measures the amount of active transport. Thus, less lithium would be transported in these conditions than sodium, in spite of greater cell concentration and greater diffusion in deep cell layers.

MAETZ: Have you used this type of kinetic approach to study the effects of neurohypophysial hormones which presumably occur at the apical face of the frog skin epithelium?

MOREL: All the SCC data I have discussed were in fact obtained in the presence of oxytocin (25 mU/ml) added to the serosal medium. Up till now

---

Keynes, R. D. & Swan, R. C. (1959) The permeability of frog muscle fibres to lithium ions. *J. Physiol. (Lond.) 147,* 627–638.

we have only performed a few preliminary experiments in which paired pieces of the same epithelia, one of each pair treated with oxytocin and the other not, were compared; these experiments suggest that oxytocin increased the size of the Na cellular pools without affecting their turnover rate; this could be accounted for by an increase in Na permeability of the outer border in response to oxytocin.

USSING: This is really a remark that refers to the beginning of your talk. The isolated epithelium preparation seems to be exceedingly useful. But there you cited me as the performer of the method for the first time. In the paper mentioned I refer to experiments by Dr. Bernhard Andersen in our laboratory where he used pressure to separate the epithelium. In his hands actually it is possible to get just as high potentials and currents as in the method you are using.

LEAF: I would just like to ask Dr. Morel about a point I might not have clearly understood. When you have 2.5 mM lithium on the outside in the steady state, the lithium accumulates with a concentration in the isolated epithelium which is almost ten times greater than the medium concentration. When you reverse the situation in your experiments and remove the lithium at the outside and replace it with choline, was the forward and reverse diffusion of lithium symmetrical or was there a difference in the two directions; because if it is the same, it seems to be a little bold to argue that there is an active process in one direction.

MOREL: They were almost symmetrical, and I completely agree with you. But the fact that such chemical gradients can be achieved is really a question that remains puzzling.

LOEWENSTEIN: I should like to make a short comment on the potentiality of feedback control in transport epithelia with junctional communication. In typical transport epithelia, large cell populations (in some, the entire population) are interconnected by junctions; the cell ensemble, rather than the single cell, is the unit as far as the molecular species are concerned that are transported through nonjunctional membrane. So control of transport can conceivably be concerted throughout the entire cell ensemble. Unlike systems controlled by extracellular means of communication, e. g., humorally controlled systems, the connected cell ensemble has a finite volume and hence is capable of self regulation on the basis of simple clues of chemical concentration. The key element here is the sharp diffusion barrier bounding the entire cell system, namely the nonjunctional membranes and the junctional seals. The boundaries not only make the cell ensemble a system of finite volume, but they also introduce an element of assymmetry into the intracellular chemical field against which time-dependent concentration parameters can be referred. These features of interconnected cell systems are potentially very important for possible feedback control.

Relation Between Structure and Function in Frog Skin[1]

C. L. Voûte & S. Hänni[2]

The rather broad subject of this meeting has been narrowed in our laboratory by focussing on the process of active transport of sodium in amphibian epithelia and more specifically to the epithelium of the frog skin. Our approach to this topic has been guided right from the beginning by the question whether or not there could be a further gain of knowledge in epithelial transport physiology by joining the disciplines of the morphologist to conventional approaches used in transport physiology. We were guided by fortune to do our first steps together with Professor Ussing, whose benevolent criticism represented an integral part of earlier and todays presentation.

Fundamentally, without changing the conventional methods of both physiologists and morphologists, with respect to active sodium transport, we were able to demonstrate so far the following functional-morphological relations in the epithelium of the frog skin:

— When comparing a fully shorted piece of skin with an identical piece under open circuit conditions, one can observe in the shorted epithelium a swelling of the first living cell layer just underneath the stratum corneum, which we named thereafter the 1st reactive cell layer. We tentatively interpreted this observation that this cell layer alone might be responsible for the asymmetric bioelectric phenomena observed in the frog skin (Voûte & Ussing 1969).

— Also we could demonstrate that there was a linear relationsship between the volume of the interspace system as well to an applied hydrostatic

1 Supported by grant 3.343.70 of the Swiss National Foundation for Scientific Research.
2 From the Laboratory of Experimental Nephrology, Department of Medicine, University of Basel, Switzerland.

»Transport Mechanisms in Epithelia«, Munksgaard, Copenhagen.

pressure gradient and the magnitude of the epithelial short circuit current (Voûte & Ussing 1970).

As both these observations could be in favor of our interpretation, however without being a proof, we tried to relate the volume of the first reactive cell layer fixed under constant short circuit conditions directly with the magnitude of the corresponding transport rate. Using conventional fixation techniques however, we were unable to pin down a parallelism between degree of swelling in this layer and the magnitude of short circuit current. We were aware that speed of fixation had always been the most critical part in our method and that this point became utmost important when trying to study correlations in rapidly reversible phenomena.

Thus we were looking for a technique which would be more promising for our purpose: Speed of fixation had to be practically immediate without losing too much structural preservation. The technique recently described by Eckert (1969) seemed to be suited best for this purpose.

Our report will serve two aims: A presentation and discussion of the new method and the presentation of the first results with the help of this method, dealing with experiments used for a reevaluation of the old question: The degree of swelling in relation to active sodium transport in the various layers of the epithelium of the fully shorted frog skin. As the volume of the extracellular space has been shown to parallel transport in a previous communication (Voûte & Ussing 1970) it will not be discussed.

All experiments in this series were done during summer. The frogs, *Rana temporaria,* were kept as usual in shallow tap water at 4°C. After removal of the skin, the ventral part was divided into four quadrants, clamped in special ring holders and positioned between two lucite half chambers containing identical Ringer's solution on both sides. Fig. 1 demonstrates our set up which has been used for the last two years in our laboratory. Four identical chambers are placed in parallel, each chamber being provided with a separate pair of electrical circuits as well as circulation-aeration devices. The half chambers are rectangular and can easily be moved back and forth in order to facilitate rapid removal and freezing of the tissue. After the usual equilibration time of 1 hr. under open circuit, the skins were short circuited for 30 min. Then an inside positive pressure gradient of 30 cm H_2O was applied for 10 min. in order to get good expansion of the epithelial interspace system. By this, morphometric evaluation of the various parts in the epithelium will be largely facilitated and improved. A pressure gradient of this magnitude does not usually interfere with active sodium transport in the frog skin. Skins not exhibiting steady values of SCC under pressure were discarded. Thereafter the tissue pieces were rapidly removed and dropped into propane cooled to liquid nitrogen temperature. Without emerging the tissue small pieces of skin were transferred into a precooled lyophilizer at a temperature of − 60°C. Lyophilization was carried out over a period of 4 hours in a vacuum of 10^{-4} Torr. The temperature was then brought to room temperature without releasing the vacuum and the specimens were fixed for another hour in osmium vapors. Thereafter the specimens were rapidly transferred at atmospheric pressure into preevacuated Epon and infiltrated again in a vacuum of 10^{-4} Torr overnight. Polymerisation, cutting and staining was done according to previous reports.

For morphometric evaluation, three random sections of 1 μ thickness cut at a 90° angle to the skin surface and stained with toluidine blue were used. The morphometric methods applied are the ones described by Weibel & Elias (1967) and are based on stereological principles by Holmes (1927) and Delesse (1847). All microscopic work was done in a light microscope, measurements at a total magnification of 3200 x and photography at an objective magnification of 100 x.

With the help of Fig. 2, we will try to illustrate which parts of the epithelium were evaluated morphometrically for this report. We will also use this picture to depict the excellent preservation of the tissue at the level of the light microscope. From top to bottom, the various layers can easily be seen: The cornified layer, the first reactive cell layer, which here appears swollen, and below the cells of the stratum granulosum, spinosum and germinativum. In order to avoid as many errors as possible the volume of the 1st RCL was not measured and expressed as volume per cell, but as absolute volume over a length of 100 μ for each counted section. The lower cells were measured individually, fifty cells per section and expressed as average volume per cell.

It is worthwhile to note some aspects of the tissue at a higher magnification, dealing with volume measurements only. Even when using the best obtainable way of freezing to liquid nitrogen temperature, ice crystal formation is so far inavoidable. Fig. 3 demonstrates this phenomenon very nicely: On top we have parts of a cornified cell with its nucleus and underneath a cell of the 1st RCL. Desmosomal attachments join both cells across the subcorneal space. Note specially in the lower cell, large amounts of ice

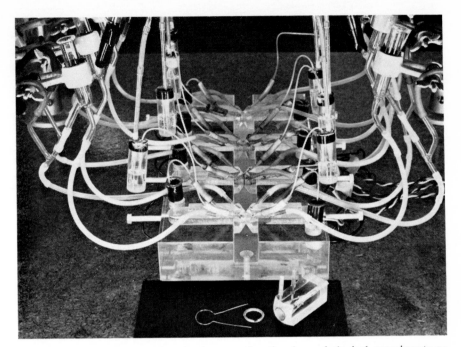

Fig. 1. Experimental set-up for simultaneous functional-morphological experiments on frog skin.

Four parallel chamber pairs held by a massive lucite block. Each pair with separate bioelectric circuits and circulation-aeration devices. Measuring surface 1 cm². In front: Ring holder for rapid deep freezing.

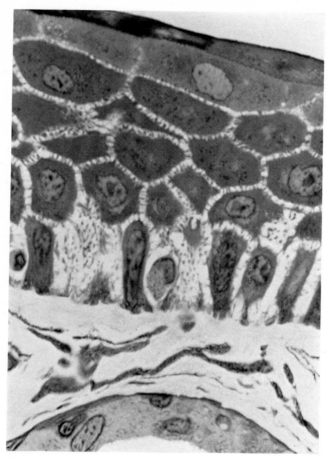

Fig. 2. Cross section of an epithelium of a frog skin under constant short circuit conditions and an inside positive pressure gradient of 30 cm H_2O. – From top to bottom:
Stratum corneum, swollen 1st RCL, interspace system separating the cells of the str.
granulosum, spinosum and germinativum. – Tissue frozen and processed according to
the described method. – Lightmicrograph of toluidine blue stained 1 μ thick section.
Magnification 1800 x.

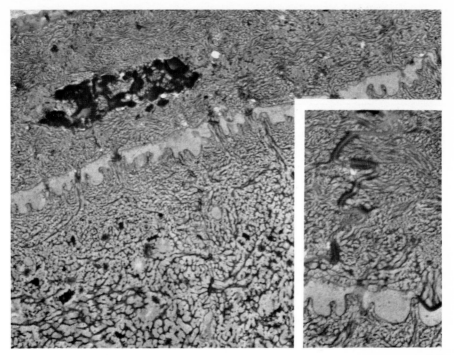

Fig. 3. Low magnification electron micrograph showing stratum corneum on top and a cell of the 1st RCL underneath separated by the subcorneal space. Note rather heavy ice crystal formation and damage. Magnification 12'000 x.

Insert: Desmosomal attachments between two cells of the str. corneum, rather well preserved. Magnification 20'000 x.

crystal holes distorting the basic tonofibrillar ultrastructure of the cytoplasm. Mitochondria appear merely as black masses with unrecognizable substructure. In the insert, however, junctional complexes between two cornified cells may be seen with rather nice preservation. In this layer tonofibrills are mostly well preserved.

The opinions, about which step in the processing is the most critical and responsible for this ice crystal formation, are highly controversial. Most experts believe that rapidity of freezing is of utmost importance. As we are merely beginners with this method and as ultrastructural preservation is not of prime importance to this paper, we may leave it to that and turn to our results at the level of the light microscope.

Fig. 4 illustrates the results of the average volume of the epithelial cells below the 1st RCL. Their volume in the abscissa is plotted against the corresponding value of short circuit current in the ordinate. Even without statistical analysis, one may easily see that there is no significant correlation between the two variables and that there is roughly no volume change for the full range of transport rates, the regression line being almost vertical.

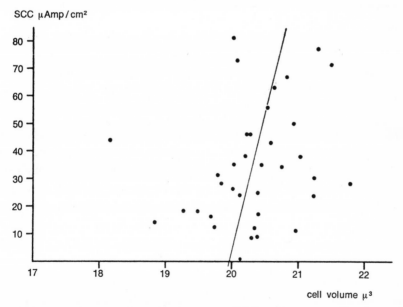

Fig. 4. Relation between short circuit current (SCC) and average cell volume of the lower epithelial cells (without 1st RCL).
Statistic: N 35 P<0.1 R 0.32
  Volume = 0.117 E – 01 × SCC + 19.95

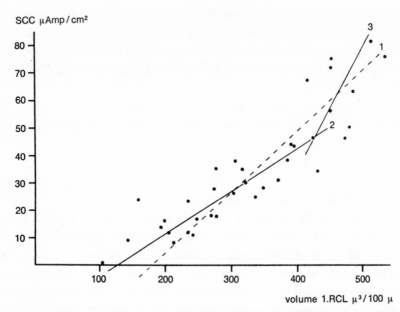

Fig. 5. Relation between short circuit current (SCC) and volume of the 1st RCL (per 100 μ measured).

Statistics: – Curve 1: N 35   P<0.001   R 0.89
                     Volume = 0.468 E 01 $\times$ SCC + 170.13
     – Curve 2: N 28   P<0.001   R 0.86
                     Volume = 0.637 E 01 $\times$ SCC + 129.32
     – Curve 3: N 13   P<0.01   R 0.75
                     Volume = 0.216 E 01 $\times$ SCC + 325.72

    The second parameter we analysed morphometrically for this study was the volume of the first reactive cell layer, the results of the counts were plotted against transport rates (Fig. 5). Here the picture is altogether different: We see that there is a highly significant positive correlation between the rate of active sodium transport and the volume of the first reactive cell layer. If we consider the relation to be linear for all values, we get regression line 1, which however does not seem to be the best representative, also when considering the cell volume at zero transport rate.

    This zero transport cell volume was measured in an experiment in which sodium was substituted by choline in the outside bathing solution. Splitting now the values for SCC at about 50 μAmps/cm² and analysing the higher and lower data separately, we get curves 2 and 3. If, in the higher ranges of SCC this non linear behaviour of the correlation could be shown to hold in further series of experiments, this might be taken to mean that at transport rates of 50 μAmps/cm^2 and above further cell volume expansion has come close to the limits of elasticity, and that a further volume increase would inevitably call for cell damage, an observation regularly encountered in skins submitted to double current values of SCC (Voûte & Ussing 1968).

Summarizing our morphometric results paralleling asymmetric sodium transport in a series of fully shorted epithelia of frog skin, we may make the following statement:

In an isosmotic environment the total epithelial volume depends on the rate of asymmetric active sodium transport in a positive linear relation. This volume is composed of at least two variable *dependent* parameters, the volume of the 1st RCL and the volume of the extracellular space, and at least two variable *independent* parameters, the volume of the cells of the stratum corneum and the volume of all cells not belonging to the 1st RCL. The stratum corneum has been measured in a previous report and found not to change under varying transport rates. (Voûte & Ussing 1968).

Before discussing the implications drawn from these results we might briefly recall our comments of the swelling phenomenon formulated previously. Any cellular volume change has to go *pari passu* with a change in the total amount of osmotically active particles in that cell. Electrolytes are in this respect the most probable actors as they represent the only rapidly exchangeable particles involved. If we assume, according to the ionic permeability characteristics predicted by the Koefoed-Johnsen-Ussing model 1958, that the outward facing boundary of the compartment is permeable to sodium, but less so to chloride; the inward facing one, however, more permeable to chloride than to sodium, whenever an inward going current is passed through this cell layer, there will be an accumulation of electrolyte (NaCl) between the boundaries and an equivalent amount of water will follow the osmotic gradient. That the sodium pump in most living cells seems to be the major factor determining a new equilibrium under constant short circuit conditions is the most probable explanation, but to what degree other factors interfere with this mechanism is beyond our present knowledge.

On the basis of these results, it seems intriguing to reevaluate an old argument. What is the role of the 1st RCL as compared to the role of the lower cells in transepithelial sodium transport? Is the main job done by the first cell layer or is the epithelium behaving also in this respect as a true syncytium? Former and present results seem to support our first opinion being in favor of the 1st RCL doing the whole or at least the major job (Voûte & Ussing 1968). One could argue that this cell layer is swelling according to SCC because it is the sole layer trespassed by all sodium transported or to be transported, wherever one wants to locate the responsible sodium pump. This so far seems reasonable. We should however be able to find at least some sort of relation between cell volume of the lower cells and short circuit current. We showed this not to be the case and at this level the extracelular volume only is transport dependent.

Fixation artefacts or artefacts due to other parts in the method may be excluded with a high degree of certainty if one stays at the light microscopic level. Freezing is immediate, and qualitatively the same results are found as previously with conventional methods.

A last but not least point to be discussed is the role of the cell junctions, the desmosomal low or high resistance couplings in this special case of totally shorted epithelia. It seems probable, always under the assumption that our explanation is correct, that the cell junctions between the 1st RCL and the lower cells have to exhibit under these bioelectric conditions a high resistance at least towards the sodium ions carrying the current. Would a partial depolarisation due to shorting of this layer trigger an increase in resistance of the junctions between the 1st RCL and the lower cells?

On the other hand a high resistant epithelium under open circuit conditions, in fact, behaves as a true syncytium as net sodium transport is blocked by the high electric gradient and thus no partial depolarisation of the 1st RCL would occur. There would be equipolarity throughout the epithelial cells, an idea which would be consistent with our morphological observations in highly resistent epithelia under open circuit conditions – a high degree of homogeneity in all epithelial cells including the ones of the 1st RCL. However, conversely, it could be the appearance of an epithelium exhibiting no net sodium transport, a fact which has been shown to be true with the help of flux measurements.

Thus, in summary, these data give further support to our previously formulated hypothesis that in the epithelium of the frog skin asymmetric active sodium transport giving rise to the conventionally used bioelectric measurements is mainly or solely performed and controlled by the first living cell layer just underneath the stratum corneum, the 1st reactive cell layer.

REFERENCES

Delesse, M. A. (1847) Procédé mécanique pour déterminer la composition des roches. – *C. R. Acad. Sci. (Paris) 25,* 544–545.

Eckert, H. (1969) Distribution of soluble substances at the ultrastructural level. In *Autoradiography of diffusible substances,* ed. Roth, J. & Stumpf, W. E., pp. 321–333. Academic Press, New York & London.

Holmes, A. H. (1927) *Petrographic methods and calculation.* Murby & Co., London.

Koefoed-Johnsen, V. & Ussing, H. H. (1958) The nature of the frog skin potential. Acta physiol. scand. *42,* 298–308.

Voûte, C. L. & Ussing, H. H. (1968) Some morphological aspects of active sodium transport. – The epithelium of the frog skin. *J. Cell. Biol. 36,* 625–638.

Voûte, C. L. & Ussing, H. H. (1970) Quantitative relation between hydrostatic pressure gradient, extracellular volume and active sodium transport in the epithelium of the frog skin. *(R. temporaria). Exp. Cell. Res. 62,* 375–383.

Weibel, E. R. & Elias, H. (1967) *Quantitative methods in morphology.* Springer, Berlin-New York.

ACKNOWLEDGEMENT

We would like to express our gratitude to Dr. Eckert and Mr. Browning from Sandoz Ltd. for their technical assistance when introducing the new freezing method. Our thanks go also to Mr. E. Wyss from the Office of Statistics, Basel.

DISCUSSION

ZADUNAISKY: Have you considered the possibility that this first layer of cells become more permeable during short circuiting and that additional flow of water occurs as a consequence of short circuiting more than as a primary mechanism.

VOÛTE: I do not know what short circuiting does to the two selective cell membranes in series concerning their resistance. Short circuiting, however, usually leads in so called good skins to a small resistance increase of the epithelium. Whether this is due to the cellular swelling and compression of the lateral interspaces at the level of the 1st RCL is certainly a possibility but unproven so far. – As to the water equilibrating osmotically with transported sodium it is difficult to separate the two moieties: The one really following sodium from out to in and the one being recirculated from inside. Here we are probably dealing with large seasonal variations.

WHITTEMBURY: In relation to the possibility of explaining the swelling of the reactive cell layer. When you short circuit the skin the cell interior becomes electronegative (Whittembury 1964, Rawlins et al. 1970). So you speed up entry of Na into the cell from the outside solution, as compared to the open circuit condition. In agreement with your observations, Mateu et al. (1970) studied the electrical connection between cells. They found that cells communicated more horizontally than in the vertical direction. The electrical communication between electrodes with tips situated less than 30 micrones apart but in cells of different layers was much smaller than that between electrodes with tips situated within the same layer but 150 μ apart. They also found a greater connection between cells of the first reactive layer.

VOÛTE: I am glad to hear this. From our morphological observations there would be horizontal equipolarity in the cells of the 1st RCL, differing, however, largely from the polarity of the lower cells.

Whittembury, G. (1964) Electrical potential profile of the toad skin epithelium. *J. Gen. Physiol. 47,* 795–808.

Rawlins, F., Mateu, L., Fragachán, F. & Whittembury, G. (1970) Isolated toad skin epithelium: transport characteristics. *Pflügers Arch. 316,* 64–80.

Mateu, L., Giménez, M., Callarotto, R. & Windhager, G. (1970) Intercellular communication is preferentially horizontal in toad skin epithelium. *Biophys. J. 10.* Biophys. Soc. Abstract. 34a.

USSING: Am I not right when I say that you got zero potential by replacing sodium with choline in the outside medium?

VOÛTE: Yes. Zero potential and transport rate.

USSING: In other words some of the experiments have been done on pieces of the same skin with sodium or choline chloride outside?

VOÛTE: Yes.

USSING: Because otherwise my question would have been: Is this a proof that large cells transport sodium faster; or is it a proof that the entry of sodium makes the cells swell. Of course, both interpretations could have been right except for your experiments with choline.

VOÛTE: Yes. We were curious to see whether cells which do not transport would still fit into the curve with respect to their volume to give us a sort of zero transport value.

USSING: That was very fortunate.

SACHS: If you voltage clamp your skins at different voltages not just the short circuit current conditions and you take a group of skins with similar short circuit currents, do they all fall in the same line in terms of that correlation?

VOÛTE: I couldn't answer that question. We haven't done this because it would be difficult or impossible to extrapolate voltage clamp currents to net sodium transport merely with the help of bioelectric measurements. One would need bidirectional flux measurements and this would be highly impractical for our purpose.

ZADUNAISKY: I would like to go back to this question of passing currents through complicated membranes. If one takes a tissue like the cornea that has a very thick layer of connective tissue between two epithelia, of which the endothelium is very permeable and passes currents of the same order that you pass across the frog skin, the cornea tends to swell. If one passes the current in the other direction the cornea tends to shrink. The connective tissue is 300–400 μ thick and these effects occur mostly in the connective tissue. My contention from these experiments is that this application of current produces an electroosmotic effect, changing the volume of the stroma. I would like to know if besides looking at the epithelium or at the reactive cells, did you look at the collagen of the frog skin?

VOÛTE: This question has been dealt with in a previous paper (Voûte & Ussing, 1970a). Although we have found a linear relation between sodium transport and extracellular volume of the epithelium proper we believe that this relation would be valid for the extracellular volume of the interstitium as well if one permits it to equilibrate for a longer period. As a matter of fact reversing the current will stop active sodium transport (Voûte & Ussing, 1970b) and will by this lead to a shrinking of the extracellular space. At least in frog skin it does and according to your observations this seems to hold for cornea as well.

LINDEMANN: I should like to mention with respect to Dr. Zadunaisky's comment that even in skins which are practically dead, current flow can at a low pH cause volume changes and electro-endosmotic blister formation between epithelium and connective tissue. This is old work done by Amberson & Klein (1928).

TOSTESON: Speaking of pH, I would like to know what the present measurements indicate about the magnitude of the potential difference between the inside and the outside of these cells as a function of short circuit current. I ask particularly because for each different value of the membrane potential there is a different distribution of protons. The more negative the cytoplasm becomes with respect to the extra-cellular fluid, the more acid it becomes. This, of course, could be one way in which the swelling could appear. Do you have any information on that?

VOÛTE: No, I haven't.

MEARES: When you short circuit the whole skin then of course you have not short circuited the first reactive cell layer, because there will be an intrinsic potential profile within the skin, so that across the first reactive layer there will be a potential, and that potential would lead to a concentration distribution. If you choose circumstances in which the short circuit current is changed then it may well be – in fact it almost certainly will be – that the concentration profile in the first reactive cell layer will be changed. Thus in

Voûte, C. L. & Ussing, H. H. (1970a) Quantitative relation between hydrostatic pressure gradient, extracellular volume and active sodium transport in the epithelium of the frog skin *(R. temporaria) Exptl. Cell Res. 62,* 373–383.
Voûte, C. L. & Ussing, H. H. (1970b) The morphological aspects of shunt-path in the epithelium of the frog skin *(R. temporaria). Exptl. Cell Res. 61,* 133–140.

addition to Dr. Zadunaisky's point that there can be an electro-osmotic transfer brought to change the volume, the change in the concentration profile can lead to a straight osmotic effect, an osmotic swelling effect. I think that the critical condition of short circuit for the whole skin is not necessarily such a significant condition in relation to the first reactive cell layer.

USSING: I should answer that question. The virtue of the preparation is that the resistance in the interspaces is quite low, so that the potential in the inside solution would be very close to the potential in the interspaces right down to the first reactive cell layer, therefore I don't think the interpretation is so difficult, as it might seem. Also the cell membranes – since they are probably of lipid nature with no large pores – do now allow large volume flows as a response to potential changes. So I think these volume changes must be somehow related to the actual amounts of osmotically active substances in the cells. The fact that the choline experiments where apparently no cation entered from the outside led to the smallest cell volumes sort of indicates that the main reasoning is correct. With respect to resistance I can tell that the running of short circuit currents through a frog skin does not lead to a drop in cell resistance. On the contrary, there usually is a slight increase which, however, is reversible within seconds when the current is broken.

ZADUNAISKY: But this system in the connective tissue is also reversible.

USSING: Yes, but that is something entirely different. May I also comment to Dr. Tosteson's question about the pH. I think there is every reason to believe that the hydrogen ion is among the ions which the cells handle by active transport, so one cannot say that one will get the Donnan change in the cell pH.

TOSTESON: I agree that the presence of active proton transport would complicate matters. Nonetheless, the membrane potential would operate in parallel with any proton pump to influence the steady state distribution of protons.

Amberson, W. R. & Klein, H. (1928) The influence of pH upon the concentration potentials across the skin of the frog. *J. Gen. Physiol. 11,* 823–841.

USSING: We would like some measurements. While I am talking: was it correct, Dr. Morel, that you found the sodium content at the isolated epithelium increased during short circuiting?

MOREL: Yes.

USSING: That really would fit in quite nicely with Dr. Voûte's observations. So the only question is: where it is located, and these experiments might suggest that most of it is in the first reactive cell layer.

ORLOFF: What about ouabain?

USSING: It has been done.

ORLOFF: Does the swelling occur only in the first reactive cell layer, or don't you know?

VOÛTE: We have done very few experiments with ouabain. Therefore my answer will be a purely qualitative one. Besides a drastic reduction of the extracellular volume the first changes can be observed at the level of the 1st RCL. These cells swell, become prenecrotic as judged by their irregular appearance of the cytoplasmic matrix (due to a detachment of the cell membrane from tonofibrills). Finally a complete detachment of a swollen to necrotic 1st RCL from the underlying cells can be observed.

WINDHAGER: Is there any new information on potassium in the outer layers of epithelial cells during short-circuiting of the skin?

VOUTE: No.

Cation Selectivities in Frog Skin

Tomuo Hoshiko

It was during my first visit to Copenhagen some eighteen years ago that Professors Ussing and Koefoed-Johnsen were formulating the now famous model of frog skin (Koefoed-Johnsen & Ussing 1958). That model has played a key role in shaping much of the subsequent investigative work, not only on frog skin, but also on almost every other epithelial tissue. The model arose out of their demonstration of sodium and potassium selective properties of the two surfaces of the frog skin. Today I will try to summarize some of our work on cation selectivity of frog skin, which may provide some clues to the molecular mechanisms.

Although on first thought, ion selectivity seems to be quite a clear cut concept, further reflection reveals that the term can and has been used in many different ways. All are aimed at establishing a selectivity order for some tissue. First, there are operationel definitions: the effect of ion substitution on open circuit potential, on clamp current at various fixed potentials, on isotope fluxes, on intracellular ionic composition, on secretory activity. Second, there are mechanistic or quasi-thermodynamic definitions: equilibrium potential, partition coefficients, permeability coefficients.

In order to estimate the selectivity order, common practice has been to measure the relative values of some functional index with and without the presence of the ions in question. Thus, the term selectivity applies to a process or functional parameter and does not necessarily imply a specific molecular process.

Nevertheless, our intuitive sense is that membrane ion selectivity involves some static or equilibrium partition among the ions or some dynamic sieving or flow-separation process. The permeability coefficient of the constant field theory incorporates both concepts – a partition coefficient and mobility term.

Department of Physiology, Case Western Reserve University, School of Medicine, Cleveland, Ohio 44106 U. S. A.

»Transport Mechanisms in Epithelia«, Munksgaard, Copenhagen.

Eisenman (1961) presented a theory of equilibrium partition of ions between solution and membrane. Partition of ions was postulated to result from free energy differences in coulombic attraction among ions of different charge density competing for association with a fixed charge or induced dipole site in the membrane.

An alternative theory based on a suggestion of Mullins (1956) was elaborated on by Lindley (1967). It constitutes an extension of the theory of ionic solution and treats the partition of ions due to differences in free energy of solvation in water, as opposed to solvation in rigid pores of different size in that relatively low dielectric medium, the lipid matrix of the cell membrane.

Neither theory appears adequate in their present forms, since they are unable to predict several selectivity orders of alkali metal ions exhibited, for example, by frog skin. Thus detailed experimental information on ion selectivity of frog skin presents restraints upon theoretical formulations for ion selectivity mechanisms.

Ion selectivity order in frog skin has been estimated by substituting one ion for another. Simultaneously, some parameter of membrane function is observed. Lindley and I have used the change in open circuit potential as the index for estimating the ionic selectivity order in isolated frog skin (Lindley & Hoshiko 1964).

The advantage of the ion substitution method is that the test conditions are different only to the extent that the ions are different. Thus the osmotic pressure and ionic strength of the bathing solution is unchanged. The tissue is not consumed in the measurement and the step can be reversed or repeated as long as the tissue can survive.

In the presence of a non-permeating anion, and at constant total cation concentration, the change in potential with a change in cation concentrations is a simple function of their permeability ratio. Experimentally, the open circuit potential of frog skin exhibits such behavior, in response to substitution of other cations for sodium at the outer border, when that solution is of relatively low ionic strength. The inside or corium surface of frog skin exhibits such behavior in response to substitutions for potassium, when the inside bathing solution contains 60 m Eq/L sodium sulfate solution. The change in potential can be related to the concentration change by the constant field equation of Hodgkin & Katz (1949). By a least squares procedure, at is possible to estimate the relative permeability $\alpha_{AB} = P_A/P_B$ which best suits each ion. Table I gives the permeability ratios for the alkali metal cations at the inside and outside surfaces of *R. pipiens* and *R. catesbeiana* at pH 8 (Lindley *et al.* 1967).

Table 1.

	Permeability Ratio		Bullfrog R. Catesbeiana	Leopard Frog R. Pipiens
At Outside Surface	α Li	Na	.301	.433
	α Cs	Na	.044	.092
	α Rb	Na	.068	.075
	α K	Na	.050	.074
At Inside Surface	α Rb	K	.740	.714
	α Cs	K	.229	.258
	α Li	K	.130	.153
	α Na	K	.102	.092

The relative permeabilities at the outer surface for K, Rb and Cs are not different statistically. This does not mean that these three cations enjoy equal selectivity. On the contrary, later experiments indicate not only that there is a definitive order of selectivity, but also that the orders are different for bull frog as against leopard frog.

Instead of obtaining an ordering from the relative permeabilities, the ions can be ordered according to their effects on the open circuit potential. In some experiments the three cations, K, Rb and Cs were presented in random order, in succession, with no intermediate exposure to a wholly Na medium. In such cases, the open circuit potentials were compared directly and the orders were established according to the descending magnitude of the potentials, i. e., the higher the potential, the higher the permeability.

When the test cation periods were alternated with periods in full sodium concentration, the change in potential was the basis for ordering the test cations: now the lower the change in potential, the higher the permeability. Each set of test exposures generated one ranking order out of the total of six possible.

What relationship may exist between the six possible orders? One approach is to postulate that some determinative parameter, such as field strength, or pore diameter, causes the relative permeabilities of the three cations to rise and fall as that parameter increases in magnitude. If the dependence of the three ionic permeabilities on the determinative parameter are out of phase, the rank order of the relative permeabilities would change, and under some circumstance, each of the six selectivity orders would appear. In the living membrane, given sufficient sensitivity of the relative permeabilities to the determinative parameter, it is conceivable that each order may be represented as the parameter value itself fluctuates randomly about some mean value. Moreover the relative frequencies of the six possible orders would be distributed about some favored order. Otherwise, given six possible orders, if each order were equally probable, the mean frequency of appearance of any order would be 1/6.

Fig. 1 shows a plot of the frequency at each selectivity order for the outer surface of leopard frog and bullfrog skins. These results represent almost 400 individual determinations of selectivity orders in 146 pieces of skin. The 6 selectivity orders could occur if the relative permeabilities rose and fell along the hypothetical determinative parameter, as represented in the lower graph. For example, at pH 6–9 the order which most skins of *R. pipiens* exhibit is Na $>>$ Cs $>$ Rb $>$ K, while that of *R. catesbeiana* is Na $>>$ Rb $>$ K $>$ Cs. This happens to agree with the order of the relative permeabilities given in Table I.

The analysis of the responses at the two borders in terms of the Ussing

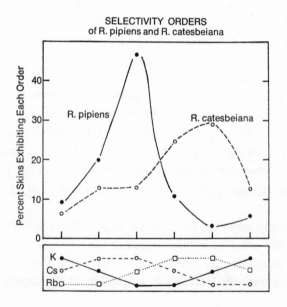

Fig. 1.

Koefoed-Johnsen model so far is straight forward. The responses at the two surfaces should be simply additive and not interact. However, careful experiments demonstrate significant interaction, (Leb *et al.* 1965). In general, the lower the total potential, the smaller is the response to a given change in concentration. Thus, the response to change in potassium at the inner border is smaller at a low sodium concentration in the outside bathing solution and larger at a high concentration.

These effects have been interpreted in terms of parallel nonselective shunt paths. Anatomical evidence in favor of such paths have been presented by Farquhar & Palade (1965) and by Voûte & Ussing (1970), while physiological evidence was given by Ussing & Windhager (1964). The shunt path would allow current flow and an "IR drop" across the internal resistance of, for example, the sodium diffusion path. The measured potential change with change in sodium concentration would not be the change in sodium equilibrium potential alone, but would include the voltage drop across the sodium diffusion barrier. Thus, the apparent selectivity may be the result of a pair of parallel ion paths, one highly selective and the other much less selective.

Numerous factors may alter or condition the apparent selectivity. The internal state of the animal strongly affects selectivity: the breeding cycle, the nutritional state, the moulting cycle. The periodic shedding of the stratum corneum in the moulting cycle is accompanied by a drastic fall in selectivity at the outer border (Hviid Larsen 1971).

Several external conditions have been studied. Ionic strength drastically inhibits the sensitivity of isolated skin to changes in outside sodium concentration. At high ionic strength, the non-selective shunt leakage paths are increased and conductance increases. Thus it is not clear whether ion selectivity of the cellular surface is altered.

The field strength theory of ion selectivity postulates that fixed charge in the membrane determines ion selectivity. These fixed charges are ionized or ionizable groups within the membrane such as proteins. The amphoteric nature of proteins allows us to alter the nature of the charged groups in the membrane by changing the pH. The pH effect at the outer border has been investigated many times and all agree that the skin potential decreases at outside pH's below about 5 or less. Below pH 4.5, the skin potential is irreversibly depressed. Amberson estimated the isoelectric point to be about pH 5.2.

We chose to compare selectivities at a number of pH's and found that the selectivity orders fall into two groups: those at pH 6 or above; and those at pH below 5, actually at pH 4.5. The data at a given pH could not be fitted by the constant field equation and thus we were unable to estimate the relative permeabilities, α. Therefore, I am presenting the data as histograms, as before.

In Fig. 2, the solid line connects the selectivity orders at pH 6 and above for *R. pipiens* and is the curve shown in Fig. 1. The broken line connects the selectivity orders at pH 4.5. Here we see that the selectivity order has shifted from the order Na $>>$ Cs $>$ Rb $>$ K at above pH 6 to a Na $>>$ Rb $>$ Cs $>$ K order at the more acid pH.

These two orders correspond to none of the orders predicted by Eisenman nor those predicted by Lindley. However the possibility remains that the observed orders are the result of two types of pathways: one highly selective for sodium and one not selective for sodium. This postulate of more than one type of selectivity is reinforced by the failure to describe the data with the constant field equation. Such a postulate possibly could be tested through quantitative comparisons of the ionic permeabilities of each ion.

The sodium selective outer border of frog skin is reminiscent of the

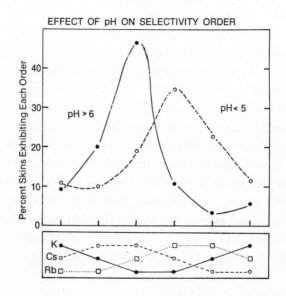

Fig. 2.

conductile membrane of nerve and muscle in the excited state or the excited junctional membrane of the neuromuscular junction. Such membranes are reported to be selective not only for sodium but also to hydrazinium and guanidinium ions (Tasaki *et al.* 1965). Therefore, these two ions were tested at the outer border of frog skin at pH 6.

In Fig. 3, change in potential is plotted on the abscissa as increasing proportions of the sodium bathing the outer surface is replaced by the test cation. Neither guanidinium nor hydrazinium ion could maintain the potential. This result shows that the sodium-selective outer border of frog skin is different from the sodium selective excited state of the conductile membrane of nerve or muscle.

Numerous other states of the membrane remain to be studied. For example, the selectivity order during stimulation with novobiocin (Johnston & Hoshiko 1971) appears to be unchanged. The remarkable and long stimulation of both current and potential depends upon an increased sodium permeability at the outer border. But the apparent selectivity order appears to be the same. Thus, novobiocin has a quantitative and not qualitative effect. Also the novobiocin effect points out another feature of the sodium selective membrane - namely that the sodium selective membrane exhibits saturation behavior.

This property was emphasized by Kirschner (1955) almost 20 years ago, and attributed to the active transport carrier. The saturable carrier concept

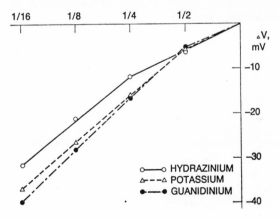

EFFECT OF ORGANIC CATIONS USED AS SODIUM SUBSTITUTES
AT OUTSIDE SURFACE OF FROG SKIN.

Fig. 3.

has proved very helpful although the question of whether this step is "active" (i. e., requires direct expenditure of metabolic energy) is still open. In other words, a process akin to facilitated diffusion if not active transport appears to be responsible for the selectivity exhibited at the outer border. Moreover, this selectivity process appears unaffected by novobiocin stimulation although the overall rate of sodium transport is enhanced.

In summary, the frog skin has been a useful preparation for the experimental study of ion selectivity, particularly as a test of theoretical models of ion selectivity. However, it is clear that the ion selective behavior of frog skin is complicated by the presence of more than one type of pathway for ion permeation and by the saturation behavior of the sodium path. Since even cell membrane systems such as the isolated node of Ranvier exhibit "leakage current", the complete modelling of ion selective behavior may require dealing with "leakage" or "shunt paths", as an integral part of the molecular scheme as well as an intercellular pathway.

More broadly speaking, ion selectivity is only one of a whole gamut of related processes located at the cell surface, such as neurohumoral and pharmacological receptor systems or cell recognition systems in development. Although the molecular steps may differ in specific detail, many of the same processes possibly are involved and I believe this perspective may be helpful in future work.

REFERENCES

Eisenman, G. (1961) On the elementary atomic origin of equilibrium ionic specificity. In *Membrane Transport and Metabolism,* ed Kleinzeller, A. & Kotyk, A., pp. 163–179. Academic Press, New York.

Farquhar, M. G. & Palade, G. E. (1965) Cell junctions in amphibian skin. *J. Cell. Biol. 26,* 263–291.

Hodgkin, A. L. & Katz, B. (1949) The effect of sodium ions on the electrical activity of the giant axon of the squid. *J. Physiol. 108,* 37–77.

Hviid Larsen, E. (1971) The relative contributions of sodium and chloride ions to the conductance of toad skin in relation to shedding of the stratum corneum. *Acta physiol. scand. 81,* 254–263.

Johnston, K. H. & Hoshiko, T. (1971) Novobiocin stimulation of frog skin current and some metabolic consequences. *Amer. J. Physiol. 220,* 792–798.

Kirschner, L. B. (1955) On the mechanism of active sodium transport across the frog skin. *J. cell. comp. Physiol. 45,* 61–87.

Koefoed-Johnsen, V. & Ussing, H. H. (1958) The nature of the frog skin potential. *Acta physiol. scand. 42,* 298–308.

Leb, D. E., Edwards, C., Lindley, B. D. & Hoshiko, T. (1965) Interaction between the effects of inside and outside Na and K on bullfrog skin potential. *J. gen. Physiol.* *49*, 309–320.

Lindley, B. D. (1967) Membrane solvation as a basis for ionic selectivity. *J. theor. Biol. 17*, 213–228.

Lindley, B. D., Bartsch, G. E. & Eberle, B. J. (1967) Determination of relative permeabilities for the constant field equation from experimental data. *Math. Biosci. 1*, 515–543.

Lindley, B. D. & Hoshiko, T. (1964) The effect of alkali metal cations and common anions on the frog skin potential. *J. gen. Physiol. 47*, 749–771.

Mullins, L. J. (1956) The structure of nerve cell membranes. In *Molecular Structure and Functional Activity of Nerve Cells,* ed. Grenell, R. G. & Mullins, L. J., pp. 123–154. Academic Press, New York.

Tasaki, I., Singer, I. & Watanabe, A. (1965) Excitation of internally perfused squid giant axons in sodium-free media. *Proc. nat. Acad. Sci. (Wash.) 54*, 763–769.

Ussing, H. H. & Windhager, E. E. (1964) Nature of shunt path and active sodium transport path through frog skin epithelium. *Acta physiol. scand. 61*, 484–504.

Voûte, C. L. & Ussing, H. H. (1970) The morphological aspects of shunt path in the epithelium of the frog skin *(R. temporaria). Exp. Cell Res. 61*, 133–140.

108

DISCUSSION

SCHULTZ: I would doubt that the shunt pathway seriously influences trans-epithelial diffusion potentials resulting from a change in the ionic composition of the solution on either side of the skin. The data of Ussing & Windhage (1964) suggest that the conductance of the shunt pathway across frog skin is less than one-tenth of that of the transcellular pathway. Under these circumstances, the likelihood that diffusion potentials arising within the shunt will significantly affect transepithelial potential differences is very remote.

TOSTESON: The shunt pathway doesn't necessarily have to be intercellular. It can also be in the membrane.

HOSHIKO: The conductance of the shunt path may be expected to be a function of ion concentration, and hence its effects would be expected to be minimized at lower concentrations.

MOREL: I have two additional questions to ask you. What was the anion you used in the outside medium and secondly: how did you estimate the change in potential?

HOSHIKO: First of all, we almost always used sulphate. The other thing is, we rather arbitrarily took the potential as being the difference from the time of the first change – just before the change – to the potential 15 minutes after the change.

SACHS: Were your data corrected for diffusion potential changes?

HOSHIKO: No.

SACHS: Comparing the PD changes in the intact tissue to PD changes with microelectric measurements, have these been carried out in frog skin? And compared the selectivity series with microelectrodes versus the intact tissue?

HOSHIKO: No.

TOSTESON: I would like to make a comment and ask one question. The comment has to do with your initial discussion of the variable definition of selectivity and the problems associated with trying to make mechanistic deductions from selectivity ratios. I strongly support the point that it is

Ussing, H. H. & Windhager, E. (1964) Nature of shunt path and active sodium transport path through frog skin epithelium. *Acta physiol. Scand. 61*, 484–504.

difficult to make reliable mechanistic deductions from selectivity ratios. The main reason for this difficulty is that the selectivities are determined by small differences between very large numbers. From studies of neutral cation carriers like valinomycin, we have learned that complexation involves the complete dehydration of the ion. Therefore, the free energy of complexation depends, among other factors, on the difference between the free energy of hydration (which has values of about −75 and −95 Kcal per mole for K^+ and Na^+ respectively) and the free energy of complexation of the naked ion in vacuo which must have comparably large values. Selectivity depends on differences between these large energies for K^+ and Na^+. Since a difference in free energy of 1.4 Kcal per mol represents a selectivity ratio of 10, the difficulties of specifying the origin of the differences are obvious. My question has to do with the experiments on ionic strength. How do you distinguish betwen ionic strength and the absolute concentration of sodium?

HOSHIKO: Theoretically the potential change depends upon the ratio of the initial and final concentrations and not on their absolute values.

KEYNES: I wasn't quite clear just how strong the evidence is that hydrazine doesn't behave as a sodium substitute.

HOSHIKO: The thing is: it may perhaps be a little more like sodium than like potassium, but it certainly is very far from being like sodium. In other words, if we substitued sodium for hydrazinium the potential change was very large. Almost as large as for − say − potassium.

KEYNES: The interesting point that Dr. Hille has been making about nerve membranes is that hydrazine and hydroxylamine, which are just about the same size as the sodium ion and one water molecule, are pretty good sodium substitutes as of course lithium is, but methylamine which is virtually the same size, too, doesn't substitute at all. It would be quite interesting somehow to see if the same kind of consideration applies in frog skin.

HOSHIKO: We haven't tried methylamine, but I would suspect it would be in the same class as hydrazinium and guanidinium and potassium, in other words not a very good sodium substitute.

LINDEMANN: Apparently you assumed in your treatment that the sodium permeability throughout the sodium concentration range is constant. Is that true?

HOSHIKO: According to the constant field theory that is possible.

LINDEMANN: I should like to challenge that, and shall do so at some length in my paper. Another problem: When using the Goldman equation you only include the sodium and the potassium terms. When one adds the terms for anions – and I have done that using your data – then one gets quite different α-values. The selectivities may still turn out similar to the one you got, but the α-values are certainly different when the anions are taken into account.

HOSHIKO: Yes, that would certainly be true. I think that our idea was that the sulphate permeabilities are in the order of 10^{-7} or something like that and are sufficiently low to be ignored in this correlation.

LINDEMANN: Are they much lower than those of potassium which you included?

HOSHIKO: I think that it is somewhat lower. It could bear reconsideration, though.

ZADUNAISKY: With respect to the comment of Dr. Tosteson, it was clear to him that ions cross naked through the membranes. Since this is a very old problem in permeability studies, maybe he could verify for me if this is the definite situation at least in the case of artificial membranes and the ions do traverse the membranes without water of hydration, and the old controversy over hydrated and non hydrated radius of the ion has been definitely resolved.

TOSTESON: In the case of transport of monovalent cations across bilayers or biological membranes as complexes of neutral carriers like valinomycin and its analogues, the ions certainly are transported without water. One point of evidence is that the proton spectra of the complexes reveal no resonances characteristic of water. A second point is that X-ray diffraction patterns from crystals of the K complexes with monactin and valinomycin are consistent with structures that contain no water. Molecular models of the complexes are consistent with this conclusion.

GENERAL DISCUSSION

LEAF: Dr. Voûte, may I ask about this new technique for studying the morphology in your preparation of the tissue? Do you have evidence that it is superior to a direct fixation of the tissue *in situ,* let us say with 1 percent gluteraldehyde. It seems to me that cutting the tissue into bits and dropping it into a liquid nitrogen may include some opportunity for artifacts which could be avoided by fixation in the experimental conditions.

VoÛTE: With respect to superiority of any fixation method: If there would be a really superior one to all others we could omit all discussions. One thing however has to be remembered: The fixation technique applied has to be adapted to the experimental question. In our case we have shown that even with osmium we had trouble to get good quantitave correlation between cell volume and transport in spite of keeping the skin shorted all through fixation. This specific experimental condition is a rapidly reversible situation and may be compared to a stretched spring which through fixation will slowly loose its elasticity. So, for us, rapidity was of first order importance, whereas preservation of ultrastructural details could be neglected. – As to my personal opinion concerning a 1 percent glutaraldehyde solution as fixative my answer is the following: Glutaraldehyde is a very slow fixative, much slower than osmium and in most cases (frog skin) it doesn't even abolish the bioelectric activity. This is just the reason why it can be used as fixative for ultrastructural enzyme-histochemistry (the only indication for us to use it) but also a sign that it is a poor fixative, especially if one tries to correlate simultaneously unstable functional and morphological parameters. – For a critical discussion of the various freezing techniques I would like to refer to the book by Roth and Stumpf (1969).

MAETZ: The first question is directed to Dr. Voûte. Have you any evidence that neurohypophysial hormones induce swelling under short circuit conditions in frog skin?

VoÛTE: We have done a few experiments not included in this series. There we had an increase in current which ran roughly parallel with the increase of the extracellular space. But we did not measure cell volume that time, so we don't know if there is a quantitatively significant correlation (Voûte & Ussing 1970).

MAETZ: My second question is directed to Dr. Hoshiko. It concerns the problem of shunt path *vs* cellular path selectivity. Have you compared the selectivity pattern before and after addition of amiloride outside, a drug

Voûte, C. L. & Ussing, H. H. (1970) The morphological aspects of shunt-path in the epithelium of the frog skin (*R. temporaria*). *Exptl. Cell Res. 61*, 133–140.

which presumable blocks the cellular path at least for some cations (Na^+ and Ca^{++}, for example)?

HOSHIKO: I should say that my remarks are all simply speculative because we haven't really pursued that question of what nature the shunt pathway might be. But of course your suggestion is good. Also perhaps this could be done by simply eliminating sodium.

We have studied the kinetics of radiosodium washout and the intracellular ion contents in the isolated epidermis (Hoshiko & Parsons 1972 a, b). These methods allowed us to make four independent types of estimates of the extra-cellular space and two independent estimates of the intracellular sodium pool. The results agree with each other and the methods promise to be powerful tools.

EFFECT OF EXTERNAL SODIUM
ON TISSUE AND CELL CONTENT

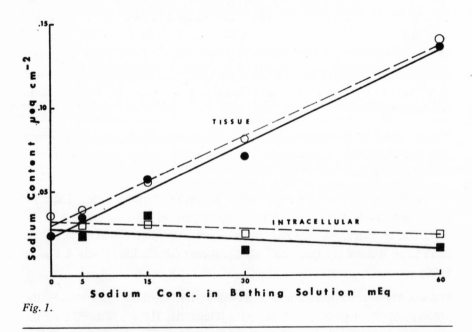

Fig. 1.

Hoshiko, T. & Parsons, R. H. (1972a) Isolation of large sheets of frog skin epidermis. *Experientia* (Basel) *28,* 795–796.

Hoshiko, T. & Parsons, R. H. (1972b) Isolation and properties of large sheets of frog skin epidermis. *IV Internat. Biophysics Congr.,* Moscow, 3, 144. Abs.

The washout kinetics of sodium, sulfate and inulin at the outer border were studied. This approach allowed estimation of extracellular spaces and of the intracellular sodium pool.

The split skin or epidermis preparation was incubated in bathing media with sodium concentrations from 0 to 60 mM/1, with and without potassium at 5 mM/1. Tissue Na contents were obtained and are plotted in Fig. 1, against the sodium concentration in the medium. The solid lines and symbols indicate potassium was present, while the broken lines and open symbols show its absence. The uppermost pair of lines and the circles represent total sodium analyzed in the tissue, whereas the lower pair of lines and the squares represent measured mannitol space. The straight lines were drawn from linear regression lines fitted to the data points.

The slopes for the regression line for tissue sodium can be interpreted to be the volume of the extracellular fluid space. In this experiment, the mannitol spaces were measured in each tissue, giving two estimates of the ECF space by the static method – from the slope of the regression line and from the mannitol space. Likewise the intercept of the upper regression curve estimates the intracellular sodium pool; all the lower points represent the same pool obtained by subtracting off the sodium in the mannitol space.

These estimates of the extracellular spaces and sodium pools obtained by the kinetic and static methods are compared on Table I. Looking down the middle column labelled ECF space, the sodium, sulfate and inulin spaces,

Table I. Extracellular fluid space and intracellular sodium pool estimated by kinetic and static methods.

Method	ECF Space μl cm^{-2}	Cell Na Pool μEq cm^{-2}
KINETIC		
Na Washout	0.93 $\pm$ .10 (14) (outside only)	0.030 $\pm$ .004 (14)
Sulfate Washout	0.72 $\pm$ .12 (10) (outside only)	–
Inulin Washout	0.33 $\pm$ .05 (8) (outside only)	–
STATIC		
Tissue Na vs [Na]$_o$	1.8 $\pm$ .3 (60)	0.023 $\pm$ .002 (60)
Tissue Na – ECF Na	–	0.024 $\pm$ .002 (63)
Mannitol Space	1.84 $\pm$ .13 (64)	–

as obtained from the washout method, are the top three figures. The slope of the line relating tissue sodium to bathing solution sodium concentration gave a value for the extracellular fluid of 1.8 μl. The last figure in the middle column is the mannitol space itself with a value of 1.84 μl. By comparison, the extracellular space at the outer border measured kinetically is exactly half. This finding is consistent with the results of another series of static measurements in which the mannitol spaces of the two surfaces were made separately and found to be equal. The right hand column shows the intra-cellular sodium pools measured from the washout experiment at the top, 0.03 μEq cm^{-2}, next is the sodium pool measured as the intercept of the tissue sodium curve at 0.023 μEq cm^2; at the bottom is the value calculated after correcting for the sodium in the mannitol space: 0.024 μEq/cm^2. These values agree and are not significantly different. Thus, the kinetic and static methods corroborate each other.

These results illustrate the usefulness of the split skin preparation and as well suggest need for great caution in interpreting results of isotope kinetics in whole skin preparations. The fact that the intracellular sodium pool is unaffected by the extracellular sodium levels raises the question of how the active transport rate is controlled. Finally, the split-skin preparation is so simple and effective, I would suggest that no frog skinner can afford to be without the technique as a routine tool.

LEAF: I will just say that the washout method is probably quite unreliable and in toad bladder at least mannitol will penetrate all the cells of the tissue.

Delayed Changes of Na - Permeability in Response to Steps of (Na)$_o$ at the Outer Surface of Frog Skin and Frog Bladder

B. Lindemann & U. Gebhardt[1]

Understanding epithelial transport requires a listing of tasks and properties of both apical and latero-basal membranes. To obtain this listing the membranes have to be studied separately. One way to study the apical membrane without getting additive signals from the latero-basal one, is to change ionic concentrations of the apical solution so briefly, that only the outermost membrane has time to respond to the change. Cell interior and latero-basal membranes remain in their steady states.

METHOD

For such experiments we have developed a 'fast flow' chamber (Gebhardt *et al.* 1971, Lindemann *et al.* 1972), in which membrane areas of up to 1 cm^2 can be mounted to form the bottom of a shallow flow channel (1 mm high). The channel is fed from an electro-magnetic flow switch, which permits flows of up to 50 ml/s through the crossectional area of 1 × 17 mm. When switching between different solutions, the 0.1 ml volume above the membrane can reach its new composition within 20–25 ms.

The flow program used is shown in the upper part of Fig. 1. The outside of the mounted abdominal frog skin (r. esculenta, summer animals) is pre-equilibrated with P-solution flowing slowly. On arrival of a trigger signal this solution is accellerated and after a pre-set time replaced by a test-solution (T) of different composition. At the end of the chosen exposure time, T is replaced by P which continues to flow fast for a third pre-set period

1. From the 'Abteilung für Membranforschung an Epithelien', 2. Physiologisches Institut, 665 Homburg, Saar, W-Germany. Supported by the DFG through SFB 38, project C$_1$.

»Transport Mechanisms in Epithelia«, Munksgaard, Copenhagen.

8*

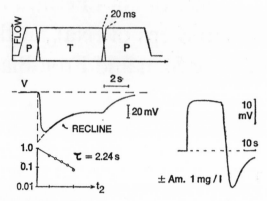

Fig. 1. Upper part: flow program for voltage response shown below. P-solution: 80 mM K-gluconate, 1 mM Ca-gluconate, 5 mM Tris-sulphate, pH 6.0. T-solution: 80 mM Na-gluconate, rest like in P-solution.

Lower left: corresponding voltage response to the P-T-P cycle described above (rana esculenta). The recline has a time constant of approximately 2 s.

Right diagram: Response of frog skin (rana esculenta) to addition and removal of 1 mg/1 of Amiloride. Exposure time to Amiloride containing T-solution was 34 s. P and T solution contained 100 mM Na-gluconate, 1 mM Ca-gluconate, 5 mM Tris-sulphate, pH 7.2. This experiment was performed together with Mr. W. Zeiske.

of time. Exposure times can be varied between 50 ms and 10 s. The only change of composition which shall concern us here is a substitution of K by Na or Na by K. The anion concentration is kept constant and poorly penetrating anions (gluconate, occasionally sulphate) are used. The inner bathing medium is Na_2SO_4-Ringer's.

UP-DOWN ASYMMETRY AND RECLINE

In frog skin, the apical (outward facing) membrane is predominantly Na-permeable. When replacing K by Na in the outer solution, the outer membrane surface becomes rapidly more negative (hyperpolarization), a change which we plot downward. When reversing the concentration change, the membrane depolarizes (plotted upward). These voltage responses can be fitted with diffusion equations (Crank 1956), to compute the nominal thickness δ of unstirred layers in a way already used by Kidder *et al.* (1964).

With hyperpolarizing voltage changes, we found nominal δ values between 10 and 15 μ. About half of this thickness is made up of free solution, the other half of organic material (mainly the str. corneum). Consistently larger

δ values (14–20 μ) were computed from depolarizing voltage changes. Thus, after correcting for the logarithmic relationship between concentration and membrane potential, we found that the membrane responded to a decrease of $(Na)_o$ more slowly than to an increase (up-down asymmetry, Fuchs et al. 1972).

Furthermore, in many skins the rapid hyperpolarization is followed by a depolarizing recline, which is approximately exponential with a time constant of one to several seconds (Fig. 1, lower left). The recline, which ends in a plateau or a very slow further depolarization, will be the main topic of this communication. We shall show that it is caused by a fall of Na-permeability, which also explains the slow speed of response to a decrease of $(Na)_o$.

CONTROL EXPERIMENTS

Our first concern was, that the recline was an artefact caused by transient changes of hydrostatic pressure. However, the pressure changes were no larger than 0.02 atm. Also, switching between identical solutions both with and without Na gave voltage responses of negligible magnitude. Furthermore, experiments involving urea showed, that we are not dealing with an osmotic response of cellular volume, nor did the osmotic gradient across the outer membrane seem to be of importance. The voltage response to outward going current pulses was monotonous, showing that the recline is not a voltage dependent phenomenon (Lindemann et al. 1972). The recorded potentials are too large to be explained by interdiffusion of Na and K in the unstirred layers. The relatively small role of K ions in the recline-phenomenon can best be shown by changing $(Na)_o$ together with its anion at zero $(K)_o$: the recline still occurs.

RESPONSE OF BLADDER EPITHELIUM

To study the possible role of the str. corneum in these transients more directly, we did fast flow experiments with frog bladder, which does not have a cornified cell layer. The speed of response was faster in this tissue. After replacing $(K)_o$ with Na, maximal hyperpolarization was reached in 50 ms, followed by a depolarizing recline of about 50 ms time constant (Fig. 2). Clearly therefore, any nonlinear properties of the str. corneum cannot be made responsible for the recline phenomenon.

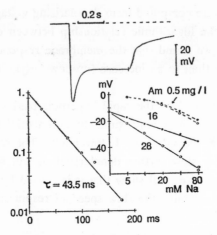

Fig. 2. Upper curve: response of bladder epithelium (rana esculenta). Solutions as in Fig. 1, but pH 7.2.

Lower left: time constant of recline approximately 0.05 s.

Lower right: Nernst plot of peaks (open circles) and plateaus (dots) observed with various $(Na)_o$ – values of T-solution. Slopes indicated as mV per 10 fold change of $(Na)_o$. The difference in slopes disappeared when P and T solutions contained 1.9 μM Amiloride (dashed curves).

RECLINE DURING WASHOUT OF NA-BLOCKER

When a partial blockage of the Na-permeability (P_{Na}) of the outer membrane was produced with Amiloride in low concentrations (3.3 μM in T-solution), and the blocker was then washed away without changing external electrolyte concentrations, an overshooting hyperpolarization with a recline lasting more than 10 s was observed during washout (frog skin, Fig. 1, right diagram). Apparently, the outer membrane responds in a similar way to a removal of the Na-blocker and to an increase of $(Na)_o$. Any diffusion effects at electrode surfaces or in unstirred layers (of organic or anorganic structure) are excluded in this experiment.

MOSAIC MEMBRANE HYPOTHESIS

The resting potential of a membrane containing Na-specific and K-specific channels can be written

$$V_o = E_{Na} R_K / (R_{Na} + R_K) \quad + \quad E_K R_{Na} / (R_{Na} + R_K),$$

where E_{Na} and E_K are the Nernst-potentials and R_{Na} and R_K are the integral resistance of the specific channels. Suppose a Na-K substitution experiment is done and V_0 as well as all E and R values are plotted against log $(Na)_0$. Then E_{Na} will be a straight line of neg. slope while E_K will at small $(Na)_0$ have a small positive slope which, however, increases with increasing $(Na)_0$. When all $(K)_0$ has been substituted by Na, E_K will formally be infinite. Since V_0 is a weighted sum of E_{Na} and E_K, it can be a non-monotonous function of $(Na)_0$ in such a substitution experiment, provided that R_{Na} and R_K are functions of $(Na)_0$ and $(K)_0$, which do not too much counteract the effect of E_K at high $(Na)_0$. A favourable situation would be that R_K is relatively constant at small $(K)_0$ but decreases at larger $(K)_0$, while R_{Na} decreases at small $(Na)_0$ and saturates at large $(Na)_0$ (compare Lindemann 1971, Fuchs *et al.* 1971, Gebhardt *et al.* 1972, Heckmann *et al.* 1972). Then V_0 will at small $(Na)_0$ hyperpolarize (become negative) with increasing $(Na)_0$, go through a minimum and depolarize again as $(Na)_0$ increases further. In a substitution experiment of increasing $(Na)_0$ all values on the concentration scale would successively be run through, and the time course of V_0 would show a negative peak followed by a depolarizing recline.

The model predicts that the steady state voltage – concentration curve has a minimum. This allows a rough check without even knowing the exact concentration dependence of R_{Na} and R_K. Plotting peak and plateau values of the voltage responses to increases of $(Na)_0$ in a Nernst diagram gave straight lines for both functions (Fig. 2, frog bladder). Also, the above model would predict a hyperpolarizing voltage peak for a substitution of $(Na)_0$ by K, which is not found. Therefore, the mosaic membrane effect cannot be responsible for the recline phenomenon.

SHUNT CURRENT CHARGING LATERO-BASAL MEMBRANE?

The Nernst-plot of Fig. 2 gave slopes of 28 and 16 mV per tenfold change of concentration, for peaks and plateaus, respectively. The relatively small slope values made us suspect that the epithelia had large paracellular shunt conductances. Can the voltage recline be explained by circular current flow through paracellular shunts?

For simplicity let us consider one layer of epithelial cells shunted by paracellular pathways. Its equivalent circuit is schematically shown in the lower left hand corner of Fig. 3. Index 1 denotes the apical, 2 the latero-basal membrane, s the shunt. When the battery voltage E of the apical membrane is suddenly increased (because $(Na)_0$ has been stepped up), additional inward current will flow through the cells and change the charge on both membrane capacitors. In frog skin, the latero-basal capacitance (C_2) can be 50 × larger than the apical one (Smith 1971). The inward current decreases as C_2 is slowly charged up and a recline of the transepithelial voltage results. Its magnitude is indicated to the right of the circuit diagram.

When the same experiment is done while the transepithelial voltage is clamped to zero, current flow through the shunt will be zero at all times. The short circuit current must then be expected to show a recline which is somewhat faster (Fig. 3, upper diagram and equations) than the voltage recline expected under open circuit conditions (middle diagram).

The equations of Fig. 3 allow solving for all circuit elements. Records of short-circuit current and open-circuit voltage are shown on the lower left and values computed from these records on the lower right of the figure. The value of R_s was found surprisingly low and that of R_2 surprisingly large. Furthermore, the time constant of the current recline was found larger than that of the voltage recline, in contrast to expectations. Both time constants are so large that C_2 values of 1000–3000 $\mu F/cm^2$ are computed from them. These numbers are 20–60 times larger than the expected values for the latero-basal membrane of frog skin (compare Smith, 1971). Thus charging of the latero-basal membrane cannot explain the recline-phenomenon under discussion.

SHUNT CURRENT LOADING CELLS WITH Na?

Which physical process activated by circular current flow would be expected to have a time constant of several seconds? We have considered the possibility that this process is a change of cellular composition, particularly an increase of cellular Na-concentration, $(Na)_c$. It is true that Na-influx will be very small in the presence of poorly penetrating anions, but only at zero current flow. If a paracellular shunt permits circular current flow, $(Na)_c$ may increase even in a fast flow experiment. However, the current flowing can be no larger than the short circuit current (typically 50 $\mu A/cm^2$), which would increase $(Na)_c$ by no more than 1 mM each second, if the cells were only 5 μ thick (left part of Fig. 4). This is a small change, but may be significant if $(Na)_c$ is very low at the beginning of the Na-exposure.

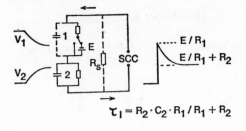

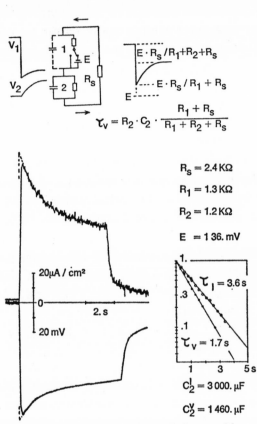

Fig. 3. Middle: equivalent network for one cell layer with a paracellular shunt resistance R_s: open circuit conditions. Hyperpolarization due to increase of $(Na)_o$ is simulated by switching in battery E.

Top: same network for short circuit conditions.

Lower left: recording of open circuit voltage (below) and short circuit (upper trace) for a P-T-P cycle (skin of rana esculenta). Both recordings show a recline.

P-solution: 80 mM K-gluconate, 0.1 mM Na-gluconate, 1 mM Ca-gluconate, 5 mM Tris-sulphate, pH 7.2.

T-solution: all K replaced by Na. Circuit component values on the right were computed using the equations shown above. The change of R_1 during exposure with Na was neglected.

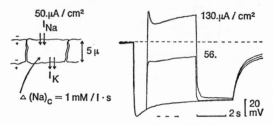

Fig. 4. Left diagram: expected maximal change of cellular Na-concentration during flow of inward current.

Right diagram: 3 responses to the P-T-P cycle of Fig. 2 (skin of rana esculenta). The almost square depolarizing voltage deflections were produced with inward going current pulses. After termination of the smaller current pulse, the voltage returned to the plateau value of the control trace, showing that the increase of (Na)$_c$ due to current flow did not significantly affect the recorded voltage.

The Na-loading hypothesis would predict that the recline amplitude increases (i. e. that the plateau level shifts further towards zero voltage) after a pulse of inward current drawn during Na-exposure from an external source. This did not prove to be the case, however, as shown in the right part of Fig. 4. After passing 56 μA for more than 3 s, the plateau level was found unchanged. After passing 130 μA for the same time, the plateau level was depolarized by an amount which was almost insignificant compared with the 20 mV recline observed with this preparation. Furthermore, when passing outward current sufficiently large to prevent Na-influx completely for the first 4 s of Na-exposure, we found the voltage to return to the plateau value of the control immediately after termination of the current pulse. Therefore, the recline, observed on our time-scale, cannot be explained by an increase of cellular Na-concentration caused by current flow.

It should be mentioned, that an electroneutral Na exchange mechanism at the apical membrane, for instance a Na-H exchange, could increase (Na)$_c$ and would not be influenced directly by passing inward or outward current through the membrane. However, the Na influx due to this mechanism – if it exists – does not exceed the short circuit current (compare Garcia Romeu *et al.* 1969). Therefore, additional Na, injected with inward current of 50 μA/cm², should have a synergic and equally noticeable effect with respect to an increase of (Na)$_c$. Since this synergic effect is not found, Na influx by exchange will have been insignificant also.

RESISTANCE INCREASE DURING RECLINE

We found a number of frog skin preparations, where the Nernst-plot of peak potentials observed after increasing $(Na)_o$ gave slopes close to the theoretical maximum of 58 mV per tenfold change (see Nernst-plot at lower left of Fig. 5). Nevertheless, these preparations showed large reclines. This observation excludes the presence of significant paracellular shunts. A shunt would already be operative while the Na-concentration at the outer

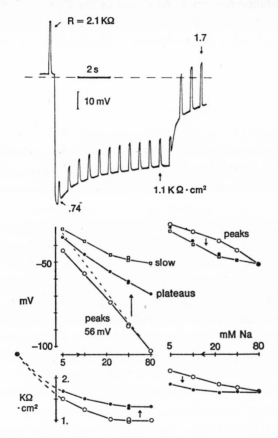

Fig. 5. *Upper diagram:* voltage and resistance response to a P-T-P cycle as in Fig. 3 (skin of rana esculenta). The narrow deflections were caused by brief inward going current pulses of 15 $\mu A/cm^2$.

Lower diagram: voltage and resistance change plotted against log $(Na)_o$ in a Na-K substitution experiment with the same preparation. P-solution of left plots: 80 mM K-gluconate, 1 mM Ca-gluconate, 5 mM Tris-sulphate, pH 7.2. P-solution of right plots: ditto, but all K replaced by Na. Arrows and abscissa indicate direction of P-T step.

membrane surface increases, and prevent the potential from reaching the theoretically maximal value.

Arguing by exclusion of other possibilities, the recline seems to be a response of the Na-selective membrane itself, formally described as a permeability change. An increase of K- or anion-permeabilities would decrease the membrane resistance, shunt the Na-battery and depolarize the membrane. A decrease of Na-permeability would also cause depolarization, but increase membrane resistance. As shown in the upper part of Fig. 5, an increase of resistance is found during the depolarizing recline (it is also found with outward going current pulses). Therefore, the recline must be caused by a decrease of P_{Na}.

EFFECT OF PRE-INCUBATION MEDIUM

In the left Nernst-plot of Fig. 5, $(Na)_o$ was increased from zero in P-solution to larger concentrations with a set of T-solutions (note arrow on abscissa). The resistance in P-solution was about 3. KOhm cm². When stepping $(Na)_o$ up to several final concentrations as rapidly as possible, peak responses gave almost the Nernst slope, indicating that the apical membrane was predominantly Na-permeable immediately after increasing $(Na)_o$. Resistances during peak time were low and independent of concentration between 20 and 80 mM $(Na)_o$ (conductance saturation, see Gebhardt et al. 1972). The peaks were followed by depolarizing voltage changes, almost stabilizing at what on our time-scale looked like 'voltage plateaus'. At the same time, the membrane resistance increased significantly. The resistance-concentration curve plotted for plateaus showed the same 'conductance saturation' as the one belonging to peak potentials. The depolarizing voltage recline together with an increase of resistance is interpreted as being caused by a lowering of P_{Na} in response to an increase of $(Na)_o$.

It should be noted, that a further depolarization is usually seen when exposure times are made longer. Very small Nernst slopes were obtained when solutions were changed slowly by pouring T-solution over the preparation from a beaker and exposing for several minutes (curve marked 'slow' on the plot).

In the right Nernst plot, Fig. 5, P-solution contained 80 mM $(Na)_o$, which was lowered (arrow on abscissa) rapidly (circles) or slowly (squares). The resistance in P-solution was larger than the saturation levels obtained when stepping (Na)$_0$ up (compare to the left diagrams) because P_{Na} was small after long time Na-exposure. Poor Nernst slopes were obtained when (Na)$_0$ was lowered. The voltage-time course showed small hyperpolarizing reclines accompanied by a decrease of resistance, indicating an increase of P_{Na} in response to a decrease of (Na)$_0$.

Erroneous results must be expected from fitting the Goldman function to such data in order to compute permeability ratios. This treatment requires the permeabilities to be independent of concentration, which is not the case at least in our experiments.

UP-DOWN ASYMMETRY

The change of P_{Na} does provide a qualitative explanation of the 'up-down asymmetry' reported earlier (Fuchs *et al.* 1972, Lindemann *et al.* 1972). After increasing (Na)$_0$ the hyperpolarizzing voltage change looks fast because it involves the more sensitive part of the Nernst-curve (low (Na)$_0$) and a large P_{Na}. In its final phase the hyperpolarization may be retarded by the beginning decrease of P_{Na}. If this occurs, the peak potential cannot quite reach the Nernst value. A subsequent drop af (Na)$_0$ (T-P switch) results in a relatively slow depolarization because the concomitant increase of P_{Na} provides a hyperpolarizing tendency as long as (Na)$_0$ is still larger than (Na)$_c$.

At the end of the second P-period, the potential frequently levelled off at a value more than the control value of the first P-period (Figs. 2, 3, 4, 5). At the same time the resistance remained smaller than that of the first period (Fig. 5). We have suggested that in this situation the outer membrane surface is for a while supplied with Na from a slowly exchanging compartment, which was loaded during the T-period (Gebhardt *et al.* 1972). This effect would also contribute to the slowness of the depolarization after removal of Na from the outer solution.

LOCAL CONTROL OF P_{Na}

It is generally known, that P_{Na} of apical membranes in amphibian skin and bladder epithelia is under hormonal control (Fuhrmann & Ussing 1955). According to current thinking, vasopressin increases P_{Na} by raising the cellular cyclic-AMP concentration (Orloff & Handler 1967). Aldosterone appears to use the same mediator (Orloff 1973), which may affect the membrane directly or indirectly. Another level of regulation is demonstrated by the experiments reported here. P_{Na} responds directly to the Na content of the outside solution. Our data corroborate the findings of Cereijido *et al.* 1964 and Biber & Curran 1970, who computed from isotope fluxes that P_{Na} decreases with increasing $(Na)_o$. We show, however, that when $(Na)_o$ is suddenly increased from very small values, P_{Na} remains at the high level belonging to the previously low concentration for some hundred milliseconds, then decreases with a time constant of one to several seconds (frog skin). This fairly slow change may suggest the involvement of metabolic reactions.

Although the cellular Na-concentration is very unlikely to have changed in our experiments, we did not rule out the possibility that $(Na)_c$, when it is allowed to change at constant $(Na)_o$, can affect P_{Na} too.

ACKNOWLEDGEMENT

We are grateful to Mr. W. Ruffing for excellent technical assistance and to Mr. G. Ganster for servicing the electronics.

REFERENCES

Biber T. U. L. & Curran, P. F. (1970) Direct measurement of uptake of sodium at the outer surface of the frog skin. *J. gen. Physiol. 56*, 83–99.

Cereijido, M., Herrera, F. C., Flanigan, W. J. & Curran, P. F. (1964) The influence of Na-concentration on Na-transport across frog skin. *J. gen. Physiol. 47*, 879–893.

Crank, J. (1956) *The Mathematics of Diffusion.* Oxford University Press, London, p. 45.

Fuchs, W., Gebhardt, U., Nacimiento, N. A. & Lindemann, B. (1971) Concentration dependence of small signal membrane conductance in the outer resistive membrane of frog skin. *Abs. 25th Int. Cong. Physiol. Sci.*, Munich, p. 189.

Fuchs, W., Gebhardt, U. & Lindemann, B. (1972) Delayed voltage responses to fast changes of $(Na)_o$ at the outer surface of frog skin epithelium. In *Passive Permeability of Cell Membranes* series Biomembranes, Vol. 3, Ed., Kreuzer, F. & Slegers, J. F. G., pp. 483–498 Plenum Press, N. Y., London.

Fuhrman, F. A. & Ussing, H. H. (1955) A characteristic response of the isolated frog skin potential to neurohypophysical principles and its relation to the transport of sodium and water. *J. cell. comp. Physiol. 45*, 89–102.

Garcia Romeu, F., Salibian, A. & Pezzani-Hernández, S. (1969) The nature of the *in vivo* sodium and chloride uptake mechanisms through the epithelium of the Chilean frog *Calyptocephalella gayi. J. gen. Physiol. 53*, 816–835.

Gebhardt, U., Fuchs, W. & Lindemann, B. (1971) A rapid flow method to study the surface permeability of frog skin. *Abs. 25th Int. Cong. Physiol. Sci.*, Munich, p. 198.

Gebhardt, U., Fuchs, W. & Lindemann, B. (1972) Resistance response of frog skin to brief and long lasting changes of $(Na)_0$ and $(K)_0$. In *Role of Membranes in Secretory Processes* ed. Bolis, L. pp. 1–17. North Holland Publ. Co., Amsterdam.

Heckmann, K., Lindemann, B. & Schnakenberg, J. (1972) Currentvoltage curves of porous membranes in the presence of pore blocking ions. I: Narrow pores containing no more than one moving ion. *Biophys. J. 12*, 683–702.

Kidder, G. W., Cereijido, M. & Curran, P. F. (1964) Transient changes in electrical potential difference across frog skin. *Amer. J. Physiol. 207*, 935–940.

Lindemann, B. (1971) Electrical excitation of the outer cell membrane in frog skin epithelium. In *Electrophysiology of Epithelial Cells*, Symposia Medica Hoechst, ed., Giebisch, G., pp. 53–86. Schattauer, Stuttgart.

Lindemann, B., Gebhardt, U. & Fuchs, W. (1972) A flow-chamber for concentration-step experiments with epithelial membranes. T.-I.-T. J. *Life Sci. 2*, 15–26.

Orloff, J., Handler, J. S. & Stoff, J. S. (1973) Permissive Effect of Aldosterone on the Action of Vasopressin. In *Transport Mechanisms in Epithelia* ed. Ussing, H. H. & Thorn, N. A., pp. 205–210, Munksgaard, Copenhagen.

Orloff, J. & Handler, J. S. (1967). The role of adenosine 3', 5' – phosphate in the action of antidiuretic hormone. *Amer. J. Med. 42*, 757–768.

Smith, P. G. (1971) The low frequency electrical impedance of the isolated frog skin. *Acta physiol. scand. 81*, 355–366.

DISCUSSION

KEYNES: Have you tried what happens to your various time constants if you alter the temperature? With your interpretation of the sodium permeabilities, I think that one would expect to find very large temperature coefficients.

LINDEMANN: Yes, it would be interesting.

KEYNES: Whereas with some of the alternative explanations the coefficients might be much smaller.

LINDEMANN: That is right, they could be smaller.

KEYNES: I am extremely interested in this because I made a voltage clamp system for looking at the transients with the same kind of objectives as you, though we have done very much less work with it. One thing we did was to see what happens when you put lithium outside; and it seems that Li makes quite large differences to the current transients that you get when you voltage clamp. It was turning some of the transients upside down. I have only done one single experiment and I am completely incapable of interpreting it at the moment; but there were certainly some clearly marked effects.

WHITTEMBURY: When you increase the sodium concentration, does the permeability go up or down?

LINDEMANN: It goes down.

WHITTEMBURY: Would there be any possibility that instead of changing permeability you have changes in cell volume or something of that sort. Can you rule that out?

LINDEMANN: The prerequisite for the change of cell volume would be an increase of cellular sodium. And we do not find evidence for that at this time.

HOSHIKO: Do you observe this recline phenomenon with low polar cation concentration, let us say 10–20 mM?

LINDEMANN: When you go from let us say 0.1 to 10 mM? Yes, it is also there, but it is smaller.

HOSHIKO: Perhaps this problem is related to the saturation phenomenon that is observed when you look at the fluxes as a function of the concentration.

LINDEMANN: Yes, I would like to remind you of my slide no. 5. (Fig. 5). Between 25 and 80 mM sodium there is no change of resistance with concentration. This may be the representation of a saturation phenomenon. The change of resistance with time in this situation (arrow) is from one saturation level to another.

TOSTESON: It is a very elegant story, but on one point I would like clarification. Even without the change in sodium permeability which you propose, the relationship between time and voltage following the step from a low to a high sodium concentration and following the step from a high to a low sodium concentration would not be expected to be symmetrical.

LINDEMANN: That is right. The famous logarithmic relationship between V and C always makes the voltage response after an increase of concentration look faster, because when C increases the more sensitive part of the Goldman function is used for the initial, fast change of concentration.

TOSTESON: Exactly.

LINDEMANN: But there is an additional difference, and this additional difference in speed expressed itself in my data as a difference in apparent thickness of unstirred layers. I have suggested, however, that the real reason is a delayed change of P_{Na}.

SACHS: I understand that your model would be some sort of feed-back mechanism between the sodium penetration through the membrane and its permeability to sodium. If that were the case you would expect that the resistance change or the recline, as you call it, would be some function of the sodium concentration.

LINDEMANN: That's right.

SACHS: From your data it didn't seem that way.

LINDEMANN: Fig. 5 again: When stepping $(Na)_o$ up, there is a large change of resistance during the recline at high sodium concentrations. When the $(Na)_o$ step is smaller, voltage and resistance reclines are also smaller.

SACHS: And that includes the PD change as well?

LINDEMANN: Yes.

CURRAN: With regard to the different calculated unstirred layer frequencies going up and going down, are there definite differences?

LINDEMANN: Yes.

CURRAN: Do you have an explanation?

LINDEMANN: Yes, I think it is what we might call the up-down asymmetry in excess of what is explained by the logarithmic relationship.

CURRAN: That is right, because with the logarithmic calculation you essentially are assuming a constant permeability.

LINDEMANN: When the sodium concentration at the membrane rises, the initial steep increase of voltage can be slowed down in its final phase and even reverse (recline) because the Na-permeability drops. When later on $(Na)_o$ is decreased, the resulting depolarisation is slowed down by an increase of P_{Na}, which provides a hyperpolarising tendency. This makes the unstirred layer look thicker in the case of decreasing $(Na)_o$.

MEARES: I would like to say something about this unstirred layer thickness. The Nernst stagnant layer model is an artificial representation of the situation in which you have gradients. When you look into the hydrodynamics of this you find that the apparent thickness of the unstirred layer – if everything else, rates of stirring, viscosity, is kept constant – is inversely proportional to the third power of the diffusion coefficient of the substance which is transported and being studied. In your experiments you start with potassium, the faster diffusing in the aqueous medium of the two ions outside, and replace it suddenly with sodium, so that the unstirred layer now concerns the diffusion of sodium, a slower ion, for which we should expect to find a thicker layer. In fact, in comparison with the final replacement of sodium by potassium, potassium being a faster ion, we would expect to find a thinner layer. I think you found the reverse of that. Thus the effect is possibly larger even than you have interpreted it to be, because you probably have not taken into account this hydrodynamic difference.

LINDEMANN: I did not take this effect into account. However, when the experiment is done with sodium gluconate steps, not substituting but stepping anions and cations together, then one can use a single diffusion coefficient which does not change for up and down movements. One still gets this appreciable difference. I think it is qualitatively a safe thing, although one might argue about 1 μ or so of thickness of unstirred layer (Fuchs et al. 1972).

Fuchs, W., Gebhardt, U. & Lindemann, B. (1972) Delayed voltage responses to fast changes of (Na)o at the outer surface of frog skin epithelium. In: *Passive Permeability of Cell Membranes,* Series 'Biomembranes', Vol. 3, ed. Kreuzer, F. & Slegers, J. F. G. pp. 483–498, Plenum Press, N. Y. London.

Effect of Amiloride, Cyanide and Ouabain on the Active Transport Pathway in Toad Skin

Erik Hviid Larsen

INTRODUCTION

According to the two-membrane hypothesis (Koefoed-Johnsen & Ussing 1958), the active sodium transport through the amphibian skin is a two step process: (1) a passive movement of sodium across an outer sodium selective barrier (denoted the outward facing membrane); (2) an active and therefore metabolically dependent extrusion of sodium across the lateral and basal membranes (denoted the inward facing membrane) of the transporting cells. Thus, in principle, the primary site for regulation of the active transport may be located at one of three levels: At the *outward facing membrane* by regulation of its passive sodium permeability, in the *energy metabolism* by regulation of the supply of ATP necessary for the working of the sodium/potassium pump, and in the active *sodium/potassium pump* itself by regulation of its transport capacity.

Substances reducing the rate of active sodium transport by acting at one of these three levels are for example (1) amiloride which inhibits the passive entry of sodium, (2) cyanide which prevents formation of ATP, and (3) ouabain which inhibits the sodium/potassium pump in a non-competitive way.

Although the target of these inhibitors are located at quite different levels of the transepithelial active sodium transport, the present study shows, that they all cause a reduction of the sodium conductance of the outward facing membrane.

The Zoophysiological Laboratory A, August Krogh Institute, 13 Universitetsparken, 2100 Copenhagen Ø, Denmark.

»Transport Mechanisms in Epithelia«, Munksgaard, Copenhagen.

MICROELECTRODE MEASUREMENTS IN TOAD SKIN

In order to establish the possible effect of an inhibitor on the outward facing membrane, membrane potential and -resistance may be followed during its application.

Table I summarizes the overall transport characteristics of the isolated skin of *Bufo bufo* bathed in SO_4-Ringer's solution on both sides. As these observations agree well with previous findings from other laboratories (Ussing & Windhager 1964, Whittembury 1964, Cereijido & Curran 1965, Rawlins *et al.* 1970), only a few comments are needed.

The transepithelial potential difference (E_t) and the short circuit current (I_{sc}) were measured as described elsewhere (Hviid Larsen 1970). The values are significantly lower than those of the skin of *Bufo marinus* reported by Whittembury (1964), who used skins only if their E_t was greater than 80 mV and I_{sc} was greater than 25 μA/cm². In the present study, such a selection was not made, the range of E_t being 47 – 126 mV and that of I_{sc} 8 – 59 μA/cm².

The epithelium of the skin was impaled from the outside perpendicular to the surface by a micropipette electrode mounted on a Leitz micromanipulator. The electrodes were filled with 2.5 M KCl and selected to have a tip potential less than 5 mV and a resistance between 5 and 15 MΩ, *i. e.* a tip diameter of approximately 0.2 μm (Lassen *et. al* 1971). The individual impalements were accepted if neither the electrode resistance nor the tip potential were significantly changed after withdrawal of the electrode. Intracellular localization of the tip of the microelectrode was established when the recorded potential was more than 4 mV positive with respect to the

Table I. Bioelectric parameters of the isolated toad skin bathed in SO_4-Ringer's solution[1]. The values are given as mean $\pm$SEM for 59 impalements in 11 skins. See text for explanation.

E_t mV	E_0 mV	I_{sc} μA/cm²	E_0^{sc} mV	R_t kΩcm²	R_0 kΩcm²
69.7 $\pm$2.4	35.5 $\pm$2.3	19.3 $\pm$1.7	–15.2 $\pm$0.8	4.49 $\pm$0.24	3.32 $\pm$0.23

[1] Composition of the SO_4– Ringer's solution used throughout this study: 113.5 mM Na^+, 1.9 mM K^+. 0.9 mM Ca^{++}, 57.4 mM SO_4^-, 2.4 mM HCO_3^- and 57 mM sucrose. pH = 8.2 when aerated with atmospheric air.

outer solution in the non-shortcircuited skin and significantly negative under short circuit conditions (Whittembury 1964).

Without any exception the potentials fulfilling the above mentioned criteria were appreciably smaller than the simultaneously recorded E_t. Their mean value is given in the column denoted E_o of Table I. The mean value of the corresponding negative potentials recorded during shortcircuiting is given in the column denoted E_o^{sc}. These negative potentials range from -5mV to -28 mV.

The resistance of the skin (R_t) and the apparent resistance between the intracellular recording site and the outer solution (R_o) was obtained by passing an inward going current through the skin and recording the voltage changes. In this procedure, the current passing through the cells is not known, and therefore R_o is not a measure of the membrane resistance; in skins with a significant paracellular shunt, R_o is an underestimate of the resistance of the outward facing membrane. The mean values of R_t and R_o are given in the last two columns of Table I.

The Bioelectric Response to Amiloride, Cyanide and Ouabain

In the experiments reported in this section the intracellular potentials were recorded from the same cell throughout the observation period. The effect of amiloride and cyanide was studied in 5 and 4 skins, respectively, and was repeated several times as the response appeared to be completely reversible. The effect of ouabain was studied in 2 skins. Representative examples are given here.

Figs. 1 and 2 depict the typical bioelectric response when amiloride or cyanide is added to the solution bathing the outside of the skin. In both cases, the decrease in short circuit current is accompanied by a depolarization of the skin as well as of the outward facing membrane. The increase in skin resistance is accounted for by an increase in the apparent resistance between the interior of the impaled cell and the outer solution. Note, that these changes of E_o, E_t and I_{sc} are completely reversible. This shows that continuous recording from the same cell for such a long period does not in itself damage the impaled cell or affect the overall transport characteristics of the skin.

Amiloride reduces the transepithelial active sodium transport by inhibition of the sodium entrance through the outward facing membrane without directly

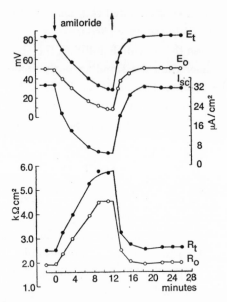

Fig. 1. The bioelectric response of toad skin to amiloride. In the control periods the outer chamber was perfused with SO_4-Ringer's solution. In the period bracketed by arrows the perfusate in addition contained 10^{-6} M amiloride. Volume of outer solution: 2 ml; rate of perfusion: 2 ml/sec.
Symbols are explained in the text.

affecting the active extrusion of sodium across the inward facing membrane (Dörge & Nagel 1970, Nagel & Dörge 1970). The parallel increase in R_t and R_o agrees well with this fact. Application of cyanide depresses ATP formation which in turn is expected to inhibit the active sodium/potassium pump located at the inward facing membrane. Nevertheless, these experiments suggest an effect of cyanide similar to that of amiloride.

In order to see whether inhibition of the active transport mechanism proper contributes to the development of the response observed, the effect of ouabain was studied (Fig. 3). Obviously, when ouabain is added to the inner solution, the response is similar to that of amiloride and cyanide. Again, inhibition of active sodium transport is followed by an increased skin resistance and by a corresponding increase in the apparent resistance between the intracellular recording site and the outer solution.

On the basis of these observations, it is tempting to conclude that the sodium conductance of the outward facing membrane is affected by cyanide and ouabain treatment. However, the increase in resistance between the

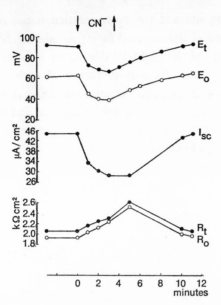

Fig. 2. The bioelectric response of toad skin to cyanide applied to the outer solution as described in the legend to Fig. 1. Cyanide concentration of the perfusate: 10^{-3} M.

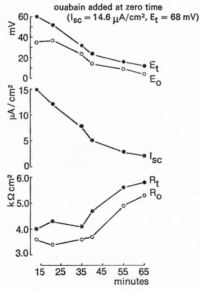

Fig. 3. The bioelectric response of toad skin to 10^{-4} M ouabain applied to the inner solution. Microelectrode impalement of the epithelium was successful about 10 min. after addition of ouabain.

Symbols are explained in the text.

intracellular recording site and the outer solution is not necessarily due to an increased resistance of the outward facing membrane. Still another possibility must be considered. It might be due to an increased resistance of a paracellular shunt pathway. As a consequence of such an effect, a larger fraction of the current is shunted through the cells, resulting in a larger voltage change between the tip of the microelectrode and the outer solution. This would in effect show up as an increased resistance between the intracellular recording site and the outer solution.

MEASUREMENT OF SHUNT CONDUCTANCE AND CALCULATION OF SERIAL
CONDUCTANCE AND EMF

In order to see whether the shunt pathway is affected by these various inhibitors, the overall shunt conductance was measured by means of tracers. The method is based on a simultaneous determination of sodium and sulphate outflux in continuously short circuited skins. Assuming that these ions are transported through the shunt according to the flux-ratio equation (Ussing 1949), the outflux values can be recalculated in terms of sodium and sulphate conductances by means of the following formula (Hodgkin 1951):

$$g_j = (z_j^2 F^2 M_j)/(RT)$$

where g_j is the conductance of the ion species, j, and M_j is the passive flux of j in the short circuited skin; z, F, R and T have their usual meaning. It is fair to assume, that sodium and sulphate are the only ions contributing significantly to the shunt due to their relatively high concentrations in the bathing solutions. Accordingly, the sum of the sulphate conductance and the passive sodium conductance is a reasonable estimate of the overall shunt conductance of the skin (g_{sh}).

A simultaneous determination of the total skin conductance (g_t) measured electrically permits calculation of the conductance of the active transport pathway (g_{ser}) and of the electromotive force of the skin (EMF) by means of the following formula (Ussing & Windhager 1964):

$$g_t = g_{sh} + g_{ser} \quad \text{and} \quad EMF = I_{sc}/g_{ser}$$

According to the two-membrane hypothesis, g_{ser} is determined by the sodium conductance of the outward facing membrane and the potassium conductance of the inward facing membrane. The EMF is the potential difference across the skin when the net transport of ions through its individual membranes is zero. Provided the outward and inward facing membranes behave like a perfect sodium electrode and a perfect potassium electrode, respectively, the EMF is given by (Koefoed-Johnsen & Ussing 1958):

$$EMF = (RT)/(F) \{ \ln(Na)_o/(Na)_c + \ln(K)_c/(K)_i \}$$

where the indices refer to the activities in the outer (o), cellular (c) and inner (i) compartment. Recent experiments carried out in our laboratory (Hviid Larsen, unpublished results) show that, even in the case of amiloride and cyanide treated skins, this equation is a valid approximation. Therefore, in the present study, where $(Na)_o$ and $(K)_i$ are kept constant, changes in EMF are assumed to reflect changes in the cytoplasmic Na:K-ratio in the transporting cells.

The results shown in Table II confirm that the total skin conductance decreases significantly due to treatment with each of the three inhibitors

Table II. Total skin conductance (g_t), shunt conductance (g_{sh}), conductance of the active transport pathway (g_{ser}), and the electromotive force (EMF) of the isolated toad skin bathed in SO_4 –Ringer's solution[1].

	g_t	g_{sh}	g_{ser}	EMF
	10^{-5} mho/cm^2			mV
AMILORIDE[2]	6.1 ±0.6	3.6 ±0.3	2.5 ±0.5	205 ±20
CONTROL	17.8 ±1.1	3.4 ±0.5	14.4 ±0.9	124 ± 2
CN$^-$ [3]	5.5 ±0.6	1.7 ±0.1	3.7 ±0.5	42 ± 4
CONTROL	11.5 ±0.7	2.0 ±0.2	9.6 ±0.6	119 ± 4
OUABAIN[4]	14.1 ±0.7	15.2 ±0.3	(6.2 ±0.9)	(31 ± 7)
CONTROL	21.0 ±1.0	7.9 ±0.4	13.1 ±1.3	120 ± 4

[1] All values given are mean ±SEM (n = 8). [2] 10^{-6} M. [3] 10^{-3} M. [4] 10^{-4} M.
In each experiment two pieces of skin were isolated from the same toad. The inhibitor was added to one of the skins, whereas the other served as control. g_{sh} was calculated on the basis of flux measurements performed during the steady state period following addition of the various inhibitors.

(first column). The second column reveals, that the shunt conductance is unaffected by treatment with amiloride and cyanide. Thus, the decrease in total skin conductance must be accounted for by a decrease in the conductance of the active transport pathway. The significant increase in EMF due to amiloride treatment (last column) is consistent with the view, that this diuretic exclusively plugs the sodium channels of the outward facing membrane without any effect on the active step at the inward facing membrane. Consequently, the continuous working of the pump decreases the sodium activity of the intracellular compartment and, as a result, the EMF increases. In contrast, cyanide treatment decreases the EMF, illustrating that the sodium activity of the intracellular compartment increases and the potassium activity decreases, confirming the assumption of a depressed pump activity.

The interpretation of the ouabain experiments apparently is not this simple. Not only does the shunt conductance increase, it even tends to exceed the total skin conductance. This apparent paradox can be explained by means of Fig. 4, showing the effect of ouabain on the sodium and sulphate outfluxes. Only, the sodium outflux is stimulated by ouabain, the sulphate outflux being unaffected. This finding is most readily explained by

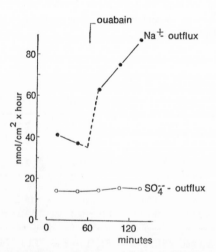

Fig. 4. The effect of 10^{-4} M ouabain on SO_4^--outflux and Na^+-outflux in short circuited toad skin. 40 min. after addition of $^{22}Na^+$ and $^{35}SO_4^-$ to the inner solution, samples from the outer solution were drawn at 30 min. intervals. Ouabain was added to the inner solution at the time indicated by an arrow.

assuming that binding of ouabain converts the sodium/potassium pump to a sodium/sodium exchange mechanism. A similar effect of ouabain on sheep red cells has been reported by Tosteson & Hoffman (1960).

However, as the sulphate outflux remains constant in the ouabain treated skins, this provides good evidence that the true shunt conductance also remains constant and therefore, the effect of ouabain on the total skin conductance is due to a reduction of the conductance of the active transport pathway. Accordingly, g_{sh} obtained in the control skins can be used in calculating g_{ser} and EMF of the ouabain treated skins. The estimates obtained in this way are given in parenthesis in the respective columns of Table II. The remarkably low EMF corresponds well to the expected increase in the cytoplasmic Na:K-ratio.

LOCALIZATION OF THE CYANIDE AND OUABAIN EFFECT

Whether inhibition of energy metabolism affects the properties of the outward facing membrane has not yet been settled, although the effect of various inhibitors on oxidative metabolism has been studied with respect to their overall inhibition of the transepithelial active sodium transport. However, as pointed out by Ussing: "The most striking feature is that although such poisoning stops current and the active sodium transport, the skin does not die, as evidenced by the fact that the skin resistance goes up, whereas in dead skins it drops to nearly zero." (Ussing *et al.* 1960). Further experimental analysis of this is presented above.

Without doubt, striking similarities exist between the effect of amiloride on the one hand and that of cyanide and ouabain on the other. In all three cases inhibition of active sodium transport is accompanied by an increased skin resistance and a parallel increase in the apparent resistance between the interior of the epithelial cells and the outer solution. Moreover, the increased skin resistance is completely accounted for by an increased resistance of the active transport pathway. This provides direct evidence that treatment of the skin with amiloride, as well as cyanide and ouabain reduces the sodium conductance of the outward facing membrane.

With respect to amiloride, this conclusion is entirely consistent with previous findings obtained by means of different experimental approaches. Nevertheless, with respect to cyanide and ouabain, the following objection must be considered: The conclusion is based on the tacit assumption that the outward facing membrane alone impedes ion movement from the outer solution towards the interior of the impaled cells. Obviously, this is correct if the impaled cells belong to the outermost cell layer of stratum granulosum.

Localization of the tip of the microelectrode was not attempted. However, the mean value of the intracellular potentials amounts to about 50 % of the simultaneously recorded E_t, indicating that the majority of intracellular potentials recorded in the present study belong to cells located below stratum granulosum (Ussing & Windhager 1964, Rawlins *et al.* 1970). Therefore, the question arises whether the impaled cells themselves contribute to the overall transport of sodium, and to what extent the cells of different strata are electrically coupled.

The concept of vertical electrical coupling between the epithelial cells of the amphibian skin was introduced by Ussing & Windhager (1964 see also Ussing 1965). Their experimental evidence supporting this hypothesis was the potential profile of the skin increases discontinuously from the outside in 2, 3 or more plateaus located to cell layers following one another towards the serosa side of the epithelium, suggesting a serial arrangement of cells interconnected by low resistance pathways. When $MgSO_4$ replaces Na_2SO_4 of the outer bath, the intraepithelial electric gradient reverses, and so does the direction of the net sodium movement, further supporting the view that the discontinuous potential profile is linked to the direction of intraepithelial sodium transport.

A discontinuous electrical potential profile consisting of three plateaus was also recorded in the present work, but a systematic study of the profiles in skins treated with the inhibitors was not performed. On the other hand, it turns out that the functional significance of the profile observed is a controversial point. Fig. 5 serves to clarify this.

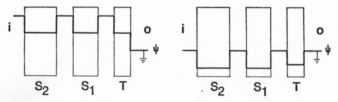

Fig. 5. Simplified diagram of the amphibian skin illustrating the electrical potential profile (Ψ) in the case the epithelium consists of one asymmetric cell layer (T) in series with two layers of symmetric cells (S_1 and S_2). *Left hand panel:* Open circuit conditions. *Right hand panel:* Short circuit conditions.

Suppose that the epithelium consists of an asymmetric, transporting cell layer (T) in series with functionally symmetric cells (S_1 and S_2) across which net transport of ions does not take place, *i. e.* the sodium ions actively transported across T bypass S_1 and S_2 via the low resistance extracellular pathway. The diagrams show the electrical potential profile (Ψ) of this arrangement during open circuit and short circuit conditions.

The first potential step in the left hand panel is expected to be determined mainly by the sodium equilibrium potential across the outward facing membrane of T, whereas the following steps are determined mainly by that of potassium. From this simple diagram a number of important conclusions can be drawn:

(1) Whatever the microelectrode impales T, S_1 or S_2, a positive potential recorded under open circuit conditions becomes negative during shortcircuiting, although the outward facing membrane of T is the only one across which the potential difference reverses. This reversal is due to an ohmic voltage drop owing to the increased sodium ion movement across this membrane. Of course, a slight depolarization of the inward facing membrane of T might occur, the size of which depends on its potassium conductance, beforehand assumed high.

(2) Any voltage signal developed across the outward as well as the inward facing membrane of T will be recorded unattenuated from S_1 and S_2 provided the shunt between the T cells is negligible.

(3) The potential profile of the whole epithelium under open circuit conditions will appear as increasing intracellular plateaus from the outside if P_K/P_{Na} of the inward facing membrane of T and of the membranes of S_1 and S_2 decreases in that order. Thus, a discontinuous increasing potential profile does not necessarily imply vertical electrical coupling between the cell layers. However, the potential profile recorded in the $MgSO_4$ experiments is not as easily explained by this simplified model.

As far as indirect evidence of electrical coupling is concerned, the study of Smith working in Ussing's laboratory showed that, under certain circumstances, the frog skin can be equivalated by a simple two-condensor model with a capacitance of the internal condensor 65 times that of the external one, the latter being about 1 $\mu F/cm^2$ in accordance with values normally reported for biological membranes. The functional interpretation of this goes a long way by assuming that all of the membranes facing the intercellular spaces of the epithelium make up the inward facing membrane, and that the cells communicate with one another through intercellular bridges of low resistance (Smith 1971). In this context it must be mentioned that neighbouring cells indeed are morphologically interconnected by several desmosomes (Farquhar & Palade 1965). Furthermore, the isolated skin bathed in Ringer's solution free of metabolites maintains an appreciable short circuit current often for more than 10 hr. This seems inconsistent with the hypothesis that all of the work is done exclusively by the outermost cell layer of stratum granulosum 10–20 μm thick.

Thus, it is assumed that the epithelium of the skin functionally behaves as a syncytium, and hence that intracellular potential recording from the epithelium does take place from transporting cells.

Consequently, the present results support the hypothesis that the sodium conductance of the outward facing membrane decreases as the sodium/potassium pump located to the inward facing membrane is inhibited, and that this control is mediated by the cytoplasmic Na:K-ratio of the transporting cells.

It cannot be excluded that inhibition of oxidative metabolism more directly affects the sodium entrance to the transporting cells. It is worth noting that the resistance increases immediately upon cyanide application whereas, in a number of experiments, the increase in the resistance following ouabain treatment showed a short time lag as compared to the decrease in short circuit current (e. g. Fig. 3). Further experiments are needed in order to evaluate the significance of this observation.

Finally, very recently, Biber showed that the saturable component of the sodium influx across the outward facing membrane in frog skin was significantly decreased when the skin was treated with ouabain or oxygen depleted by N_2-aeration of the bathing solutions (Biber 1971). The hypothesis put foreward above agrees fairly well with Biber's conclusion based on his direct experimental approach.

ACKNOWLEDGEMENT

The author wishes to express sincere thanks to Professor H. H. Ussing for many stimulating discussions. Thanks are also due to Miss Lone Christensen for her assistance in the laboratory and to Mr. Ole Bengtson for his skilfull technical advice. The work was supported by a grant from the Danish Natural Science Research Council.

REFERENCES

Biber, Th. U. L. (1971) Effect of changes in transepithelial transport on the uptake of sodium across the outer surface of the frog skin. *J. gen. Physiol. 58*, 131–144.

Cereijido, M. & Curran, P. F. (1965) Intracellular electrical potentials in frog skin. *J. gen. Physiol. 48*, 543–557.

Dörge, A. & Nagel, W. (1970) Effect of amiloride on sodium transport in frog skin. II. Sodium transport pool and unidirectional fluxes. *Pflügers Arch. ges. Physiol. 321*, 91–101.

Farquhar, M. G. & Palade, G. E. (1965) Cell junctions in amphibian skin. *J. Cell Biol. 26*, 263–291.

Hodgkin, A. L. (1951) The ionic basis of electrical activity in nerve and muscle. *Biol. Rev. 26*, 339–409.

Hviid Larsen, E. (1970) Sodium transport and d. c. resistance in the isolated toad skin in relation to shedding of the stratum corneum. *Acta physiol. scand. 79*, 453–461.

Koefoed-Johnsen, V. & Ussing, H. H. (1958) The nature of the frog skin potential. *Acta physiol. scand. 42*, 298–308.

Lassen, U. V., Nielsen, A.-M. T., Pape, L. & Simonsen, L. O. (1971) The membrane potential of Ehrlich ascites tumor cells. Microelectrode measurements and their critical evaluation. *J. Membr. Biol. 6*, 269–288.

Nagel, W. & Dörge, A. (1970) Effect of amiloride on sodium transport in frog skin. I. Action on intracellular sodium content. *Pflügers Arch. ges. Physiol. 317*, 84–92.

Rawlins, F., Mateu, L., Fragachan, F. & Whittembury, G. (1970) Isolated toad skin epithelium: Transport characteristics. *Pflügers Arch. ges. Physiol. 316*, 64–80.

Smith, P. G. (1971) The low-frequency electrical impedance of the isolated frog skin. *Acta physiol. scand. 81*, 355–366.

Tosteson, D. C. & Hoffman, J. F. (1960) Regulation of cell volume by active cation transport in high and low potassium sheep red cells. *J. gen. Physiol. 44*, 169–194.

Ussing, H. H. (1949) The distinction by means of tracers between active transport and diffusion. *Acta physiol. scand. 19*, 43–56.

Ussing, H. H. (1965) Transport of electrolytes and water across epithelia. *Harvey Lect. 59*, 1–30.

Ussing, H. H., Kruhøffer, P., Hess Thaysen, J. & Thorn, N. A. (1960) The alkali metal ions in biology. In *Handbuch der Experimentellen Pharmakologie*, ed. Eichler, O. & Farah A., p. 125, Springer-Verlag, Berlin, Göttingen, Heidelberg.

Ussing, H. H. & Windhager, E. E. (1964) Nature of shunt path and active sodium transport path through frog skin epithelium. *Acta physiol. scand. 61*, 484–504.

Whittembury, G. (1964) Electrical potential profile of the toad skin epithelium. *J. gen. Physiol. 47*, 795–808.

DISCUSSION

ORLOFF: Some years ago Dr. Leaf argued that removal of potassium from the pump surface decreases the permeability of the mucosal surface to sodium. Have you examined this possibility?

HVIID LARSEN: No, I haven't.

LEAF: I might say that ouabain also does this, and I would ask the question of whether you think that the permeability at the outer surface may actually be controlled by the potassium concentration in the cell rather than the sodium.

HVIID LARSEN: Yes, this is a possibility.

LEAF: I should add that your studies are much more elegant than ours.

SACHS: I noticed on your slides that you had an absolute value for R_o, which was the membrane resistance. How did you derive that absolute value with your technique?

HVIID LARSEN: Actually, this is not an absolute value of the resistance of the outer membrane. It is determined by passing an inward directed current through the skin and measuring the voltage deflection between the intracellular recording site and the outer solution. As we do not know how large a fraction of this current passes through the membrane itself, it is not possible to calculate the membrane resistance itself. This was the reason why it is necessary to see whether the shunt of the skin is affected.

SACHS: Is it not a function of three resistances, the shunt resistance, the resistance of the luminal facing membrane and the resistance of the serosal facing membrane?

HVIID LARSEN: Yes. Of course a *decrease* of the resistance of the inward facing membrane alone might be reflected in an increase of the parameter denoted R_o. However, such a mechanism can easily be disregarded in the present study, where the total skin resistance as well as the overall serial resistance was shown to increase.

LOEWENSTEIN: Couldn't you measure resistance with a bridge arrangement with the same electrode?

HVIID LARSEN: This we have considered. But the method is of no help, because the current passing between the microelectrode and the outer reference electrode is expected to be partially shunted through the inward facing membrane and out via the paracellular pathway. Thus, even in this case the measured resistance depends on more than the resistance of the outward facing membrane.

KEYNES: This comment is possibly relevant to Dr. Lindemann's paper as well. Can we be certain that there isn't any rectification taking place at the two faces of the cell? I tried doing some voltage clamp studies at different steady potentials. Although the resistance seemed to be fairly linear over the normal range, there was quite a large increase when you started clamping at reversed potentials.

HVIID LARSEN: I have no comment on this, because I have not studied the effect of clamping at various levels out of the normal range on the resistance of the skin.

KEYNES: Quite a lot of other cells do show very marked rectification. I think this is something one wants to be on one's guard against.

LOEWENSTEIN: I think I can answer this for some epithelial cells, one can get easily into, other than skin cells. These cells do not rectify; they behave like nice ohmic resistors: salivary glands, liver, kidney and lens cells.

HVIID LARSEN: Gastric mocusa?

LOEWENSTEIN: That we have not tried.

WHITTEMBURY: How sure can you be that you, with the tip of your microelectrode, were in the first reactive cell layer?

HVIID LARSEN: I am not sure of that.

WHITTEMBURY: If it were in the second would this change your interpretation?

HVIID LARSEN: As discussed in the text, intracellular potential recording probably took place from cell layers located below stratum granulosum. Thus, my interpretation rests on the assumption, that these cells take part in the overall transport of sodium, and that the main barrier to sodium ion movement from the outer solution into the impaled cells is the outward facing membrane of the first reacting cell layer. Argumentation favouring this view is given in the last part of the paper as far as untreated skins are concerned. To my knowledge studies on vertical electrical coupling between the epithelial cells in CN^- or ouabain treated skins have not yet been performed. On the other hand, Kristensen & Schousboe (1968) showed that during *anaerobic* conditions all of the glycogen content of the whole epithelium was used as energy source for the remaining active sodium transport. This indicates, that also during anaerobic conditions where the cytoplasmic Na:K-ratio is expected to increase, the epithelium functionally behaves like a syncytium. Still, this only gives circumstantial evidence for the correctness of my interpretation.

LINDEMANN: Do I remember correctly that after adding amiloride the change recorded with the microelectrode was less than the total change?

HVIID LARSEN: Yes, in the experiment depicted in Figure 1 the decrease of E_o is less than the decrease of E_t. Although, in other experiments, the decrease of E_o fully accounted for the decrease of E_t, you must speculate whether the actual observation is due to a leak around the microelectrode rather than a depolarization of the inward facing membrane. I think, that the complete reversibility of the response indicates that the impaled cell is not seriously damaged, favouring the last interpretation. The mechanism of a depolarization of the inward facing membrane might then be explained on the basis of the significant decrease of the intracellular sodium activity.

KEYNES: There was one thing about that slide that puzzled me slightly; this is the slowness of the effect of amiloride. On the few occasions that I have tried it, I was rather impressed to find that it reduced the short circuit current to zero as fast as I could change the solution, a time constant of a second or two.

HVIID LARSEN: Yes, this applies to frog skin. I have made the same observation. Definitely, the response in toad skin is much slower. This was a consistent finding.

CRABBÉ: Dr. Hviid Larsen, did you examine the possible effects of temperature changes on the permeability of the outermost cells barrier?

Kristensen, P. & Schousboe, A. (1968) The role of glycolysis in energy production in the isolated skin of the brown frog (*Rana temporaria L.*). *Biochim. biophys. Acta. 153*, 132–137.

HVIID LARSEN: No.

CRABBÉ: I raise the question becauce amphibia exposed to low temperature (which is the way we often store them) don't seem to swell unbearably; so one would be tempted to think that when one decreases the metabolic rate by just lowering temperature there would be an adjustment involving sodium entry, maybe via a mechanism such as the one you have examined here.

HVIID LARSEN: This is an interesting suggestion.

Anion Transport Across Frog Skin

Poul Kristensen

INTRODUCTION

It has early been shown that frogs are able to absorb chloride from very dilute solutions (Krogh 1937). Later work showed that this is an active process (Jørgensen et al. 1954). More recent work shows that an active transport system is operating in isolated skins of *Rana pipiens* and *Rana esculenta* (Martin & Curran 1966). A similar system for Cl⁻ transport has been observed in toad bladders bathed in potassium free media (Finn et al. 1967). In *Rana temporaria* skins chloride transport is of such a magnitude (Kristensen 1972) that it could not be measured in chloride Ringer's solution because of the large background of passive fluxes, so it was measured in sulphate Ringer's solution containing small concentrations (5mM or less of chloride). This chloride transport was not inhibited by ouabain (Kristensen 1972) and is therefore different from the chloride transport system in the skin of *Leptodactylus ocellatus* (Zadunaisky et al. 1963).

A working hypothesis with respect to how chloride crosses the frog skin *(R. temporaria)* has been published (see Fig. 1). According to this, chloride is transported actively from the mucosal side across the outward facing membrane of the outermost living cell layer into the cells, from where it diffuses passively to the interspace system. The main transepithelial passive chloride movements occur via the so called tight seals.

The major arguments for this hypothesis were the following (Kristensen 1972):

1) The graphically determined influx pool of chloride is greater than the efflux pool, and is reduced to the same size as this under the influence of Diamox®, which is an inhibitor of chloride transport in this tissue; 2) The efflux varies with the transepithelial potential difference as would be expected

Institute of Biological Chemistry A, University of Copenhagen.

»Transport Mechanisms in Epithelia«, Munksgaard, Copenhagen.

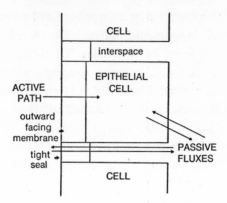

Fig. 1. Schematical representation of the working hypothesis. Only the outermost living cell layer is shown.

from the integrated flux equation (Goldman 1943–44); 3) The net influx was not influenced by variations in the potential difference across the skin; 4) The presence of an active sodium transport is not required for the maintenance of chloride transport.

In this report the simple concentration dependence of chloride is described. Further, determinations of apparent pool sizes for chloride at low and high concentrations are compared to the apparent pools of sulphate, and the results are found to be in agreement with the previously published working hypothesis (Kristensen 1972). As sulphate in these experiments is considered to move passively across the tissue, some control experiments had to be carried out with this ion. It was found that the flux ratio does not vary with the potential as would be expected for a divalent ion. However, no arguments in favour of an active transport system or exchange diffusion system for the sulphate ion could be found. A similar behaviour of sulphate was found in experiments carried out with a synthetic cation exchange membrane.

METHODS

The frogs *(R. temporaria)* were stored at about 4°C and killed immediately before use. The abdominal skins were dissected off and divided into two laterally symmetrical halves; each of these were mounted between two half chambers, in which they were bathed in Ringer's solution on both sides.

The skins could be shortcircuited or clamped at fixed potentials as described under the individual experimental series. The bathing solutions were bubbled with atomspheric air on both sides. To obtain flux ratio measurements, fluxes were measured in opposite directions across the two halves, which were otherwise treated identically. Chloride fluxes were measured with Cl^{36} and sulphate fluxes with S^{35} labelled sulphate. The isotopes were obtained from the Danish Atomic Energy Commission (Risø). The composition of the Ringer's solutions were as follows: Chloride Ringer: 111 mM NaCl, 2.4 mM $NaHCO_3$, 1.0 mM $CaCl_2$ and 2.0 mM KCl. Sulphate Ringer: 56.5 mM Na_2SO_4, 2.4 mM $NaHCO_3$, 2.0 mM K_2SO_4, and 1.0 mM $CaSO_4$. In the calculations of the individual experiments the amounts of carrier accompanying the isotopes are taken into account. Higher concentrations of chloride in sulphate Ringer's solution were obtained by mixing the two Ringer's solutions in the appropiate ratios.

The apparent pools were determined by adding isotope to one side of the skin and following the appearance on the other side as a function of time (Andersen & Zerahn 1963, Harvey & Zerahn 1969) The pool is determined as the vertical distance between the steady state part of the obtained curve and a line parallel to this through the origin. Chloride and sulphate pools were determined simultaneously on the same skin piece, as the two isotopes could be distinguished from each other by the Packard Spectrometer.

In the experiments concerning sulphate fluxes across an ion exchange membrane, an Amfion C. 100 membrane (American Machine and Foundry Company, Springdale, Conn.) used. The potential difference was obtained as a diffusion potential resulting from a concentration difference of NaCl across the membrane. Sucrose was added to obtain osmotic balance.

RESULTS AND DISCUSSION

If chloride combines with a carrier, as it was postulated in the working hypothesis, it should be expected that the active transport of chloride obeys saturation kinetics. That this is the case can be seen in Fig. 2, in which influx is plotted as a function of chloride concentration in the mucosal bathing solution. The obtained curve is clearly composed of a saturating component and a diffusion component following Fick's law. In order to estimate the K_m value, a permeability constant for the skin was calculated from the fluxes at 11 mM and 57 mM Cl^-. From this the passive fluxes at

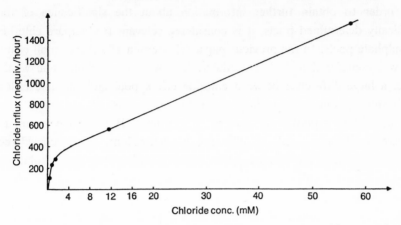

Fig. 2. Chloride influx as a function of chloride concentration. The curve is clearly composed of a saturating and a linear component.

the low concentrations could be calculated and subtracted from the influx values.

The reciprocal of the net fluxes plotted against the reciprocal of the concentration gave straight lines, enabling an estimation of the concentration of chloride giving half maximal net transport for the skin piece. Values from six experiments are given in Table I. The magnitude of these values indicates that the described transport may be of significance for the frogs under natural conditions, as the chloride concentration in fresh waters often is in the range of 1 mM. Further, the fact that chloride transport obeys saturation kinetics makes it unlikely that chloride transport occurs as a result of solvent drag. Further support for the carrier hypothesis is found in the fact that bromide and isothiocyanate inhibit chloride transport (Kristensen 1972).

Table I: K_M-values for chloride transport across the isolated frog skin.

Exp. no.	K_M (mM)
1	0.24
2	0.40
3	0.40
4	0.49
5	0.27
6	0.28

In order to obtain further information about the significance of the graphically determined pools, it is considered relevant to compare chloride and sulphate pools. In the previous paper (Kristensen 1972) the small efflux pool was considered to be extracellular. If it were intracellular one would expect a large difference between chloride efflux pool and sulphate efflux pool, resulting from the large difference between the ions.

By definition, a pool is an amount of substance. But when we want to compare pools measured at different concentrations, it is considered permissible to use a corrected pool, which is defined as the graphically determined pool divided by the concentration of the compound for which the pool is determined. The corrected pool will thus be a volume of solution occupied by the amount of substance in the pool, if the concentration is equal to that in the bathing solution. This is clearly not the case with respect to intracellular pools and may even be wrong for intercellular pools, although the approximation in the latter case may be better. On the other hand, the use of corrected pools is the only possible means by which pools measured at different concentrations of ions can be compared.

Table II shows the corrected pools under different conditions. It appears that chloride influx pool at low chloride concentration is larger than any other measured pool (P = 0.001–0.005), whereas no other pools are significantly different from each other (P-values for the various comparisons ranging from 0.1–0.7).

Table II: Apparent corrected pools for chloride and sulphate. Averages of 5–10 experiments.

Ionic species	Conc. of the ion mM	Influx pool $\mu l/5\ cm^2$	Efflux pool $\mu l/5\ cm^2$
Cl^-	3.09	3.03	0.43
Cl^-	115	0.38	0.39
SO_4^-	0.21	0.26	0.42
SO_4^-	57	0.21	0.25

The results indicate that passive chloride fluxes occur through the same path as movements of sulphate. This is in agreement with the working hypothesis, if the solution in the extracellular space equilibrates faster with the bathing solution than with the intracellular space.

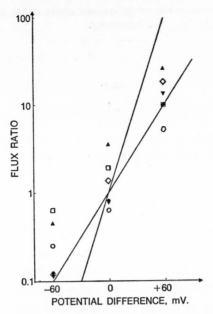

Fig. 3. Flux ratio of sulphate as a function of transepithelial potential difference. The steeper line indicates the slope expected for a divalent ion, the other that for a monovalent ion.

In the preceding discussion it was assumed that sulphate behaves as a passive ion when crossing the frog skin. It is therefore relevant to study the behaviour of sulphate in more detail.

When sulphate flux ratios are measured at different potential difference values, Fig. 3 shows that sulphate does not follow the flux ratio equation (Ussing 1949) for a divalent anion. There may be several explanations for this:

1) Sulphate crosses the frog skin in combination with one cation and thus behaves as a monovalent ion or ion pair.

2) Sulphate movements occur partially via an exchange mechanism or active transport.

3) Sulphate fluxes are disturbed by water movements due to electroosmosis through negatively charged pores.

Possibility (2) may be considered unlikely because it can be shown that in the range of sulphate concentrations 0.2 to 60 mM no evidence of

Table III: Permeability constants for influx and efflux of sulphate as a function of sulphate concentration. Average of 9 experiments.

Conc. of sulphate	$P \times 10^8$ (cm/sec.)	
	Influx	Efflux
20 mM	3.8 (a)	4.0 (c)
52 mM	4.6 (b)	3.6 (d)

When applying Student's t-test it was found that a is significantly smaller than b, and d smaller than c. In both cases $P < 0.001$.

saturation kinetics is present. On the contrary it may be shown that the permeability constant for influx increases with increasing sulphate concentration, while the constant for efflux decreases with increasing concentration (Table III). This phenomenon may not be a function of the sulphate concentration but rather of the chloride concentration, which varies from nearly full chloride Ringer concentration to 1/10 of that.

The explanation may be the existence of a standing potential gradient in the interspace system resulting from the sodium transport, either due to a potassium diffusion potential or to an electrogenic sodium transport. It is conceivable that such a potential will tend to slow down outward sulphate movements and to increase inward diffusion. Such a potential gradient may be more pronounced in sulphate solutions because it may be more easily short circuited by the chloride ions. A similar effect on the sulphate permeability would be excerted by a standing concentration gradient in the extracellular space just below the tight seals. In this case water movements through channels in the tight seals could influence the sulphate fluxes.

Possibility (3) cannot be excluded. An estimation of the water flux required to change the sulphate fluxes to the observed extent can only be calculated if the length of the pores and their cross sectional area are known. It is however worth mentioning that passive chloride movements obey the flux ratio equation (chloride Ringer's solution). An explanation for this difference between chloride and sulphate might be that the latter, because of the double negative charge, will have to travel a more tortuous path through the rate limiting structure. The time during which a water flow could exert its action might thus be considerably longer for sulphate than for chloride. As the difference between the permeability constants for chloride and sulphate in

Table IV: The variation of sulphate flux ratio with the potential difference across a cation exchange membrane.

Exp. no.	P. D. mV	Fluxes (neq./h) 5 cm²	Flux ratio
A	+56	0.51	10.2
	—60	0.05	0.098
B	+60	0.40	6.7
	—58	0.06	0.15
C	+55	0.40	8.0
	—58	0.05	0.125

many experiments is a matter of a factor of three, this explanation seems less likely.

Possibility (1) requires that the structure through which the diffusion takes place contains negative charges or dipoles (Frizzell & Schultz 1972, Wright et al. 1971) so that double negatively charged ions are prevented from entering this structure. A few experiments were therefore carried out with a synthetic cation exchange membrane. The results are shown in Table IV. If we assume symmetrical behaviour of this membrane, the flux ratios for the three experiments may be calculated. Disregarding the small deviations from 58 mM seen in the experiments, we obtain the values shown in Table IV, last column. It may be concluded that sulphate also behaves more like a monovalent than like a divalent ion, in this case.

It may be concluded that the experiments with sulphate do not prove that the apparently smaller pool is extracellular, but the results are still in agreement with this idea. Although the present method of measuring pools gives results which are difficult to interpret, it may be concluded that the difference obtained between measurements made from the two sides of the skin indicates that the flux ratio varies with time in the beginning of the experimental period. This must lead to the conclusion that chloride follows two paths through the skin.

REFERENCES

Andersen, B. & Zerahn, K. (1963) Method for non-destructive determination of the sodium transport pool in frog skin with radiosodium. *Acta physiol. scand.* 59, 319–329.

Finn, A. L., Handler, U. S. & Orloff, J. (1967) Active Chloride Transport in the isolated toad bladder. *Am. J. Physiol. 213*, 179–184.

Frizzell, R. A. & Schultz, S. G. (1972) Ionic conductances of extracellular shunt pathway in rabbit ileum. Influence of shunt on transmural Na transport and electrical potential differences. *J. gen. Physiol. 39*, 318–346.

Goldman, D. E. (1943–44) Potential, impedance and rectification in membranes. *J. gen. Physiol. 27*, 37–60.

Harvey, W. R. & Zerahn K. (1969) Kinetics and route of active K-transport in the isolated midgut of *Hyalophora cecropia*. *J. exp. Biol. 50*, 297–306.

Jørgensen, C. B., Levi, H. & Zerahn, K. (1954) On active uptake of sodium and chloride ions in anurans. *Acta physiol. scand. 30*, 178–190.

Kristensen, P. (1972) Chloride Transport across isolated frog skin. *Acta physiol. scand. 84*, 338–350.

Krogh, A. (1937) Osmotic regulation in the frog (R. esculenta) by active absorption of chloride ions. *Scand. J. Physiol. 76*, 60–78.

Martin, D. W. & Curran, P. F. (1966) Reversed potentials in isolated frog skin: II Active transport of chloride. *J. cell. comp. Physiol. 67*, 367–374.

Ussing, H. H. (1949) The destinction by means of tracers between active transport and diffusion. The transfer of iodide across the isolated frog skin. *Acta physiol. scand. 19*, 43–56.

Wright, E. M., Barry, P. H. & Diamond, J. M. (1971) The mechanism of cation permeation in rabbit gallbladder. Conductances, the current-voltage relation, the concentration dependence of anion-cation discrimination, and the calcium competition effect. *J. Membr. Biol. 4*, 331–357.

Zadunaisky, J. A., Candia, O. A. & Chiarandini, D. J. (1963) The origin of the short-circuit current in the isolated skin of the South American frog *Leptodactylus ocellatus*. *J. gen. Physiol. 47*, 393–402.

DISCUSSION

CURRAN: I was not entirely clear as to how you estimated the pools you were talking about.

KRISTENSEN: The pools were measured in the same way as the one Harvey & Zerahn (1969) used for the *cecropia* midgut. You add radioactive chloride to one side and you get a curve which becomes a straight line sooner or later. And then by extrapolation you get something which may represent an apparent pool, an amount of substance at least, with which the isotope has to equilibrate before the flux is constant.

LINDEMANN: Could the chloride transport be a secondary phenomenon due to the large sodium transport?

KRISTENSEN: It doesn't seem to. We have done experiments with amiloride and ouabain, and in both cases chloride is stille actively transported. We have also done it with sodium free solutions on the outside. I don't know what the reason is for that but we still have a rather normal chloride transport.

ZADUNAISKY: I want to express to Dr. Kristensen how happy I am that he has confirmed now that in the skins of European frogs there is also transport of chloride ions that can be demonstrated *in vitro* similar to our old demonstration of a chloride transport in the same direction as sodium in the skin of the *Leptodactylus ocellatus*. Did you try the effect of copper on the ion fluxes?

KRISTENSEN: I tried it but it was very inconclusive up until now, becauce it varies a lot with the copper concentration. It is something we are going to work with. Sometimes it increases and sometimes it decreases. It is very difficult to handle.

ZADUNAISKY: And did you get negative potentials very easily?

KRISTENSEN: No, not in the *Rana temporaria*. Whatever I do I cannot get a negative potential and a negative current in the *temporaria*. I haven't tried other species.

Harvey, W. R. & Zerahn, K. (1969) Kinetics and rout of active transport in the isolated midgut of *Hyalophora cecropia*. *J. exp. Biol. 50*, 297–306.

TOSTESON: Could the monovalent cation that is being transported with sulfate be protons? Have you studied the effect of pH?

KRISTENSEN: No, we haven't done the experiments yet.

SCHULTZ: Has carbonic anhydrase been identified in the frog skin preparation you have used?

KRISTENSEN: There is carbonic anhydrase almost anywhere so it would also be in the frog skin, but whether it has any relevance to the transport I cannot say. I tried to do some experiments some time ago with changing of the bicarbonate concentration, and I couldn't really see any clear effects. It has to be repeated in the future, because, like in the fish gills, there may be some effect of bicarbonate. But up to now we haven't seen it.

MAETZ: Very recently, in our laboratory Dr. M. Istin in collaboration with Dr. F. Garcia Romeu have reinvestigated the problem of carbonic anhydrase activity in frog skin. It is presumed that toxic secretion from frog skin may inhibit the enzyme in ground tissue. Indeed, whole frog skin extracts were shown to inhibit frog red cell carbonic anhydrase. To avoid such interference, homogenates of isolated frog skin epithelia were prepared and submitted to Sephadex column chromatography in order to separate fractions containing the inhibitor from those containing the enzyme. No enzyme containing fractions were found, however. It remains possible that the enzyme is an insoluble form. Already such forms of carbonic anhydrase have been discovered in other tissues (see Istin & Girard 1970).

HOSHIKO: We tested the effect of pH on anion selectivity. And in the alkaline range the bromide ions seemed to be preferred, but around pH 4.5 chloride seemed to be more permeable. What was the time constant on that chloride curve? It is in minutes, isn't it?

KRISTENSEN: Yes, it is 2–3 minues for the active one, and a little longer for the passive, 7 minutes, as far as I remember.

Istin, M. & Girard, J. P. (1970) Carbonic anhydrase and mobilisation of calcium reserves in the mantle of lamellibranchs. *Calc. Tiss. Res. 5*, 247–260.

GENERAL DISCUSSION

MAETZ: There was a question raised about cell feed-back in response to increased sodium content of the epithelial cell. Dr. Lindemann, could you please comment on the experiments which were performed in your laboratory by Dr. U. Katz concerning toads adapted for a month in high salinity waters. If I remember well from my discussion with Dr. Katz, there were drastic adaptive changes in the sodium permeability of the outside facing membrane as well as changes in the shunt pathway leakiness.

LINDEMANN: These experiments were done by Dr. Uri Katz. He took Isralian toads which can adapt to high salinity water. He found that the sodium permeability of the outer membrane was drastically reduced after adaptation. At the same time the response to amiloride was very small. He didn't perform any measurements of cellular concentration. Therefore we cannot really evaluate whether the effect is partly due to a high cellular sodium concentration.

MAETZ: This has been done by Dr. Ferreira in Portugal on the same animal. The sodium content of the cells goes up (unpublished experiments).

LINDEMANN: That would be expected.

LOEWENSTEIN: May I make a general comment. As an outsider to the field of epithelial transport, it seems to me that for analysis at the cell level, the frog skin, which has ben so fruitfully exploited by Dr. Ussing and many other participants at this symposium, presents some difficulties because the cells are small and multilayered. I should like to point out a skin preparation which may be useful for transport work at the cellular level: the skin of Amblystoma and Triturus larvae. Its cells are large and single-layered. I used these skins with Penn (Loewenstein & Penn 1967) to show coupling between skin cells. We had no difficulties in visualizing the cells *in vivo* or in impaling them with microelectrodes.

KIRSCHNER: We have used them.

USSING: They have not much of a potential as far as I know or have they?

Loewenstein, W. R. & Penn, R. D. (1967) Intercellular communication and tissue growth. II. Tissue Regeneration. *J. Cell. Biol. 33,* 235–242.

KIRSCHNER: It is just like that of the frog. A little bit lower becauce there is probably a shunt through the gills.

SCHULTZ: Dr. Lindemann, in your experiments did you always replace a sodium-free solution with one containing sodium? Have you replaced a sodium-free, potassium solution with one containing, say, 20 mM sodium and then restored the full sodium concentration? Do you still see the overshoot and redrive under these conditions?

LINDEMAANN: Yes, we did, and we still see it.

Intercellular Pathways for Water and Solute Movement Across the Toad Bladder

Donald R. DiBona, Mortimer M. Civan

(communicated by Alexander Leaf)

INTRODUCTION

In 1963, Ussing reported that when urea is added to the outer medium bathing isolated frog skin, making the solution hypertonic, the transepithelial electrical resistance falls. This observation has been subsequently extended to a wide variety of solutes (Ussing 1965, 1966, 1969, Biber & Curran 1968, DiBona *et al.* 1969b, Franz *et al.* 1968, Lindley *et al.* 1964) and to other epithelia, including the urinary bladder of the toad (Urakabe *et al.* 1970).

Recently, deformations of the apical junctions of toad bladder and skin have been noted to occur when the resistance of these preparations was lowered by making the mucosal bathing medium hypertonic with sucrose (DiBona 1972) and other solutes (Wade & DiScala 1971).

This communication reports the results of a more detailed study of the relationship between the osmotically-induced morphologic and resistance changes. The findings show that solutes which reduce the tissue resistance also produce "blistering" in the apical intercellular junctions, that the two effects are reversible with similar time courses, and that they are independent of active sodium transport. We conclude that the impenetrability imputed to the tight apical junctions is more apparent than real, that their structure may be reversibly altered, and that intercellular channels or potential channels exist which provide a pathway for the passive movement of solutes and water across the epithelium.

Laboratory of Renal Biophysics and the Department of Medicine, Massachusetts General Hospital, and the Departements of Anatomy and Medicine, Harvard Medical School. Supported in part by grant HE-06664 from the National Heart and Lung Institute and by Grant-in-Aid 71–847 from the American Heart Association. Dr. Civan is an Established Investigator (70–184) of the American Heart Association.
Please address correspondence to Alexander Leaf, M. D., Massachusetts General Hospital, Boston, Massachusetts 02114 U. S. A.

»Transport Mechanisms in Epithelia«, Munksgaard, Copenhagen.

METHODS

The electrical resistance across paired half bladders was measured so as to exclude the effects of resistance changes in the bathing media produced by altering the composition or concentrations of bathing solutions. Measurements were made in conventional lucite chambers with the hemibladder separating mucosal and serosal media. A nylon mesh supported the bladder and a slight hydrostatic pressure applied from the mucosal side kept the bladder immobilized against the nylon mesh so as to avoid physical distortion or injury to the tissue. The transephithelial potential was clamped at 0 mv (short-circuited state) except for 9 sec. intervals each 30 sec. when the transepethelial potential was increased to 12 mv with serosa positive to mucosa in order to determine the tissue resistance. The composition of the bathing media could be altered at will; the resulting resistance changes recorded.

At the appropriate time, the tissues were fixed by the simultaneous additions of suitable volumes of 50 % glutaraldehyde to mucosal and serosal solutions to provide a final glutaraldehyde concentration of 1 % and were allowed to stand for 15–30 minutes before removal. Rectangles of tissue excised from the chambers were immersed in 1 % glutaraldehyde in phosphate buffer. Tissue samples were post-fixed in osmium tetroxide and embedded in epoxy as previously described (DiBona *et al.* 1969a) but one half of each sample was also stained *en bloc* with uranyl acetate (Farquhar & Palade 1963). Sections were cut with a Reichert OmU2 multramicrotome and examined with a Philips EM-200 electron microscope.

RESULTS

Tissue resistance began to fall within 30 seconds of the addition of a variety of solutes to the mucosal medium resulting in tonicities as high as 672 mOsm/kg water. Although qualitatively similar results were observed following the addition of NaCl, KCl, urea, mannitol, sucrose or raffinose to the mucosal medium, the magnitude of the effect depended on the concentration and molecular size of the solute applied. For example, raffinose was effective when dissolved in Ringer's solution to a final mucosal tonicity of 650 but not at 430 mOsm/kg water, whereas addition of KCl to a final tonicity of only 400 mOsm/kg water significantly decreased the tissue resistance.

Subsequent electron microscopic examination of the fixed tissues consistently demonstrated a bullous enlargement or "blistering" of the space within the apical junctions in every preparation in which the resistance had been lowered by mucosal hypertonicity, but none in the control tissues which had been exposed only to isotonic solutions.

Fig. 1* illustrates an experimental tissue at low power and demonstrates both the junctional blistering and the associated consistent finding of strict closure of the lateral intercellular spaces. Fig. 2* shows the apical intercellular junction as it appears in toad bladder from control preparations bathed on both surfaces by isotonic solutions. The appearance is similar to the "tight" apical junctions which have been described in other epithelia (Brightman & Reese 1969).

By contrast application of hypertonic mucosal solutions resulted in alteration of a majority of junctional profiles as previously described (DiBona 1972) and as shown in Figs. 2b, 3 and 4*. Considerable variability in the size and appearance of the blisters was noted, even within single preparations. Oblique sectioning of the tissue demonstrates that the junctional alteration is not continuous but is composed of multiple, focal deformation. For quantitative comparison of results the fraction of all the observed junctional profiles containing at least a single blister has been determined for each tissue.

The correlation between the osmotically-induced decrease in resistance and the appearance of blistering was strengthened by observations on the reversibility of the two effects. In three paired experiments, a decrease to 50 % of control transepithelial resistance was produced by addition of mannitol to raise the tonicity of the mucosal medium to 450 to 467 mOsm/kg water. Draining the mucosal medium and replacing it with isotonic Ringer's solution resulted in a return of resistance in the ensuing 23 to 26 minutes to 95 $\pm$ 6.5 % of its initial value. Subsequent examination demonstrated blisters in only 9 $\pm$ 4 % of the junctional profiles in those bladders as compared with an incidence of blistering of 89 $\pm$ 3 % of the junctional profiles from bladders from controls in which the hypertonicity of the mucosal medium induced by mannitol had not been reduced.

The structural changes in the junctional complexes were shown not to be dependent upon active transport of sodium. Sodium transport was prevented by replacing the Ringer's solution with a sodium-free choline Ringer's solu-

* Figs. 1, 2, 3, 4 see figure insert opposite p. 168.

11*

tion and adding KCl to increase the tonicity of the mucosal medium in the experimental tissues. Subsequent electron micrographs demonstrated blistering in 90 $\pm$ 1 % of observed junction profiles from the experimental side and no blistering whatsoever on the control side. In the presence of sodium in the bathing media, but with active sodium transport largely inhibited by ouabain (10^{-3}M), mucosal hypertonicity caused with KCl or NaCl produced the same fractional reduction in resistance and the same degree of blistering as observed in the absence of the inhibitors. With Amiloride (10^{-4}M) as inhibitor, marginally significant changes in both effects were noted.

The effect of molecular size of the solute used on the osmotically induced changes was examined with KCl, mannitol and raffinose. KCl at a mucosal medium tonicity of some 400 mOsm/kg water caused twice the reduction in resistance as mannitol at some 450 mOsm/kg. Similarly, mannitol at this concentration was slightly more effective in reducing resistance than was a still higher concentration of raffinose ($\sim$ 650 mOsm/kg of water).

Subsequent examination of the tissues by electron microscopy revealed a similar high degree of blistering in both the mannitol and KCl treated tissues although the resistance drop was nearly twice as large with KCl as with mannitol.

When the size of the blisters was measured without prior knowledge of the treatment, however, the degree of disruption of junctions was found to be clearly greater with KCl than with mannitol. The KCl-treated preparation also showed 18 % of the observed junction profiles to contain more than a single blister while, following mannitol, only 6 % showed multiple deformations.

Furthermore, comparison between mannitol- and raffinose-treated tissues demonstrated a marked difference in the frequency of blistered junctions; 92 $\pm$ 2 % of the junctures was blistered by mannitol, as opposed to 18 $\pm$ 4 % raffinose. No morphological change, other than the deformation of the junctional complexes, correlated with the fall in resistance produced by mucosal hypertonicity.

Although mucosal hypertonicity was regularly associated with both the appearance of junctional blisters and a fall in resistance, it was unclear whether these effects were determined by the tonicity of the mucosal medium *per se* or by the osmotic gradient established across the preparation.

To resolve these possibilities, the serosal medium was replaced by a Ringer's solution diluted with an equal volume of distilled water but normal Ringer's solution served as mucosal bathing solution on the experimental side; this preparation was compared with control tissue exposed on both surfaces to the diluted Ringer's solution.

The short-circuit current was little affected by these treatments but the experimental side showed a fall in resistance whereas the control tissues with dilute Ringer's solution bathing both sides showed a slight rise in resistance. Upon microscopic examination, the epithelial cells of both experimental and control tissues appeared swollen to comparable degrees. Blisters were observed in 90 % of the apical junctions observed from the experimental tissue, but in none of those from control tissue.

Also comparison of the effects of mucosal hypertonicity alone versus hypertonicity of both mucosal and serosal medium with NaCl resulted in a larger resistance drop in the former than in the latter. Whereas no blisters were seen in the preparation exposed to hypertonic sodium chloride on both surfaces, the hypertonic mucosal medium alone produced the usual high incidence of junctional blisters (91 % of the apical junctions examined). Thus, it is the presence of a transepithelial osmotic gradient rather than hypertonicity *per se* which is responsible for the observed effects.

DISCUSSION AND CONCLUSIONS

Addition of a variety of solutes to an initially isotonic mucosal medium reduces the electrical resistance across the toad bladder and simultaneously induces bullous deformation (blisters) of the most apical intercellular junctions (generally called "tight junctions") of the mucosal epithelium. This phenomenon is reversible, as shown by the effects of adding and then removing mannitol from the mucosal medium. The changes in resistance and the blistering of intercellular junctions is not dependent on active sodium transport, as they occur in the absence of sodium in the media or when sodium transport has been inhibited by either ouabain or amiloride.

Since the same decrease in resistance and blistering was produced by diluting the serosal medium rather than making the mucosal medium hypertonic (fig. 5*), it is clearly the osmotic gradient rather than mucosal hypertonicity *per se* which causes the effects. When both mucosal and serosal tonicities were equally elevated, no blistering was seen. The direction of the osmotic gradient is critical; dilution of the mucosal medium (DiBona & Civan 1972)

* see insert opposite p. 168.

or addition of solute to the serosal medium (DiBona 1972) does not induce blisters and increases, rather than decreases, the tissue resistance.

The effects are apparently not dependent on the changes in cell volume; application of an osmotic gradient across the bladder with a higher solute concentration of the mucosal than of the serosal medium reduces resistance and causes blisters whether the osmotic forces tend to swell the cells (serosal hypotonicity) or to shrink the cells (mucosal hypertonicity). Closure of the lateral intercellular spaces beneath the junctional complexes accompanied the resistance change and blistering. Finally, the size of the solute molecule was shown to be important; smaller solutes, e.g., KCl, were more effective than larger molecules, e.g. raffinose, in causing these effects.

The simplest explanation for the observed findings is that the solute molecules diffuse into the intercellular junctional complexes from the more concentrated mucosal medium. They then draw water osmotically from the adjacent cells or lateral intercellular spaces into the junctions causing the appearance of blisters. At the same time water is drawn into the cells from the lateral intercellular spaces in accordance with the osmotic gradients, thus closing the lateral intercellular spaces. Since the apical junctional complexes normally provide a major resistance barrier to the movement of water and solutes between the cells, their deformation with blistering causes a lowering of resistance and a pathway for the passive movement of water and solute across the toad bladder.

Although the apical junctions have been found to be an impassable barrier for molecules as large as peroxidase (Masur *et al.* 1971), smaller solutes, as used in these studies, do penetrate into them. Thus these junctions provide a limited, but perhaps adjustable pathway for the movement of water and solutes across the bladder, a conclusion consistent with recent results obtained with other epithelia (Boulpaep 1967, Frizzell & Schultz 1972, Frömter & Diamond 1972, Mandel & Curran 1972).

Movement of ions across the junctional complexes and through the lateral intercellular spaces is, therefore, likely to constitute at least one of the parallel shunt pathways previously hypothesized on the basis of electrophysiologic evidence (Civan *et al.* 1966, Ussing & Windhager 1964). For this reason, it is suggested that the terms "tight junction" and "*zonula occludens*" (Farquhar & Palade 1965) based on purely anatomic evidence and perhaps evoking the erroneous concept of an infinite resistance of these junctional complexes to water and solutes, should be replaced by the term "limiting junction" (DiBona 1972) which more correctly describes the probable physiologic function of this structure.

REFERENCES

Biber, T. U. L. & Curran, P. F. (1968) Coupled solute fluxes in toad skin. *J. gen. Physiol. 51,* 606–620.

Boulpaep, E. L. (1967) Ion permeability of the peritubular and luminal membrane of the renal tubular cell. In *Symposium über Transport und Funktion Intracellularer Elektrolyte,* ed. F. Kruck, 98–107. Urban and Schwarzenberg, Munich.

Brightman, M. W. & Reese, T. S. (1969) Junctions between intimately apposed cell membranes in the vertebrate brain. *J. Cell Biol. 40,* 648–677.

Civan, M. M., Kedem, O. & Leaf, A. (1966) Effect of vasopressin on toad bladder under conditions of zero net sodium transport. *Amer. J. Physiol. 211,* 569–575

DiBona, D. R. (1972) Passive pathways in amphibian epithelia: Morphologic evidence for intercellular route. *Nature,* New Biology. *238,* 179–181.

DiBona, D. R. and Civan, M. M. (1972) Osmotically-induced conductance changes in toad bladder under physiologic conditions. *Int. Union Pure and Applied Biophysics IVth Cong.*

DiBona, D. R., Civan, M. M. & Leaf, A. (1969a) The anatomic site of the transepithelial permeability barriers of toad bladder. *J. Cell Biol. 40,* 1–7.

DiBona, D. R., Civan, M. M. & Leaf, A. (1969b) The cellular specificity of the effect of vasopressin on toad urinary bladder. *J. Membrane Biol. 1,* 79–91.

Farquhar, M. & Palade, G. E. (1963) Junctional complexes in various epithelia. *J. Cell. Biol. 17,* 374–412.

Farquhar, M. and Palade, G. E. (1965) Cell junctions in amphibian skin. *J. Cell Biol. 26,* 263–291.

Franz, T. J., Galey, W. R. & Van Bruggen, J. T.: (1968) Further observations on asymmetrical solute movement across membranes. *J. gen. Physiol. 51,* 1–12.

Frizzell, R. A. & Schultz, S. G. (1972) Ionic conductances of extracellular shunt pathway in rabbit ileum. Influence of shunt on transmural sodium transport and electrical potential differences. *J. gen. Physiol. 59,* 318–346.

Frömter, E. & Diamond, J. (1972) Route of passive ion permeation in epithelia. *Nature (Lond.) 235,* 9–13.

Lindley, B., Hoshiko, T. & Leb, D. E. (1964) Effects of D_2O and osmotic gradients on potential and resistance of the isolated frog skin. *J. gen. Physiol. 47,* 773–793.

Mandel, L. & Curran, P. F. (1972) Response of the frog skin to steady-state voltage clamping: I. The shunt pathway. *J. gen. Physiol. 59,* 503–518.

Masur, S. K., Holtzman, E., Schwartz, I. L. & Walter, R. (1971) Correlation between pinocytosis and hydroosmosis induced by neurohypophyseal hormones and mediated by adenosine 3', 5' – cyclic monophosphate. *J. Cell Biol. 49,* 582–594.

Urakabe, S., Handler, J. S. & Orloff, J. (1970) Effect of hypertonicity on permeability properties on the toad bladder. *Amer. J. Physiol. 218,* 1179–1187.

Ussing, H. H. (1963) Effects of hypertonicity produced by urea on active transport and passive diffusion through the isolated frog skin. *Acta physiol. scand. 59,* (Suppl. 213) 155–156.

Ussing, H. H. (1965) Relationship between osmotic reactions and active sodium transport in the frog skin epithelium. *Acta physiol. scand. 63*, 141–155.

Ussing, H. H. (1966) Anomalous transport of electrolytes and sucrose through the isolated frog skin induced by hypertonicity of the outside bathing solution. *Ann. N. Y. Acad. Sci. 137*, 543–555.

Ussing, H. H. (1969) The interpretation of tracer fluxes in terms of membrane structure. *Quart. Rev. Biophys. 1*, 365–376.

Ussing, H. H. & Windhager, E. E. (1964) Nature of shunt-path and active sodium transport path through frog skin epithelium. *Acta physiol. scand. 61*, 484–504.

Wade, J. B. & DiScala, V. A. (1971) The effect of osmotic gradients on the ultrastructure of the zonulae occludentes in toad bladder epithelia. *11th Ann. Meeting of the Am. Soc. Cell Biol.* Abs. 622.

ACKNOWLEDGEMENTS

The technical assistance of Miss Bonnie Lord is gratefully acknowledged. This work was supported in part by grant HE 06664 from the National Heart and Lung Institute and by Grant-in-Aid 71–847 from the American Heart Association. Dr. Civan is an Established Investigator (70–148) of the American Heart Association.

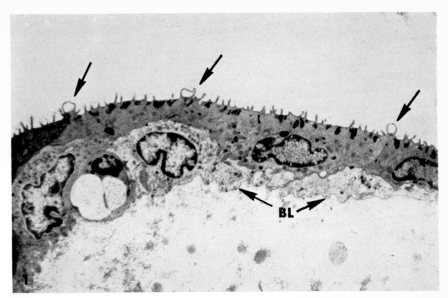

Figure 1. Low power view of toad bladder epithelium fixed after mucosal solution was made hypertonic with mannitol (final osmolality = 467 mOsm/kg water.) Lateral margins of the epithelial cells are tightly opposed to one another everywhere, except within the most apical cell-cell junctions which are grossly deformed into prominent »blisters« (Bl, arrows) projecting beyond the mucosal surface. X *3,500.*

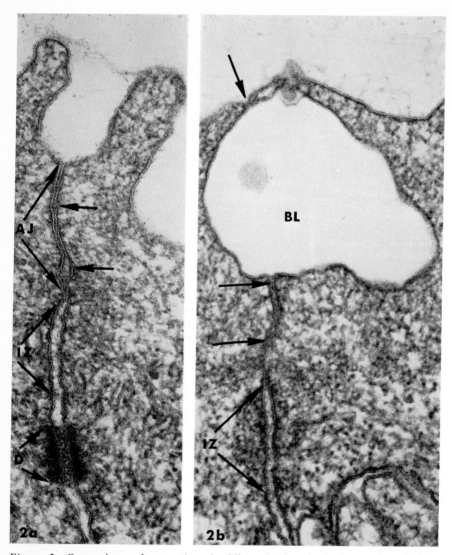

Figure 2. Comparison of normal and »blistered« junctions. 2a. Apical limiting junction (AJ) between two granular cells as it appears when the tissue is exposed only to isotonic solutions and where the preparation is *en bloc* stained with uranyl acetate. It is clear that within this zone the adjacent cell membranes only make periodic contact, revealing small clear spaces at several positions (e. g. unlabelled arrows). lZ, intermediate zone or *zonula adhaerens;* D, desmosome. X 100,000.

2b. »Blistered« junction produced by elevation of mucosal tonicity to 385 m0sm/kg water with KCl. Above the intermediate zone (lZ) the cells make close contact at the unlabelled arrows, but a prominent blister (BL) or bullous enlargment of the extracellular space within the junction itself is also present. X 100,000.

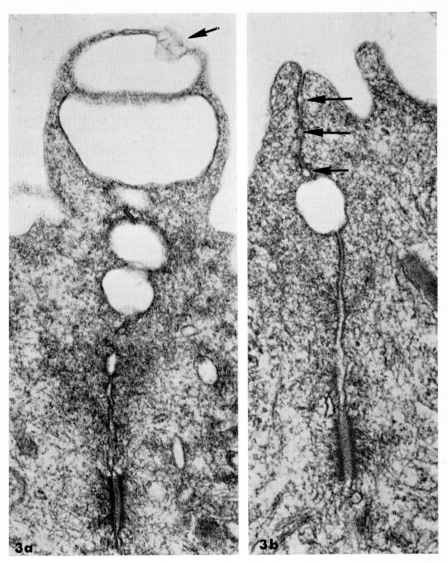

Figure 3. Junction profiles after mucosal application of hypertonic mannitol (3a) (to a final osmolality of 467 mOsm/kg water) and raffinose (3b) (to a final osmolality of 650 mOsm/kg water). In *Figure 3a,* multiple blisters are observed even along the length of a single junction profile; at the arrow, a membrane disruption is also present suggesting that, with time, the junction may be at least partially disrupted. *Figure 3b* illustrates a single prominent blister and three sites (arrows) where additional deformation may be being initiated. 3a, X *42,500;* 3b, X 57,500.

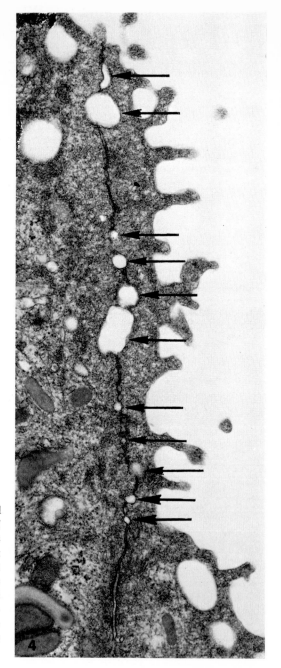

Figure 4. Oblique section of apical junction fixed after exposure of mucosal surface to a hypertonic solution (brought to 467 mOsm/kg water with mannitol). This section allows observation of an extensive region of the apical junction and confirms that the blistering (arrows) is focal rather than continuous around the periphery of the cell. *X 31,000.*

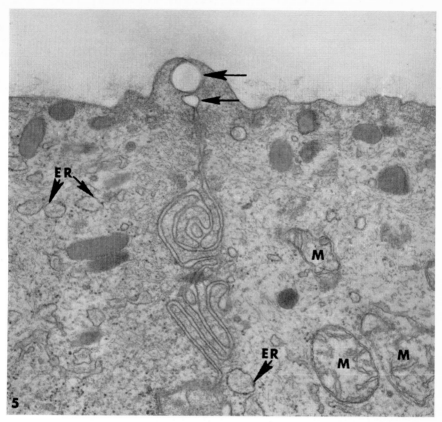

Figure 5. Preparation of bladder epithelium fixed after transepithelial resistance was reduced by dilution of the serosal bath (to a final concentration of 146 mOsm/kg water). As in experiments with mucosal hypertonicity, apical junctions are deformed (arrows) and lateral cell margins are tightly closed. However, in this case, the epithelial cells are swollen as is evidenced by the swollen profiles of mitochondria (M) and saccules of the rough endoplasmic reticulum (ER). X *31,000.*

DISCUSSION

SKADHAUGE: Did you actually observe whether the permeability to the hypertonic agent or others, perhaps smaller molecules was correlated to the degree of blistering?

LEAF: We are making only electrical measurements here, obviously. Whatever ion is carrying the current would be the one that must be penetrating more readily in the presence of the blistering than in the absence of blistering. All the solutes that we are studying produced the blistering so I'm not saying that the solutes are the ones that are carrying the current, any other small ions should do this. It's rather interesting that although the horseradish peroxidases' molecular size is 18,000 it fails to penetrate the tight junctions, the smaller solutes we studied all managed to penetrate. This evidenced by the morphological changes and the resistance change.

SCHMIDT-NIELSEN: I have a question to either Dr. Leaf or Dr. Skadhauge: The only places that I know where hyperosmotic reabsorption occurs naturally and physiologically is in the fish intestine or in the bird intestine, where the solution on the mucosal side is hypertonic to the blood side. And I wonder whether any electronmicroscopy has been done in these two tissues?

LEAF: DiBona and Civan have examined only toad bladder.

SKADHAUGE: Not so far as I know.

MOREL: It would be interesting to know whether the tight junction is open or not in the thin limb of the descending loop of Henle in the kidney.

KEYNES: What is the normal range of osmotic gradient one actually has in the bladder?

LEAF: The toad produces a very dilute urine. We measured values of 50 mOsm in the bladder with 220 mOsm in plasma and they can go up to an isotonic fluid but normally the urine does not go to hypertonic values.

WHITTEMBURY: As far as I understand, when the toad is dehydrated for a very long time, the toad can have a rather concentrated urea solution in the lumen of the bladder.

LEAF: That's right. If the toad gets very dehydrated the osmolality of urine in the bladder can go up to 300 or 400 and at that time plasma osmolality will also be correspondingly increased. The osmolality of the urine can be almost entirely due to urea since the bladder and the kidney can reabsorb the sodium. In other words this condition is characterized by a very low sodium in the urine.

ORLOFF: What path does water traverse under these circumstances?

LEAF: Probably largely across the cell.

ORLOFF: Do other solutes induce blister formation as well?

LEAF: I don't know.

HOSHIKO: We made observations by looking on the effect on potential in frog skins some time ago (Lindley et al. 1964) and found that in series of polyalcohols the small molecules were not effective and the large molecules were not and there was a sort of maximum effect with adonitol, a 5 carbon polyalcohol. I was wondering if there were similar effects on blister formation?

LEAF: No, here the smaller the solute, the more effective it is in producing a higher percentage of the junctions of these formations and blisters of larger size.

HOSHIKO: But raffinose also was not quite as effective.

LEAF: That was a point I was making. Raffinose I recall has a 6 Å molecular radius. Even at concentrations over 600 mOsm it only produced junctional deformations in 18 percent of the observed junctions and the resistance block was much less than with lower concentrations of mannitol.

KIRSCHNER: I'd like to go back to the first question that was asked: I don't know the answer, but it suggests something very interesting to me. There are some animals, hypotonic regulators, which normally face the situation that you just described. Sea water is half molar NaCl and the animals have blood concentrations pretty much like the frog. Now pretty clearly those tight junctions can't be blistered because the fish is normally found in sea water. What is interesting to me is that several of us have tried to perfuse the vascular system in order to have control of solutions on both sides. And

---

Lindley, B. D., Hoshiko, T., & Leb, D. E. (1964) Effects of D$_2$O and osmotic gradients on potential and resistance of the isolated skin. *J. Gen. Physiol. 47*, 773–793.

nobody since Keyes has been successful in double perfusing a fish, fresh water or marine, for very long. What happens is that, while perfusing with Ringer solutions with appropriate colloid osmotic pressure, the preparation turns very leaky and it gets progressively more leaky until it is behaving like a better filtration apparatus than the glomerulus. It would be very interesting to look at EM of these perfused marine fish gills and see whether the tight junctions had opened. What is it in the blood in the normal fish that keeps the situation from developing?

LEAF: Maybe the fish gill has an entirely different structure for its tight junctions.

MAETZ: There has been a question about possible effects of neurohypophysial hormones on reversed osmotic water flow across frog bladder. More than 10 years ago, in collaboration with Dr. J. Bourguet, we observed an increased reversed flow under neurohypophysial stimulation (unpublished experiments). Such an effect does not necessarily signify that the hormone acts on the mucosal border of the epithelial cell. It may also act by loosening the tight junction. I must add that I have in my files an electronmicrograph made on such preparations by Dr. N. Carasso. We found fantastic blisters in the region of the tight junction but we did not know how to interpret them!

ZADUNAISKY: I wonder if the mechanism of formation is the same as the one produced by application of hydrostatic pressure in the inside of frog skin, because if one applies the pressure before the big separation of the epithelium occurs, one can see in the electronmicroscop the small blisters.

LEAF: Windhager or Ussing may want to comment on that, but I don't think it's the same thing you get when you increase the pressure on the serosal side. There you actually get separation of the lateral intercellular space whereas I think the tight junctions remain intact. And the same thing occurs in renal tubules. If you overhydrate the animal one sees a separation between the basal lateral portion of the intercellular space but the tight junctions I don't think are affected.

WHITTEMBURY: Rawlins in our laboratory, when he designed this preparation of the isolated epithelium, made some electronmicrographs during application of hydrostatic pressure in the outward direction. He used thorium dioxide as an extracellular marker. He found particles of about 20 Å in radius flowing along the socalled tight junctions of the outermost layer and of the layer below in the outer spaces. So the tight junctions also become a little leaky when hydrostatic pressure is applied (Rawlins *et al.* (1970, Figs. 6 and 7). I would like to add that Erlij & Martinez-Palomo (1972) have reported formation of vacuoles by hypertonic urea, which they believe are independent of the openings of tight junctions.

Rawlins, F., Mateu, L., Fragachán, F. & Whittembury, G. (1970) Isolated toad skin epithelium: transport characteristics. *Pflügers Arch. 316,* 64–80.
Erlij, D. & Martinez-Palomo, A. (1972) Opening of tight junctions in frog skin by hypertonic urea solutions. *J. Membrane Biol. 9,* 229–240.

Insulin, Glucagon and Active Sodium Transport: from Man to Amphibia - and back

J. Crabbé

Very soon after the discovery of insulin, 50-odd years ago, the hormone was reported to bring about a transient lowering of plasma potassium concentration in man and in laboratory animals (Harrop & Benedict 1923, Briggs *et al.* 1924). This finding began to be amenable to interpretation when Zierler observed that hyperpolarization of isolated rat striated muscle cells results from exposure of the preparation to insulin (1957). This electrical effect may be ascribed to a stimulation of active outward sodium transport across the cell membrane, as recently described by Moore who studied the isolated frog sartorius muscle (in publication).

Tissues carrying out transcellular sodium transport are of course similarly equipped with a sodium "pump", and it was therefore not unreasonable to assume that insulin could stimulate this process in such specialized preparations as well. Actually, Herrera *et al.* (1963) reported that, upon introduction of insulin in the solution bathing frog skin *(Rana pipiens)* examined by the short-circuit current technic of Ussing and Zerahn (1951), sodium transport increased when the solutions were enriched in calcium or magnesium. A stimulating influence of insulin can be demonstrated when the hormone comes into contact with the inside surface of the ventral skin of the toad, *Bufo marinus,* incubated in frog Ringer's fluid* (André & Crabbé 1966). (Fig. 1).

Toad colon mucosa is another structure capable of efficient sodium transport: it reacts also to insulin, and in terms of sensitivity, an effect can be observed with hormone concentrations smaller than 10 mU/ml (Fig. 2)

(Endocrine Unit, Depts of Physiology and Medicine, University of Louvain (U.C.L.) Medical School, B-3000 Louvain, Belgium)

* made out of NaCl, 115 mM; $KHCO_3$, 2.5 mM; $CaCl_2$ 1 mM.

»Transport Mechanisms in Epithelia«, Munksgaard, Copenhagen.

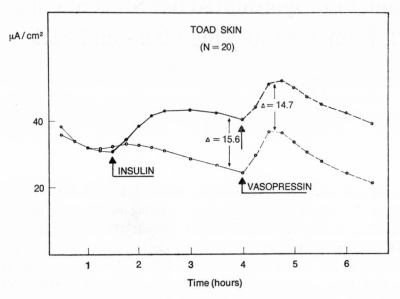

Fig. 1. Influence of antidiuretic hormone and insulin on active sodium transport by the ventral skin of *Bufo marinus*

After 90 min of incubation of paired pieces of skin, insulin was added to the solution of the inner surface of one preparation (final concentration: 125 mU/ml) while the matched one was treated with the excipient. Posterior pituitary extract (Pitressin®, 125 mM/ml) was added 2½ hours later to the solution on the inside of both preparations of each pair, and incubation continued for an analogous period of time.

which are insufficient to elicit a response on toad skin. With the urinary bladder of *Bufo marinus,* this hormonal effect (Herrera 1965) is less regularly obtained, although the tissue can respond, as treatment of the animal with insulin results in stimulated sodium transport by the bladder to an extent at least as marked as when skin and colon are examined (Table I.). The reason for the discrepancy between this insulin effect observed *in vivo* vs. *in vitro,* is obscure.

The mechanism of the stimulating action of insulin on sodium transport awaits elucidation. Interaction of the hormone with some "receptor" located at the basal cell border would constitute the first step (Ehrlich *et al.* 1972). This factor does not seem to require the integrity of sulphhydryl groups, as indicated by failure on the part of N-ethymaleimide to block the effect of insulin on short-circuit current (François & Crabbé 1969).

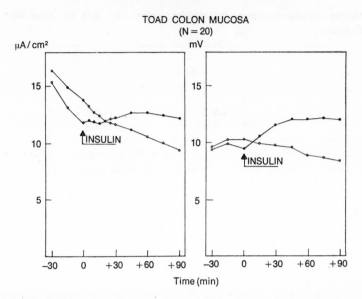

Fig. 2. Stimulation of sodium transport by the mucosa of toad colon exposed to insulin.

The mucosa of the colon of *Bufo marinus* was dissected free from the underlying layers, and divided in 2 pieces incubated simultaneously according to Ussing and Zerahn (1951). After 30 min, insulin, 5 mU/ml, was introduced in the solution on the inside (serosa) of one preparation of each pair and incubation continued for 90 min.

The thought is entertained of late that insulin interferes with the activation of adenyl cyclase (Jefferson *et al.* 1968, Hepp & Renner 1972, Krug & Cuatrecasas 1972). Yet insulin fails to modify the amplitude of the response of toad skin to the natriferic effect of antidiuretic hormone (Fig. 1), while the latter presumably acts by stimulation of the adenyl cyclase system (Orloff & Handler 1964).

On the other hand, inhibitors of nucleic acid, and of protein synthesis were devoid of influence on the electrical response of toad skin to insulin. Fig. 3 summarizes data obtained with actinomycin D and puromycin; additional experiments, carried out with colon mucosa exposed to cyclo-heximide, yielded identical results. Such data are in keeping with the observation that glucose uptake by the isolated rat diaphragm was normally stimulated by insulin despite addition of these inhibitors to the medium (Eboué-Bonis *et al.* 1963).

Table I. Active sodium transport by bladder, colon and skin of *Bufo Marinus* treated with insulin.

Tissue examined	Untreated toads	Toads treated with insulin
Bladder	11.7 ± 1.8 (10)	33.6 ± 5.4 (12)
Colon	9.5 ± 1.9 (6)	13.6 ± 3.5 (6)
Skin	12.3 ± 1.8 (6)	16.7 ± 1.3 (6)

1 Unit insulin per 10 g. body weight was injected subcutaneously 6 hours prior to sacrifice to 6 toads which had been maintained for a few days in dilute saline. Six control animals were injected with saline. The means stand for the electrical activity during the first hour of incubation *in vitro* (current was read every 15 min).

The effect is observed even when hypoglycemia resulting from insulin is prevented by repeated injection of glucose to the animal. For these preparations, short-circuit current is a quantitative reflection of net, active transepithelial sodium transport.

Concerning the site of action of insulin on sodium-transporting amphibian epithelia, tissues such as toad bladder, colon, and skin are currently understood as transporting sodium in the following way: the ion first has to move inward, across the apical pole of the epithelial cells of these preparations; this process supposedly is a "downhill" one, thermodynamically speaking, at least when the solution on the outside (*i. e.* in the lumen of bladder and colon) is rich in sodium; still it would require interaction of the ion with specialized carrier molecules, the existence of which is postulated at the apical membrane. The "uphill", endergonic movement of sodium ascribed to a sodium "pump", occurs in turn at the cell borders facing interspaces and basement membrane. Obviously, these two steps are possible sites of action of insulin.

Data discussed elsewhere (Crabbé, 1972a, 1972b) led to a confirmation of the conclusion formulated by Herrera (1965) according to whom insulin stimulates the active energy-requiring sodium extrusion mechanism presumably located at the non-apical border(s) of the cells specialized in net transepithelial sodium transport.

Such a conclusion is congruent with the hypokalemic effect of insulin observed *in vivo* as the latter can be taken as a consequence of a stimulation of the sodium "pump" in several tissues. But the experiments conducted on isolated amphibian epithelia would acquire more biological significance, if a role of insulin could be documented for mammalian sodium-

TOAD SKIN

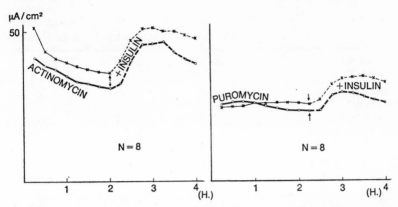

Fig. 3. Lack of influence of actinomycin D and puromycin on the response of the isolated ventral skin of *Bufo marinus* to insulin.

Matched fragments of toad skin were incubated according to Ussing and Zerahn (1951). After one hour, actinomycin D (10^{-5}M) or puromycin (5.10^{-5}M) were introduced in the solution on the inside of one preparation of each pair. Two hours later, both membranes of each pair were treated with insulin (125 mU/ml added on the inside) and incubation proceeded for another 2 hours.

transporting organs such as the kidney. It was indeed reasoned that, should insulin act on the sodium "pump", the function of tissues specialized in transcellular sodium movement would most likely be influenced, on the basis that such tissues have in common with cells at large the sodium "pump" mechanism.

Because of the metabolic reactions attending hypoglycemia which results from insulin administration, it is almost impossible to establish whether insulin administered systemically exerts an effect on sodium reabsorption by organs such as the kidney. Therefore insulin was added to the blood perfusing dog kidneys studied in the isolated state (Nizet *et al.* 1971); in order to eliminate animal-to-animal variations, both kidneys of each animal were perfused concomitantly, in 15-odd instances, with one preparation serving as a reference for the matched one treated with the hormone. Amounts of the latter were such that insulinemia ranged between 80 and 500 μU/ml – which values correspond to those encountered in several clinical states. As appears from Table II, this experimental

Table II. Influence of insulin on excretion of sodium, potassium and water by the isolated perfused dog kidney.

	Control	+ Insulin	Statistical significance of differences
Renal plasma flow (ml/100 g/min)*	284	290	–
Glomerular filtration rate (ml/100 g/min)*	41.6	41.6	–
Haematocrit (p. 100)	48.1	48.2	–
Plasma sodium (mEq/1)	155.2	154.2	–
Plasma potassium (mEq/1)	4.5	4.6	–
Post-glomerular plasma protein concentration (g/100 ml)	8.2	8.0	–
Plasma glucose (mg/100 ml)	152	153	–
Sodium excretion (μEq/100 g/min)*	233	191	$P<0.001$
Potassium excretion (μEq/100 g/min)*	128	98	$P<0.001$
Water excretion (ml/100 g/min)*	2.6	2.2	$P<0.001$

* Weight of the kidney was taken into account (mean weight was identical for both kidneys of each pair: 31.7 g). The parameters indicated above were determined for 3 consecutive 30-min periods for each pair of perfused kidneys (N = 14 pairs), after half an hour being allowed for »stabilization« of the preparations.

protocol made it possible to demonstrate a sodium-retaining effect of insulin on the nephron.

It was attempted to move one step further and to identify in man a condition associated with sodium retention at the renal level, which could be ascribed to insulin. It is well known that, during its initial phase, starvation leads to sodium wastage while sodium almost disappears from urine upon refeeding with carbohydrates. This manifestation is especially well documented in obese patients undergoing fast for therapeutic reasons (Kolanowski *et al.* 1970). This phenomenon was interpreted as the consequence of levels of insulin in circulating blood being very low during fast, while a large outpouring of insulin would occur upon subsequent administration of carbohydrates. More recent studies indicate however that insulin cannot be the only factor involved. Glucagon, in all likelihood, also comes into play as

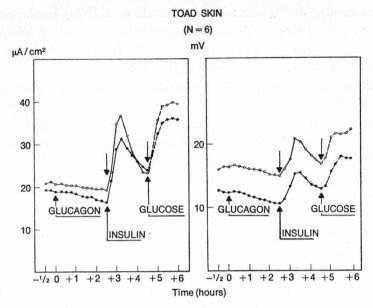

Fig. 4. Lack of effect of glucagon on the response of aldosterone treated toad skin to insulin and to glucose.

Pairs of the ventral skin of *Bufo marinus* were set up for incubation overnight in the presence of aldosterone, $5 \cdot 10^{-8}$M, in the open-circuit state. The following morning the hormonal effect was definite since, for untreated preparations handled similarly, short-circuit current averages 10 μA/cm^2 (Crabbé *et* al. 1971). Then the solutions were renewed and the preparations were short-circuited for one hour prior to addition of glucagon, 2.5 μg/ml, to the solution bathing the inner surface of one membrane of each pair. Thereafter both preparations were treated first with insulin, 125 mU/ml (2½ hours later), later with glucose, 10 mM (4½ hours later).

a natriuretic substance (Kolanowski & Crabbé 1971, Saudek *et al.* 1972). It should be kept in mind that glucagon in blood is elevated during fast while carbohydrates inhibit glucagon secretion (Foá 1968).

Thus, amphibian epithelia were again resorted to so as to examine whether glucagon interferes with sodium transport in baseline conditions or at least when hormonal stimuli are applied. To date, however, it has not been possible to bring about *in vitro* any effect of glucagon on the sodium-transporting activity of such preparations in baseline conditions. Further, no attenuation of the response usually elicited by hormones stimulating sodium transport resulted, as summarized in Fig. 4 in the case of toad skin stimulated with aldosterone since this steroid hormone is actually secreted in increased amounts during starvation (Kolanowski *et al.* 1970). Furthermore, insulin and glucose brought about the large stimulation of sodium transport usually observed under such circumstances (André & Crabbé 1966; Crabbé *et al.* 1971) irrespective of the presence of glucagon.

Additional experiments meant to evaluate the response to antidiuretic hormone of glucagon-treated vs. control skin preparations also led to negative results: a mean short-circuit current ($\mu A/cm^2$) of 43.4 was recorded for the hour following introduction of this hormone when no glucagon was used, as against 38.3 when it was present at a concentration of 2.5 $\mu g/ml$. When the response to antidiuretic hormone was evaluated – which requires account be taken of baseline activity – the decrease in the presence of glucagon, of 4.2 $\pm$ 2.6 (S. E.), proved not significant, statistically ($P > 0.1$, N = 10).

Insofar as studies on amphibian epithelia help shed light on some aspects of the function of the mammalian nephron, these data render it unlikely that the natriuretic effect of glucagon is due to inhibition of the response to sodium-retaining hormones such as aldosterone, as has been recently suggested (Boulter *et al.* 1972). The natriuretic activity of glucagon, the physiological significance of which deserves consideration, may result instead from hemodynamic fluctuations of the renal level (Levy & Starr 1972).

ACKNOWLEDGEMENTS

These studies were supported financially by grants from the Fonds National de la Recherche Scientifique and the Fonds de la Recherche Scientifique Médicale (Belgium).

Insulin and Glucagon were gifts from Novo Industri A/S; Aldosterone was generously supplied by CIBA, and Pitressin by Parke-Davis.

REFERENCES

André, R. & Crabbé, J. (1966) Stimulation by insulin of active sodium transport by toad skin: influence of aldosterone and vasopressin. *Arch. int. Physiol. Biochem.* 73, 538–540.

Boulter, P., Spark, R. & Arky, R. (1972) Starvation: dissociation of renin and aldosterone and the insensitive nephron. *J. clin. Invest.* 51, 12a.

Briggs, A. P., Koehig, I., Doisy, E. A. & Weber, C. J. (1924) Some changes in the composition of blood due to the injection of insulin. *J. biol. Chem.* 58, 721–730.

Crabbé, J. (1972a) Hormonal influences on transepithelial sodium transport: aldosterone vs. insulin. *J. Steroid Biochem.* 3, 229–235.

Crabbé, J. (1972b) L'influence de l'aldostérone et de l'insuline sur le transport du sodium à travers la membrane cellulaire. *Schweiz. med. Wschr. 102*, 1001–1008.

Crabbé, J., Decoene, A. & Ehrlich, E. N. (1971) Some characteristics of the response of the ventral skin of the toad *Bufo marinus*, to aldosterone *in vitro*. *Arch. int. Physiol. Biochem. 79*, 805–808.

Ehrlich, E. N., Siddique, F. & Rubenstein, A. H. (1972) Biologic action and tissue binding of insulin in ventral toad skin. Abstr. IV int. Congress End., Washington, *Excerpta med., Int. Congress Series 256*, 168.

Eboué-Bonis, D., Chambaut, A. M., Rolfin, P. & Clauser, M. (1963) Action of insulin on the isolated rat diaphragm in the presence of Actinomycin D and Puromycin. *Nature Lond. 199*, 1183–1184.

Foá, P. P. (1968) Glucagon. *Ergebn. Physiol. 60*, 141–219.

François, B. & Crabbé, J. (1969) Interaction between isolated amphibian skin and insulin. *Arch. int. Physiol. Biochem. 77*, 527–530.

Harrop, G. A., Jr. & Benedict, E. M. (1923) The role of phosphate and potassium in carbohydrate metabolism following insulin administration. *Proc. Soc. exp. Biol. Med. (N. Y.) 20*, 430–431.

Hepp, K. D. & Renner, R. (1972) Insulin action on the adenyl cyclase system: antagonism to activation by lipolytic hormones. F. E. B. S. letters. *20*, 191–194.

Herrera, F. C. (1965) Effect of insulin on short-circuit current and sodium transport across toad urinary bladder. *Amer. J. Physiol. 209*, 819–824.

Herrera, F. C., Whittembury, G. & Planchart, A. (1963) Effect of insulin on short-circuit current across isolated frog skin in the presence of calcium and magnesium. *Biochim. biophys. Acta (Amst.) 66*, 170–172.

Jefferson, L. H., Exton, J. H., Butcher, R. W., Sutherland, E. W. & Park, C. R. (1968) Role of adenosine 3',5'-monophosphate in the effects of insulin and anti-insulin serum on liver metabolism. *J. biol. Chem. 243*, 1031–1038.

Kolanowski, J. & Crabbé, J. (1971) Rôle du glucagon dans la rétention du sodium observée en cas d'administration de glucose à l'obèse soumis au jeûne total. *J. Physiol. (Paris) 63*, 243 A.

Kolanowski, J., Pizarro, M. A., de Gasparo, M., Desmecht, P., Harvengt, C. & Crabbé, J. (1970) Influence of fasting on adrenocortical and pancreatic islet response to glucose loads in the obese. *Europ. J. clin. Invest. 1*, 25–31.

Krug, F. & Cuatrecasas, P. (1972) Modulation of adenyl cyclase activity by insulin. Abstr. IV int. Congress End., Washington, *Excerpta. med., Int. Congress Series 256*, 158.

Levy, M. & Starr N. L. (1972) Mechanism of glucagon-induced natriuresis in dogs. Kidney int. *2*, 76–84.

Moore, R. D. The effect of insulin upon active sodium efflux from frog striated muscle. Submitted for publication.

Nizet, A., Lefebvre, P. & Crabbé J. (1971) Control by insulin of sodium, potassium and water excretion by the isolated dog kidney. *Pflügers Archiv ges. Physiol. 323*, 11–20.

Orloff, J. & Handler, J. S. (1964) The cellular mode of action of antidiuretic hormone. *Amer. J. Med. 36*, 686–697.

Saudek, C. D., Boulter, P. R., Stark, R. F. & Arky, R. A. (1972) Glucagon: the natriuretic hormone of starvation? *Clin. Res. 20*, 556.

Ussing, H. H. & Zerahn, K. (1951) Active transport of sodium as the source of electric current in the short-circuited isolated frog skin. *Acta physiol. scand.* *23*, 110-127.

Zierler, K. L. (1957) Increase in resting membrane potential of skeletal muscle produced by insulin. *Science 126*, 1067–1068.

DISCUSSION

SCHULTZ: Is there any evidence that these increases in short circuit current are in fact attributable to an increase in sodium transport?

CRABBÉ: The relationship between short circuit current and net sodium transport still holds when insulin is acting: this has been established by the conventional bidirectional sodium flux measurements.

SACHS: Was the concentration of potassium in the sodium bathing medium ever varied in your experiments?

CRABBÉ: No. It was kept constant. When experiments were performed with Ringer's fluid containing 25 mEq/L potassium (instead of 2.5) there was little, if any, repercussion on the amplitude of the response to insulin.

SACHS: You don't think the mechanism would be an increased access of potassium to the phosphorylated protein?

CRABBÉ: We don't have any experimental evidence that would lead us to believe that this is critical.

LEAF: This past year I had the privilege of working with Brian Ross and Franklin Epstein with the perfused isolated rat kidney and we were able to examine the effect of substrates added to the perfusion fluid. Although fatty acids, including short chain fatty acids, butyrate, are the best substrates for kidney cortex slices in that they will maintain the highest rate of oxygen consumption, if one perfuses the rat kidney with butyrate as the metabolic substrate there is rather poor reabsorption of sodium. Some 87 percent of the sodium was reabsorbed with butyrate as the sole substrate. When one adds a small amount of glucose to the perfused kidney, the reabsorption of sodium increases to 98 percent on the average in a large series, so I will suggest that the available substrate is perhaps more pertinent than the hormonal relationship during fasting. The effect of glucose cannot be substituted by any increase in the amount of butyrate or fatty acids added to the substrate in the perfused preparation.

CRABBÉ: One piece of data that might be relevant here is what one obtains when epinephrine is given to people in this sodium losing phase of a fast. Epinephrine obviously, apart from increasing glycemia to a certain extent, also raises fatty acid concentration; this, however, failed to revert the sodium retention brought about by carbohydrates (Kolanowski *et al.* 1972).

LEAF: Of course epinephrine was reported years ago to cause increase in blood glucose if there is glycogen in the liver.

CRABBÉ: The fatty acid concentration in blood did increase in those patients whom I am talking about and, as I said, this failed to modify the sodium excretion pattern*).

LEAF: What I am suggesting is that it is the utilization of fatty acids rather than glucose which may cause the losses. Drs. Hynie and Sharp investigated the specificity of the adenyl cyclase in the isolated toad bladder membrane and they found that insulin is not acting on the adenyl cyclase which is affected by vasopressin.

CRABBÉ: But did they conduct experiments to examine whether insulin would interfere with the amplitude of the response to vasopressin?

LEAF: No.

Kolanowski, J., De Gasparo, M., Desmecht, P. & Crabbé, J. (1972) Further evaluation of the role of insulin in sodium retention associated with carbohydrate administration after a fast in the obese. *Europ. J. clin. Invest. 2,* 439–444.

*) When glucagon was injected together with glucose, in fasting obese patients, the plasma level of free fatty acids dropped as much as when glucose was injected alone; yet, the sodium retaining effect of glucose was abolished.

Regulation of Transepithelial Sodium Transport by Aldosterone

I. S. Edelman

In the last decade, a general scheme has been formulated on the molecular mechanisms that mediate the action of aldosterone on active Na^+ transport across epithelia (Fig. 1). In brief, the proposed events are: 1) migration of d-aldosterone from the capillary into the cytoplasm of the target cell; 2) binding of the native steroid to an aporeceptor; 3) temperature-dependent

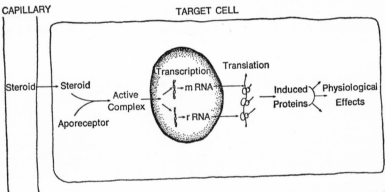

Fig. 1. General model proposed for the mechanism of action of aldosterone (Feldman *et al.* In press).

transformation of the steroid-aporeceptor complex into an active steroid-receptor complex; 4) migration of the active complex into the nucleus; 5) second stage modification of the active complex and binding to specific genomic sites; 6) activation or derepression of transcription which results

* From the Cardiovascular Research Institute and the Departments of Medicine, and of Biochemistry and Biophysics of the University of California School of Medicine. San Francisco, California 94122.

»Transport Mechanisms in Epithelia«, Munksgaard, Copenhagen.

Table I. Distribution of ³H-aldosterone in toad bladder epithelial cells*

Competitive Steroid (100:1)	Grains/Nucleus	Grains/Cytoplasm
None	2.66	0.79
Estradiol-17β	2.75	0.96
9α-Fluorocortisol	0.84	0.45

* Toad bladders were exposed to ³H-aldosterone (5.3×10^{-8} M) for 30 min with or without a competitive steroid (5.3×10^{-6} M). Radioautographs of single cell layer epithelial scrapings were prepared by the dry mount technique. Results are the average of 200 cells counted per section. (Bogoroch & Edelman, unpublished results).

in induction of the synthesis of specific proteins; and 7) augmentation of Na^+ transport by the action of the aldosterone-induced proteins (AIP). Evidence has been presented suggesting that this general pattern applies to the action of most, if not all, steroid hormones (Raspé 1970).

The purpose of this report is to present some of the experimental findings that bear on the applicability of this hypothesis to the mechanism of action of aldosterone.

Existence of Nuclear and Cytoplasmic Aldosterone-Binding Proteins

Crabbé (1963) used the short-circuit current (scc) technique of Ussing and Zerahn (1951) in studies on the *in vitro* action of aldosterone in the isolated toad bladder system. At the time of the maximal effect on scc, 100% (± 5%) of the labelled aldosterone was recovered as the unmetabolized, native steroid. Similar results were obtained by Kliman (Quoted by Sharp & Leaf 1966). That native d-aldosterone acted by binding to a receptor system was indicated by high resolution radioautographs of the toad bladder: ³H-aldosterone was preferentially placed in target cell nuclei in contrast to the random distribution of the inactive steroid, ³H-progesterone (Edelman *et al.* 1963, Porter *et al.* 1964).

The specificity and sites of binding of ³H-aldosterone were examined by addition of competitive steroids to the incubation media (Bogoroch & Edelman, unpublished observations). The potent mineralocorticoid, 9α-fluorocortisol, at 100-fold excess inhibited both nuclear and cytoplasmic uptake of ³H-aldosterone. Estradiol-17β (an inactive steroid in the toad bladder system) however, had no effect on the distribution of ³H-aldosterone

Table II. Steroidal specificity for ^{3}H-aldosterone binding sites in renal chromatin

No.[a]	Competing Steroid[b]	Relative Specific activity %
24	None	100
6	17α-Isoaldosterone (10:1)	106 ± 10
9	Progesterone (100:1)	106 ± 18
6	17β-Estradiol (100:1)	89 ± 12
10	Cortisol (100:1)	51 ± 10
8	Spirolactone (10,000:1)	21 ± 3
6	9α-Fluorocortisol (100:1)	11 ± 0.9
12	d-Aldosterone (10:1)	20 ± 3

a Number of experiments.
b Molar ratio of competing steroid to ^{3}H-aldosterone is given in parentheses. (Swaneck *et al.* 1970).

(Table I). These results implied the existence of mineralocorticoid-specific cytoplasmic and nuclear binding sites. Sharp *et al.* (1966) and Ausiello & Sharp (1968) used a competitive steroid displacement method and inferred the existence of a saturable pool of aldosterone-specific binding sites in the cytoplasmic and nuclear fractions of the toad bladder epithelium.

The most complete documentation of the properties of the aldosterone receptor system has been derived from studies on the rat kidney. In adrenalectomized rats, a saturable pool of renal nuclear receptor sites was found by conventional cell fractionation methods after infusion of varying amounts of ^{3}H-aldosterone *in vivo* (Fanestil & Edelman 1966a). The nuclear receptor was subsequently isolated, partially purified and identified as a heat-labile protein (Herman *et al.* 1968). The existence of renal cytoplasmic aldosterone-binding proteins was also demonstrated in this study. The cytoplasmic and nuclear binding proteins had corresponding affinities for a variety of steroids and the affinities of these steroids for the ^{3}H-aldosterone binding sites were in direct proportion to their potencies, either as mineralocorticoid agonists or antagonists.

Further evidence that these aldosterone-binding proteins were a part of the physiological receptor system was provided by studies with the physiologically inert stereoisomer, 17α-isoaldosterone. Excess cold 17α-isoaldosterone had no effect on the formation of the cytoplasmic or nuclear ^{3}H-aldosterone complexes nor did ^{3}H-17α-isoaldosterone bind to renal cytoplasmic or nuclear fractions. The nuclear receptor system has been resolved into two species; a Tris-soluble 3S and a chromatin-bound 4S form (Herman *et al.* 1968, Swaneck *et al.* 1970).

If the chromatin-bound aldosterone complex regulates gene expression as part of the Na^+ transport control system, the steroidal specificity of the chromatin-bound form should relate to physiological potency. As shown in Table II, the hierarchy of competitive affinity of various steroids for the ^{3}H-aldosterone chromatin-bound complex conforms to their agonist or antagonist potencies. The existence of cytoplasmic and nuclear aldosterone-receptor complexes directed attention to the question of the interrelationships among these forms and their distinct physiological roles.

Renal Cytoplasmic Receptors

In the adrenalectomized state, administered aldosterone may occupy receptor sites that serve glucocorticoid as well as those that serve mineralocorticoid functions, since both types of steroids are eliminated from the circulation. Recently, three types of adrenal steroid receptors have been identified in rat renal cytosol fractions on the basis of differential affinities for aldosterone, dexamethasone and corticosterone (Feldman *et al.*, 1972). The equilibrium dissociation constants (K_{diss}) of the primary steroid receptor complexes and the relative potencies of key steroids for each type are shown in Table III. The Type I receptor has been identified as

Table III. Corticosteroid receptors in rat kidney: Affinities and relative binding potencies of various steroids

	Type I	Type II	Type III
Optimal steroid	Aldosterone	Dexamethasone	Corticosterone
K_{diss}*	$5 \times 10^{-10}M, 37°C$	$5 \times 10^{-9}M, 25°C$	$3 \times 10^{-9}M, 25°C$
% Relative potency†			
aldosterone	100	20	<1
DOC	85	20	25
corticosterone	2	40	100
dexamethasone	2	100	<1

* Determined by Scatchard analysis for the optimal steroid.
† Determined by competitive binding assay against tritiated optimal steroid. The potency of unlabelled optimal steroid is taken as 100 %. (Feldman *et al.* 1972).

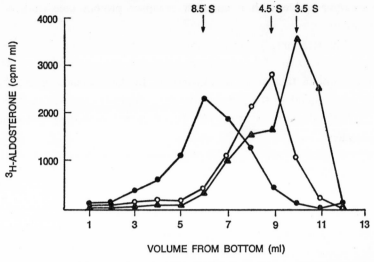

Fig. 2. The effect of high salt and Ca^{++} on the glyceral density gradient patterns of cytosol labelled in vitro with ^{3}H-aldosterone. Adrenalectomized rat kidneys were homogenized in 0.1 M Tris-HCl, pH = 7.2. The cytosol fraction was prepared by centrifugation at 19,000 $\times$ G for 15 min and labelled by incubation with 1 $\times$ 10^{-8} M ^{3}H-aldosterone for 30 min at 0° C. Aliquots were layered on 13 to 34 % glycerol gradients containing 0.1 M Tris-HCl pH = 7.2, and 1.6 $\times$ 10^{-9} M ^{3}H-aldosterone, and: a) 1.5 mM EDTA (–●–); or b) 1.5 mM EDTA and 0.4 M KCl (–o–); or c) 1.5 mM EDTA, 0.4 M KCl, 6 mM CaCl$_2$ (–▲–). One ml fractions were collected and filtered through G-50 Sephadex to remove free ^{3}H-aldosterone. (Marver *et al.* 1972).

serving mineralocorticoid functions on the basis of selective preference for d-aldosterone and DOC.

As shown in Fig. 2, sedimentation analysis in glycerol density gradients revealed three states of aggregation of the cytoplasmic aldosterone-receptor complex, with sedimentation coefficients of 8.5S, 4.5S, or 3.5S, depending on the ionic strength and Ca^{++} concentration of the medium (Marver *et al.* 1972). On the basis of gel filtration analysis, Ludens & Fanestil (1971) estimated the molecular weight of the renal cytosol (100,000 X g supernatant) aldosterone complex in low salt media to be greater than 1.5 X 10^6 Daltons and in high salt media (0.3 M KCl) to be $\sim$ 50,000 Daltons.

Analysis of the intranuclear complexes indicated that the Tris-soluble species sedimented at 3S and the chromatin-bound species at 4S (Marver *et al.* 1972, Swaneck *et al.* 1970). The shifts in S values induced by K$^+$ salts as well at the differences noted between the extranuclear and intranuclear forms are compatible with a complex receptor protein consisting of two or more subunits.

A possible relationship between various forms may be formulated as follows: *In vivo* (0.16 M K$^+$ salt concentrations) the aporeceptor is a 4.5S oligomeric protein that binds d-aldosterone. In the presence of Ca^{++}, the complex is activated to the 3.5S form which migrates into the nucleus and binds to chromatin as a 4S form. This postulate was examined in a series of further studies.

Conversion of Cytoplasmic Receptor to Nuclear Receptor

Adrenalectomized rats were injected with ^{3}H-aldosterone and pairs of kidneys were removed at various intervals. Labelling of the cytoplasmic complex was maximal in less than two min., whereas the intranuclear species attained maximum concentrations at 10 min. (Fig. 3). To resolve the sequence of events further, kidney slices were prepared from adrenalectomized rats and incubated in ^{3}H-aldosterone at 25° C. The 4.5S cytosol complex was generated in 15 min., followed by appearance of the 3S Tris-soluble nuclear complex over a 40 min. period and finally formation of the 4S chromatin-bound complex over a 1 to 2 hour period (Marver *et al.* 1972). The same sequence of events was found in rat parotid gland slices, another well-characterized target tissue for aldosterone (Funder *et al.* 1972).

The validity of the concept that the cytoplasmic complex is the precursor of the nuclear complexes was tested in a series of reconstitution experiments with pre-labelled renal cytosol fractions and unlabelled nuclear or purified chromatin fractions (Marver *et al.* 1972, Marver & Edelman, unpublished observations). After mixing ^{3}H-labelled renal cytosol with unlabelled nuclei, there was progressive depletion of the cytosol receptor content, an early appearance of the Tris-soluble nuclear species and later formation of the chromatin-bound species (Fig. 4). Nuclear uptake accounted for 60% of the receptor content lost from the cytosol donor during incubation. Although binding of ^{3}H-aldosterone to the cytosol was evident at 0° C, nuclear transfer did not occur until the complex was warmed to 25° C. These findings imply that aldosterone first binds to an aporeceptor to form in inactive complex which is converted to an active donor form by a temperature-dependent process.

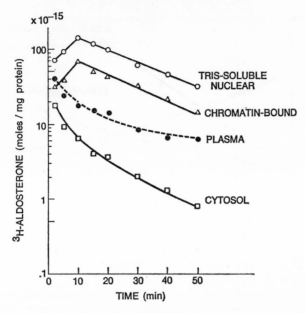

Fig. 3. Time-course of formation of renal ³H-aldosterone-receptor complexes *in vivo*. The labelled cytosol (–☐–), Tris-soluble nuclear (–o–), and chromatin-bound (–△–) fractions were prepared as described under Methods. Adrenalectomized rats were injected with 2.6×10^{-10} moles of ³H-aldosterone with or without 2.6×10^{-8} moles 9α-fluorocortisol and sacrificed 2 to 50 min after injection. The renal fractions were corrected for non-specific labelling. The concentration of ³H-aldosterone in plasma, both free and bound, is shown as a dashed line (– –●– –). The receptor content in each fraction was corrected for variations in plasma ³H-aldosterone concentration by a factor which was the ratio of the plasma concentration in that sample to the average concentration of all of the plasma samples at that time point. Each point represents an average of five experiments. (Marver *et al.* 1972).

Similar reconstitution experiments with unlabelled purified renal chromatin fractions and pre-labelled cytosol fractions yielded equivalent results to those obtained with nuclear fractions, i.e., temperature-dependent transfer process of ³H-aldosterone from cytosol to nucleus (Table IV)

In addition, cross-over experiments with kidney and spleen cytosol and nuclear fractions indicate that mineralocorticoid specificity resides in the donor cytoplasmic complex. Thus, the available information supports the model shown in Fig. 1 on the receptor mechanisms involved in the formation of the chromatin-bound aldosterone-receptor complex. Evidence has also been obtained implicating regulation of gene expression in the action on Na⁺ transport.

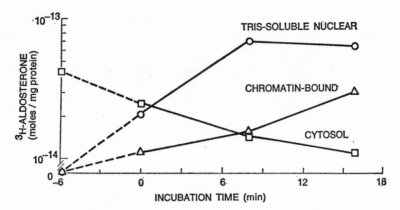

Fig. 4. Time-course of formation of nuclear ³H-aldosterone-receptor complexes in reconstituted mixtures of cytosol and nuclear fractions. Cytosol fractions were prelabelled with ³H-aldosterone (1.3 × 10⁻⁸ M), with and without 9α-fluorocortisol (1.3 × 10⁻⁶ M) for 30 min at 0° C. The labelled cytosol fractions were mixed with washed renal nuclear fractions and incubated for various times at 25° C. The "zero-time"values represent measurements obtained on mixtures of labelled cytosol and unlabelled nuclei that were processed immediately after mixing at 25° C. An extrapolated "true zero" point of –6 min has been included to take into account binding developed during the 6 min required to process the mixture. The results are the average of 5 experiments. (Marver *et al.* 1972).

Table IV. Temperature dependence of ³H-aldosterone binding in rat renal cytosol and nuclear fractions*

Temperature	Cytosol	Tris-Soluble	Chromatin-Bound
	$\times 10^{-15}$ moles/mg protein		
0° C	22.5	19.0	2.1
25° C	31.4	28.4	12.9
0° C/25° C	71.7 %	66.9 %	16.3 %

* Adrenalectomized rat kidney slices were incubated at either 25° C or 0° C for 1 hr in 5.2 × 10⁻⁹ M ³H-aldosterone with or without 5.2 × 10⁻⁶ M d-aldosterone. The cytosol, Tris-soluble nuclear, and chromatin-bound fractions were prepared as under Methods. The quantity of ³H-aldosterone bound has been corrected for non-specific labelling. The results are the mean of five experiments. (Marver *et al.* 1972).

Aldosterone Regulation of Transcription

That induction of DNA-dependent RNA synthesis mediated mineralocorticoid action was first proposed independently by Edelman *et al.* (1963) and

Williamson (1963). In the isolated toad bladder system, inhibition of nuclear RNA synthesis with actinomycin D or of protein synthesis with puromycin abolished the augmentation of Na^+ transport by aldosterone (Edelman *et al.* 1963, Fanestil & Edelman 1966b). In the rat kidney, *in vivo,* actinomycin D impaired the anti-natriuretic but not the kaliuretic effect of parenteral aldosterone (Williamson 1963, Fimognari *et al.* 1967). Recently, Lifschitz *et al.* (1973) obtained similar results in the adrenalectomized dog; actinomycin D inhibited the anti-natriuretic response to aldosterone completely, had a lesser effect on the kaliuretic response, and no effect on the urinary acidification response. In addition, incorporation of precursors into epithelial RNA was enhanced by aldosterone in the isolated toad bladder (Porter *et al.* 1964, Rosseau & Crabbé 1968) and in the adrenalectomized rat kidney (Fimognari *et al.* 1967, Forte & Landon 1968).

Sharp and his colleagues (Vancura *et al.* 1971, Sharp & Komack 1971), did not find an aldosterone-induced increase in the incorporation of labelled precursors into particular species of RNA (density gradient separation) of toad bladder and questioned the induction hypothesis. Rosseau and Crabbé (1968, 1972) also found no change in labelling of specific classes of toad bladder RNA after aldosterone, but confirmed increased synthesis of unfractionated nuclear RNA. These negative results may be a consequence of ribonuclease activity commonly present during fractionation of RNA into the major classes (Joel & Hagerman 1969) and to the insensitivity of the methods used to detect changes in messenger RNA (mRNA) in particular.

Majumdar & Trachewsky (1971) recently exploited a reconstituted ribosomal system to explore the effect of aldosterone on polysomal mRNA content and on translational capacity of polyribosomes. Renal cortical ribosomes prepared from aldosterone-treated adrenalectomized rats incorporated $\sim 22\%$ more ^{14}C-phenylalanine into protein in translating endogenous mRNA compared to control preparations. Poly-uridine directed incorporation was also $\sim 25\%$ greater than the controls in ribosomes prepared from aldosterone-treated rats. These results imply augmentation by aldosterone of translational capacity as well as an increase in polyribosomal-bound mRNA. The former effect may reflect a bifunctional role in steroidal regulation of gene expression: enhanced ribosomal RNA synthesis (rRNA) as well as a shift in or augmentation of mRNA synthesis.

Newer information lends credence to the inference of a dual role for aldosterone in the regulation of gene expression. Specific eukaryotic nuclear polymerases have been identified: Polymerase I is localized to the nucleolus and transcribes genes that code for rRNA. Polymerase II is localized to the nucleoplasm and transcribes genes coded for heterogeneous RNA (HnRNA), the putative precursors of mRNA.

Chu & Edelman (1973) identified Polymerase I (cordycepin-sensitive) and Polymerase II (α-amanitin-sensitive) in nuclear fractions of the rat kidney as had been described earlier in rat liver nuclei (Roeder & Rutter 1969). RNA Polymerases I and II are activated by Mg^{++} and Mn^{++} with some degree of selectivity. Liew et al. (1972) reported that in renal nuclei from adrenalectomized rats, prior injections of aldosterone enhanced the activities of the Mg^{++}-activated enzyme by 117% and of the Mn^{++}-activated enzyme by 107 % compared to untreated controls. Chu & Edelman (1973) used α-amanitin to discriminate more precisely between Polymerase I and Polymerase II activities and found that physiological doses of aldosterone increased the ratio of nucleolar to nucleoplasmic polymerase activity: The time-course of this shift correlated with the change in urinary Na^+ and K^+ excretion. Thus, the findings described by Majumdar & Trachewsky (1971), Liew et al. (1972), and Chu & Edelman (1973) implicate both rRNA and mRNA synthesis in the action of aldosterone on the kidney.

The relevance of transcriptional control mechanisms to the action of aldosterone was assessed further with cordycepin (3'-desoxyadenosine) in the isolated toad bladder system. This adenosine analogue impairs nucleolar polymerase activity and the processing of HnRNA to mRNA by addition of polyadenosine sequences but does not inhibit the synthesis of HnRNA (Darnell et al. 1971). In the isolated toad bladder system, cordycepin inhibited incorporation of ^{3}H-uridine into epithelial RNA by $\sim$ 50% and the aldosterone-induced rise in scc by more than 90% (Fig. 5, Chu & Edelman, 1973).

That cordycepin did not impair steroid-independent Na^+ transport via a non-specific pathway was indicated by the following controls: a) the Na^+ transport response to vasopressin was not inhibited by cordycepin. b) In substrate-rich toad bladders, basal Na^+ transport was only slightly depressed by cordycepin, and c) Stimulation of Na^+ transport by addition of substrate to the media of toad bladders previously deprived of substrate was unimpaired by cordycepin.

A plausible hypothesis of the mechanism of action of aldosterone can be constructed on the basis of these and related findings. Aldosterone may

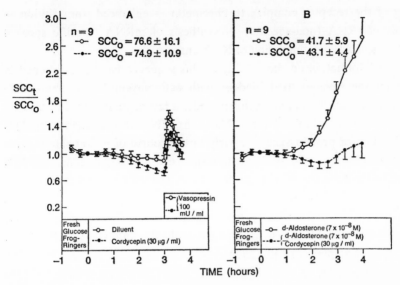

Fig. 5. Effect of cordycepin on aldosterone stimulation of Na⁺ transport in the toad bladder. Pairs of hemibladders were pre-incubated in steroid-free glucose (10 mM)-frog Ringers for 15 hr. The media were then exchanged for fresh glucose (5 mM)-frog Ringers solution. In the experiments shown in Panel A, cordycepin (30 μg/ml) was added to the serosal and mucosal media of one af each pair (– –●– –) and vasopressin (100 mU/ml) was added to the serosal media of both hemibladders at 3 hr. In the experiments shown in Panel B, aldosterone (7×10^{-8}) M) was added to the serosal and mucosal media of both hemibladders and cordycepin (30 μg/ml) was added to the serosal and mucosal media of one of each pair (– –●– –). SCC_t/SCC_0 denotes the short-circuit current at time "t" divided by that at time zero. Each point and vertical line represents the mean ± s.e.m. "n" denotes the number of pairs of hemibladders. SCC_0 denotes the absolute short-circuit current at time zero and is given as the mean ± s.e.m. (Chu & Edelman, 1973).

act at two classes of gene sites to increase both the capacity for protein synthesis by enhancement of Polymerase I activity, and the formation of specific and less easily detected classes of mRNA that dictate the synthesis of the physiologically active proteins.

Transcription as the Initiating Event in the Action of Aldosterone

In essence the hypothesis displayed in Fig. 1. proposes a linear sequence of events consisting of binding of aldosterone to stereospecific receptors →

13*

binding of the receptor complex to chromatin → enhanced transcription → synthesis of rRNA (non-specific) + synthesis of mRNA (specific species) → synthesis of AIP → stimulation of Na^+ transport.

Further information on the validity of this sequence has been obtained in studies on the isolated toad bladder with actinomycin D, an irreversible inhibitor of transcription, and cycloheximide, a reversible inhibitor of translation (Lahav *et al.* submitted for publication). This linear sequence predicts that inhibition of protein synthesis early in the course of action should have little or no effect on the subsequent response if the translational block is removed at the time newly transcribed mRNA becomes available for translation.

This prediction was tested by fractionating the latent period of the Na^+ transport response of the toad bladder with the inhibitors. At 22° to 24° C, the latent period of mineralocorticoid action is 75 to 90 min. in the isolated toad bladder (Porter 1970). If the latent period represents the time required to synthesize and translate mRNA, inhibition of protein synthesis during the latent period and removal of the cycloheximide just before the end of the latent period, should not lengthen the latent period appreciably.

Herman (Quoted in Edelman 1969) tested this prediction by adding cycloheximide to the media of pairs of hemibladders, one of each pair of which received aldosterone at the same time. Seventy-five min. later, the cycloheximide was washed out. As predicted, the Na^+ transport response was not delayed nor reduced in magnitude. A similar study in which cycloheximide was added 15 min. before aldosterone, yielded the same result (Lahav *et al.,* submitted for publication). These results imply that the dependence of the mineralocorticoid effect on protein synthesis does not develop until the end of the latent period.

The inference that the latent period is determined largely by the time needed for RNA production was assessed by adding actinomycin D to the media of isolated toad bladders either 15 min., 30 min. or 60 min. after the addition of aldosterone (Lahav *et al.,* submitted for publication). Sixty min. after aldosterone, actinomycin D had no effect on the duration of the latent period or the initial rate of increase in Na^+ transport compared to the uninhibited response. The increase in Na^+ transport, however, was of short duration and the scc began to decline towards baseline values 150 min. after the addition of aldosterone. Similar patterns were seen when the inhibitor was added earlier in the latent period, i. e., at 15 or 30 min.

Table V. The effect of varying the period of uninterrupted RNA synthesis on the mineralocorticoid response*

Time of Addition of Actino-mycin D (min)	Inital Slope			% Increase in scc at t = 150 min (peak)		
	Treated	Control	Treated/ Control Ratio	Treated	Control	Treated/ Control Ratio
15	0.11	0.29	0.38	11	35	0.31
30	0.31	0.40	0.78	33	46	0.72
60	0.48	0.50	0.96	56	72	0.78

* The initial slope is the rate of rise in scc during the first 30 min after the end of the latent period. "Treated" denotes hemibladders exposed to actinomycin D (5 μg/ml) at the times indicated in the first column. The peak response was calculated as the difference in normalized scc between 75 and 150 min in each group of hemibladders. The peak response after actinomycin D was at t = 150 min. (Lahav et al., submitted for publication).

Two measurements of the response were made to quantify the results: 1) the initial rate of increase in scc at the end of the latent period $\equiv$ initial slope, and 2) the percent increase in scc at the peak of the response (t = 150 min.). As shown in Table V, the longer the exposure to actinomycin D, the lower the initial slope and the less the % increase in scc.

These results are compatible with the hypothesis that aldosterone-induced RNA (presumably mRNA) made during the latent period is pooled and translated synchronously at the end of the latent period. The increase in scc at any one time would then reflect the magnitude of the increase in the pool of AIP.

REFERENCES

Ausiello, D. A. & Sharp, G. W. G. (1968) Localization of physiological receptor sites for aldosterone in the bladder of the toad, Bufo marinus. Endocrinology 82, 1163–1169.

Bogoroch, R. & Edelman, I. S. Unpublished observations.

Chu, L. H. H. & Edelman, I. S. (1973). Cordycepin and α-amanitin: Inhibitors of transcription as probes of aldosterone action. I. Membrane Biol., 10, 291–310.

Crabbé, J. (1963) The Sodium Retaining Action of Aldosterone. Presses Acad. Euro-péennes S. C. Brussels. pp. 75–82.

Darnell, J. E., Philipson, L., Wall, L. & Adesnik, M. (1971) Polyadenylic acid sequen-ces: Role in conversion of nuclear RNA into messenger RNA. Science 174, 507–510.

Edelman, I. S. (1969) Intracellular action of aldosterone on sodium transport. In *Renal Transport and Diuretics,* ed. Thurau, K. & Jährmarker, J., pp. 139–151. Springer Verlag, New York.

Edelman, I. S. Bogoroch, R. & Porter, G. A. (1963) On the mechanism of action of aldosterone on sodium transport. The role of protein synthesis. *Proc. nat. Acad. Sci. (Wash.) 50,* 1169–1177.

Fanestil, D. D. & Edelman, I. S. (1966a) Characteristics of the renal nuclear receptors for aldosterone. *Proc. nat. Acad. Sci. (Wash.) 56,* 872–879.

Fanestil, D. D. & Edelman, I. S. (1966b) On the mechanism of action of aldosterone on sodium transport: Effects of inhibitors of RNA and of protein synthesis. *Fed. Proc. 25,* 912–916.

Feldman, D., Funder, J. W. & Edelman, I. S. (1972) Subcellular mechanisms in the action of adrenal steroids. *Amer. J. Med. 53:* 545–560.

Fimognari, G. M., Porter, G. A. & Edelman I. S. (1967) Induction of RNA and protein synthesis in the action of aldosterone in the rat. *Amer. J. Physiol. 213,* 954–962.

Forte, L. & Landon, E. J. (1968) Aldosterone-induced RNA synthesis in the adrenalectomized rat kidney. *Biochim. biophys. Acta (Amst.) 157,* 303–309.

Funder, J. W., Feldman, D., Edelman, I. S. (1972) Specific aldosterone binding in rat kidney and parotid. *J. Steroid Biochem. 3,* 209–218.

Herman, T. S., Fimognari, G. M. & Edelman, I. S. (1968) Studies on renal aldosterone binding proteins. *J. biol. Chem. 243,* 3849–3856.

Joel, P. B. & Hagerman, D. D. (1969) Extraction of RNA from rat uterus. *Biochim. biophys. Acta (Amst.) 195,* 328–339.

Lahav, M., Dietz, T. & Edelman, I. S. Further studies on the role of RNA and protein synthesis in the action of aldosterone on sodium transport. Submitted for publication.

Liew, C. C., Liu, D. K. & Gornall, A. G. (1972) Effects of aldosterone on RNA polymerase in rat heart and kidney nuclei. *Endocrinology 90,* 488–495.

Lifschitz, M. D., Schrier, R. W. & Edelman, I. S. (1973) Effect of actinomycin D on aldosterone-mediated changes in electrolyte excretion. *Amer. J. Physiol. 224,* 376–380.

Ludens, J. H. & Fanestil, D. D. (1971) Studies on cytosol aldosterone binding macromolecules. *Biochim. biophys. Acta (Amst.) 244,* 360–371.

Majumdar, A. P. N. & Trachewsky, D. (1971) Protein synthesis by ribosomes from rat kidney cortex: Effect of aldosterone and bilateral adrenalectomy. *Canad. J. Biochem. 49,* 501–509.

Marver, D. & Edelman, I. S. Unpublished observations.

Marver, D., Goodman, D. & Edelman, I. S. (1972) Relationships between renal cytoplasmic and nuclear aldosterone-receptors. *Kidney International 1,* 210–223.

Porter, G. A. (1970) Temperature-dependence of sodium transport in the isolated toad bladder. *Biochim. biophys. Acta (Amst.) 211,* 487–501.

Porter, G. A., Bogoroch, R. & Edelman, I. S. (1964) On the mechanism of action of aldosterone on sodium transport: The role of RNA synthesis. *Proc. nat. Acad. Sci. (Wash.) 52,* 1326–1333.

Raspé, G. (Ed.) (1970) *Advances in the Biosciences 7,* Schering Workshop on Steroid Hormone Receptors. Pergamon Press-Vieweg, Oxford.

Roeder, R. G. & Rutter, W. J. (1969) Multiple forms of DNA-dependent RNA polymerase in eukaryotic organisms. *Nature (Lond.) 224,* 234–237.

Rousseau, G. & Crabbé, J. (1968) Stimulation by aldosterone of a rapidly labelled RNA fraction in toad bladder tissue. *Biochim. biophys. Acta (Amst.) 157,* 25–32.

Rousseau, G. & Crabbé, J. (1972) Effects of aldosterone on RNA and protein synthesis in the toad bladder. *Europ. J. Biochem. 25,* 550–559.

Sharp, G. W. G. & Komack, C. L. (1971) The effects of aldosterone on toad bladder measurements of ^{3}H-uridine incorporation into RNA. *Biochim biophys. Acta (Amst.) 247,* 66–73.

Sharp, G. W. G. & Leaf, A. (1966) Mechanism of action of aldosterone. *Physiol. Rev. 46,* 593–633.

Sharp, G. W. G., Komack, C. L. & Leaf, A. (1966) Studies on the binding of aldosterone in the toad bladder. *J. clin. Invest. 45,* 450–459.

Swaneck, G. E., Chu, L. L. H. & Edelman, I. S. (1970) Stereospecific binding of aldosterone to renal chromatin. *J. biol. Chem. 245,* 5382–5389.

Ussing, H. H. & Zerahn, K. (1951) Active transport of sodium as the source of electric current in the short-circuited isolated frog skin. *Acta physiol. scand. 23,* 110–127.

Vancura, P., Sharp. G. W. G. & Malt, R. A. (1971) Kinetics of RNA synthesis in toad bladder epithelium: Action of aldosterone during latent period. *J. clin. Invest. 50,* 543–551.

Williamson, H. E. (1963) Mechanism of the antinatriuretic action of aldosterone. *Biochem. Pharmacol. 12,* 1449–1450.

DISCUSSION

No discussion followed this section, since Dr. Edelman due to sickness was unable to deliver his lecture.

GENERAL DISCUSSION

LOEWENSTEIN: Dr. Leaf, have you any information about what portion within the junction complex you are looking at? Where does the blistering occur? My second question is: Is the resistance changed within the cells themselves under the effect of your hypertonic solutions and the third question is: are the cells electrically coupled?

LEAF: I can only answer one of these question. The site where DiBona saw the changes at the apical portion of the tight junctions was above the desmosome. It is in the part which is considered to be the zonula occludens, the tightest portion of the tight junction.

LOEWENSTEIN: Isn't there a gap-type junction there?

LEAF: That is a little lower down.

LOEWENSTEIN: That is not separated?

LEAF: That didn't seem to change much. As far as the resistance of the cells we don't have any independent measurements. No experiments have been done with electrodes within these cells under these circumstances. I don't know whether this affects the intercellular coupling.

WHITTEMBURY: I think I can answer some of your questions, Dr. Loewenstein. We have some electronmicroscopic observations (see Whittembury, this Symposium) on perfused amphibian kidney under conditions which somehow resemble those that Dr. Leaf has used. We have been using soluble lanthanum as extracellular marker. We tried to precipitate it with divalent anions (Whittembury & Rawlins 1971) somehow following Dr. Ussing's idea

Whittembury. G. & Rawlins, F. A. (1971) Evidence of paracellular pathway for ion flow in the kidney proximal tubule: electronmicroscopic demonstration of lanthanum precipitate in the tight junction. *Pflügers Arch. 330*, 302–309.

of barium and sulphate, that were published some years ago (Ussing 1970, 1971). We have observed under control conditions that lanthanum goes through the junctions in the proximal tubule of the kidney and then if one puts urea in the luminal phase to produce an effect similar to what has been reported by Dr. Leaf, there's more lanthanum going through those junctions in the first place. In the second place the length of the intercellular space shrinks and in the third place, and this somehow answers your question about the cell resistance, there is some lanthanum going into the cells, so it might be that the cell resistance drops to some extent under hypertonic solutions on the luminal side. But this is of course an indirect answer.

LOEWENSTEIN: How much potassium is there in your potassium medium? How high is the concentration?

LEAF: The concentrations were increased from 230 mOsm/kg by addition of about 230–150 mM potassium salt.

LOEWENSTEIN: Under these circumstances one may expect, from experience with other types of epithelial cells, a substantial change in conductance. Another question: Is there any information available on freeze-fractured material where one can see a little bit better what junctional complex you are dealing with?

MAETZ: I wonder whether the toad bladder submitted to hyperosmotic solutions (on the mucosal side) and submitted to neurohypophysial hormones becomes leaky to inulin. Other epithelia have been shown to be or to become permeable to inulin. For instance, the gill of teleosts and elasmobranchs has been shown by A. Masoni and P. Payan to be permeable to inulin (unpublished) in confirmation of earlier work by Lam (1969).

In fact, branchial clearance of inulin is equal to if not higher than renal clearance. Also mammalian intestine becomes permeable to inulin under condition of volume expansion by isotonic saline loading (see Humphreys &

Ussing, H. H. (1970) Tracer studies and membrane structure. Alfred Benzon Symposium II, In: *Capillary Permeability*, (Crone, C. & Lassen, N. A., eds.), Munksgaard, pp. 654–656.

Ussing, H. H. (1971) Introductory remarks. *Phil. Trans. roy. Soc. Lond. B. 262*, 85–90.

Lam, T. J. (1969) Evidence of loss of C^{14}-inulin via the head region of the threespine stickleback *Gasterosteus aculeatus*, form trachurus. Comp. Biochem. Physiol. *28*, 459–463.

Earley 1971). It has been suggested that this leakiness is the result of a loosening of the tight junction.

CURRAN: Dr. Leaf, could you tell us what happens to the sodium transport under the conditions of the experiment you were describing?

LEAF: The sodium transport measured by the short circuit current is hardly affected.

CURRAN: Wouldn't you expect a decrease in net transport when you markedly lowered a shunt resistance?

LEAF: I should have mentioned that in all these experiments compensations were made for the changes in the resistance of the solutions themselves and solute concentration, respectively. Perhaps Dr. Ussing would like to add a comment.

USSING: Yes, it is true that in frog skin we have the same experience that although the resistance drops, still the short circuit current remains relatively constant. I must say, though, that we have a few cases where we did see an initial slight stimulation.

As to Dr. Maetz's question about inulin I can say at least for the frog skin with moderate increases in the osmotic pressure on the outside there is no increase in the passage of inulin, but maybe we didn't punish the tissues as hard as we should.

TOSTESON: I would like to ask about a point that Dr. Leaf and Dr. Orloff discussed in passing after Dr. Leaf's paper. The question was about the effect of vasopressin on this phenomenon. I think it is very interesting in a general way that the mucosal membranes of these epithelial cells are so tremendously impermeable to water. To a red cell and bilayer man the idea that there could be a sustained difference in the solute concentration in the mucosal fluid and the cells for any appreciable period of time is strange. The high water permeability of these membranes, ca. 10^{-3} cm sec^{-1}, precludes such an occurrance. I'd like to ask people who work on these systems whether there has been any serious attempt to estimate quantitatively the water

Humphreys, M. H. & Early, L. E. (1971) The mechanism of decreased intestinal sodium and water absorption after acute volume expansion in the rat. *J. Clin. Invest. 50*, 2355–2367.

permeability of these mucosal membranes and also to invite discussion about what could be the physico-chemical basis for this really quite amazing phenomenon. We know now quite a bit about the water permeability of bilayers, and the water permeability of this mucosal membrane must be quite another thing.

LINDEMANN: I should like to make a brief comment on the effect of potassium on the membrane resistance of the mucosal border. When sodium is present in the outer solution and potassium is added, the membrane resistance actually increases. So I would suggest that in your situation there is a decrease of total resistance due to the opening of the shunt pathway but the resistance in the membrane itself increases.

LEAF: Yes, I think one would expect this. Civan found that a very large portion of the conductance across the tissue is accomplished by the sodium ion, partly going through the transport channels. And if you keep sodium away and put potassium there, obviously conductance through the transport pathway disappears.

LINDEMANN: Even when (with frog skin) you leave the sodium there in constant concentration and add potassium, surprisingly there is an increase of resistance, and I suppose that this might be explained by a partial blockage of the sodium channels by potassium.

LEAF: I believe bladder would behave the same way. There is some competition.

WINDHAGER: In urea treated frog skin the microelectrode potentials remained the same at the time when the short circuit current was also the same, roughly. So it would indicate that the membrane resistance did not change.

LINDEMANN: I confirmed this observation with our fast flow method. When we add urea to the outer solution of frog skin in the presence of sodium, the membrane resistance does not change. However, when we do the same experiment with sucrose instead of urea, adding 300 mMol/l sucrose, we find sometimes at first a small increase of the resistance due to the addition of sucrose. When removing the added sucrose from the outer solution after an exposure time of less than 10 seconds, there is often a very clear decrease of resistance.

ZADUNAISKY: I would like to address a question to the workers who have made hypertonicity on outside of the skin or the mucosal side of the bladder.

Does this hypertonicity affect calcium movements or calcium concentration? Would that explain some of the permeability changes?

LINDEMANN: Fishman & Macey (1968) have tried to correlate calcium concentrations in the outer solution and resistance changes of the skin, expecting that the resistance would drop when they decreased the calcium concentration. It dropped by 5–30 % within 2–10 seconds when the outside solution contained 10 mM EDTA.

SKADHAUGE: About the question Dr. Schmidt-Nielsen raised early in the discussion of the possible effect of the hypertonicity in the intestines which in nature receive hyperosmotic fluids. I think, although we do not have any direct measurements on it, that it can be answered at least for the fish. It appears that there is a lower limit to the hypertonicity, which makes the tight junction leaky. In toad skin, for example, the phenomenon ranges from 400 to 50 mOsm, but there is only real good effect from 200 mOsm hypertonic and upward. Even in the little cyprinodont *Aphanius dispar,* which can live in and drink a 4000 mOsm solution with a plasma osmolality of 500 mOsm, the greater part of the intestine has an osmolality less than 100 mOsm higher than that of plasma. The answer thus seems to be that the intestines quickly dilute the incoming hypertonic fluids to an osmolality where the tight junctions do not become untight. In coprodeum of some birds, however, much higher hypertonicity may be found, and here the possible effect should be measured.

KIRSCHNER: But the gills certainly can't.

Fishman, H. M. & Macey, R. I. (1968) Calcium effects in the electrical excitability of 'split' frog skin. *Biochim. Biophys. Acta 150,* 482–487.

Permissive Effect of Aldosterone on the Action of Vasopressin

J. Orloff, J. S. Handler & J. S. Stoff

Vasopressin is known to exert its physiologic effects in toad bladder and other responsive epithelial structures via the intermediacy of cyclic AMP (Orloff & Handler 1962, 1967). The hormone accelerates the conversion of ATP to cyclic AMP by increasing the activity of the enzyme adenylate cyclase. Incubation of toad bladder with cyclic AMP increases the transport of sodium and the permeability to water of the tissue as does vasopressin.

Aldosterone, on the other hand, increases sodium transport in the toad bladder (Crabbé 1961), but does not alter its permeability to water. Although the effect of aldosterone on sodium transport is not thought to involve cyclic AMP, both permeability responses of the tissue to vasopressin which do, are enhanced to a significant degree by the steroid (Handler *et al.* 1969). This permissive effect of aldosterone appears to be mediated by a steroid dependent decrease in the rate of degradation of cyclic AMP.

PERMISSIVE EFFECT OF ALDOSTERONE

Studies were performed on paired hemibladders of *Bufo marinus*. The water permeability response to vasopressin of each bladder was first measured within 2 hours of removal from the toad. Following this, one hemibladder of each pair was incubated for 18–20 hours in Ringer solution containing 2×10^{-7}M aldosterone (or an equivalent amount of dexamethasone); the other was incubated for a similar length of time in Ringer solution without steroid. At the end of this period the response to vasopressin was re-examined. As can be seen in Figure 1, no difference in response was detected 2 hours after removal of the bladder from the toad. In contrast, the response to

The Laboratory of Kidney and Electrolyte Metabolism, National Heart and Lung Institute, Bethesda, Maryland. 20014, U.S.A.

»Transport Mechanisms in Epithelia«, Munksgaard, Copenhagen.

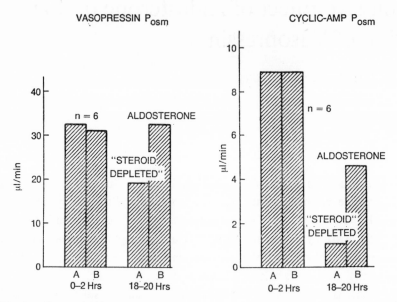

Fig. 1. Effects of aldosterone on osmotic water flow response (P_{osm}) of toad bladder to vasopressin and cyclic AMP.

vasopressin was significantly reduced in steroid depleted hemibladders, whereas the response of those incubated with aldosterone was maintained. Analogous results were obtained with cyclic AMP as well as with theophylline, an inhibitor of the degradation of the nucleotide (not illustrated).

Fig. 2 illustrates a similar pattern with respect to sodium transport. Once again the effects of vasopressin and cyclic AMP were considerably greater in steroid treated bladders than in depleted controls. The findings with theophylline were similar. On the basis of these results we concluded initially that aldosterone affected a step or steps in the permeability processes after the generation of cyclic AMP. We further postulated that aldosterone conceivably acted by increasing the sensitivity of the tissue or the permeability processes to cyclic AMP.

ALDOSTERONE AND CYCLIC AMP ACCUMULATION

In order to determine the role of aldosterone in the accumulation of cyclic AMP, paired bladders were subjected to conditions analogous to those employed in the physiologic studies just described. During the final 15 minutes of the 18–20 hours of incubation with or without steroid, the tissues

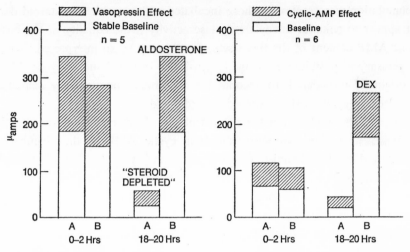

Fig. 2. Effect of steroids on short circuit current response of toad bladder to vasopressin and cyclic AMP.

were challenged with vasopressin. The epithelial cells were then removed and their cyclic AMP content measured employing the technique of Gilman (1970). As reported elsewhere (Stoff *et al.* 1972), vasopressin induced a considerably greater increase in the cyclic AMP content of steroid treated bladders than in depleted controls. The pertinent results are illustrated in Table I.

Note first that the cell content of cyclic AMP in the basal state, i. e. prior to challenge with vasopressin, was unaltered by 18 hours of incubation of the bladders with or without steroid. Clearly aldosterone of itself does not induce a detectable change in the net accumulation of cyclic AMP.

Vasopressin on the other hand, significantly increased the cyclic AMP content of cells from all three groups of bladder segments. The extent of

Table I. Cylic-AMP in toad bladder epithelial cells

		18 Hours	
	Fresh Control	Steroid Depleted	Aldosterone $(2 \times 10^{-7}M)$
		pmoles/mg Protein	
Basal	8.95 ± 1.03	7.19 ± 1.48	7.67 ± 2.08
Vasopressin (25 mU/ml)	24.4 ± 4.39	12.2 ± 0.84	$125. \pm 41.2$

accumulation was greatest in those incubated with steroid. The steroid does not appear to enhance adenylate cyclase activity if the equality of the basal cyclic AMP content of the three sets of cells may be so interpreted. Nor is it necessary to postulate a change in sensitivity of the tissue to cyclic AMP, as originally considered, to account for the greater permeability responses to exogenous cyclic AMP of steroid treated bladders.

The likeliest explanation for the permissive effect of aldosterone on the responses to vasopressin and exogenous cyclic AMP is that incubation with steroid in effect diminishes the degradation of cyclic AMP within the cell.

The aldosterone (or dexamethasone) effect on cyclic AMP accumulation is specific in that testosterone, a steroid which does not amplify the physiologic responses to vasopressin in toad bladder, did not increase the vasopressin elicited net accumulation of cyclic AMP (Table II).

Table II. Effect of steroids and ADH on toad bladder

18–24 Hours of	C-AMP Concentration pmole/mg Protein
Depletion	16.36 ± 1.62
Testosterone	11.82 ± 2.48
Aldosterone	472 ± 81.1
Dexamethasone	534 ± 121.7

Steroids 2×10^{-7}M;
ADH 25 mu/ml for 15 min. at end of 18–24 Hours.

Cyclic AMP, as far as is known, is degraded to a physiologically inactive nucleotide, 5'AMP, in all tissues studied (Butcher et al. 1962). Thus, although it is unlikely that the steroid interacts directly with the enzyme with a resultant inhibitory effect, as do methyl xanthines for example, an assumption subsequently proved (Stoff et al. unpublished), we (Stoff et al. 1972) thought it conceivable that cyclic nucleotide phosphodiesterase activity might be reduced in steroid treated tissues.

To test this hypothesis, bladders were again incubated for 18–20 hours with and without aldosterone. Cells were removed at the end of this time and cyclic nucleotide phosphodiesterase activity measured employing a modification of the method of Murad et al. (1970).

Most of the diesterase activity was present in the supernatant fraction of homogenized epithelial cells. The crude enzyme, as in many other tissues, appears to have 2 Km's for cyclic AMP (Jard & Bernard 1970). Aldosterone significantly decreased phosphodiesterase activity of the cell supernatant fraction at both high and low substrate concentrations. Preliminary results of a large series of assays employing 6.75×10^{-6}M cyclic AMP are illustrated in Table III. Though the magnitude of the effect is small, the difference in activity of the $700 \times$ g supernatant fractions is highly significant (Stoff et al. unpublished).

Table III. Phosphodiesterase activity in toad bladder epithelial cells

Homogenate Fraction	C-AMP Hydrolyzed nmoles/mg. protein/10 min.		
	Depleted	Aldosterone 2×10^{-7}M	$\triangle$
$700 \times$ G Supernatant $n = 16$	2.61	2.34	$-.29 \pm .07$

Although we tentatively have concluded that this is the effect of aldosterone for which we were searching, we are not certain. The assay is based on the assumption that all of the label from the 5'AMP derived from the degradation of cyclic AMP is recovered for radioassay after snake venom treatment. This may not be the case. Were aldosterone to stimulate the transformation of 5'AMP to a degradation product that is not recovered in our final sample, the result could be interpreted erroneously as indicative of a change in diesterase activity. Such an effect of the steroid, however, would not account for the enhanced accumulation of cyclic AMP in response to vasopressin. It would merely eliminate inhibition of phosphodiesterase activity as the mechanism.

CONCLUSION

In view of the uncertainty of the interpretation of the phosphodiesterase studies, it is not profitable to speculate on the mechanism of the aldosterone-dependent decrease in cyclic AMP degradation. We do know that the resultant increase in cyclic AMP accumulation in response to vasopressin, which accounts for the amplification of the physiologic response to the peptide, requires several hours of incubation with steroid to be expressed, suggestive of a role of protein synthesis in its genesis. We are aware of one other study employing dexamethasone in which a significant reduction in the diesterase activity of rat hepatoma cells in tissue culture was induced by the steroid (Manganiello & Vaughan 1972).

Since presentation of these results, we have repeated the studies employing a modification of our original method that assured recovery of all of the radioactivity from 5' AMP in the final samples. Unequivocal depressin of true phosphodiesterase activity in cells of steroid treated bladders was demonstrated.

REFERENCES

Butcher, R. W. & Sutherland, E. W. (1962) Adenosine 3', 5'-phosphate in biological materials. I. Purification and properties of cyclic 3', 5'-nucleotide phosphodiesterase and use of this enzyme to characterize adenosine 3', 5'-phosphate in human urine. *J. biol. Chem. 237*, 1244–1250.

Crabbé, J. (1961) Stimulation of active sodium transport by the isolated toad bladder with aldosterone *in vitro. J. clin. Invest. 40*, 2103–2110.

Gilman, A. G. (1970) A protein binding assay for adenosine 3':5'-cyclic monophosphate. *Proc. Nat. Acad. Sci. (Wash.) 67*, 305–312.

Handler, J. S., Preston, A. S. & Orloff, J. (1969) Effect of adrenal steroid hormones on the response of the toad urinary bladder to vasopressin. *J. clin. Invest. 48*, 823–833.

Jard, S. & Bernard, M. (1970) Presence of two 3'–5'-cyclic AMP phosphodiesterases in rat kidney and frog bladder epithelial cell extracts. *Biochem. biophys. Res. Commun. 41*, 781–788.

Manganiello, V. & Vaughan, M. (1972) An effect of dexamethasone on adenosine 3', 5'-monophosphate activity of cultured hepatoma cells. *J. clin. Invest. 51*, 2763–2767.

Murad, F., Manganiello, V. & Vaughan, M. (1970) Effects of guanosine 3', 5'-monophosphate on glycerol production and accumulation of adenosine 3', 5'-monophosphate by fat cells. *J. biol. Chem. 245*, 3352–3360.

Orloff, J. & Handler, J. S. (1962) The similarity of the effects of vasopressin, adenosine 3', 5'-phosphate (cyclic AMP) and theophylline on the toad bladder. *J. clin. Invest. 41*, 702–709.

Orloff, J. & Handler, J. S. (1967) The role of adenosine 3', 5'-phosphate in the action of antidiuretic hormone. *Amer. J. Med. 42*, 757–768.

Stoff, J. S., Handler, J. S. & Orloff, J. (1972) The effect of aldosterone on the accumulation of adenosine 3':5'-cyclic monophosphate in toad bladder epithelial cells in response to vasopressin and theophylline. *Proc. Nat. Acad. Sci. (Wash.) 69*, 805–808.

DISCUSSION

SACHS: There are probably more than one diesterase in the tissue and presumably the diesterase of interest is the one of the low Km type, and also for the diesterase there are activators and inhibitors present in the tissue. Have you had any look at that?

ORLOFF: There may or may not be two diesterases present in toad bladder. Evidence favoring the former view has come from Morel's laboratory and is consistent with studies in other tissues. However, as far as I know no chemical separation of two distinct enzymes has been accomplished. We also have observed two »apparent« Km's in studies in which diesterase activity was measured over a wide range of substrate concentration. The differences in the slopes of the curves relating the reciprocal of activity and substrate concentrations at the extremes of concentration may be indicative of two enzymes, but it is certainly not conclusive evidence. In any case the V max's of the putative low and high Km enzyme are decreased in preparations from steroid treated tissue. In answer to your second question, we have looked for inhibitors of diesterase in steroid treated tissue without success.

ULLRICH: In some diesterase preparations imidazol augments the diesterase activity, Have you observed something like this in your preparations?

ORLOFF: No.

SCHULTZ: Have you examined the possibility that aldosterone increases the intracellular concentration of other cyclic nucleotides that use the same phosphodiesterase and secondarily increases cyclic AMP concentration by competitively inhibiting the phosphodiesterase?

ORLOFF: No, we have not.

CRABBÉ: When one brings about the effect of aldosterone on fresh tissues (it is more difficult but in a couple of hours one can get it) the reaction of such aldosterone stimulated tissue to ADH appears to be of normal amplitude, both in terms of sodium transport and water permeability (Crabbé 1961, 1964). For such »acute« experiments, do you see potentiation as could

Crabbé, J. (1961) Stimulation of active sodium transport by the isolated toad bladder with aldosterone *in vitro. J. clin. Invest. 40,* 2103–2110.

Crabbé, J. (1964) Effects of corticosteroids on sodium transport by, and permeability to water of, a living membrane. *J. Proc. 2nd Int. Cong. Nephrol,* eds. Vostal, J. & Richet, G., pp. 548–551.

be demonstrated after a long preincubation?

ORLOFF: We have observed potentiation after three to six hours of incubation with steroid.

CRABBÉ: With fresh tissue you report an increase in cyclic AMP concentration with ADH from 9 to 24 pmoles/mg protein.

ORLOFF: That is a 15 minute point. It's considerably higher in 30 minutes.

CRABBÉ: With your substrate depleted tissue, after 18 hours of incubation, don't you eventually in the presence of aldosterone get higher cyclic AMP concentrations than when fresh tissue was exposed to ADH? If so, how do you interpret this? Dr. Edelman, years ago, commented that by using these long preincubation periods, one gets rid of whatever aldosterone is present *in vivo* and therefore one starts from a better base-line to demonstrate the steroid hormone effect; one would then expect aldosterone just to reproduce at that stage what's taking place *in vivo*. Yet, you get, with ADH after 18 hours of incubation, in the presence of aldosterone, a concentration of cyclic AMP considerably higher than when ADH is added to a fresh preparation.

ORLOFF: In my view this is not surprising. The term »fresh« tissue is unfortunate and I'm sorry that it was used on the slides. We have no idea of the steroid content of these cells. It may be close to that of the so-called depleted cells and is certainly considerably less than that of cells incubated with steroid for 18 hours or more. Nor do I think the extraordinarily high concentrations of cyclic AMP in the latter occur *in vivo*. Our experiments were designed to maximize the changes. We have demonstrated, under artificial circumstances (*in vitro* incubation), a qualitative correlation between steroid-induced enhancement of the effects of ADH on both the permeability response of the bladder and the accumulation of cyclic AMP by its cells.

CRABBÉ: And what happens when you treat your tissues with aldosterone?

ORLOFF: There is no detectable effect of aldosterone alone on the basal cyclic AMP content of the epithelial cells.

MAETZ: I wonder whether you have noticed whether protein synthesis inhibitors interfere with this enhancement effect.

ORLOFF: Yes, cyclohexamide inhibits the effect.

SACHS: I have two questions – one in relation to the effect of aldosterone on phosphodiesterase. Can you tell any difference of the effect of theophylline or papaverin on aldosterone treated bladders?

ORLOFF: We have not tested papavarine.

SACHS: This is not in the Ussing chamber preparation. Is the sensitivity to theophylline altered after aldosterone?

ORLOFF: Yes. The permeability responses to theophylline as well as to exogenous cyclic AMP are greater in steroid treated bladders than in depleted paired controls.

SACHS: Does aldosterone change glycolosis in the toad bladder?

LEAF: Well, I'll just say that surely in the presence of aldosterone certain substrates are effective in stimulating sodium transport. These compounds are: glucose, pyruvate, lactate on one hand and also aceto acetate and β-hydroxybutyrate.

Effects of Aldosterone on Frog Skin

Robert Nielsen

INTRODUCTION

It is well established that aldosterone induces a moult in the isolated frog skin (Nielsen 1969, Eigler 1970), and in skins from hypophysectomized toads (Hviid Larsen 1970). The moult in the isolated frog skin is accom-

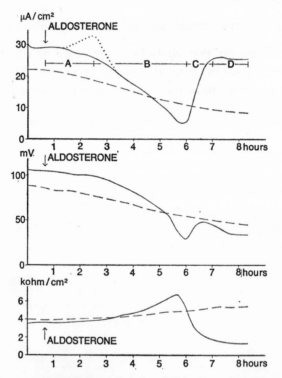

Fig. 1: Effect of $7 \cdot 10^{-7}$ M aldosterone on the short-circuit current and potential across the frog skin.

————————— aldosterone treated skin half.

— — — — — — control skin half.

Institute of Biological Chemistry A. Universitetsparken 13, DK 2100 Copenhagen, Denmark.

»Transport Mechanisms in Epithelia«, Munksgaard, Copenhagen.

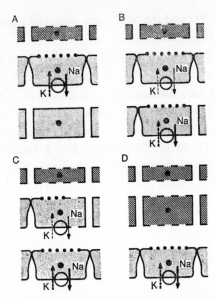

Fig. 2: Diagram of the 3 outermost cell layers of the isolated frog skin. A to D correspond to the 4 periods described in Fig. 1., B the inhibition period, C the activation period.

panied by characteristic changes in the electrical potential (PD), the short-circuit current (SCC) and the trans-epithelial resistance. Fig. 1 shows the effect of aldosterone on these parameters. The graph depicting the SCC after the addition of aldosterone is divided into 4 parts (periods A-D).

In period A, which normally lasted from 1 to 4 hr., an activation of the SCC was observed in about 20 % of the experiments (indicated by the dotted line). In the rest of the experiments there was no significant difference between the control skin half and the aldosterone treated skin half in this period. When this initial activation of the SCC was observed, it commenced 40–90 min. after the addition of the aldosterone. A lag time of the same order was observed for the activation of the SCC in the toad bladder by aldosterone (Crabbé 1961, Sharp & Leaf 1964, Porter & Edelman 1964).

In the inhibition period (period B), an inhibition of the SCC and the PD were observed together with an increase in the resistance. At the end of period B it was rather easy to remove the s. corneum. If the s. corneum is removed at the end of this period one observes an immediate activation of the SCC and the PD (Nielsen 1969). In the activation period (period C), an activation of the PD and the SCC were observed together with a decrease in the trans-epithelial resistance.

In period D the SCC was higher in the aldosterone treated skin half as compared to the control skin half. If the s. corneum which was released during the inhibition period was removed in period D, one observed a further activation of the SCC and a decrease in the resistance. Therefore, one might conclude that the activation observed in period D was due to a disintegration of the s. corneum. However, because of data I have obtained during the investigation of the effect of polyene antibiotics on the aldosterone induced moult (Nielsen 1972a), I have proposed the model shown in Fig. 2.

THE MODEL

The designations A to D of Fig. 2 show a diagram of some of the morphological changes which are supposed to take place in the 3 outermost cell layers during the aldosterone induced moult. A to D on Fig. 2 also correspond to the 4 periods described above. (Fig. 1).

The top layer is the s. corneum. According to the two membrane hypothesis (Koefoed-Johnsen & Ussing 1958, Ussing 1965) the outward facing membrane of the next layer, the s. granulosum, is selectively but passively permeable to Na^+. The outward facing membrane is separated from the inward facing membrane by tight seals (Ussing 1965, Farquhar & Palade 1966). The inward facing membrane is permeable to K^+ but impermeable to free Na^+. Thus, the major Na^+-flux across the inward facing membrane takes place via the $(Na^+–K^+)$-pump.

In the inhibition period, Fig. 2B, the outermost layer of the s. granulosum starts to turn into a new cornified layer, and the underlying cell layer starts forming new tight seals. Since the cornified layer consists of dead cells it is most likely that the activity of the $(Na^+–K^+)$-pump decreases in that layer of the s. granulosum which starts to turn into a cornified layer.

In the activation period (Fig. 2C) the former first layer of the s. granulosum becomes leaky, therefore the Na^+ would now diffuse more easily to the new transporting cell layer which would result in an increased Na^+-transport. In period D (Fig. 2D) the skin is back in its normal transport situation but with two cornified layers. Thus, one of the reasons why aldosterone activated the Na^+-transport in period C and D might be that the transport was performed by a new layer of cells.

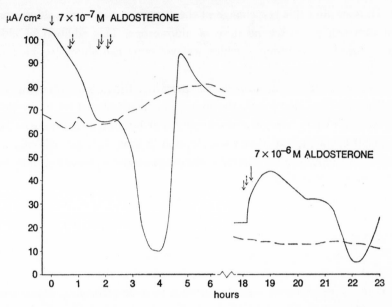

Fig. 3: Effect of two subsequent additions of aldosterone on the short-circuit current across the isolated frog skin. The arrow pair shows when the Ringer's solution was changed.

———————— Aldosterone treated skin half.
– – – – – – – – Control skin half.

According to this model, the outermost layer of the s. granulosum was transformed into a new cornified layer during the moult. If this is so, it might be possible to induce a second moult in the isolated frog skin, after it had accomplished the first moult. In order to investigate this, experiments as that shown on Fig. 3 were carried out.

At time zero aldosterone was added to both sides of the skin. After 40 min. of incubation the aldosterone was removed by renewing the Ringer's solution. The change of the Ringer's solution was repeated again after about 100 and 125 min. of incubation. Fig. 3 shows that the moult was accomplished 5 hr. after the addition of the aldosterone. 18½ hr. after the removal of the aldosterone, the Ringer's solution was renewed again. This change of the Ringer's solution caused an immediate activation of the SCC in the aldosterone treated skin half, whereas it had no effect on the control.

Previous experiments have shown that this activation was due to a breakage of the s. corneum, which was loosened but not removed during the previous moult. 10 min. after this last change of the Ringer's solution the aldosterone treated skin half got a second dose of aldosterone. This addition of aldosterone induced a second moult which was accomplished about 4½ hr. after the addition.

In experiments (data not given here) where the Ringer's solution was not changed during the experiment, the first moult was induced by a low dose of aldosterone ($3 \cdot 10^{-9}$ M) and the second moult by a high dose ($7 \cdot 10^{-6}$ M). In about 50 % of the experiments performed in that way, the SCC did not recover spontaneously after the inhibition period in the second moult, but not until the cornified layer had been broken by applying a pressure gradient across the skin or removed by a piece of cotton wool.

Histological examinations of the skin halves showed that the control skin half had one cornified layer, whereas the skin half which had moulted twice had three cornified layers. Since it was possible to induce two moults in series in the isolate frog skin, we can conclude that the former outermost layer of the s. granulosum was converted into a new cornified layer. Furthermore, the isolated frog skin formed a new Na^+-selective outward facing membrane during the aldosterone induced moult.

ALDOSTERONE INDUCED CHANGES IN THE OUTWARD FACING MEMBRANE

It has been demonstrated by Koefoed-Johnsen & Ussing (1958) and Ussing (1965) that the PD across the isolated frog skin responds as a more or less ideal Na^+-electrode to a change in the Na^+ concentration of the outside solution. In order to investigate when the new Na^+-selective outward facing membrane was formed, I have studied the changes in the Na^+-selective membrane during the aldosterone induced moult. The investigation was carried out by observing the voltage changes across the skin in response to a Na^+ concentration step in the outside solution. The interpretation of such experiments would be more difficult, if secondary changes of the cellular composition take place. In order to minimise such changes, I have used the rapid-flow method described by Lindemann et al. (1972). With this method it is possible to change the outside concentration so fast and so briefly, that the electrical response to a concentration step only will be a response of the Na^+-selective membrane, the layers in front of the Na^+-selective membrane and the extracellular shunt pathway.

The experiments were performed in the chamber shown in Fig. 4. The

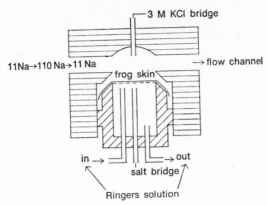

Fig. 4: Flow-chamber. The outer chamber consists of a flow channel 1 mm high and 1 cm broad. The inner chamber had a nylon mesh to support the skin. The inside solution was circulated by a peristaltic pump.

inside of the skin was bathed with Ringer's solution containing 110 mM NaCl. The inside solution was circulated by a peristaltic pump. The outside of the skin faced a flow channel (1 mm high and 1 cm broad). The outside solution could be changed very fast, by an electromagnetic valve system, from Ringer's solution containing 11 mM NaCl to Ringer's solutions containing 110 mM NaCl and back to Ringer's solution containing 11 mM NaCl. In order to keep the Cl⁻ concentration and the osmolarity constant during the experiment the Ringer's solution with 11 mM NaCl contained also 99 mM choline chloride.

The flow rate of the outside solution was 3 ml/min, however when a concentration step experiment was performed the flow rate was 30 ml/sec. Conductivity measurements showed that it took 10 msec to half-way complete the exchange of the outside solution when the flow rate was 30 ml/sec. The change in the electrical potential across the skin during this fast 10 fold increase and decrease in the Na^+ concentration was measured with a double beam oscilloscope.

The oscilloscope tracing (Fig. 5) shows the result obtained when an Amfion C-100 cation exchange membrane was mounted in the fast flow chamber and exposed to a fast (11,110,11 mM) change in the Na^+ concentration. Before the fast changes in the Na^+ concentration were carried out the membrane was brought into a steady state with respect to the outside solution. The steady state PD across the Amfion C-100 membrane was 13

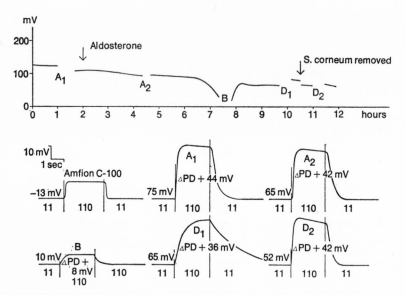

Fig. 5: Effect of aldosterone on the Na+-selectivity of the isolated frog skin. The upper curve shows the PD across the skin during the incubation with 110 mM NaCl Ringer's solution on both sides of the skin. The letters A to D indicate when the Na+ concentration step experiments were carried out. The lower part shows the corresponding changes in the PD across the skin. The 110 mM NaCl Ringer's solution consisted of, 110 mM NaCl, 2.5 mM KCl, 1 mM CaCl₂, and 5 mM Tris, pH- 8.0. The 11 mM NaCl Ringer's solution consisted of 11 mM NaCl, 99 mM Choline chloride, 2.5 mM KCl, 1 mM CaCl₂ and 5 mM Tris, pH- 8.0.

mV (inside negative) when the outside solution was 11 mM NaCl Ringer's solution, and the inside was 110 mM NaCl Ringer's solution.

Fig. 5 shows that the PD across the Amfion C-100 membrane decreased very fast to 0 mV when the Na concentration on the outside was changed from 11 mM to 110 mM, and increased again to 13 mV when the outside solution was changed back to 11 mM NaCl Ringer's solution. The response of the Amfion C-100 membrane to the increase of the Na+ concentration from 11 mM to 110 mM was found to be half completed (t $\frac{1}{2}$) within 30 msec. From the half time 30 msec. (and taking into account the logarithmic response of the membrane) the apparent thickness of the unstirred layer between the flowing solution and the membrane was calculated to be 8.5 μ.

Crank's equation (1956) for diffusion in unstirred layers was used in these calculations. Furthermore it was assumed that the diffusion coefficient for the Na^+ was constant and equal to $0.5 \cdot 10^{-5} cm^2/sec$ (Lindemann *et al.* 1972).

The upper part of Fig. 5 shows the PD across the skin during the incubation with 110 mM NaCl Ringer's solution on both sides of the skin. The letters A to D show when the Na^+ concentration step experiments were carried out; the lower part of Fig. 5 shows the corresponding changes in the PD across the skin. The concentration step (11–110–11 mM Na^+) experiments were performed as described above for the Amfion C-100 membrane.

The PD with 11 mM NaCl Ringer's solution on the outside and 110 mM NaCl Ringer's solution on the inside was in experiment A_1-75 mV, in A_2-65 mV, in B-10 mV, in D_1-65 mV and in D_2-52 mV. The increase in the PD across the skin ($\triangle$ PD) when the Na^+ concentration of the outside solution was changed from 11 mM to 110 mM was in experiment A_1-44 mV, in A_2 -42 mV, in B-8 mV, in D_1-36 mV and in D_2-38 mV.

The response ($\triangle$ PD) of the skin when the Na^+ concentration of the outside solution was changed from 11 mM to 110 mM was found to be half completed within-72 msec in experiment A_1, -72 msec. in A_2, -400 msec. in D_1, -and 66 msec. in D_2. From these half times the apparent thickness of the unstirred layer between the flowing solution and the Na^+-selective membrane was calculated: the thickness of the layer in experiment A_1 was 13μ, in A_2- 13μ, in D_1- 31μ, and in D_2- 12.5μ.

From these data it appears, that there was no great difference in the thickness of the unstirred layer before the aldosterone induced moult and after the s. corneum was removed (cf. fig. A_1 and D_2). The same result was obtained (data not shown here) when the s. corneum was removed at the very end of the inhibition period (period B, Fig. 1). If aldosterone had only induced the release of the s. corneum, one would expect that the thickness of the unstirred layer would decrease with about $4-5\mu$ (the thickness of the s. corneum). However, since such a decrease in the unstirred layer was not observed, one might suggest that a new cornified layer was formed during the inhibition period.

These data support the hypothesis shown in Fig. 2 which suggest that a new cornified layer was formed during the inhibition period. The data obtained by Lindemann *et al.* (1972) also support this hypothesis. By using spontaneous moulting frogs skins Lindemann *et al.* (1972) conclude: "The increase in speed was not very convincing, (in other words, the decrease in thickness of the unstirred layer was not very convincing), however, since the total deflection was small". The $\triangle$ PD found by Lindemann et al. (1972) before and after the removal of the s. corneum was about 10 mV for a 40 times change in the Na$^+$-concentration, whereas the data presented here show a $\triangle$PD of about 40 mV for a 10 times change in the Na$^+$-concentration.

The thickness of the unstirred layer after the s. corneum was loosened, but not removed, was 31μ (c. f. Fig. 5D), whereas the thickness after the removal of the cornified layer was 12.5μ. The difference between these two thicknesses (31–12.5) 18.5μ should give the thickness of the cornified layer. However, this figure was too high, which is probably due to the formation of a subcorneal space during the moult (Voûte et al. 1969).

The fact that the $\triangle$PD caused by the change in the Na$^+$ concentration at the outside decreases in the inhibition period (cf. Fig. 5B) might be due to: 1. an increase in the shunt permeability or 2. a decrease in the Na$^+$ selectivity of the outward facing membrane of the outermost layer of the s. granulosum. However, previous experiments (Nielsen 1969) have shown that the permeability to Na$^+$ and Cl$^-$ does not increase in the inhibition period; the same appears to be the case for sucrose (Nielsen 1972b). Thus, one can conclude that the observed decrease in the inhibition period of $\triangle$PD was caused by a decrease in the Na$^+$ selectivity of the outward facing membrane.

If one compares the slope of the $\triangle$PD curves in Fig. 5A$_1$, 5A$_2$ and 5D$_2$ during the exposure of the outside to the 110 mM NaCl Ringer's solution, one observes that the decrease of the $\triangle$PD was enhanced after the removal of the s. corneum. This decrease in the $\triangle$PD in normal frog skins could be inhibited by amiloride (Lindemann 1973). Since amiloride acts by inhibiting the passive Na$^+$ influx (Baba et al. 1968, Nielsen & Tomlinson 1970), one might suggest that the enhanced decrease in the $\triangle$PD was caused by an increased permeability for Na$^+$ across the outward facing membrane.

But why should an increase in the Na$^+$ permeability of the outward facing membrane cause a decrease in the $\triangle$PD? This might be due to one of the following two reasons or to a combination of both. The increased permeability caused the Na$^+$ concentration in the cells to increase; accordingly the Na$^+$ gradient and the PD across the outward facing membrane would decrease. In the toad bladder it has been shown that the cellular concentration of Na$^+$ increased after the addition of aldosterone (Leaf & MacKnight 1972, Handler et al. 1972). On the other hand it has been shown by Cereijido *et al.* (1964) and by Lindemann in this symposium (Lindemann 1973) that the apparent Na^+ permeability of the outer membrane was decreased by an increase in the Na^+ concentration of the outside solution. A decrease in the Na^+ flux into the cells across the Na^+ selective membrane would because of shunt increase the attenuation of the PD.

CONCLUSION

It has been shown that aldosterone induces a moult in the isolated frog skin. The moult can be repeated by adding a second dose of aldosterone. The moult was accompanied by characteristic changes in the Na^+-transport (Fig. 1). In the inhibition period the s. corneum was detached from the underlying cell layer, and the outermost layer of the s. granulosum was converted into a new cornified layer. The Na^+ selectivity of the outermost layer of the s. granulosum decreased in the inhibition period. The removal of the loosened s. corneum caused an increase in the Na^+ permeability of the outward facing membrane. The late activation of the Na^+ transport (Fig. 1 period D) was due to 1. the Na^+ transport was performed by a new cell layer and 2. the outward facing membrane of this cell layer was more permeable for Na^+.

REFERENCES

Baba, W. I., Lant, A. F., Smith, A. J., Townshend, M. M. & Wilson, G. M. (1968) Pharmacological effects in animals and normal human subjects of the diuretic amiloride hydrochloride (MK-870). *Clin. Pharmacol. Ther.* 9, 318–327·

Cereijido, M., Herrera, F. C., Flanigan, W. J. & Curran, P. F. (1964) The influence of Na concentration on Na transport across frog skin. *J. gen. Physiol.* 47, 879–893.

Crabbé, J. (1961) Stimulation of active sodium transport by the isolated toad bladder with aldosterone *in vitro. J. clin. Invest.* 40, 2103–2110·

Crank, J. (1956) *The Mathematics of Diffusion,* p. 45. Oxford Univ. Press, London.

Eigler, J. (1970) The effect of aldosterone on sodium transport and transepithelial water flux through isolated ventral skin of *Rana temporaria. Pflügers Arch. ges. Physiol.* 317, 236–251.

Farquhar, M. G. & Palade, G. E. (1966) Adenosine triphosphatase localization in amphibian epidermis. *J. Cell Biol.* 30, 359–379.

Handler, J. S., Preston, A. S. & Orloff, J. (1972) Effect of aldosterone on the sodium content and energy metabolism of epithelial cells of the toad urinary bladder. *J. steroid Biochem.* 3, 137–141.

Hviid Larsen, E. (1970) Sodium transport and D. C. resistance in the isolated toad skin in relation to shedding of the stratum corneum. *Acta physiol. scand.* 79, 453–461.

Koefoed-Johnsen, V. & Ussing, H. H. (1958) The nature of the frog skin potential. *Acta physiol. scand. 42*, 298–308.

Leaf, A. & MacKnight, A.D.C. (1972) The site of the aldosterone induced stimulation of sodium transport. *J. steroid Biochem. 3*, 237–245.

Lindemann, B., Gebhardt, U. & Fuchs, W. (1972) A flow-chamber for concentration-step experiments with epithelial membranes. T.-I.-T. *J· Life Sci. 2*, 15–26.

Lindemann, B. (1973) Delayed Changes of Na-Permeability in Response to Steps of (Na)o at the Outer Surface of Frog Skin and Frog Bladder. Alfred Benzon Symposium V. *Transport mechanisms in epithelia*, H. H. Ussing & N. A. Thorn (ed.), pp. 115–127. Munksgaard, Copenhagen.

Nielsen, R. (1969) The effect of aldosterone *in vitro* on the active sodium transport and moulting of the frog skin. *Acta physiol. scand. 77*, 85–94.

Nielsen, R. & Tomlinson, R. W· S. (1970) The effect of amiloride on sodium transport in the normal and moulting frog skin. *Acta physiol. scand. 79*, 238–243.

Nielsen, R. (1972a) The effect of polyene antibiotics on the aldosterone induced changes in sodium transport across isolated frog skin. *J. steroid Biochem. 3*, 121–128.

Nielsen, R. (1972b) Unpublished findings.

Porter, G. A. & Edelman, I. S. (1964) The action of aldosterone and related corticosteroids on sodium transport across the toad bladder. *J. clin. Invest. 43*, 611–620.

Sharp, G.W.G. & Leaf, A. (1964) Biological action of aldosterone *in vitro*. *Nature (Lond.) 202*, 1185–1188.

Ussing, H. H. (1965) Transport of electrolytes and water across epithelia. *Harvey Lect. 59*, 1–30.

Voûte, C. L., Dirix, R., Nielsen, R. & Ussing, H. H. (1969) The effect of aldosterone on the isolated frog skin epithelium (*R. Temporaria*) A morphological study. *Exp. Cell Res. 57*, 448–449.

DISCUSSION

WINDHAGER: What was the diffusion coefficient that you assumed in the calculation for the unstirred layer?

NIELSEN: 0.5×10^{-5}.

WINDHAGER: That is quite close to the free diffusion coefficient.

NIELSEN: Yes. Since I did not know what the diffusion coefficient for sodium was in the cornified material, I have used that from water.

WHITTEMBURY: Have you studied the effect of ADH?

NIELSEN: No.

LINDEMANN: I would like to say that I agree with you. In electronmicrographs one can sometimes see two cornified layers: Is there any morphological difference between a moult induced by aldosterone and a natural one?

NIELSEN: The only thing I can say is: When I have a spontaneous moult in the chamber, the bioelectrical changes are just the same as in aldosterone induced ones. But I have no histological data on that.

HVIID LARSEN: In Fig. 2, showing the three outermost cell layers of the frog skin during the aldosterone induced moult, you propose that an extracellular shunt is opened (part C of the figure). I would like to ask whether the very small voltage response to changing the sodium concentration of the outer bath during this period could be due to a shunting effect through these paracellular leaks instead of being due to a loss of the Na-selectivity of the outward facing membranes themselves.

NIELSEN: The very low voltage response to changing the sodium concentration refer to Fig. 2 part B and not C (the inhibition period). In the inhibition period the resistance was still high and furthermore no increase in the passive chloride and sodium flux took place in this period. Thus, it is very unlikely that the low voltage response caused by the change in the sodium concentration was due to an increase in the shunt permeability. However, the somewhat lower PD (period C and D Fig. 2) and $\triangle$ PD (Fig. 5 D_1 and D_2) observed latter in the experiment was probably, at least partically, due to an increase in the shunt permeability; both the passive sodium and sucrose permeability did increase in these periods.

HVIID LARSEN: Isn't it true that if you increase the paracellular shunt this would in effect show up as a decreased sodium selectivity of the outward facing membrane?

LINDEMANN: We have done similar experiments to the ones just described and I got the impression that the sodium selectivity is lost at a time when the resistance is still up. Therefore I would not think that this is due to an ordinary shunt.

LEAF: After the moult occurs do you then have the skin as one layer of cells thinner than before the aldosterone or is there concomitant synthesis of another layer?

NIELSEN: There is no increase in the DNA synthesis and my impression is that the aldosterone increases the turnover in already existing cell layers but it does not induce a formation of new cell layers.

CRABBÉ: If you preincubate the tissue without aldosterone for, say 15–18 hours, do you get exactly the same results?

NIELSEN: Yes, we do.

VOÛTE: Speaking about preincubation of skins for *in vitro* aldosterone experiments I would like to report about some preliminary results we obtained on spring frogs kept at 4° C. Getting tired of the very irregular appearance of these skins with respect to mitochondria rich cells (MR cells, see also Voûte *et al.* 1972) and their possible involvement in aldosterone effect in epithelia, we decided to preincubate some skins for 24 h at 4° C in aerated frog Ringer's in order to get rid of all endogenous steroid effect. And we were lucky enough to find that not only the MR cells had almost completely disappeared but that the average electric resistance of these skins was much higher than in fresh preparations of spring frogs. When adding aldosterone to these preincubated skins *in vitro* (see exptl. protocol) already after 30 minutes the MR cells had become normal in number and shape and the aldo response of the preparations was absolutely normal (Nielsen 1969, Voûte *et al.* 1969). When in another chamber (same skin) an identical piece was treated with aldosterone in the presence of actinomycin D (6 μg/ml) we got, as reported by Nielsen (1969), no aldosterone effect with respect to bioelectric changes and moulting. Looking now at the biopsies (for timing see protocol) after 30 minutes there was not much difference to see in number

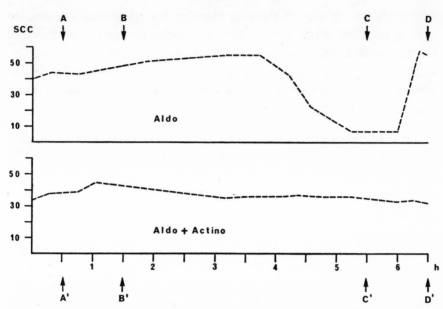

For this *in vitro* experiment on preincubated skins a double set of four chambers (see fig. 1) was used (8×1 cm² from an identical skin). A B C D are the biopsies with Aldosterone, A' B' C' D' the ones with Aldo + Actinomycin D. For reasons of simplicity only the SCC tracings (uAmps/cm²) of the last two chambers (D and D') are given. The upper tracing shows the typical Aldo-response which is inhibited in the lower chamber by Actinomycin D.

and shape of MR cells as compared to aldosterone alone: The morphogenetic effect of aldosterone, the differentiation of a normal epithelial cell to an MR cell was not blocked by actinomycin D. From thereon, however, in later biopsies, these cells, in spite of being present, didn't show any sign of

Voûte, C. L., Hänni, S. & Ammann, E. (1972) Aldosterone induced morphological changes in amphibian epithelia *in vivo*. *J. Steroid Biochem. 3*, 161–165.

Nielsen, R. (1969) The effect of aldosterone *in vitro* on the active sodium transport and moulting of frog skin. *Acta physiol. scand. 77*, 85–94.

Voûte, C. L., Dirix, R., Nielsen, R. & Ussing, H. H. (1969) The effect of aldosterone on the isolated frog skin epithelium (R. temporaria). A morphological study. *Exptl. Cell. Res. 57*, 448–449.

Voûte, C. L. & Ussing, H. H. (1970a) Quantitative relation between hydrostatic pressure gradient, extracellular volume and active sodium transport in the epithelium of the frog skin (*R. temporaria*) *Exptl. Cell Res. 62*, 375–383.

Voûte, C. L. & Ussing, H. H. (1970b) The morphological aspects of shunt-path in the epithelium of the frog skin (*R. temporaria*) *Exptl. Cell Res. 61*, 133–140.

further changes as seen in the aldo biopsies: No cytoplasmic changes, no signs of secretion or lake formation, and of course no sign of detachment of the stratum corneum. In addition more and more MR cells showed peculiar nuclear and cytoplasmic changes, signs of degeneration and finally disintegration and disappearance. In the undifferentiated normal epithelial cells no changes could be observed to occur.

KEYNES: I listened rather enviously to the last two papers, because the preparation we work on at Babraham is sheep rumen epithelium. We took sheep and adrenalectomized them so as to get them nicely aldosterone-depleted, remembering that within a very few hours this gives you substantial effects both on salivary sodium-potassium balance, as has been shown by Blair-West *et al.* (1964) and on kidney function in conscious animals as we have tried. When we give aldosterone back by infusion, within an hour you get restoration of normal kidney and of salivary gland function; but when we took rumen epithelium from these aldosterone-depleted animals and incubated it with aldosterone for long periods, we couldn't demonstrate any difference at all between the depleted controls and the test epithelia – completely negative results. Of course we didn't think to look at the amount of cornified epithelium in our preparations.

NIELSEN: The cornified layers are in some cases the rate limiting step for the sodium influx in the isolated frog skin.

KEYNES: Rumen epithelium normally has a rather thicker cornified layer than frog skin. The structure of rumen epithelium is rather like the pictures we saw yesterday with first a cornified layer, then the layer that pumps, and then lots of cells which seem to have quite wide gaps in between. We have more cornified layers in rumen epithelium than in frog skin. This may be significant.

FRÖMTER: Do you expect if you continue to give aldosterone shocks that the frog could come out with a single layered skin epithelium?

NIELSEN: I have done that but you can't do it on a good frog, because then you have to incubate for days. The first experiments I did in the spring were

Blair-West, J. R., Coghlan, J. P., Denton, D. A., Goding, J. R. & Wright, R. D. (1964) The effect of adrenal cortical steroids on parotid salivary secretion. In: Salivary Glands and their Secretions (Sreebny, L. M. & Meyer, J., ed.), Pergamon, Oxford, London, pp. 253–279.

on rather skinny frogs. What happened when I induced the second moult was that the preparation died. I looked at it in the microscope and the reason was that there were only two cell layers. So the preparation died because both the sodium transporting cell layers were converted into cornified layers.

GENERAL DISCUSSION

MOREL: This morning's session dealt with hormonal control of transport and permeability in epithelial cells of amphibia. In this context, I would like to take the liberty of opening the generel discussion by reporting results which were recently obtained in our laboratory by S. Jard and M. Bernard. They will suggest that the generally accepted model of cell control by cyclic AMP, as illustrated by Dr. Orloff's first slide, may be an oversimplified view of the situation. The model implies a linear sequence of reactions from the hormone-receptor interaction, via cAMP production up to the final biologi-cal effect. I would like to suggest the presence of negative retroaction exerted by the output of the system and controlling the actual level of cAMP within the cell.

S. Jard and M. Bernard measured both the time course of the biological effect (change in P. D. or change in S. S. C.) and the time course of the cAMP cell concentration in isolated frog skin epithelia in response to stimu-lation by isoproterenol. Roughly similar observations were made after oxy-tocin stimulation. Cyclic AMP concentration within the structure reached its peak at about 1 min. after catecholamine addition up to a maximum about 8 times higher than the baseline; it then gradually dropped to a steady state level which was about $1^1/_2$–2 times higher than the control level. This be-ginning and subsequent development of the drop in concentration mirrored the time course of the biological response (increase in S. C. C.). In order to test the hypothesis that the drop in cAMP concentration may have resulted from a negative feed-back exerted by the biological response, similar experi-ments were performed under conditions where no final effect could be elicited, that is, by omitting sodium from the outside medium. In this condi-tion, the cAMP concentration curve within the cells was completely different; it started in the same way, but instead of dropping after one min or so, it continued to increase and tended towards very high steady state levels, as predicted by the hypothesis. We don't know at present the mechanism of the retroaction observed, nor if it results in cyclase inhibition or phosphodiesterase stimulation. But the physiological meaning of such a negative feedback control could be to maintain the cell permeability changes induced by hormones within the limits required for normal functioning of the target cells.

I therefore wonder if the so-called permissive action of aldosterone might have something to do with such a cellular feed-back mechanism, whatever it may be. My question to Dr. Orloff is: I understand that you measured cyclic AMP contents, in your experiments, 15 min after adding vasopressin; did you look at what happened after shorter incubation times? In addition, I would like to ask the audience if anybody has other indications suggesting that the biological response to hormones can in some way limit cyclic AMP production or concentration in the epithelial cells.

ORLOFF: We have measured the cyclic AMP content at earlier times and have found that it rises rapidly, reaches a maximum at about 30 minutes and is still elevated to the same level at 60 minutes. The tissues had been incubated in Ringer's solution containing NaCl and thus our results in toad bladder cells differ from yours in skin. Mendoza *et al.* (1972) have obtained similar results in whole bladder in that they too found that the cAMP content at 30 and 60 minutes did not differ despite a decline in the short-circuit current response to vasopressin at the 60 minute point. The late fall off in cAMP content that you have observed has been noted in other tissues, as you know, but the interpretation is quite different.

MOREL: In fat cells for instance.

ORLOFF: Yes, in the fat cell hormone-induced increases in cAMP content reach a maximum and then decline; yet the physiologic response continues at the same rate (Manganiello *et al.* 1971). Whether this is indicative of negative feed-back as a physiologic control mechanism or of an overshoot phenomenon with respect to cAMP production and accumulation (now more fashionably referred to as "receptor reserve"), I'll let you decide.

Mendoza, S. A., Murad, F., Handler, J. S. & Orloff, J. (1972) Refractoriness of toad bladder to stimulation of sodium transport. *Am. J. Physiol. 223*, 104–109.
Manganiello, V. C., Murad, F. & Vaughan, M. (1971) Effects of lipolytic and antilipolytic agents on cyclic 3', 5'-adenosine monophosphate in fat cells. *J. Biol. Chem. 246*, 2195–2202.

OSCHMAN: Berridge & Prince (1972) in Cambridge have clearly demonstrated a negative feed-back relationship between cyclic AMP and calcium in the insect salivary gland. It appears that as the cyclic AMP level goes up, calcium is released, probably from mitochondria, and this feeds back on the adenyl cyclase to reduce the rate of cyclic AMP production.

TOSTESON: A simple reason for this result would be by a reduction in the substrate concentration. It is known that the active transport consumes ATP. Have you measured the ATP concentration as a function of time during this period. Have you studied the effect of ouabain?

MOREL: No, the ATP concentration was not measured, but the amount of ATP which is used for cyclic AMP production is negligible compared with that used for the ATPase transport.

TOSTESON: Of course one wouldn't expect the ATP to disappear entirely but the steady state ATP concentration might be reduced when rate of transport is increased.

MOREL: Yes, but as far as ATP concentration is concerned, when the cyclase activity is measured on membrane preparations, it works with a very low ATP concentration so I don't think that ATP cell concentration could control or limit cyclic AMP production under physiological conditions.

CRABBÉ: Dr. Morel, I would concur with you in thinking that the amount of ATP available would not set a limit, at least here, to what takes place. You remember what was reported last summer by Dr. De Weer (1972) who studied the rate of efflux of sodium from perfused squid axon, monitoring the concentration of ATP; one has to reduce the concentration drastically, that is with respect to the physiological state, to start noticing a change in the rate at which the preparation pumps sodium.

TOSTESON: I can only comment that there is a good deal tissue-to-tissue variation in the relationship between the rate of active transport and the concentration of ATP. One unusual fact is that the effective Km for the

Berridge, M. J. & Prince, W. T. (1972) The role of cyclic AMP and calcium in hormone action. *Adv. Insect Physiol. 9*, 1–49 (refer to Fig. 11).

De Weer, P. (1972) Discussion to »A study on the effect of aldosterone on the extra-mitochondrial adenine-nucleotide system in rat kidney« Kirsten, E., Kirsten, R. & Salibian, A. *J. Steroid Biochem. 3*, 178.

enzyme system studied in an isolated state is generally (certainly in red cells) very much lower than the Km for active transport in a functioning cell.

KEYNES: Can I comment on Dr. Crabbé's point? Although the sodium efflux from the squid axon is fairly sensitive to the ATP level, what happens is that it switches from a sodium-potassium exchange system to a sodium-sodium exchange system at a low ATP level.

MAETZ: I want to ask just one question about these experiments by Dr. Jard. Isn't there simultaneously an increase of water permeability by isoproterenol? Probably the experiments were done without an osmotic gradient. However, if you suppress sodium transport you may still have the osmotic effect.

MOREL: There is another long story which has not been discussed up to now and that is the dissociation of the two hormonal effects. These experiments were done under conditions where there was no osmotic gradient either in the presence or in the absence of sodium in the outer medium. I cannot tell you what happens when only the osmotic effect is the final effect of the hormone. That was not achieved up to now.

SACHS: Are there any changes in the cyclic AMP content in the bathing solution in your experiments?

MOREL: There was no cyclic AMP added. If there was any leakage of cyclic AMP out of the cell I'm afraid the resulting concentration will be much too low to be measured.

SACHS: Because with the absence of sodium pump and water flow, if there is leaking cyclic AMP out of your cells you will also reduce that.

MOREL: That could explain it.

LEAF: There may be a serosal side which is surprisingly permeable.

SACHS: This movement is in the serosal direction.

ULLRICH: Dr. Morel, didn't Dr. Jard in your laboratory observe the formation of a phosphoprotein and can you comment on that?

MOREL: A phosphokinase is present in this structure – as in any structure in which cyclic AMP has been involved – which is specifically stimulated by cyclic AMP. What kind of reaction this protein kinase controls physiologically is completely unknown. The Km of this kinase for cAMP is, as in other structures, of the order of $10^{-8}-10^{-9}$; it is therefore difficult to visualize how this kinase acts as a regulatory factor, since even the base-line concentrations of cyclic AMP in the whole structure are higher than its Km for cyclic AMP. This leads to the problem of the compartmentalization of cyclic AMP in the structure.

ULLRICH: And is this phosphokinase in the cytoplasmatic fraction, or is it in a particulate fraction?

MOREL: Part is apparently bound to the particulate fraction, I don't know what information you have Dr. Orloff on the problem in the bladder?

ORLOFF: All our results have been negative. We have demonstrated cyclic dependent kinase activity, but no effects of hormone or other experimental manipulations which could be considered fruitful.

LEAF: If one has an osmotic gradient of 150 mOsm solute across the bladder, this is equivalent to a hydrostatic pressure of four atmospheres, but in the absence of ADH, we recorded something between 2 or 4 μl water/cm² per hour across the tissue. If one adds ADH the value of course goes way up to about 200 μl/cm²/hr. The actual basis of this I don't know. One could produce this degree of impermeability on calculation with a 50 Å layer of structured water considering the diffusion coefficient for water in ice is about 10^{-11} cm/sec as compared with 10^{-5} for water in water, but whether that is the basis or not I have no clue. I'd like to ask you what is the permeability of a lipid barrier-like membrane to water. Is it a comparable figure?

TOSTESON: The permeability of lipid bilayers to water is about 10^{-3} cm · sec^{-1}.

KEYNES: Dr. Bangham was telling me the other day that he got some very dramatic differences in water permeability of lysosomes depending on the nature of the phospholipids. I don't remember what the actual chemical differences were, but he got orders of magnitude effects on water permeability of bilayers.

TOSTESON: I believe that I was referring to the work of Krasne et al. (1971). They showed that the conductance of bilayers made with glyceryl distearate

---

Krasne, S., Eisenman, G. and Szabo, G. 1972. Freezing and melting of lipid bilayers and the mode of action of nonactin, valiomycin and gramicidin. *Science, 174,* 412.

or glyceryl dipalmitate and exposed to monactin with KCl in the bathing solutions decreased by 5 orders of magnitude when the temperature was reduced below the point where the hydrocarbon chains became frozen. It is highly likely that this procedure would produce a comparable reduction in the diffusion coefficient of water in the hydrocarbon chains and thus in the permeability of the membrane to water.

BERLINER: Was the temperature studied?

TOSTESON: Have you tried to warm the thing and see whether there is a sudden break in the water permeability.

USSING: In the first place, when the frog skins are kept cold, the water permeability drops and drops and gets very low indeed, so as if there is some kind of a phase change at low temperature. I may bring up two points: one is that I seem to remember that high cholesterol contents reduce the permeability, is that not true?

TOSTESON: Yes.

USSING: I should also like to remind you of the fact that amphotericin and analogues act only on the outside of the amphibian skin and bladder. As far as I know they have no effect when applied to the inside. So there must be a chemical difference between these two membranes. It may have to do with the contents of cholesterol.

TOSTESON: That is an excellent point worth investigating further.

OSCHMAN: With respect to Tosteson's and Ussing's discussion, the apical plasma membranes of toad bladder and of a variety of other epithelia are structurally different from the lateral and basal plasma membranes. They look different in ordinary electronmicroscope preparations, and there is also a very distinct fuzzy glycocalyx or carbohydrate coat on the apical plasma membrane, clearly indicating that there is much more to the membrane than lipid.

TOSTESON: J. D. Robertson and his colleagues at Duke (Vergara et al. 1969) have observed an hexagonal array, probably due to these glycoproteins, on the surface of mammalian bladder. I wonder whether that is a common

Vergara, J., Longley, W. & Robertson, J. D. (1969) A hexagonal arrangement of subunits in membrane of mouse urinary bladder. *J. Mol. Biol. 46*, 593–596.

feature of membranes that have very low water permeability. Has any such structure been observed in amphibian bladders or skin?

LEAF: There is a fuzzy coat on the mucosal surface of toad bladder cells. This could perhaps be a glycocalix, organizing a layer of impermeable structured water or simply be an artifact of fixation. No difference in the fuzzy coat has been described in the presence or absence of neurohypophyseal hormones but quantitation of differences in the amount of fuzz present would be difficult by microscopy.

Can we ask whether you can produce a black membrane by any combination of lipids, which will have as low an osmotic or hydraulic permeability to water as we have found for toad bladder in the absence of neurohypophyseal hormones?

TOSTESON: The temperature at which hydrocarbon chains freeze depends on the chain length and the degree of unsaturation. The shorter the chain length and the greater the degree of unsaturation, the lower the freezing temperature. For membranes made of 1:1 glyceryl dipalmitate and glyceryl distearate, the freezing temperature is about 40° C.

KEYNES: If you ask what epithelium in a mammal needs to be highly water impermeable, then the obvious answer is the urinary bladder. I think Dr. Bangham says that the lipid composition of that epithelium does agree with his findings on liposomes, but I may be mistaken about it.

LINDEMANN: The beautiful work by Marian Hicks and co-workers at the Middlesex Hospital, Medical School, London, on mammalian bladder has shown that the thick apical membrane of the transitional epithelium is asymmetric. On the outer surface is a hexagonal lattice of subunits ("doughnuts") with a center to center spacing of 140 Å. Each doughnut is made up of 12 globular particles, which may well be there to structure water and keep the water permeability low. The rotation of the doughnuts with respect to the lattice is normally 19 degrees. When the rotation is changed, the water permeability increases markedly (Hicks & Ketterer 1970). I would like to ask whether anybody has tried or heard of similar studies done with frog skin or the toad bladder.

Hicks, R. M. & Ketterer, B. (1970) Isolation of the plasma membrane of the luminal surface of rat bladder epithelium, and the occurrence of a hexagonal lattice of subunits both in negatively stained whole mounts and in sectioned membranes. J. Cell. Biol. 45, 542–553.

Further Observations on the Isolated Rat Gastric Mucosa

C. Adrian M. Hogben and Dorothy R. Karal

INTRODUCTION

The isolated rat gastric mucosa has become an important experimental preparation for elucidating the movement of Na^+ and K^+ (Sernka & Hogben 1969, Sernka 1969; the latter having several appendices not published in the former; Hogben 1972), as well as prompting scrutiny of the metabolic pathways necessary to support H^+ secretion (Sernka & Harris 1972). To provide a firmer basis for effectively utilizing this preparation we have undertaken the several studies reported here.

The origin of the short-circuit current, which was previously incompletely understood, has been clarified. Further evidence is presented establishing that the active absorption of Na^+ displayed by the isolated mammalian gastric mucosa can be suppressed by a reasonable $[H^+]$ at its mucosal surface, corresponding to the expectations for the secreting stomach in the intact animal.

By use of ion substitution, we have sought to determine the extent to which the active secretion of H^+ is dependent on the concomitant active secretion of Cl^-, the active absorption of Na^+ and the active absorption of K^+. Within the constraints of the experimental design, the active transport of H^+ is not coupled to any of the other transmucosal transport processes and may be quite independent of them.

Further observations indicate that the carbonic anhydrase inhibitor, methazolamide, in extremely high concentration, 10 mM, does not abolish the short-circuit current as it does in the isolated frog gastric mucosa.

Exogenous lactate can serve as a substrate for support of H^+ secretion and the generation of the short-circuit current (Isc), although with the concentra-

Department of Physiology and Biophysics, University of Iowa. Iowa City, Iowa 52240, U. S. A.
Supported by Grant AM 05848 from the National Institutes of Health.

»Transport Mechanisms in Epithelia«, Munksgaard, Copenhagen.

tions used it was less effective than glucose for H^+ secretion. This finding is in contrast with the report by others that exogenous pyruvate was essentially ineffective (Sernka & Harris 1972).

In the course of improving the viability of the isolated mucosa, we conducted a study of the temperature dependence of H^+ secretion, the generation of Isc and the preservation of the mucosal resistance (R). Some evidence is provided that cortexone acetate (DOCA, deoxycorticosterone) might help to sustain ion transport. A previous report showed that histamine failed to enhance the spontaneous secretion of H^+ (Sernka 1969); our studies showed the cholinergic agent carbachol to be similarly ineffective.

METHODS

Male white rats weighing 300–350 grams (suppliers: Simonsen, Minn. or Sasco Co.) were anesthetized with ether or halothane (the latter for experiments of Tables V–VII and Fig. 3). Stomachs were removed and separated along the greater and lesser curvatures into reasonably symmetrical halves (McCabe et al. 1969). Serosal muscle and connective tissue were removed by scissors and forceps from the gastric body, the mid-glandular portion of which has the secreting parietal cells. These mucosal halves were mounted in identical lucite flux chambers each having an aperture of 1 cm². Approximately 20 minutes elapsed between anesthesia and mounting the dissected mucosae. Except as noted subsequently, temperature was maintained at 36° C, by thermostatically controlled heat exchanger plates.

Both surfaces of the mucosa were bathed by 12 ml of reasonably iso-osmotic saline solutions approximating 300 milliosmols. They were exposed to the control solutions for an hour to reach a quasi-stable potential difference (PD) before definitive observations were obtained. Table I provides the composition of the normal or control solutions as well as those used in ion substitution experiments.

For the experiments presented in Table III, the mucosal solutions were removed hourly and titrated to pH 7.0 at room temperature to determine the rate of H^+ secretion. In other experiments, the mucosal solution was stat titrated to pH 5.6 by approximately 25 mM NaOH while being gassed by 100 % O_2.

Table I. Composition of Bathing Solutions

	Serosal Solutions			Mucosal Solutions					
	Control	Na$_2$SO$_4$	NaCH$_3$SO$_4$	Control	Choline Cl	Na$_2$SO$_4$	Choline SO$_4$	Choline SO$_4$ s̄ K$^+$	Choline CH$_3$SO$_4$
Na$^+$	153.5	240.0	141.0	152.5	–	244.0	–	–	–
K$^+$	5.0	5.0	5.0	5.0	5.0	5.0	5.0	–	–
Ca^{++}	2.5	2.5	2.5	2.5	2.5	2.5	2.5	2.5	2.5
Mg^{++}	1.0	1.0	1.0	1.0	1.0	1.0	1.0	–	–
Choline$^+$	–	–	–	–	152.5	–	204.0	211.0	154.5
Cl$^-$	151.0	–	–	161.0	161.0	–	–	–	–
HCO$_3^-$	10.0	10.0	10.0	–	–	–	–	–	–
HPO$_4^-$	1.0	1.0	1.0	–	–	–	–	–	–
Lactate$^-$		2.5	2.5	–	–	2.5	2.5	2.5	2.5
SO$_4^-$	–	235.0	6.0	–	–	250.0	210.0	211.0	–
CH$_3$SO$_4^-$	–	–	130.0	–	–	–	–	–	154.5
Osmolality milliosmols s̄ glucose	289	293	251	289	287	293	298	–	–

Serosal solutions gassed by 5 % CO_2, 95 % O_2; mucosal by 100 % O_2. Glucose, 28 mM, was added to all solutions.

For the experiments cited in Table III, in order to approach isotopic steady state, radioisotopes were added 30 minutes before the first flux period. Transmucosal flux has been designated by M.

Except for experiments presented in Fig. 1, the mucosae were voltage-clamped to the short-circuit state. The values given for the short-circuit current (Isc) and, except for the experiments of Table VII, mucosal conductance (G) have not been corrected for the interbridge resistance of 41 ± 6 $\Omega.cm^2$ between the potential sensing bridge and calomel cell pair. The correction is not inconsiderable, given the high mucosal conductances encountered.

Ionic movements and Isc have been given a negative sign when they correspond to electron flow from the serosal to mucosal surface, or anion movement in the same or cation movement in the opposite direction. A positive sign designates the reverse flow. Otherwise procedures are those which have been previously reported (Hogben 1972, Sernka & Hogben 1969).

Influence of Temperature

Although it might seem obvious to some, the selection of 36° C to maintain the isolated mucosa might not have been the best choice. We followed the H^+ secretion, Isc and G of mucosae held at temperatures ranging from 25–36° C, Table II. In this table, we give both our consolidated data for each temperature step as well as the paired differences of smaller numbers.

The overall data suggest that 33° C might be better than 36° C in favoring H^+ and Isc, with a comparable decline of mucosal resistance at either temperature. However, our experience is that rats taken at different times may differ much more than would be predicted from the variance of a group. We do not have any explanation for the group to group variation. Consequently, it would be improper to conclude which temperature between 30–36° C is optimal.

Paired analysis suggests that H^+ secretion is just barely significantly greater at 36° C than at 33° C and the higher value for Isc is not statistically significant. It would be meaningless to analyze activation energies for H^+ secretion and generation of the Isc in a system as complex as the mucosa.

Table II. Effect of Temperature

	25° C	28° C	30° C $\mu Eq.cm^{-2}hr^{-1}$	33° C	36° C
H^+	1.51±.06	2.97±.01	2.83±.04	3.12±.14	2.79±.14
% 135–180'/0–45'	99	101	84	94	78
I_{sc}	−4.20±.06	−6.73±.24	−7.20±.12	−9.01±.31	−8.32±.16
% 135–180'/0–45'	100	104	85	79	77
n	9	12	22	19	19
Terminal R $\Omega.cm^2$	129±9	133±5	128±6	106±4	102±4
		$\bar{X}$±SE			

Comparison of Paired Differences

	28–25 °C	30–28° C	33–30° C	36–33° C
$\triangle H^+$	+1.21±.21	−0.12±.17	+1.24±.49	+0.46±.20
p	<0.005	>0.5	<0.1	<0.5
$\triangle I_{sc}$	+1.62±.67	−0.28±.86	+3.34±.69	+0.67±.72
p	<0.1	>0.5	<0.05	>0.2
n	6,6	6,6	6,6	13,13

For $\triangle I_{sc}$ a positive value signifies that the absolute magnitude was higher at the higher temperature.

Cortexone acetate

In a blind search for agents which might improve the viability of the mucosal processes, we tested, among other agents, cortexone acetate (DOCA, deoxy-corticosterone). DOCA was added in 50 μl of ethyl alcohol to the 12 ml of saline in the serosal compartment to a final concentration of 3×10^{-5} M. This exceeds the aqueous solubility of DOCA, but we do not know the extent of tissue uptake and binding. Paired analysis of one group indicated that the following values were increased in the presence of DOCA: H^+ secretion $+ 0.54 \pm .23$ μEq.cm^{-2}hr^{-1}, $<.05$; Isc $+1.00 \pm .55$ μEq.cm^{-2}hr^{-1}, $<.2$; terminal G $+1.25 \pm .52$ mmhos.cm^{-2}, $<.05$ (mean difference, its SE and p from Student's t- test, $n = 7$). Thus there is a marginal improvement in H^+ secretion and a disadvantageous increase in G.

The trend from this smaller number of paired analyses might be considered to be supported by data from the control groups with and without DOCA given in Table V, but again it should be stressed that in spite of larger numbers, comparisons between groups of rats at different times should be approached with caution.

Carbachol

In a search for an agent that might stimulate H^+ secretion above the level of spontaneous secretion, we endeavored to stimulate the isolated mucosa with the cholinergic agent carbachol [(2-hydroxyethyl) trimethyl ammonium carbamate]. With a concentration of 10 μM in the serosal solution, the rate of H^+ secretion was $0.46 \pm .07$ μEq.cm^{-2}hr^{-1}, $n = 12$.

RESULTS AND DISCUSSION

Origin of the Short-Circuit Current

The previous attempt to elucidate the short-circuit current generated by the isolated rat gastric mucosa was not entirely successful (Sernka & Hogben 1969). A comparable difficulty was encountered earlier by those studying the isolated guinea pig stomach (Shoemaker *et al.* 1966).

It was clear that the isolated rat gastric mucosa actively transported Cl$^-$ and Na$^+$ as well as secreting H^+. The failure to achieve a balance between the transport of these ions and the short-circuit current did lead to the

Table III. Origin of the Short-Circuit Current

	S→M Hour 1	2	3	$\overline{X}$	M→S Hour 1	2	3	$\overline{X}$	Δ Hour 1	2	3	$\overline{X}$
μEq.cm^{-2}hr^{-1}												
MCl$^-$	−15.8	−15.2	−14.9	−15.28±.25	10.7	10.6	10.3	10.54±.12	−5.0	−4.6	−4.7	−4.74±.42
MNa$^+$	2.4	2.7	3.0	2.69±.10	−6.1	−6.1	−6.2	−6.13±.03	−3.6	−3.4	−3.2	−3.44±.44
$\triangle$H$^+$	1.7	1.6	1.3	1.55±.14	1.6	1.4	1.0	1.34±.26	1.7	1.5	1.1	1.44±.15
I$_{sc}$	−6.8	−6.7	−7.1	−6.86±.26	−7.7	−7.7	−7.9	−7.77±.49	−7.2	−7.2	−7.5	−7.31±.30
I$_{sc}$-Σ									−0.19	−0.72	−0.82	−0.58±.59

	Initial R Ω.cm^2	Terminal PD mv	Terminal G mmhos.cm^{-2}	Wet weight mg.cm^{-2}
S→M	139±9	22±1	8.66±.30	73±5
M→S	135±5	25±2	8.63±.13	77±5

X±SE, n=6

$I_{sc} - \Sigma = I_{sc}$ (Short-circuit current) − algebraic sum of $\triangle M^{Cl^-}$, $\triangle M^{Na^+}$ & $\triangle H^+$

important and unequivocal demonstration of an active absorption of K^+. However the balance sheet achieved was ΔM^{Cl^-} -1.8, ΔM^{Na^+} -1.5, H^+ $+2.7$, ΔM^{K^+} -1.1 and an Isc of -5.1 $\mu Eq.cm^{-2}hr^{-1}$, with a resultant discrepancy of -1.7 (Sernka & Hogben 1969).

In reaching such a balance, one deals with errors compounded by the circumstance that the net fluxes are obtained from differences between the unidirectional fluxes measured across different mucosae which may not be well matched by other criteria: Isc, H^+ secretion rates, terminal G and wet weight.

In Table III we present a reevaluation, where the latter parameters were reasonably well matched for the mucosae across which the unidirectional fluxes were measured in opposing directions. The discrepancy given in Table III between the algebraic sum of ΔM^{Cl^-}, ΔM^{Na^+} and H^+ related to Isc is reduced to $-0.58 \pm .59$ $\mu Eq.cm^{-2}hr^{-1}$, which while not statistically different from zero is in the expected direction for the unmeasured previously demonstrated active transport of K^+. We conclude that the origin of the Isc generated by the isolated rat gastric mucosae is the algebraic sum of ΔM^{Cl^-}, ΔM^{Na^+}, H^+ and ΔM^{K^+} in order of their relative weight.

Prior to this reevaluation of the origin of the short-circuit current, we would have been forced to look for unlikely net ionic movement(s) or reevaluate the way in which we reach assignment of the appropriate charge transferred by H^+ secretion. As previously noted, the "free" H^+ is substantially less than that recorded by titration of the mucosal solution to pH 7.0 (Sernka & Hogben 1969). The evidence now available leads to the conclusion that the net effect of H^+ secretion and the appearance of a buffer, presumably mucopolysaccharide, is such that the H^+ secretion rate obtained by titration to neutrality is the best estimate of net H^+ transport; the vexing buffer can be considered to be released at the mucosal surface as an electrically neutral component, or, not impossibly, for the most part secreted before we begin definitive observations.

Influence of 5mM H^+ at the Mucosal Surface.

It has come as a surprise that the isolated rat gastric mucosa actively absorbs Na^+ and K^+ to an unexpected degree. The latter remains an unexplored enigma. There is a precedent for the active absorption of Na^+ by the non-secreting canine gastric mucosa in situ (Bornstein et al. 1959). It is also clear that the *in vivo* flux of Na⁺ from lumen to blood is dependent on the intra-luminal $[Na^+]/[H^+]$ ratio (Cope *et al.* 1943, Moll & Code 1962, Code *et al.* 1963). This suggests that *in vivo* there is a carrier-mediated transfer of Na⁺ if not an active absorption for which the H⁺ ion may compete. This may turn out to be more important than an idle laboratory curiosity.

The story of the feedback mechanism for gastrin has been so exciting that we may have been overlooking another local feedback mechanism at the mucosal surface involving this cation pump which might potentially raise the pH locally when appropriate.

Thus for the active absorption of Na⁺ the discrepancy between the findings that have been developed for the *in vitro* mucosa or stomach (Kitahara 1967, Kitahara & Hogben 1968, Kitahara *et al.* 1969, Sernka & Hogben 1969) and what we expect for the secreting stomach *in vivo* may be no more than that as the *in vitro* studies have been conducted the pH at the mucosal surface does not fall much below 4.

The obvious strategy would be to intentionally force the pH lower but the integrity of the isolated mucosa is such that it will not tolerate a very low pH. The prompt deterioration is indicated by a rapid decrease in mucosal resistance. In so far as it has been examined, the evidence supports the contention that a low pH, 3 to 2, at the mucosal surface inhibits the Na⁺ pump (Kitahara *et al.* 1969).

An inference that a pH of 3.8 (and throughout each hour even higher) at the mucosal surface of the isolated rat gastric mucosa also leads to an inhibition of active Na⁺ transport may in retrospect appear not to be established; compare Table 3 (Sernka & Hogben 1969) with Table III of this report. We have found that the isolated rat gastric mucosa will tolerate a pH of approximately 2.5 at its mucosal surface; compare the terminal conductances (G) given in Tables III and IV.

For the experiments of Table IV, at time zero the mucosa was exposed to both radioisotopes and the mucosal surface to a solution similar to the control mucosal solution (Table I) except that it had 5 mM H⁺ and the [Na⁺] correspondingly reduced by an inconsequential 5 mM. It should be recognized that for the first 45′ period, we may not have reached the necessary quasi-steady state for isotopic flux.

Nevertheless, we must accept the implication as one reads the row $M^{Na^+}\triangle$ that the effect of 5 mM H⁺ is not prompt, as one would expect for simple competition for a carrier, unless the critical carrier is located at the serosal

Table IV. Exposure of Mucosal Surface to 5mM H^+

		1	2	$\mu Eq.cm^{-2}hr^{-1}$ 3	4	$\overline{X}$
M^{Cl^-}	S→M	18.1 ±1.0	16.0 ±.8	14.6 ±.6	14.9 ±1.0	15.9 ±.7
	M→S	13.3 ± .9	12.2 ±.9	11.2 ±.9	10.6 ± .7	11.6 ±.8
	△	−4.9	−3.9	−3.4	−4.3	−4.1
M^{Na^+}	S→M	2.23±.27	2.52±.22	3.21±.34	4.61±.84	3.14±.41
	M→S	4.04±.26	3.82±.25	3.58±.21	3.58±.21	3.75±.22
	△	−2.21	−1.30	−0.37	⁺1.03	−0.64
I_{sc}	(S→M)	−7.5 ±.5	−6.4 ±.5	−5.4 ±.4	−4.8 ±.4	−6.0 ±.4
	(M→S)	−8.9 ±.5	−7.7 ±.6	−6.8 ±.6	−6.2 ±.6	−7.4 ±.6
	$\overline{X}$	−8.1	−7.1	−6.1	−5.5	−6.7

	Initial G	Terminal G	Initial PD	Terminal PD
	$mmhos.cm^{-2}$		mv	
(S→M)	6.9 ±.6	8.6 ±.4	29±3	12±1
(M→S)	6.6 ±.4	8.5 ±.3	33±2	18±2

$\overline{X}\pm SE$, n = 11, 45′ Periods

membrane of the epithelium. These reservations having been stated, it is apparent from Tables III and IV, that exposure of the mucosal surface to a pH of 2.5 does lead to an inhibition of the active absorption of Na^+ without having an important effect on the active secretion of Cl^-.

Independence of H^+ Secretion on Other Transmucosal Active

Transfers

We have so far identified four major transmucosal net active transport systems: Cl^-, Na^+, H^+ and K^+. There is a precedent for obligatory coupling an epithelial transport of an ion pair (Diamond 1962). For the isolated gastric mucosa, in the case of one species of frog, H^+ and Cl^- can be dissociated though the rate of H^+ secretion in the absence of Cl^- is substantially lower (Heinz & Durbin 1959, Forte *et al.* 1963, Rehm 1967) while transport is exceedingly tightly coupled in the dogfish (Hogben 1967a, Hogben *et al.* 1972).

By ion substitution in the solutions bathing the mucosa we have sought to determine the dependence of H^+ secretion on the transport of the other ions. The results are presented in Table V. The results given in Table V are amplified by Table VI because the solutions chosen to bathe the mucosa cannot be considered to be innocuous.

The dependence of H^+ secretion upon Na^+ absorption was the most thoroughly studied. In the paired experiments given in the first two rows of Table V, H^+ secretion was unimpaired by the relatively innocuous procedure, Table VI. Though only four of the six 45′ periods are presented in Tables V and VI, Student's test for 4.5 hours yielded for the control-choline Cl minus control-control: $\triangle H^+$ $+0.15 \pm .13$, $<.2$; $\triangle$Isc $-1.77 \pm .37$, $<.005$; $X \pm SE$, p; n = 7. To achieve substitution the mucosal solution was changed from the control to choline Cl twice at −5 and 0 minutes.

This series of experiments was backed up by an earlier series where the mucosal surface was exposed to either control or choline Cl alternately for four 45′ periods (7 mucosae were started with control and 7 with choline Cl to minimize any time-dependent effects). Again the paired differences, control choline Cl minus control-control, were: $\triangle H^+$ $+0.04 \pm .19$, >4.5; $\triangle$Isc $-1.83 \pm .34$, $<.001$; as above.

In the first series the exclusion of Na^+ at the mucosal surface was virtually complete; the second series has a better statistical design. In both series, it may be that H^+ secretion was greater after removal of Na^+ from the serosal surface. The Isc decreased as expected, though not to the degree perhaps expected from Table III where the $M^{Na^+}M{\rightarrow}S$ was -6.1. But perhaps energy was diverted to enhance active transport of either Cl^- and/or K^+. It is clear that H^+ secretion is not dependent on a significant Na^+ absorption from the mucosal surface.

We cannot exclude catalysis of H^+ secretion by a trace of Na^+ at the mucosal surface nor can we exclude Na^+ transport from a micro-environment, an unstirred layer or a region of confined diffusion sequestered from the bulk solution bathing the mucosal surface. At the moment there would appear to be little profit in exploring either question.

It might be noted in passing that even though there is a marginal (an order of magnitude less) transmucosal active absorption of Na^+ by the frog gastric mucosa (Kitahara et al. 1969, Sernka 1969, Flemström & Öbrink 1972), in a fraction of cases the frog gastric mucosa will continue to secrete acid in spite of the absence of Na^+ at both surfaces (Davenport 1957, Sachs et al. 1966). The isolated dogfish gastric mucosa secretes H^+ surprisingly well in the absence of Na^+ at both surfaces (Hogben et al. 1972).

Table V. Ionic Substitution

Serosal	Mucosal	H^+	I_{sc}	$\Sigma(1)$	
	Solutions		$\mu Eq.cm^{-2}hr^{-1}$		
Control	Control (2)	1.87±.30	-5.33±.62	-7.20±.89	7
Control	Choline Cl	1.94±.28	-3.19±.42	-5.13±.67	7
Control	Control (3)	2.46±.18	-7.00±.30	-9.45±.29	21
Na_2SO_4	Na_2SO_4	1.13±.12	-2.84±.37	-3.98±.35	11
Na_2SO_4	Choline SO_4	1.34±.14	-0.15±.22	-1.49±.32	12
Na_2SO_4	Choline SO_4 s̄ K^+	0.58±.11	-0.37±.29	-0.94±.35	11
$NaCH_3SO_4$	Na_2SO_4	1.96±.19	-4.76±.53	-6.72±.63	11
$NaCH_3SO_4$	Choline CH_3SO_4	1.66±.25	+0.17±.22	-1.49±.39	11

$\bar{x}±SE$, Values integrated over 3 hours

(1) Σ = unidentified ionic currents.
(2)   Paired control with mucosal choline Cl. No deoxycorticosterone.
(3)   From 3 groups (n=9,6,6). Deoxycorticosterone $3 \times 10^{-5}M$ in serosal solution.

Table VI. Time Dependence of Variables Presented in Table V

Serosal	Mucosal	H^+	I_{sc}	Σ
	Solutions		% Change from 22 to 158 Minutes	
Control	Control	71	68	69
Control	Choline Cl	63	62	62
Control	Control	86	76	72
Na_2SO_4	Na_2SO_4	62	83	77
Na_2SO_4	Choline SO_4	93	ns	92
Na_2SO_4	Choline SO_4 s̄ K^+	178	88	134
$NaCH_3SO_4$	Na_2SO_4	125	81	92
$NaCH_3SO_4$	Choline CH_3SO_4	109	12	162

Means of the ratio: (Period 135–180')/(Period 0–45')
No change of sign except Na_2SO_4/Choline SO_4 I_{sc} which
converted from –0.02 to +0.01.

Others have developed comparable information (Linde *et al.*, 1947, Rehm 1953, Moody & Durbin 1965) that H^+ secretion by the intact mammal is independent of the presence of Na^+ in the solution bathing the mucosal surface. But as in the experiments reported in this paper, one cannot rigorously exclude a leak of Na^+ from serosa to mucosa which is then recycled actively.

For the dogfish (Hogben 1971) and the frog (Sachs *et al.,* 1966), Na^+ was removed from solutions at both surfaces.

In the subsequent experiments summarized in Table V, the next important ion removed was Cl^- and it was removed from both surfaces of the mucosa. For the experiments to be considered, the substitute solution was replaced at the serosal surface four times, at -20, -15, -10 and 0 minutes; the mucosal solution twice, at -20 and -15 minutes. Unlike the previous maneuver, these drastic changes in the environment of the mucosa cannot be considered relatively innocuous, Table VI terminal G.

Unlike the previous carefully paired studies, our reference is a set of studies conducted over the span of the experiments. With the foregoing reservations in mind, elimination of extracellular Cl^- might be considered to be not terribly important; i.e. Na_2SO_4-Na_2SO_4. The additional step of removing Na^+ at the mucosal surface, Na_2SO_4-choline SO_4, did not lead to a further reduction of H^+ secretion. The further step of eliminating mucosal K^+, Na_2SO_4-choline SO_4 s̄ K^+, did substantially reduce H^+ secretion, but we may have pushed the mucosa beyond reasonable limits.

The use of methyl sulfate rather than sulfate as a substitute for chloride appears in the last two rows of Table V. There is less indication of suppression of H^+ secretion from removal of Cl^-; i.e. $NaCH_3SO_4$-Na_2SO_4. The further removal of Na^+ and K^+ from the mucosal bathing solution, $NaCH_3SO_4$-choline CH_3SO_4, leads to a modest depression of H^+ secretion but it is not impressive. Undoubtedly K^+ was leaking from cells of the epithelium.

We reach the conclusion that the secretion of H^+ is substantially if not essentially independent of the other concomitant transmucosal transport mechanisms of Cl^-, Na^+ and K^+. This should be considered in the framework of a complex epithelium. It consists of at least five cell types in a unicellular layer: the parietal or oxyntic cell, believed to be the major site of the active transport of H^+ and Cl^-; the surface epithelial cell; the chief cell; the unhappily named mucous neck cell and the amine-storing cell. There is no clue to suggest which cell(s) is(are) involved in the active transport of Na^+ and K^+.

To partially close the circle, we sought to eliminate the possibility that $SO_4^=$ is actively transported when it has been substituted for Cl^-. In Table VII the fluxes of $SO_4^=$ are shown indicating that there is not a significant net transport from serosa to mucosa, replacing the role of Cl-. In fact when the fluxes are normalized with respect to membrane conductance any appa-

Table VII. Flux of Sulfate

S→M		Hours			
	1	2	3	4	$\bar{X}$
$MSO_4^=$ $\mu Eq.cm^{-2}hr^{-1}$	3.40±.39	5.43±.51	6.88±.65	7.83±.88	5.89±.56
$MSO_4^=$ /G	0.24	0.23	0.26	0.36	0.28
G mmhos.cm^{-2}	14.11	23.93	25.98	21.87	20.81
$I_{sc}\mu Eq.cm^{-2}hr^{-1}$	-2.69±.35	-1.77±.51	-1.29±.55	-1.05±.64	-1.65±.52
M→S					
$MSO_4^=$ $\mu Eq.cm^{-2}hr^{-1}$	3.85±.40	6.19±.46	7.52±.49	7.79±.59	6.34±.45
$MSO_4^=$ /G	0.26	0.24	0.27	0.21	0.25
G mmhos.cm^{-2}	14.69	25.64	27.51	37.32	24.98
$I_{sc}\mu Eq.cm^{-2}hr^{-1}$	-2.45±.40	1.53±.60	-1.33±.65	1.34±.67	-1.66±.57

$\bar{x}\pm SE$, n=16

Both surfaces exposed to *serosal* Na_2SO_4 solution gassed by 5% CO_2, 95% O_2, for 50 minutes prior to first hourly flux period. Added $^{35}SO_4$ 60 minutes before the first hour. G measured at midpoint of each period and corrected for the mean interbridge resistance of 41 $\Omega.cm^2$.

rent difference disappears; it would appear that most of the flux of $SO_4^=$ is passive given the time course of $MSO_4^=$ and G. This was not an inconsequential study. Even though there is no indication for a significant transport of $SO_4^=$ by the mucosa of *Rana pipiens* (Kaneko-Mohammed & Hogben 1964) there is a modest secretion of $SO_4^=$ by *Rana catesbiana* (Hogben 1961, Hogben 1965). Prior study (Sernka 1969) gives comparable results for $SO_4^=$ flux using 14.6 mEq $SO_4^=$ and a high concentration of Cl$^-$ that could be expected to competitively supplant any active transport of $SO_4^=$ had it been present (Hogben 1965).

Methazolamide

There is a mysterious action of carbonic anhydrase inhibitors on that fraction of Cl$^-$ transport of the isolated gastric mucosa that has been labelled electromotive (Hogben 1955, Durbin & Heinz 1958, Kitahara & Imamura 1966, Hogben 1967b). The mystery is the extraordinary concentration of inhibitor, 10 mM, required to elicit an effect. Fig. 1 displays that a comparable effect was not achieved with 10 mM methazolamide exposed to the isolated rat gastric mucosa, even when the component of the Isc represented by Na$^+$

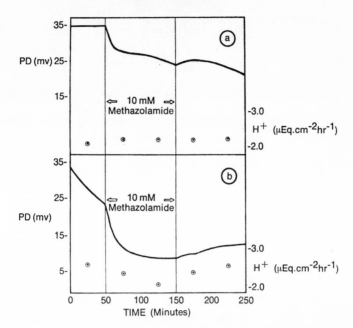

Fig. 1. The mean spontaneous PD is given as a function of time when the serosal surface of the mucosa is exposed to control saline containing 10 mM methazolamide for 100 minutes; continuous lines. The mean H^+ secretory rates, every 50 minutes, are designated by circles. (a) The mucosal surface was exposed to control mucosal saline throughout; $n = 16$. (b) At time zero the control saline at the mucosal surface was replaced by choline Cl saline (Table I) and continued throughout the remainder of the experiment; $n = 9$. (For comparison with the frog gastric mucosa see Hogben 1967b).

was presumably eliminated by bathing the mucosal surface by the choline Cl solution.

For a series in which the mucosa was exposed to 1 mM methazolamide, there was no discernible effect on the transmucosal PD. The reader should be aware that from our studies of *Rana catesbiana* the mystery is compounded because there appears to be an "escape phenomenon" with the passage of time, and that provision of CO_2, HCO_3^- at the mucosal surface diminishes inhibition (Hogben 1955, Hogben 1967b).

Support of H^+ Secretion by Exogenous Lactate

Recent work (Sernka & Harris 1972) implicates the pentose shunt as a major pathway for mobilizing energy for the secretion of H^+ by the isolated

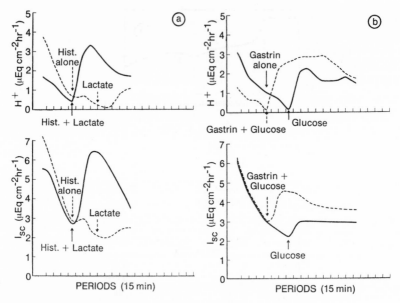

Fig. 2. The time course of the H^+ ion secreted and the Isc developed by the isolated canine gastric mucosa initially deprived of exogenous substrate. At times indicated by arrows either substrate, or substrate plus a stimulus for H^+ secretion was added. The conditions for study of the isolated canine gastric mucosa were very similar to those described here for the rat mucosa. For each substrate, curves are for a portion of mucosa from the same dog. (a) Exogenous substrate 14 mM Na lactate. (b) Exogenous substrate 5.6 mM glucose. (From Kitahara & Hogben 1968; previously unpublished figure).

gastric mucosa. They were unable to sustain H^+ secretion with exogenous pyruvate. This failure did not seem to be in accord with prior experience (Davenport & Chavré 1952, Kitahara & Hogben 1968). The latter authors were led to believe that exogenous lactate was a reasonable substrate for supporting H^+ secretion, Fig. 2.

Our current evaluation leads to the conclusion that exogenous lactate is modestly capable of supporting H^+ secretion, Fig. 3. These experiments are backed up by a prior series where the serosal surface was bathed by a control solution with 10 mM lactate, while the customary control mucosal solution had 28 mM glucose.

It should be noted that the environment of the mucosae studied by Sernka and Harris and ourselves differed in some respects. We deliberately chose a relative metabolic acidosis for our serosal extracellular solution, with

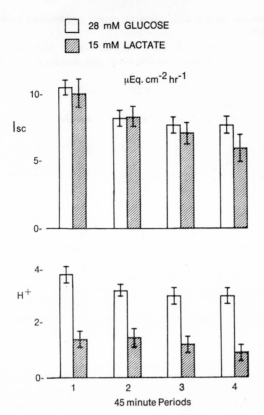

Fig. 3. A paired comparison of the mean H^+ secretory rates and the Isc developed by the isolated rat gastric mucosa when one half was exposed to glucose control salines and the other half to lactate salines. The serosal and mucosal lactate salines were identical to the control salines but for the addition of 15 mM Na lactate and the omission of 28 mM glucose. Error bars signify standard errors; n = 6.

the usual HCO_3^-/CO_2 buffer pair. In some of our experiments we continued to employ DOCA but there is nothing to suggest that this difference would be a profitable line of study. The interested reader might wish to take note of a letter in reference to the paper of Sernka and Harris (Katz 1972).

REFERENCES

Bornstein, A. M., Dennis, W. H., & Rehm, W. S. (1959) Movement of water, sodium, chloride and hydrogen ions across the resting stomach. *Amer. J. Physiol. 197*, 332–336.

Code, C. F., Higgins, J. A., Moll, J. C., Orvis, A. L., & Scholer, J. F. (1963) The influence of acid on the gastric absorption of water, sodium and potassium. *J. Physiol. (Lond.) 166,* 110–119.

Cope, O., Cohn, W. E. & Brenizer, A. G., Jr. (1943) Gastric secretion. II. Absorption of radioactive sodium from pouches of the body and antrum of the stomach of the dog. *J. clin. Invest. 22,* 103–110.

Davenport, H. W. (1957) Ion requirements for gastric acid secretion. *Proc. Soc. exp. Biol. (N. Y.) 95,* 562–564.

Davenport, H. W. & Chavré, V. J. (1952) Evidence that glycolysis contributes to acid secretion. *Amer. J. Physiol. 171,* 1–6.

Diamond, J. M. (1962) The mechanism of solute transport by the gall-bladder. *J. Physiol. (Lond.) 161,* 474–502.

Durbin, R. P. & Heinz, E. (1958) Electromotive chloride transport and gastric acid secretion in the frog. *J. gen. Physiol. 41,* 1035–1047.

Flemström, G. & Öbrink, K. J. (1972) Electrogenic properties of frog gastric mucosa: Effect of ouabain and hypoxia. In *Gastric Secretion,* ed. Sachs, G., Heinz, E. & Ullrich, K. J., pp. 189–200. Academic Press, N. Y. & London.

Forte, J. G·, Adams, P. H. & Davies, R. E. (1963) Source of the gastric mucosal potential difference. *Nature (Lond.) 197,* 874–876.

Heinz, E. & Durbin, R. P. (1959) Evidence for an independent hydrogen-ion pump in the stomach. *Biochem. biophys. Acta (Amst.) 31,* 246–247.

Hogben, C. A. M. (1955) Biological aspects of active chloride transport. In *Electrolytes in Biological Systems,* ed. Shanes, A. M., pp. 176–204. Amer. Physiol. Soc., Washington D. C.

Hogben, C. A. M. (1961) Active transport of sulfate by frog gastric mucosa. *Fed. Proc. 20,* 139. (Abstract)

Hogben, C. A. M. (1965) Introduction: the natural history of the isolated bullfrog gastric mucosa. *Fed. Proc. 24,* 1353–1359.

Hogben, C. A. M· (1967a) Secretion of acid by the dogfish, *Squalus acanthias.* In *Sharks, Skates and Rays,* ed. Gilbert, P. W., Mathewson, R. F. & Rall, D. P., pp. 299–315. Johns Hopkins Press, Baltimore, Maryland.

Hogben, C. A. M. (1967b) The chloride effect of carbonic anhydrase inhibitors. *Molec. Pharmacol. 3,* 318–326.

Hogben, C. A. M. (1971) Further observations on the response of the dogfish gastric mucosa to cation substitution. *Bull. Mt. Desert Island Biol. Lab. 11,* 36–38.

Hogben, C. A. M. (1972) The reverse "Rehm-effect". In *Gastric Secretion,* ed. Sachs, G., Heinz, E. & Ullrich, K. I., pp. 111–130. Academic Press, N. Y. & London.

Hogben, C. A. M., Brandes, M. & Stavens, B. (1972) The electrophysiological response of the dogfish gastric mucosa to extracellular potassium. *Comp. Biochem. Physiol. 42A,* 153–168.

Kaneko-Mohammed, S. & Hogben, C. A. M. (1964) Ionic fluxes of *Rana pipiens* stomach bathed by sulfate solutions. *Amer. J. Physiol. 207,* 1173–1176.

Katz, J. (1972) Letter to the editor. *Amer. J. Physiol. 223,* 739–740.

Kitahara, S. (1967) Active transport of Na^+ and Cl^- by *in vitro* non-secreting cat gastric mucosa. *Amer. J. Physiol. 213,* 819–823.

Kitahara, S., Fox, K. R. & Hogben, C. A. M. (1969) Acid secretion, Na^+ absorption, and the origin of the potential difference across isolated mammalian stomachs. *Amer. J. dig. Dis. 14,* 221–238.

Kitahara, S. & Hogben, C. A. M. (1968) Dependence of H^+ secretion of isolated canine gastric mucosa on transmucosal potential difference and exogenous substrate. *Biophys. J. 8*, A-128.

Kitahara, S. & Imamura, A. (1966) Chloride transport, acid secretion and carbonic anhydrase activity in frog stomach. *Life Sci, 5*, 215–218.

Linde, S., Teorell, T. & Öbrink, K. J. (1947) Experiments on the primary acidity of the gastric juice. *Acta physiol. scand. 14*, 220–232.

McCabe, D. R., Kent, T. H. & Hogben, C. A. M. (1969) Distribution and weights of various cell types in the rat stomach. *Anat. Rec. 163*, 555–562.

Moll, J. C. & Code, C. F. (1962) Rates of insorption of sodium and potassium from stomachs of rats. *Amer. J. Physiol. 203*, 229–231.

Moody, F. G. & Durbin, R. P. (1965) Effects of glycine and other instillates on concentration of gastric acid. *Amer. J. Physiol. 209*, 122–126.

Rehm, W. S. (1953) Electrical resistance of resting and secreting stomach. *Amer. J. Physiol. 172*, 689–699.

Rehm, W. S. (1967) Ion permeability and electrical resistance of the frog's gastric mucosa. *Fed. Proc. 26*, 1303–1313.

Sachs, G., Shoemaker, R. L. & Hirschowitz, B. I. (1966) Effects of sodium removal on acid secretion by the frog gastric mucosa. *Proc. Soc. Exp. Biol. (N. Y.) 123*, 47–52.

Sernka, T. J. (1969) Ion transport through rat gastric mucosa. Ph. D. thesis, University of Iowa.

Sernka, T. J. & Harris, J. B. (1972) Pentose phosphate shunt and gastric acid secretion in the rat. *Amer. J. Physiol. 222*, 25–32.

Sernka, T. J. & Hogben, C. A. M. (1969) Active ion transport by isolated gastric mucosa of rat and guinea pig. *Amer. J. Physiol. 217*, 1419–1424.

Shoemaker, R. L., Sachs, G. & Hirschowitz, B. I. (1966) Secretion by guinea pig gastric mucosa *in vitro*. *Proc. Soc. Exp. Biol. (N. Y.) 123*, 824–827.

DISCUSSION

KEYNES: I can't make a very complete comment, but I think I ought to report that working in Babraham early this year Trudy & John Forte (Forte *et al.* 1972) developed a new mammalian preparation of isolated gastric mucosa from baby pigs. I don't remember the actual figures, but they were finding, I think, considerably larger hydrogen and chloride fluxes in these isolated mucosae from baby pigs than the sodium movements that you have reported in other mammals. And what they found was that it is also much more readily stimulated by histamine.

HOGBEN: In our experience histamine does not augment the rate of spontaneous acid secretion by the isolated gastric mucosa or stomach wall of several mature mammals.

KEYNES: You get a nice big effect of histamine with this preparation, but only when the baby pigs are not more than about 3 days old. If you try preparations from an animal that is a week old, then it doesn't seem to work nearly as nicely. What they thought might be happening was simply that the mucosae of the pigs were getting much thicker during those first few days, and once this had happened then hypoxia intervened. Whether that conclusion still stands up, I don't know.

TOSTESON: In the case where there is hydrogen ion secretion and no short circuit current, what is the missing ion?

HOGBEN: I don't want to leave the impression that I think that the active absorption of sodium and potassium is totally a laboratory artifact. It may turn out to be one of the most important properties of the stomach. We know of one important feedback mechanism. When the pH in the stomach falls to a low value, a stimulus to further gastric secretion, the release of gastrin ceases. But that doesn't help the cells that are being insulted at the low pH. I hope that we may establish that the cation pump is functionally significant *in vivo* and it is turning over all the time. If the pH is not low, it takes up sodium and/or potassium but as the pH falls, it starts to pick up hydrogen

---

Forte, G. M., Forte, J. G. & Machen, T. E. (1972) Active ion transport by isolated piglet gastric mucosa. *J. Physiol. (Lond.) 226,* 31–32P.

ion instead of sodium-potassium and thus restores pH and protects the gastric mucosa.

ULLRICH: Did you observe an effect of bicarbonate and CO_2 pressure on your preparation?

HOGBEN: We have deliberately chosen to employ a metabolic acidosis, with a bicarbonate concentration of 10 mMol and a pCO_2 of approximately 35 mmHg. We have exposed the serosal or nutrient surface to a metabolic acidosis in the expectation that this favours hydrogen ion secretion, as it does in the intact animal and as we have been shown in the isolated gastric mucosa of the dogfish.

LINDEMANN: What is the meaning of short circuit current or the significance of it when one is dealing with such a convoluted surface as in the stomach?

HOGBEN: If two well stirred solutions are separated by a black box and a current is allowed to flow between these two solutions, the measured current, including specifically the short-circuit current, must be equal to the alge- brance sum of all net ionic movement between the solutions and through the black box. Equating the short-circuit current with the identified net ionic currents is an important if not necessary first step in characterizing the black box. A discrepancy betwen the short-circuit current and identified net ionic currents compels us to search for additional currents.

A major thrust of this symposium has been to emphasize the complex orga- nization of epithelia. At this juncture, it would be rash to approach the gastric black as being less or more complicated than any other epithelium, for instance, the frog skin. The convolution of the single cell layer of the gastric epithelium into tubules is an interesting and important morphologic feature but not necessarily less or more germane than there being several distinct cell types. If net ionic currents flow through the black box in the absence of an electrochemical potential difference, then by the definition I use, these ions are actively transported. Extension of inquiry into the mechanism of active transport does compel us to consider morphological restraints, but not less or necessarily more so than in the study of other epithelia.

KEYNES: I pressed John Forte quite hard to see whether he gets a good balance sheet in his piglet preparation, and I don't think there was a problem. I think the identified fluxes and short circuit current added up satisfactorily.

LEAF: How do you mobilize energy from the pentose phosphate shunt pathway?

HOGBEN: Since in our hands exogenous lactate does support H^+ secretion, I am not compelled to route energy transfer through the pentose shunt.

A Molecular Approach to Epithelial Conductance: Gastric Mucosa

G. Sachs, J. G. Spenney, R. L. Shoemaker & M. C. Goodall

INTRODUCTION

Transport across epithelial tissues can be divided into active and passive components. Biochemically an Na-K (Skou 1957), Ca^{++} (Martonosi 1971), and HCO_3^- ATPase (Tanisawa & Forte 1971) have been extensively studied and implicated in active transport function of various epithelia. The similarities in the hydrolytic site of these membrane bound ATPases – an acyl group, a serine, and an imidazole residue – leads to the conclusion that their functional differences may result from associated subunits. Most recently, conformational change in a part (or whole) of the molecule (Jardetzky 1966) has been suggested as the transport mechanism. In the case of Na-K ATPase assessment of intrinsic or extrinsic fluorescence (Mayer & Avi-Dor 1970), or circular dichroism (Long *et al.* In preparation) have revealed no evidence of conformational change. An alternative is the presence of a carrier or channel subunit whose mobility or structure is a function of ATPase activity. What is even more likely is that a specialized structure is present in the membrane for the passive movement of ions accompanying the actively transported species, for example, the Cl ion which accompanies the secreted H ion. Indeed, in epithelia such as frog skin or toad bladder, there also has to be a route with passive conductance properties for ion transport across the cell membrane.

Clear definition of molecular functions such as these has been possible only since the advent of artificial bilayers as a model system in which they could be studied (Mueller & Rudin 1963). The classical carrier prototype is

Departments of Physiology and Biophysics, Medicine and Division of Neurosciences, University of Alabama in Birmingham, Birmingham, Alabama 35294.
Supported by NIH AM08541, Cancer and Cardiovascular Training and Research Grants; NSF Grant GB31075; and Veterans Administration Hospital, Birmingham, Alabama 35294, and Grant by Smith Kline & French Laboratories.

»Transport Mechanisms in Epithelia«, Munksgaard, Copenhagen.

valinomycin (Lev *et al.* 1966), and channel prototype gramicidin (Hladky & Haydon 1970). It is pertinent to inquire whether this type of molecule exists in epithelial tissues and whether it can be isolated and incorporated in an active form into artificial bilayers. In this context, the intact tissue is a useful starting point for a study of the nature of the conductance pathways present.

Epithelial tissues may be classified firstly as to whether conductance is largely cellular or paracellular (Diamond *et al.* 1971). If conductance is paracellular, we would be tempted to conclude that the properties of this conductance pathway would reflect the anatomic apposition of adjacent cells, rather than the presence of isolatable entities such as channels. Tissues, however, which are tight, a reflection perhaps of the density of Ca^{++} sites on the opposing junctional membrane surfaces, will have electrical properties which are those of the cell membranes rather than those of the junctional regions.

With many assumptions, it is possible then to study the electrical properties of tight tissues in a simple Ussing chamber, and to obtain some idea of the types of molecules contributing to the electrical structure of the tissue. We can classify the types of molecules reponsible for charge transport across membranes as mobile or carrier type and fixed or channel type. In turn, these can be either neutral, such as valinomycin or gramicidin, or have formal charges such as those associated with amine or carboxyl groups. There can be anion or cation selectivity or no charge discrimination. The selectivity series can also give useful information for relating the isolated channels to the channels in the intact tissue.

Thus it is possible to establish the predominant route – cellular vs. paracellular – as well as the nature – mobile vs. fixed and charged vs. neutral – of the sites of ion permeation in the intact tissue. Extraction of the tissue, assessment of activity in artificial bilayers, and purification of activity will enable passive permeability characteristics to achieve the same molecular basis as active transport.

METHODS

A. Paracellular vs. Cellular Conductance:

As a model tight tissue, gastric mucosa would appear to be an appropriate choice, because of high resistance, high P D and extremely high H^+ concentration gradients developed, of the order of 10^4–10^6 depending on species.

However, a more quantitative definition of tightness would be desirable, and this can be done in several ways.

1. Comparison of isolated cell conductance and tissue conductance: This approach depends on the capability of isolation of viable cells from a given tissue and of making electrical measurements on these isolated cells. In the case of *Necturus* gastric mucosa, this has proved possible (Blum *et al.* 1971). Fig. 1 shows a current-voltage plot for isolated cells compared to intact tissue. The 100-fold difference in slope is readily explained by intercellular coupling in this tissue.

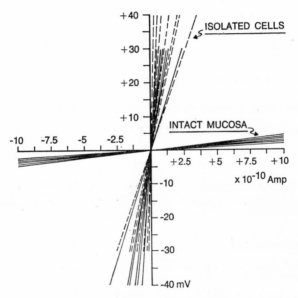

Fig. 1. Current voltage plots obtained by current sending through a microelectrode implanted either in an isolated cell, where no junctional conductance exists, or in similar cells where intercellular junctional conductance is present. This accounts for the differences in slope observed.

Since we can measure the conductance of the isolated cell in $\Omega^{-1}\ cm^{-2}$, and the tissue conductance in similar units, then the shunt conductance is simply derived as $g_{shunt} = g_{tissue} - g_{cells}$.

Using this approach for *Necturus* gastric mucosa, the shunt conductance would appear to be about equal to cellular conductance. This is to be contrasted with a tissue such as gall bladder epithelium, where the ratio is about 20.

17*

2. Direct Measurement in Intact Tissue: An alternate approach to this problem is the use of a 2-electrode system for measuring voltage decay with distance from a current sending electrode. The apparatus used for gastric mucosa is illustrated in Fig. 2. Data derived from this type of experiment are illustrated in Fig. 3 for *Necturus* antral and fundic mucosa.

Considering a disc of tissue of thickness L and radius r with the current source inserted in the center of the disc, then voltage change across a distance $\triangle$r is:

$$(1) \quad V_r = V_{(r+\triangle r)} - V_{(r)} = -IR$$

R is the resistance across $\triangle$r

$$(2) \quad R = (R_t \triangle r) / 2\pi rL$$

R_t = specific horizontal restivity

$$(3) \quad V\triangle r = - (I R_t \triangle r) / (2\pi rL)$$

$$(4) \quad dV/dr = - (R_t/2\pi L) \cdot I \cdot 1/r$$

Taking the second derivative with respect to r:

$$(5) \quad \frac{dV}{dr^2} = - \frac{R_t}{2\Pi L} \cdot \frac{1}{r} \cdot \frac{dI}{dr} + \frac{R_t}{2\Pi L} \cdot I \cdot \frac{1}{r^2}$$

Substituting from Eq. 4 for I:

$$(6) \quad \frac{d^2V}{dr^2} = - \frac{R_t}{2\Pi L} \cdot \frac{1}{r} \cdot \frac{dI}{dr} - \frac{1}{r} \frac{dV}{dr}$$

Finding an expression for dI/dr: $\triangle$ I, the current loss vertically is proportional to the surface area subject to loss and inversely proportional to the resistance of this surfaze (R_z = vertical specific resistivity).

$$(7) \quad \triangle I = - (V/R_z) \cdot 2\pi r \triangle r$$

$$(8) \quad dI/dr = - (2\pi rV) / R_z$$

and substituting into Eq. 6:

$$(9) \quad \frac{d^2V}{dr^2} = \frac{VR_t}{LR_z} - \frac{1}{r} \frac{dV}{dr}$$

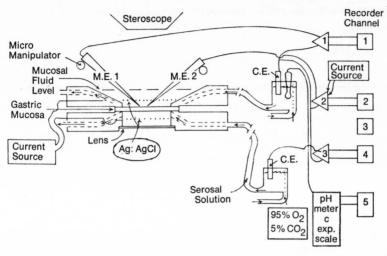

Fig. 2. A diagram of the equipment used to study coupling in intact gastric mucosa. One half of the circuit is used to send and measure current through a microelectrode, the other half to measure the P D changes in a distant cell. Simultaneous measurement of trans-tissue resistance, the luminal/serosal resistance ratio, and pH in the luminal fluid can be carried out.

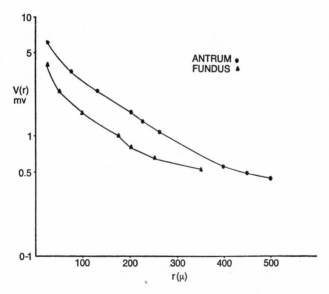

Fig. 3. The decay of voltage with distance from a current-sending microelectrode for fundus and antrum of *Necturus* gastric mucosa. Note the break in the curve for the fundus, due to the presence of a tubule.

letting $\lambda^2 = LR_Z/R_t$ and rearranging:

$$(10)\quad \frac{d^2V}{dr^2} + \frac{1}{r}\frac{dV}{dr} - \frac{1}{\lambda^2}V = 0$$

The solution to this differential equation is a zero order modified Bessel function of the second kind:

$$V = AK_0\,(r/\lambda) \text{ or } V/IK_0\,(r/\lambda) = A \text{ (a constant)}$$

Practically, the solution of the Bessel function is obtained by repetitive expansion of the Bessel function using the experimental data V_r and r and increasing values for λ. The solution is considered that value of λ for which A is most constant for all the data points.

and
$$R_t/L = 2\pi\,A/I_0$$
$$R_Z = \lambda^2\,(R_t/L) = (\lambda^2 2\pi A)\,/\,I_0$$

I_0 is the current injected.

The membrane resistances are derived from consideration of an equivalent electrical circuit:

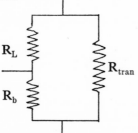

R_L = luminal cell membrane
R_b = basal cell membrane
r_{tran} = trans epithelial resistance

R_L is the positive solution of the quadratic equation:

$$(-\,R_b/R_L)\,R_L^2 + [R_Z + R_Z\,(R_B/R_L) - R_T\,]R_L + R_Z R_T = 0$$
$$R_B \text{ from } (R_B/R_L)\,R_L$$

This approach is discussed in detail for the gall bladder membrane (Frömter 1973).

However, this only holds for a simple 2-dimensional sheet such as gall bladder or antrum. The fundic mucosa is riddled with a series of indentations (i. e. tubules) which will result in an area correction term for the transepithelial resistance (2.7 fold for *Necturus,* 13 fold for dog gastric mucosa). In addition, in the microelectrode experiments, these indentations will provide a complex pathway for current flow. This is a separate case from infolded tissue where simple dimensional corrections will suffice.

A more detailed consideration of the problem shows that only between 1 in 10 or 1 in 20 surface epithelial cells make connection with the tubular indentations. The error introduced, therefore, due to the tubular parallel resistive shunt with both the transverse and vertical R can probably be ignored. The values derived for the surface epithelial system may then be regarded as correct within experimental error.

The tubular system must be considered in transepithelial measurements, and hence a separate estimate will have to be made on this part of the tissue for precise quantitation of shunt conductance. The equation, therefore, for a tissue such as amphibian gastric mucosa becomes:

$$g_{tissue} = g_{SEC} + g_{tub} + g_{shunt} \therefore g_{shunt} = g_{tissue} - g_{SEC} - g_{tub}$$

g tissue is obtained from transepithelial measurements and surface area correction for this value from histological sections, g_{SEC}, the surface epithelial conductance from microelectrode measurements on intact tissue, g_{tubule} from measurements on isolated tubules and area correction again from histological data. The difference between these terms is the shunt conductance.

Table I.

	Space Constant λ $(10^{-4}$ cm)	Trans Resistance Ω cm^2	Cell Membrane Resistance Ω cm^2
Fundus	352.9 ± 211	1882 ± 534	6433 ± 4119
Antrum	511	2254	8,130

Values for space constant, corrected transepithelial resistance, and sum of luminal and serosal membrane resistance of surface cells of fundus (n = 17) and antrum (n = 3).

For *Necturus* gastric mucosa, tubular conductance is at least equal to surface cell conductance per unit area, and is 1.7 times greater because of the area difference. Table I gives the values for the different conductive parameters of the surface epithelial cells and using these data we have:

$$g_{shunt} = 5.31 \times 10^{-4}\Omega^{-1} - 4.2 \times 10^{-4}\Omega^{-1} = 1.11 \times 10^{-4}\Omega^{-1}cm^{-2}$$

a very low shunt value relative to loose epithelia, being 1/5 of cellular conductance.

In the case of the antrum, a region of the stomach, largely devoid of indentations, the problem is comparable to that of gall bladder. Table I summarizes our findings for this tissue. Here, therefore:

$$g_{shunt} = 4.9 \times 10^{-4}\Omega^{-1}cm^{-2} - 1.2 \times 10^{-4}\Omega^{-1}cm^{-2} = 3.7 \times 10^{-4}\Omega^{-1}cm^{-2}$$

still a low value, but higher than that of fundic mucosa. Since this is about thrice that of the cellular conductance, there could therefore be some significant back leak of acid through the junctional region, with effects such as inhibition of gastric secretion.

To amplify these comments, the antrum is the site of release of gastrin. This hormone is situated deep in the tissue and its release is inhibited by acidification of antral contents. Evidently, a possible route for H^+ back diffusion is through the junctional region. An unlikely route is through the cell, since the cell pH would have to fall to low values. This may therefore be a justification for the leakiness of the antral junctions compared to the fundic junctions.

3. Intracellular $\triangle P D$ Measurements: A third approach to the problem of paracellular conductance also uses microelectrodes. A constant product KCl change in the serosal bathing solution of *Necturus* gastric mucosa results in a P D change across the whole tissue, and also across the serosal membrane of the cells. If a shunt conductance is present, then:

$$\triangle P D \text{ tissue} = \triangle P D \text{ cell.} \frac{(g_{cell})}{(g_{cell} + g_{shunt})}$$

so that if the shunt and cell conductance were equal the tissue P D change would only be half that of the cell. From Fig. 4, it can be seen that the cellular and tissue P D changes are almost equal. This results in one of two conclusions: that either there is a low shunt conductance or else the permeability properties of the shunt region are the same as that of the cell membrane. If the latter were so, then altering the ionic concentration on the other side of the tissue should have the same effects, i. e. the tissue should be symmetric. This is not the case in the gastric mucosa, hence again there is little shunt conductance in *Necturus* gastric mucosa.

Summarizing our findings, therefore, the shunt conductance of *Necturus*

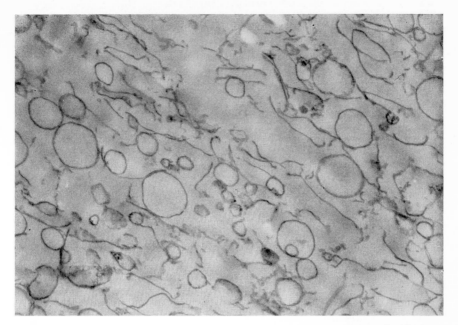

Fig. 9. Electron micrograph of membrane fraction used for channel work. (Courtesy Dr. H. F. Helander). ($\times$ 75,000)

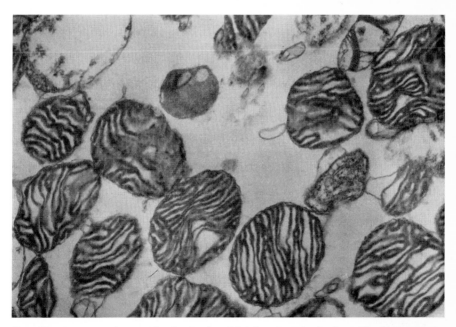

Fig. 10. Electron micrograph of mitochondrial fraction obtained at 40+% sucrose in the zonal rotor, from which channels were not extracted by Triton X-100. (Courtesy Dr. H. F. Helander). ($\times$ 75,000)

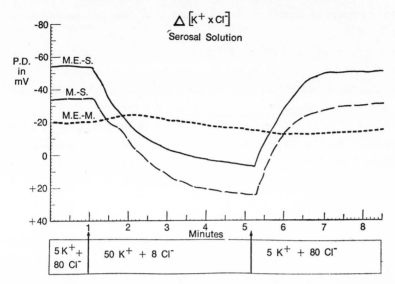

Fig. 4. The effect of a constant product KCl change on the serosal surface of *Necturus* gastric mucosa, on the trans-tissue P D, the P D from cell to serosal bathing solution, and the P D from cell to mucosal solution.

gastric fundic mucosa is almost two orders of magnitude less than that of *Necturus* gall bladder or proximal tubule. In studying the electrical properties of this tissue, therefore, in a simple Ussing chamber, we are dealing with the properties of the cell membranes.

B. Channels and Carriers, Charged and Uncharged:

It has been shown due to the work of Eisenman and his collaborators (Conti & Eisenman 1966; Eisenman *et al.* 1968) that one can distinguish the mechanism whereby ions permeate artificial membranes. Many of these considerations have been applied to the gall bladder in a series of three papers (Barry & Diamond 1971, Wright *et al.* 1971, Barry *et al.* 1971).

The advantage of the gall bladder in these studies is that permeation is controlled by the junctional region, i. e. a single membrane. This is not the case in the gastric mucosa where permeation is controlled by the luminal and basal membranes of a heterogeneous cell population. It is questionable, therefore, whether a simplistic approach can yield any useful information about any ion permeation mechanism other than the dominant one, which on the basis of chemical flux studies is Na^+ for cations, and Cl^- for anions, and on the basis of electrical consideration is K⁺ for cations and Cl⁻ for anions.

Moreover, under usual circumstances, two electrogenic pumps may complicate the picture, for H⁺ and Cl⁻. The use of unstimulated *Necturus* gastric mucosa where the H⁺ rate is zero, bathed in SO_4'' solutions where there is no anion pump gets around these two problems. With all these reservations in mind, we carried out a few simple experiments to attempt to determine the characteristics of the gastric membrane which might aid us in ebstablishing the significance of any of the materials we might succeed in isolating and incorporating in to an artificial bilayer.

1. Carrier vs Channel: From Fig. 5, the current-voltage plot for *Necturus* gastric mucosa is linear over a considerable range. This argues against a carrier system being the major current carrying mechanism in this tissue, since free carrier depletion would result in non-linearity. This finding also suggests that the rate limiting step is not at the solution-membrane interface, since here too we would expect a saturation phenomenon. Thus this finding can be taken as tentative evidence that we are dealing with fixed sites or channels. It must be pointed out that carrier systems may still be present as a minor component of the conductance system.

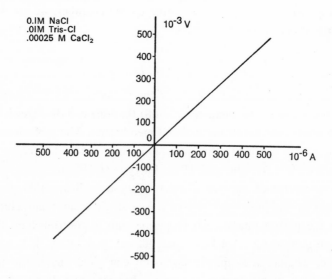

Fig. 5. Current-voltage plot obtained for *Necturus* gastric mucosa by recording voltage change obtained after 100 msec. current pulse showing linearity up to 500 mv.

2. *Charged vs Neutral Sites:* In this type of experiment the argument is that charged sites are required to satisfy electroneutrality, hence will always have the counter ion associated. Accordingly the conductance will be independent of concentration until this is high enough for co-ion to be forced into the membrane when the conductance will increase; in contrast a neutral fixed site will show initially a linear conductance concentration relationship with saturation at high concentrations.

Our findings for the *Necturus* fundic mucosa are illustrated in Fig. 6 for NaCl or LiCl, showing that gNa $>$ gLi, as found for biionic potentials and also that tissue conductance is a linear function of concentration. We can then conclude that this tissue contains neutral fixed sites or channels.

3. *Selectivity:* Fig. 7 summarizes the PD changes obtained for 10-fold changes in K^+, Na^+ and Cl^-. Both anion and cation sites are present. From other experiments the selectivity sequence for this tissue is $K^+ > Rb^+ > Na^+ > Li^+$, Cs^+ for cations and $Br^- > Cl^- > I^- > SO_4^=$ for anions, different from the free solution sequence.

Section Summary: We therefore have tentative evidence for both anion and cation selective channels present in gastric mucosa, and have some know-

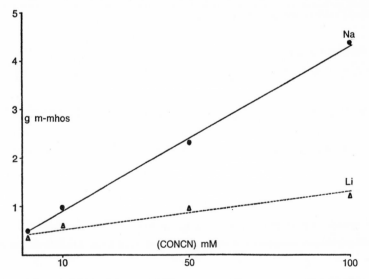

Fig. 6. A plot of the conductance of *Necturus* gastric mucosa against varying concentrations of NaCl and LiCl, showing good linearity over the concentration range studied.

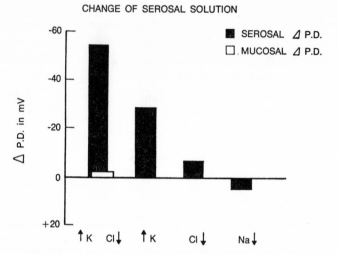

Fig. 7. A summary of the P D changes obtained across the serosal membrane of Necturus gastric mucosa in response to 10-fold changes in K^+, Cl^- and Na^+, and constant
product KCl changes.

ledge of their selectivity. An additional feature of this tissue that may be of
importance is the effect of acetyl choline (Shoemaker *et al.* 1970), which
may be interpreted as being due to the opening of nonselective channels in
the serosal membrane of the gastric cell.

C. *Isolation of Channels:*

Since we are interested in transport in the cell membrane and since other
subcellular particles may contain ion translocators, a prerequisite for further
progress is purified plasma membrane free of other subcellular fragments.
Based on our experience with purification of the gastric HCO_3^- stimulated
ATPase (Sachs *et al.* 1972), we purified the plasma membrane fraction in
a Ti XIV zonal rotor as shown in Fig. 8. Fig. 9 shows the ultra-structural
appearance of the membrane fraction and Fig. 10 of the mitochondrial fraction, see figure insert opposite page 264.

The membrane fraction was solubilized with Triton X-100, the soluble
100,000 g Triton supernatant passed over a sephadex G-200 column and
studied further in an artificial bilayer system. Indeed, this fraction induced
step changes in conductance; 3 types of channels initially were distinguishable
on the basis of their size or charge discrimination.

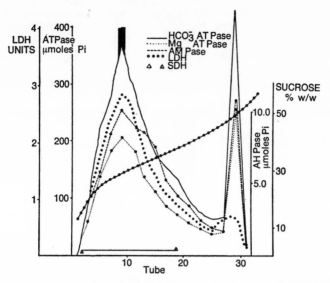

Fig. 8. Zonal pattern of 100,000g precipitate (post 5000g), showing activities of succinic dehydrogenase (SDH), lactic dehydrogenase (LDH), ATPase, HCO₃⁻ ATPase. The shaded area represents the tube used for the channel extraction.

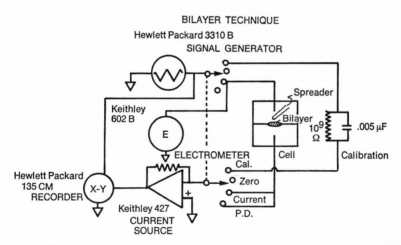

Fig. 11. Equipment used to study step conductance changes in an experimental bilayer, consisting of an X-Y recorder, a voltage source and current amplifier. An electrometer may be used in this set-up to measure 0 current voltage and provision is made for conductance and capacitance calibration.

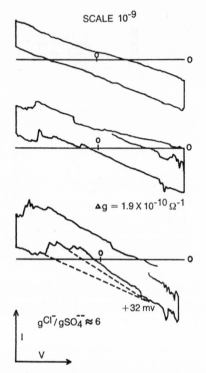

Fig. 12. The effect of addition of membrane material on the bilayer conductance. The upper figure shows the control condition. The middle figure shows the events occurring with addition of the extract, with step changes of the size (g) noted in the figures. The lowest portion shows successive events occurring with an intercept on the lines occurring at 32 mv in SO_4'':Cl^- biionic solutions showing that $gCl' > gSO_4''$.

Further fractionation on polyacrylamide gels allowed separation of at least 3 channel types corresponding to discrete protein bands; they are an anion selective channel, a cation channel and a non-selective channel.

D. *Bilayer Studies:*

The electrical method used to detect channel formation was a voltage current sweep across a bilayer formed from soy bean lecithin dissolved in decane as illustrated in Fig. 11.

Fig. 12 shows the events obtained when the gastric preparation was added to one side of the system. With this technique the slope change immediately gives us the size (conductance value) of the channel, and the intersection of

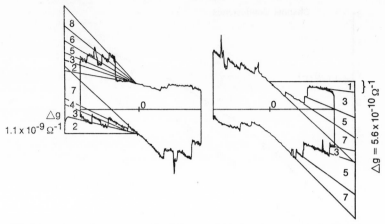

Fig. 13. Two successive sweeps of a bilayer which has incorporated the anion selective channels. This illustrates successive opening or closing of channels with increasing voltage and that the size of the steps can be related to a single unit conductance $g = 5.6 \times 10^{-10} \Omega^{-1}$.

the slopes in asymmetric solutions the biionic voltage or selectivity of the channel. It can be seen that discrete step changes occur upon addition of the extract demonstrating the presence of channels.

Fig. 13 shows two successive sweeps of a bilayer in symmetric solutions with the extract already present. It can be seen that successive increments are obtained, as well as decrements. These events occur at higher voltage; hence these channels demonstrate voltage dependence.

By measuring the conductance change of these individual steps as a function of salt concentration, as was done to obtain Fig. 14, we can see in one case the behavior to be expected for a neutral channel, and in the other the behavior of a channel with positive charge, i. e. anion selective.

Summarizing our findings, we have shown that at least 3 separable channel forming substances can be found in Triton extract of gastric membranes, with a conductance of from 2 to 5 $\times$ $10^{-10}\Omega^{-1}$, a neutral cation-selective channel, a charged anion-selective voltage dependent channel and a non-selective channel. Similar types of events can be induced in bilayers by extracts of sarcoplasmic reticulum (Szabo *et al.* In press.), and perhaps by Na^+K^+ ATPase preparations from kidney (Tosteson & Skou, personal communication), and extracts of electric organ or rat brain (Goodall & Sachs 1972). These channels are absent from similar extracts of purified gastric mitochondria.

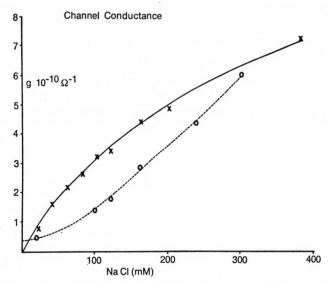

Fig. 14. The effect of varying NaCl concentration on unit channel conductance. The upper line represents the neutral cation selective channel, the lower line the smaller, voltage dependent anion selective channel.

CONCLUSIONS

This work has demonstrated the feasibility of a direct approach to the passive conductance elements occuring in epithelial tissues. Thus having established the cellular pathway of gastric mucosal conductance by direct measurements and having found suggestive evidence for a mosaic of anion and cation selective channels, we have succeeded in incorporating these channels into a bilayer. We can therefore study their properties directly, a feat which would be impossible in the high conductance situation in the intact cell membrane. We have further shown that these channels can be separated and purified and may become amenable to direct chemical and biochemical investigation.

HCl Secretion Implications:

Finally, we are tempted to speculate about the significance of the voltage dependent, anion selective channel that we have found. This we do by considering the model illustrated in Fig. 15. Here the HCO_3^- ATPase is considered to act by removing OH^- or HCO_3^- from the site of proton secretion. This system is clearly electrogenic. In the absence of any other mem-

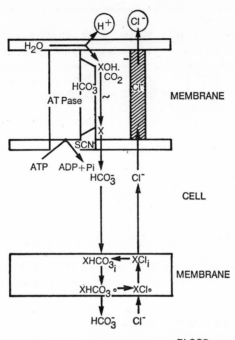

MODEL FOR HCl SECRETION

Fig. 15. A model incorporating the voltage dependent anion channel into an electrogenic HCO_3^- ATPase mechanism for H^+ secretion by the gastric mucosa.

brane conductance system a P D would build up equal to the pump EMF. In the presence of a voltage dependent Cl^- channel, in association with the ATPase site, but not necessarily part of it, the developed voltage would open the channel and Cl^- flux would occur exactly equal to the demands of the H_3^+ pump resulting in electroneutral secretion.

Similar mechanisms may be involved for other electrogenic pumps such as the Na^+K^+ ATPase.

REFERENCES

Barry, P. H. & Diamond, J. M. (1971) Theory of ion permeation through membranes with fixed neutral sites. *J. Membr. Biol. 4*, 295–330.
Barry, P. H., Diamond, J. M. & Wright, E. M. (1971) Mechanism of cation permeation in gall bladder. *J. Membr. Biol. 4*, 358–394.

Blum, A. L., Hirschowitz, B. I., Helander, H. F. & Sachs, G. (1971) Electrical properties of isolated cells of *Necturus* gastric mucosa. *Biochim. biophys. Acta (Amst.) 241*, 261–272.

Conti, F. & Eisenman, G. (1966) Steady state properties of an ion exchange membrane with mobile sites. *Biophys. J. 6*, 227–246.

Diamond, J. M., Barry, P. H. & Wright, F. M. (1971) Route of transepithelial ion permeation in gall bladder. In *Electrophysiology of Epithelial Cells*, pp. 23–28, ed. Schattauer, Stuttgart, New York.

Eisenman, G., Ciani, S. M. & Szabo, G. (1968) Some theoretically expected and experimentally observed properties of lipid bilayer membranes containing neutral molecular carriers of ions. *Fed. Proc. 27*, 1289–1297.

Frömter, E. (1973) Route of passive ion permeation in gall bladder. In Alfred Benzon Symposium V: *Transport mechanisms in epithelia*, ed. Ussing, H. H. & Thorn, N. A., pp. 492–499. Munksgaard, Copenhagen.

Goodall, M. C. & Sachs, G. (1972) Extraction of K selective channels from excitable tissue. *Nature (Lond.) 237*, 252–253.

Hladky, S. B. & Haydon, D. A. (1970) Discreteness of conductance change in bimolecular lipid membranes in the presence of certain antibiotics. *Nature (Lond.) 225*, 451–453.

Jardetzky, O. (1966) Simple allosteric model for membrane pump. *Nature (Lond.) 211*, 969–970.

Lev, A. A., Gotlib, V. A. & Buzhnisky, E. P. (1966) Cationic specificity of model bimolecular phospholipid membranes. *J. Evol. Biochem. Physiol. (USSR) 2*, 109–117.

Long, M. M., Masotti, L., Sachs, G. & Urry, D. W. In preparation.

Martonosi, A. (1971) Sacroplasmic reticulum. In *Biomembranes,* ed. Hanson, L. A, pp. 191–224. Plenum Publishing Company.

Mayer, M. & Avi-Dor, V. (1970) Fluorescence of ANS bound to ox brain NaK ATPase. *Israel J. Med. Sci. 6*, 276.

Mueller, P. & Rudin, D. O. (1963) Induced excitability in reconstituted cell membrane structure. *J. theoret. Biol. 4*, 268–280.

Sachs, G., Shah, G. Strych, A., Cline, G. & Hirschowitz, B. I. (1972) Properties of ATPase of gastric mucosa. III. Distribution of HCO_3^- – stimulated ATPase in gastric mucosa. *Biochim. biophys. Acta (Amst.) 266*, 625–638.

Shoemaker, R. L., Makhlouf, G. M. & Sachs, G. (1970) Action of cholinergic drugs on *Necturus* gastric mucosa. *Amer. J. Physiol. 210*, 1056–1059.

Skou, J. C. (1957) Influence of cations on ATPase from peripheral nerves. *Biochim. biophys. Acta (Amst.) 23*, 394–401.

Szabo, G., Eisenman, G., Laprade, R., Ciani, S. & Krasne, S. Experimentally observed effects of carriers on the electrical properties of bilayer membranes – equilibrium domain. In *Membranes,* ed. Eisenman, G. Dekker. In press.

Tanisawa, A. S. & Forte, J. G. (1971) Phosphorylated-intermediate of microsomal ATPase from rabbit gastric mucosa. *Arch. Biochem. Biophys. 147*, 165–175.

Tosteson, M. T., Tosteson, D. C. & Skou, J. C. Personal communication.

Wright, E. M., Barry, P. H. & Diamond, J. M. (1971) Mechanism of cation permeation in gall bladder. *J. Membr. Biol. 4*, 331–357.

DISCUSSION

ZADUNAISKY: Could you tell us a bit more about how you incorporated these membranes into artificial lipid black membranes?

SACHS: You can do it simply by adding the Sephadex purified ATPase fraction into the inner or the outer chamber. We use inner chamber of an artificial bilayer setup, and after having added the ATPase you form the bilayer and then look at its properties. Alternatively you can form the bilayer first and then add ATPase preparation: Now how much incorporates, obviously we don't know. At the sensitivity we are looking at here we can detect single channels incorporating into a bilayer.

SCHULTZ: Villegas (1962) has reported that the intracellular electrical potential of parietal cells, identified iontophoretically, was positive with respect to the lumen of the frog stomach. What are your observations on this point?

SACHS: In Necturus gastric mucosa we have done many many experiments and we have also identified the cells, electrophoretically. We do not find positive potential in the oxyntic cells. We inevitably find a negative level. In amphibia where we have had less experience so far we haven't been able to find this positive level that Villegas is talking about.

ULLRICH: How pure is your enzyme preparation, did you check that?

SACHS: At the moment it looks like there are 15 bands on gel electrophoresis.

FRÖMTER: Could you please repeat for me what the resistance of the cell membrane was? I think that you mentioned a figure of 800 Ohm cm^2. This value would be much smaller than the values of approximately 3000 Ohm cm^2, which have been observed in gallbladder (this symposium p. 496) and proximal tubule (Boulpaep 1971) of the same species.

SACHS: In the fundus the calculated values were about 700–800 Ohm cm^2 and the trans tissue resistance in the experiments that we showed was about

Vilegas, L. (1962) Cellular location of the electrical potential difference in frog gastric mucosa. *Biochim. Biophys. Acta, 64,* 359–367.

Boulpaep, E. (1971) Electrophysiological properties of the proximal tubule: Importance of cellular and intercellular pathways. In *Electrophysiology of Epithelial Cells,* ed. Giebisch, G. pp. 91–112 F. K. Schattauer Verlag, Stuttgart.

a 1000 Ohm cm^2. So we sum the two and there was a small shunt conductance. But that is only assuming that we can neglect the tubules, which could account for all of the shunt conductance, obviously.

HOSHIKO: Were you classifying these channels in terms of the height of the transients?

SACHS: In terms of the change of slope.

HOSHIKO: What sort of dispersion did you get with these slopes?

SACHS: If you look at the smallest channels which you have got at the high sensitivity, that was one family. And then you have two other families of slope, one 3–5 and one 2×10^{-10} reciprocal Ohms.

HOSHIKO: They were separated by orders of magnitude?

SACHS: No.

TOSTESON: What was the speed of your ramp?

SACHS: The whole sweep takes about between 30 sec to 1 min.

TOSTESON: What is the relation between the zero frequency conductance and the concentration of the material which you get from the gradient?

SACHS: This is the bilayer system. And we incorporated some channels. And instead of doing a sweep over 30 sec we do a very slow sweep.

TOSTESON: You impose 1 voltage, measure the current . . .

SACHS: and go on very slowly, right.

TOSTESON: Have you made such measurements in the presence of different concentrations of the material from the gradient?

SACHS: Yes, we can do that.

TOSTESON: What is the maximum conductance you have under those conditions?

SACHS: For a single channel?

TOSTESON: No, for the whole bilayer in Ohm^{-1} cm^{-2}.

SACHS: If you have a lot of channels present the bilayer becomes very unstable, to this sort of slow sweep. It seems to break at extreme voltages, so if we do it within a very narrow range.

TOSTESON: Extreme means very low voltages, is that right?

SACHS: No, it was $\pm$ 100 mV across the bilayer, or 80.

HOSHIKO: I was surprised to see your voltage curve for the gastric mucosa, the voltage span going up to $1/2$ V. What is the break-down voltage of the gastric mucosa?

SACHS: It must be larger than that.

HOGBEN: Which gastric mucosa? Around 60 mV a break-down occurs.

SACHS: And which gastric mucosa was that?

HOGBEN: *Rana temporaria* and *Bufo bufo*.

SACHS: Warren Rehm did the stuff in *Rana pipiens* up to 200–250 mV and with *Necturus* which didn't break down at 500.

HOGBEN: I presume that an alteration was not detected because the mucosa was challenged by a transient pulse.

SACHS: Oh yes, it wasn't clamping.

KEYNES: Are these channels of yours like gramicidin ones? Do they just stay open, or are there fluctuations all the time?

SACHS: You can get them staying on or sometimes switching off. We don't know why sometimes it will stay on so the conductance will climb up, and up and up.

KEYNES: But the story about gramicidin is that the channels only open for dimers. When the dimer form is present it is long enough to stretch all the way across the membrane and make a channel right through.

WHITTEMBURY: How many cell types do you have in the *Necturus* gastric mucosa, is there any difference in their profile of electrical potential, are they communicated?

SACHS: Yes, they are all communicated. The *Necturus* gastric mucosa has at least one cell type less than the guinea-pig gastric mucosa, because the oxyntic cell in the *Necturus* has both peptic or zymogen granules and acid secretory properties. In the guinea-pig where you have as many cell types as in the rat I think, you have coupling between each cell type. In the *Necturus* gastric mucosa you also have. We have not found, however, any significant difference in potential between the different cells we punctured.

GENERAL DISCUSSION

TOSTESON: Dr. Sachs, could I ask a question about these bilayers. You commented that when you add magnesium, the selectivity changes from being cation-selective to being anion-selective. You gave a number for chloride compared with sulphate but you did not cite a number for chloride as compared with cations. For instance if you put a 10-fold concentration ratio of sodium chloride across the bilayers in the presence of this material, what is the value of the zero current potential difference across the membrane?

SACHS: Before we added magnesium, if you had a 10-fold gradient for sodium chloride you get a PD of about 16–20 mV showing sodium selectivity. After you add magnesium, the PD will invert and you get a value which is the one calculated for a chloride/sodium conductance of about 3.

TOSTESON: And may I ask, if you add some other peak from the sucrose gradient to the solutions bathing the bilayer, do you observe changes in conductance?

SACHS: If we take some other peak, no we don't get these data at all. We don't see anything then, except what you call a garbage effect where you don't get thinning of the bilayer. If you take preparations from rat brain or electroplax, you get data which look like that but with different selectivities and different sizes.

KIRSCHNER: Does ATP have any effect?

SACHS: Yes, it does. But I don't know how to explain it. There is a potential development, as a function of the presence of ATP in symmetrical solutions and you require the ATP and the enzyme preparation to be present. You don't see the potential when the bilayer goes black. But if you watch it thin and at the silvery stage of the bilayer a potential starts to develop. If you scan the bilayer at that time there is no evidence of channels. Then if you see evidence of channels occuring under a scan and then look at the PD, the PD is shorted out. Occasionally the channels will disappear as I showed on a slide and you look at the PD again it is present. The PD is in the direction which you would expect the protons to be moving. But whether you can conclude that we get an active transport system or surface charge effect or other positive effects I can't say.

SCHULTZ: Dr. Hogben, the resistance of the rat stomach that you recorded was in the order of a 100 Ohm cm² which places it in a rather low resistance range, and yet it maintains roughly a 10,000-fold gradient for hydrogen ion. Do you expect there is a rather large leak of hydrogen ion back from the lumen?

HOGBEN: Yes, there is a large leak. This is an abnormal situation, the low resistance or high conductance you referred to is one of the artifacts of isolating the mucosa. We cannot maintain the hydrogen concentration in the mucosal chamber at 5 mMol without continually replacing the mucosal solution.

ZADUNAISKY: Dr. Sachs. If you tried to keep the ATPase activity of your Sephadex column and then incorporated it in the inhibited membrane did you get these changes or not?

SACHS: Yes, we do. If I found that thiocyanate or cyanide inhibited the PD developed in the presence of ATP, I would have had some confidence that I might have been dealing with active transport. Since these do nothing in terms of PD that's why I would hesitate very much to assume that it has anything to do with active transport. Also, the presence of thiocyanate or cyanide does not affect the appearance of these channels.

KIRSCHNER: Let me make sure I sorted this out correctly. The thiocyanate doesn't prevent the formation of channels, but it also does not block this effective ATP, which is really what you would like to.

SACHS: That's right.

MAETZ: Has something similar been done on bilayer preparations with Na-K-dependent ATPase?

SACHS: Yes, there have been papers published by Jarni *et al.* where they look at fractions coming off a zonal rotor. I think it was a rat brain membrane preparation, and they were interested mainly in the zero potential current developed as a function of the presence of ATP, that was inhibited by ouabain. If you look at their figure they have coming off their zonal rotor a peak which they said gave step-like changes in conductance of the bilayer. Similar to what we found. They did not apparently take it any further. And they were not dealing with soluble, purified ATPase preparation.

TOSTESON: I can report that we have observed rather similar effects on the electrical properties of bilayers exposed to preparations from torpedo electric organ (in collaboration with E. Schoffeniels), and with preparations of Na-K ATPase in collaboration with J. C. Skou. The problem to be resolved is whether these effects have something to do with this specific protein or with any protein which is able to interact with bilayers.

ULLRICH: Is there a difference between the HCO_3 ATPase and Na-K ATPase with respect to solubility?

SACHS: You can solubilise both but with different detergents.

Shunt Pathway, Sodium Transport and the Electrical Potential Profile Across Rabbit Ileum

Stanley G. Schultz

There is compelling evidence that, in general, two pathways are available for ion movements across *all* epithelia, a transcellular pathway which involves movements across at least two cell membranes arranged in series, and a parallel, transepithelial, extracellular ("shunt") pathway which circumvents the transcellular route. The relative roles played by each of these pathways covers a broad spectrum. One extreme is exemplified by tissues such as isolated frog skin (Ussing & Windhager 1964) and toad urinary bladder (Civan *et al*. 1966) that are characterized by high transepithelial electrical potential differences (PDs) and resistances and low hydraulic conductivities; the conductance of the shunt pathway across these tissues is but a small fraction of the total tissue conductance.

At the other extreme are tissues such as small intestine (Frizzell & Schultz 1972), proximal renal tubule (Boulpaep 1971) and gall bladder (Barry *et al*. 1971) that are capable of transporting relatively large volumes of isotonic solutions, and are characterized by low, and often negligible, transepithelial PDs, low transepithelial resistances and relatively high hydraulic conductivities.

It is now clear that the electrical characteristics of these tissues are not consequences of unusually "leaky" or expansive cell membranes but instead reflect the fact that the conductance of the shunt pathway greatly exceeds that of the transcellular route, often by more than one order of magnitude; that is, these tissues are to a large extent "self short-circuited". Epithelia such

Department of Physiology. University of Pittsburgh School of Medicine. Pittsburgh. Pennsylvania 15213.
Supported by research grants from the National Institutes of Health (AM-13744), the American Heart Association (70–633) and a Research Career Award from the U.S.P. H.S., National Institute of Arthritis and Metabolic Diseases.

»Transport Mechanisms in Epithelia«, Munksgaard, Copenhagen.

as renal distal tubule and collecting ducts, colon and salivary gland ducts appear to fall in an intermediate position between these two extremes.

The single structural feature common to this otherwise diverse collection of epithelia is the series array formed by the junctional complexes and the intercellular spaces and there is now convincing histological evidence that directly correlates the width of the intercellular spaces (c. f. Smulders *et al.* 1972) and the ultrastructure of the "tight" junctions (DiBona 1972) with tissue conductances under a variety of experimental conditions.

In addition there is compelling histochemical evidence that the tight junctions afford a pathway for the movement of small ions and some water soluble nonelectrolytes between the mucosal solution and the intercellular spaces (Ussing 1971, Whittembury & Rawlins 1971). The relative contributions of the junctional complexes and the intercellular spaces to the total shunt resistance has not been explicitly defined and appears to vary with different tissues and experimental conditions.

However, it seems reasonable to suspect that in low resistance epithelia comprising a single layer of columnar cells, where the lateral intercellular spaces may range from several hundred to several thousand Å wide, and are readily accessible to macromolecules, the tight junction normally comprises the principal resistive element of this series array. The finding that dilatation of the intercellular spaces of rabbit gall bladder above their normal width does not decrease tissue resistance supports this notion (Smulders *et al.* 1972).

In the light of these considerations, the simplest equivalent electrical circuit model applicable to all epithelia is that illustrated in Fig. 1; also illustrated are the fluxes that comprise the extracellular and transcellular routes for transepithelial movements.[1]

Clearly, a detailed understanding of the mechanism underlying transepithelial ion movements and bioelectric phenomena requires (i) a quantitative distinction between the contributions of the transcellular and extracellular routes; and (ii) an appreciation of the way in which shunt pathways influence the electrical potential profile across low resistance epithelia and, in turn, the driving forces for ionic diffusion across the limiting cell membranes and the shunt itself.

---

1. The equivalent electrical circuit model illustrated in Fig. 1 does not include the likelihood of electrical coupling between adjacent cells. Thus it must be considered a first approximation and is strictly valid only if neighboring, coupled cells have identical electrical properties.

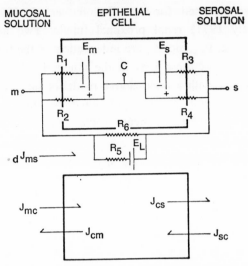

Fig. 1. Equivalent electrical circuit model of small intestine. E_m designates an electromotive force across the mucosal membrane; R_1 is the internal resistance of this battery and R_2 is a shunt resistance across the mucosal membrane. E_s, R_3 and R_4, and E_L, R_5 and R_6 are the respective parameters of the serosal membrane and the shunt pathway; m, c and s are the mucosal, intracellular and serosal electrodes, respectively. $_dJ_{ms}$ is the unidirectional diffusional flux through the shunt pathway directed from mucosa to serosa. The bidirectional fluxes across the mucosal and serosal membranes of the epithelial cell are defined in the lower portion of the figure.

In this paper, I shall briefly describe recent findings regarding the shunt pathway across rabbit ileum and discuss the implications of these findings with respect to transcellular sodium transport and the electrical potential profile across small intestine.

The Shunt Pathway

Studies on the effect of an imposed transepithelial PD on the unidirectional influxes of Na, K and Cl from the mucosal solution into the intestinal epithelium have permitted the distinction between influx across the mucosal membrane (J_{mc}^i) and diffusion into the shunt pathway ($_dJ_{ms}^i$) and have provided minimum estimates of the absolute permeabilities of the shunt pathway to these ions (Frizzell & Schultz 1972).

As shown in Table I, the shunt is cationselective and $P_K:P_{Na}:P_{Cl}=$ 1.14:1.00:0.55; interestingly, these relative permeabilities are essentially identical to those reported for rat (Frömter et al. 1971) and canine (Boulpaep & Seely 1971) renal proximal tubule. Two additional points should be noted. First, P_{Na} and P_{Cl} are independent of the transepithelial PD and the concentration of these ions in the mucosal solution. This finding is difficult to reconcile with a fixed-charge model of the shunt pathway but, instead, suggests constant products of ionic partition coefficients and mobilities which is suggestive of a "neutral-polar" pathway.

Table I. Unidirectional Influxes Across the Mucosal Membranes and into the Shunt Pathway

	J^i_{mc} (μmoles/cm²hr)	$({}_dJ^i_{ms})_{s \cdot c.}$* (μmoles/cm²hr)	P_i** (cm/hr)
Na, 140 mM	16	4.9 ± 0.9	0.035
Na, 28 mM	3	1.0 ± 0.1	0.036
Cl, 145 mM	10	2.8 ± 0.1	0.019
Cl, 50 mM	6.5	1.0 ± 0.1	0.020
K, 12 mM	2.5	0.5 ± 0.01	0.040

* $({}_dJ^i_{ms})_{s \cdot c.}$ designates the unidirectional influx of i into the shunt pathway under short-circuit conditions.

** $P_i = ({}_dJ^i_{ms})_{s \cdot c.}/[i]$

Second, unlike most cell membranes, the shunt does not markedly distinguish between Na and K; P_K/P_{Na} is only 20% lower than that which would be predicted on the basis of their free-solution mobilities. This suggests that the shunt affords a watery environment for ionic diffusion and that Na and K traverse this pathway in their hydrated forms.

It should be stressed that the finding that $P_K/P_{Na} = 1.1$, a value significantly lower than the ratio of free-solution mobilities, 1.5, cannot be attributed to steric factors. On the contrary, any steric restrictions would be expected to hinder the movement of Na (the larger of the two hydrated radii) to a greater extent than that of K. Thus, steric hinderance alone would, if anything, lead to a P_K/P_{Na} greater than 1.5. Indeed, the Renkin equation predicts a P_K/P_{Na} of 1.8 for restricted diffusion through a channel with an "equivalent pore radius" of 15 Å based on hydrated ionic radii alone. Thus, it seems evident that the anionic field that restricts the movement of Cl also prefers Na over K, a preference that is consistent with a field of high to intermediate strength (Eisenmann 1961).

The bidirectional fluxes of Na, K and Cl through the shunt under short-circuit conditions, $(_dJ_{ms}^l)$ $_{s.c.}$, in the presence of a buffered electrolyte solution containing 140 mM Na, 12 mM K and 145 mM Cl are also given in Table I. Since these fluxes, expressed in μmoles/hr, cm², are numerically equivalent to the partial ionic conductances (G_i), expressed in mmhos/cm², the total conductance of the shunt patway attributable to the movements of Na, K and Cl is 8.2 mmhos/cm². The total tissue conductance in these experiments averaged 10 mmhos/cm²; thus, at least 82% of the transepithelial conductance can be attributed to the movements of Na, K and Cl through the shunt pathway.

Further, if we assume that an additional 0.2–0.5 mmhos/cm² results from the combined movements of other ions through the shunt, the conductance of the extracellular pathway approaches 85–90% of the total tissue conductance. Thus, we estimate the resistance of the shunt pathway to be approximately 100 ohm-cm² and the resistance of the transcellular pathway to approach 1000 ohm-cm².

Diffusion Potentials and Ionic Fluxes Through the Shunt

According to the model illustrated in Fig. 1,

$$\Psi_{ms} = [(E_sR_s + E_mR_m) R_5R_L/R_t] + [E_LR_L (R_1R_m + R_3R_s)/R_t] \tag{1}$$

and

$$\Psi_{mc} = [E_mR_m(R_3R_s + R_5R_L) - (E_sR_s - E_LR_L)R_1R_m]/R_t \tag{2}$$

where

$$\Psi_{ms} = \Psi_s - \Psi_m; \ \Psi_{mc} = \Psi_c - \Psi_m; \ R_m = R_2/(R_1 + R_2); \ R_s = R_4/(R_3 + R_4)$$

$$R_L = R_6/(R_5 + R_6); \text{ and } R_t = R_1R_m + R_3R_s + R_5R_L.$$

Thus, when $R_5R_L \ll (R_1R_m + R_3R_s)$, Ψ_{ms} will be dominated by E_LR_L; that is, because of the high conductance shunt, $(E_mR_m + E_sR_s)$ is markedly attenuated and, in the presence of transepithelial ionic asymmetries, Ψ_{ms} should closely approximate the PD resulting from ionic diffusion through the permselective shunt. This prediction is borne out by the results of experiments illustrated in Fig. 2, in which (i) NaCl in the mucosal solution

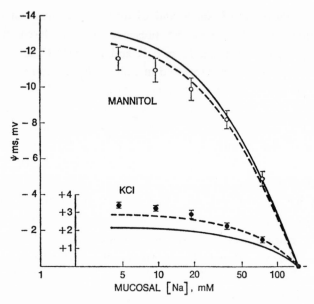

Fig. 2. Diffusion potentials across rabbit ileum. The values of Ψ_{ms} for mannitol replacement are given on the larger ordinate; those for KCl replacement are given on the smaller ordinate. All values reflect the electrical potential of the serosal solution with respect to the mucosal solution. The curves were obtained as described in the text. (Reproduced from Frizzell & Schultz (1972) by permission from the Rockefeller University Press.)

was replaced with mannitol and (ii) Na in the mucosal solution was replaced with K.

The solid curves were calculated using the Goldman-Hodgkin-Katz "constant-field" equation and the absolute permeabilities given in Table I. The dashed curves were calculated using a P_{Na} of 0.034 cm/hr (rather than 0.035 cm/hr) and provide a somewhat better fit to the experimental data.

The finding that the ionic permeabilities determined from the effect of an imposed PD on unidirectional influxes in the absence of transepithelial concentration differences adequately predict diffusion PDs under "zero current" conditions provides strong support for the internal consistency of our data and interpretations.

In addition to permitting an accurate estimate of transepithelial diffusion potentials in the presence of ionic asymmetries, the data in Table I enable us to calculate the contribution of the shunt pathway to transepithelial ionic movements and, thereby, permit a quantitative distinction between the transcellular and extracellular routes in these processes. Thus, it can be readily demonstrated that for small values of Ψ_{ms} (<25 mV), the *net* flow of i through the shunt is given by

$$_dJ^i = P_i \left\{ [i]_m \exp (z_i \mathscr{F} \Psi_{ms}/2RT) - [i]_s \exp (-z_i \mathscr{F} \Psi_{ms}/2RT) \right\} \tag{3}$$

Under physiological conditions, where $[i]_m = [i]_s$, equation (3) reduces to

$$_dJ^i = - P_i[i]_m z_i \mathscr{F} \Psi_{ms}/RT \tag{4}$$

When the mucosal and serosal solutions contain 140 mM Na, 12 mK K and 145 mM Cl,

$$_dJ^{Na} = -0.22 \Psi_{ms}; \quad _dJ^K = 0.02 \Psi_{ms}; \quad \text{and} \quad _dJ^{Cl} = 0.13 \Psi_{ms} \tag{5}$$

where $_dJ^i$ is in μEq/hr.,cm^2, Ψ_{ms} is in mV and a negative flux indicates *net* movement from serosa to mucosa.

The data given in Table I and the relations described by equations (3) and (5) completely define the contributions of the shunt to the transepithelial movements of Na, K and Cl and are both necessary and sufficient for an analysis of transcellular ion movements.

Transcellular Sodium Transport

In the presence of a normal buffer solution, the intracellular Na concentration of rabbit ileal mucosa is 40–50 mM (Schultz *et al.* 1966) and the cell interior is, on the average, 36 mV negative with respect to the mucosal solution (Rose & Schultz 1971); in the absence of sugars or amino acids, Ψ_{ms} is approximately 3–5 mV serosa positive (Schultz & Zalusky 1964).

Although many uncertainties becloud the interpretation of bulk intracellular concentrations, studies on a variety of cells and tissues suggest that the thermodynamic activity of intracellular Na is significantly lower than its bulk concentration. In particular, Lee and Armstrong (1972) have recently demonstrated, using cation-sensitive microelectrodes, that activity coefficients of cell Na and cell K in bullfrog small intestinal cells are 0.5 and approximately 1.0, respectively. Thus, the conclusion that net entry of Na into the cell from the mucosal solution, though to a large extent carrier-mediated (Frizzell & Schultz 1972), takes place down a steep electrochemical potential gradient, whereas extrusion of Na from the cell across the baso-lateral membranes into the serosal solution is an active transport process, appears inescapable.

Active transepithelial Na transport is abolished by the presence of ouabain in the serosal solution but is unaffected by the presence of ouabain in the mucosal solution (Schultz & Zalusky 1964). Further, it appears that most, if not all, of the Na-K stimulated, ouabain-sensitive ATPase activity in rat small intestinal cells is located in the baso-lateral membranes (Quigley & Gotterer 1969, Fujita et al. 1971). Thus, it seems reasonable to conclude that active Na extrusion across these membranes is coupled to active K uptake and is mediated by the ATPase mechanism that has been implicated in Na-K transport by a wide variety of animal cells. Unpublished observations from this laboratory suggest that this mechanism is solely responsible for the high intracellular K concentration in rabbit ileal cells (c.a. 140 mM).

Our findings with respect to the Na permeability of the shunt pathway have clarified our understanding of the mechanism of active Na extrusion across the baso-lateral membranes in two respects:

First, it is now clear that most, if not all, of the unidirectional transepithelial flux of Na from the serosal solution to the mucosal solution (J_{sm}^{Na}) can be attributed to simple ionic diffusion through the shunt pathway and does not involve a significant transcellular component. This was suggested nearly a decade ago by our findings regarding the effects of an imposed transepithelial PD on J_{sm}^{Na} (Schultz & Zalusky 1964, Fig. 6) and is now strongly supported by our direct measurements of the Na permeability of the shunt pathway. Thus, $_dJ_{ms}^{Na}$ does not differ significantly from J_{sm}^{Na} when Ψ_{ms} is clamped over the range $\pm$ 50 mV. Indeed, Schultz and Zalusky (1964) reported that net Na transport from mucosa to serosa was reduced from 2.7 – 3.0 μEq/hr, cm² under short-circuit conditions to zero when the tissue was clamped at 12 mV serosa positive. As indicated by equations (5), this is entirely consistent with the notion that the abolition of the net flux is attributable to the diffusional backflux through the shunt. These findings, coupled with the observation that there is rapid exchange of Na across the mucosal membrane (i. e. $J_{cm}^{Na} \neq 0$) (Schultz, et al. 1967) can only mean that the baso-lateral membranes are virtually if not totally impermeable (i.e. $J_{sc}^{Na} = 0$).

Second, there has long been considerable concern over whether or not the active Na extrusion mechanism at the baso-lateral membranes is "rheogenic" (frequently referred to as "electrogenic" but as suggested by Schwartz (1971)

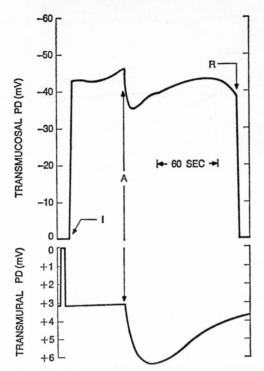

Fig. 3. The effect of 20mM L-alanine, (A) added to the mucosal solution, on the transmucosal (Ψ_{mc}) and transmural (Ψ_{ms}) PDs. All values are with reference to the mucosal solution. The return to baseline values is due to the washout of the alanine by a continuous circulation of alanine-free Ringer's solution. (Reproduced from Rose & Schultz (1971) by permission from the Rockefeller University Press.)

the term "rheogenic" or "current-generating" seems preferable) and, thereby, contributes directly to the transepithelial PD. Evidence favoring this notion follows:

Rose and Schultz (1971) demonstrated that the addition of actively transported sugar or amino acids to the mucosal solution results in a prompt depolarization of Ψ_{mc} and a simultaneous increase in Ψ_{ms} (Fig. 3) which is dependent upon the presence of Na in the bathing media (Fig. 4). The change in Ψ_{mc} was attributed to the rheogenic influx of Na across the brush border coupled to the influxes of the sugar or amino acid. However, the decrease in intracellular negativity significantly exceeded the increase in serosal positivity; the average $\triangle\Psi_{ms}/\triangle\Psi_{mc}$ was 0.3–0.4. Similar findings have been reported for bullfrog small intestine (White & Armstrong 1971).

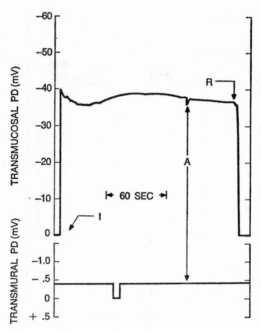

Fig. 4. Effects of 20 mM L-alanine, (A) on Ψ_{mc} and Ψ_{ms} of tissue bathed on both surfaces with a Na-free, choline Ringer's. (Reproduced from Rose & Schultz (1971) by permission from the Rockefeller University Press.)

According to the model illustrated in Fig. 1, the relation between a change in Ψ_{ms} and a change in Ψ_{mc} resulting from a change in E_m *alone* is

$$\triangle \Psi_{ms}/\triangle \Psi_{mc} = 1/[1 + (R_3 R_s / R_5 R_L)]$$

Thus, $\triangle \Psi_{ms}/\triangle \Psi_{mc} < 1$ whenever $R_5 R_L < \infty$ and the observed ratio of 0.3–0.4 could be explained in terms of a change in E_m alone if $R_3 R_s$ were *twice* $R_5 R_L$. However, this explanation was challenged by the results of experiments performed on tissues that had been poisoned with either metabolic inhibitors and/or ouabain.

As shown in Fig. 5, the addition of alanine to the mucosal solution still resulted in a significant depolarization of Ψ_{mc} but the accompanying change in Ψ_{ms} was minimal; the paired $\triangle \Psi_{ms}/\triangle \Psi_{mc}$ in a series of experiments on poisoned tissues averaged only 0.06. We suggested at that time that (i) the $\triangle \Psi_{ms}$ in response to sugars or amino acids *cannot* be attributed to a change in E_m *alone* but that an additional electromotive force that is dependent upon

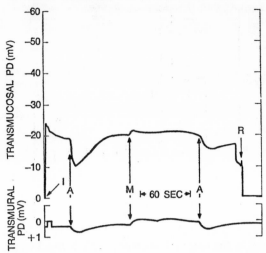

Fig. 5. Effect of 20 mM L-alanine (A) and 20 mM D-mannitol (M) on Ψ_{mc} and Ψ_{ms} of tissue bathed in a Ringer's solution containing metabolic inhibitors and ouabain. (Reproduced from Rose & Schultz (1971) by permission from the Rockefeller University Press.)

metabolic energy and inhibited by ouabain must be involved; and (ii) the value of R_3R_s/R_5R_L must be considerably greater than 2. This suggestion has now been corroborated by our direct measurements of the shunt resistance and estimates of the relative resistances of the serosal and mucosal barriers which indicate that R_3R_s is at least 6–9 times R_5R_L (600–900 ohm-cm^2 versus 100 ohm-cm^2).

Thus, we conclude that (i) because of the low resistance shunt, the change in E_m resulting from the rheogenic influx of Na coupled to the influxes of sugars or amino acids is markedly attenuated and cannot account for the overall change in Ψ_{ms}; and (ii) Na extrusion across the baso-lateral membranes is rheogenic so that the increase in the rate of Na extrusion into the serosal solution elicited by sugars or amino acids results in an increase in E_s *which makes the major contribution* to the overall $\triangle\Psi_{ms}$.

Electromotive Forces and the Electrical Potential Profile: Some Questions for the Future

As recognized by Ussing and his collaborators nearly two decades ago, a detailed understanding of transepithelial electrolyte transport requires that

19*

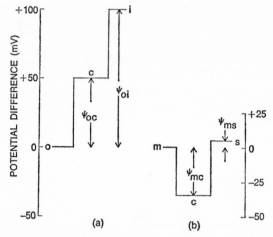

Fig. 6. (a) Schematic of the electrical potential profile across isolated frog skin; o, c and i designate the outer solution, intracellular compartment and inner solution, respectively. (b) Schematic of the electrical potential profile across isolated rabbit ileum; m, c, and s designate the mucosal solution, intracellular compartment and serosal solution, respectively. (Reproduced from Schultz (1972) by permission from the Rockefeller University Press.)

the transepithelial PD and electrical potential profile be explicable in terms of the electromotive forces (emf's) operating across the limiting membranes of the epithelial cell. For the case of the high resistance epithelia alluded to previously, $R_5 R_L \gg (R_1 R_m + R_3 R_s)$ so that equations (1) and (2) reduce to

$$\Psi_{ms} \cong E_s R_s + E_m R_m \text{ and } \Psi_{mc} \cong E_m R_m$$

Thus, transmembrane PDs must have the same polarity as the emf across the punctured membrane.

A typical electrical potential profile across high resistance epithelia is illustrated in Fig. 6a; in every reported instance, the cell interior is found to be electrically positive with respect to the outer or mucosal solution. Evidence that $E_m R_m$ may be largely attributed to the Na-gradient across the outer or mucosal membrane and that $E_s R_s$ may be attributed to the K-gradient and, perhaps, a rheogenic pump across the serosal or inner membrane, need not be elaborated upon.

In marked contrast with high resistance epithelia, the electrical potential profile across low resistance epithelia, illustrated for the case of rabbit ileum in Fig. 6b, consists of two steps in opposite directions that sum to give the transmural PD; in each instance, the cell interior is electrically negative with respect to the mucosal solution. At first sight, this would suggest that the orientation of the emf across the mucosal membrane is opposite to that found in high resistance epithelia. However, it can be readily demonstrated that the intracellular negativity may be a consequence of electrical coupling between the emf's and resistive elements across the serosal and mucosal membranes through the shunt pathway (Schultz 1972). Thus, when the tissue is bathed on both surfaces with solutions of identical composition and $E_L R_L = 0$, then, referring to Fig. 1,

$$\Psi_{mc} \quad < 0 \text{ when } E_s R_s > [(R_3 R_s + R_5 R_L)/R_1 R_m] E_m R_m$$

and

$$\Psi_{mc} \quad > 0 \text{ when } E_s R_s < [(R_3 R_s + R_5 R_L)/R_1 R_m] E_m R_m$$

When $R_5 R_L = \infty$, $\Psi_{mc} \geq 0$ and will always have the same orientation as E_m; under these circumstances, the electrical potential profile will resemble that illustrated in Fig. 6a.

On the other hand, if $R_5 R_L$ is small, Ψ_{mc} may assume negative values due to the influence of $E_s R_s$, in spite of the fact that the emf across the mucosal membrane is oriented in the opposite direction, as illustrated in Fig. 1.

In brief, the difference between the electrical potential profile characteristic of frog skin and that observed in rabbit ileum may simply be a consequence of the fact that the latter is to a large extent self-short circuited by the presence of a high conductance shunt pathway. Whether or not this is true poses a formidable challenge for future investigations. Nonetheless, it is clear that in the presence of high conductance shunt pathways, the magnitude and orientation of the intracellular PD do not provide firm grounds for deductions regarding relative ionic permeabilities, etc., across the limiting cell membranes, and the interpretation of electrophysiological studies is fraught with uncertainties.

REFERENCES

Barry, P. H., Diamond, J. M. & Wright, E. M. (1971) The mechanism of cation permeation in rabbit gallbladder: Dilution potentials and biionic potentials. *J. Membrane Biol. 4*, 358–394.

Boulpaep, E. L. (1971) Electrophysiological properties of the proximal tubule: Importance of cellular and intercellular transport pathways. In *Electrophysiology of Epithelial Cells*, ed. Giebisch, G., F. K. Schattauer Verlag, Stuttgart, pp. 91–112.

Boulpaep, E. L. & Seely, J. F. (1971) Electrophysiology of proximal and distal tubules in the autoperfused dog kidney. *Amer. J. Physiol. 21*, 1084–1096.

Civan, M. M., Kedem, O. & Leaf, A. (1966) Effect of vasopressin on toad bladder under conditions of zero net sodium transport. *Amer. J. Physiol. 211*, 569–575.

DiBona, D. R. (1972) Passive pathways in amphibian epithelia: Morphologic evidence for an intercellular route. *Nature (Lond.) 238*, 179–181.

Eisenman, G. (1961) On the elementary atomic origin of equilibrium ionic specificity. In *Membrane Transport and Metabolism*, ed. Kleinzeller, A. & Kotyk, A. Academic Press, New York, pp. 163–179.

Frizzell, R. A. & Schultz, S. G. (1972) Ionic conductances of extracellular shunt pathway in rabbit ileum. *J. gen. Physiol. 59*, 318–346.

Frömter, E., Muller, C. W. & Wick, T. (1971) Permeability properties of the proximal tubular epithelium of the rat studied with electrophysiological methods. In *Electrophysiology of Epithelial Cells*, ed. Giebisch, G., F. K. Schattauer Verlag, Stuttgart, pp. 119–146.

Fujita, M., Matsui, H., Nagano, K. & Nakao, M. (1971) Asymmetric distribution of ouabain-sensitive ATPase activity in rat intestinal mucosa. *Biochim. biophys. Acta (Amst.) 233*, 404–408.

Lee, C. O. & Armstrong, W.McD. (1972) Activities of sodium and potassium ions in epithelial cells of small intestine. *Science 175*, 1261–1264.

Quigley, J. P. & Gotterer, G. S. (1969) Distribution of (Na+K)-stimulated ATPase activity in rat intestinal mucosa. *Biochim. biophys. Acta (Amst.) 173*, 456–468.

Rose, R. C. & Schultz, S. G. (1971) Studies on the electrical potential profile across rabbit ileum. *J. gen. Physiol. 57*, 639–663.

Schultz, S. G. (1972) Electrical potential differences and electromotive forces in epithelial tissues. *J. gen. Physiol. 59*, 794–798.

Schultz, S. G. & Zalusky, R. (1964) Ion transport in isolated rabbit ileum: I. Short-circuit current and Na fluxes. *J. gen. Physiol. 47*, 567–584.

Schultz, S. G., Fuisz, R. E. & Curran, P. F. (1966) Amino acid and sugar transport in rabbit ileum. *J. gen. Physiol. 49*, 849–866.

Schultz, S. G., Curran, P. F., Chez, R. A. & Fuisz, R. E. (1967) Alanine and sodium fluxes across the mucosal border of rabbit ileum. *J. gen. Physiol. 50*, 1241–1260.

Schwartz, T. L. (1971) Direct effects on the membrane potential due to "pumps" that transfer no net charge. *Biophys. J. 11*, 944–960.

Smulders, A. P., Tormey, J. McD. & Wright, E. M. (1972) The effect of osmotically induced water flows on the permeability and ultrastructure of the rabbit gallbladder. *J. Membrane Biol. 7*, 164–197.

Ussing, H. H. (1971) Introductory remarks. *Phil. Trans. B 262*, 85–90.

Ussing, H. H. & Windhager, E. E. (1964) Nature of shunt path and active sodium transport path through frog skin epithelium. *Acta physiol. scand. 61*, 484–504.

White, J. F. & Armstrong, W. McD. (1971) Effect of transported solutes on membrane potentials in bullfrog small intestine. *Amer. J. Physiol. 221*, 194–201.

Whittembury, G. & Rawlins, F. A. (1971) Evidence of a paracellular pathway for ion flow in the kidney proximal tubule: Electronmicroscopic demonstration of lanthanum precipitate in the junction. *Pflügers Arch. ges. Physiol, 330*, 302–309.

DISCUSSION

CURRAN: Would you comment on what seems to be a discrepancy. Your data indicate that about 85–90 percent of the conductivity lies in the shunt pathway and yet if you compare the fluxes through the shunt pathways and the flux across the brush border membrane, the latter are in fact considerably higher than those through the shunts.

SCHULTZ: The fact that the unidirectional influx of sodium across the mucosal membranes, approximately 16 μEq/cm²hr, greatly exceeds the rate of active transepithelial sodium transport has perplexed both of us. The only explanations I can offer are that either a significant fraction of this influx represents exchange-diffusion and is, thus, non-conductive and/or that a large fraction enters cells or compartments that are not engaged in transepithelial sodium transport. In either circumstances, this fraction of sodium influx would not contribute to the transepithelial conductance.

WHITTEMBURY: First, I would like to say that there are great similarities between the proximal tubule of the kidney and your preparation. If you take the transtubular hydraulic permeability coefficient and you assume that water could flow in the intercellular space you come down to a figure about 5 Å of half-width of the channel which would roughly fit with the data you have (Whittembury et al. 1971). Now considering the electrogenicity you obtain for the sodium pump there seems to be also analogies (c. f. Fig. 15 of Giebisch et al. 1971). Now I noticed in the equivalent circuit diagrams that the luminal battery was with the negativity towards the inside of the cell and I would like really that you comment on the possibility that it is the other way around.

SCHULTZ: As for the orientation of the electromotive force across the mucosal membrane, the evidence bearing on this point is very indirect. This evidence suggests that the mucosal membrane is permeable to sodium but relatively impermeable to potassium, whereas the converse appears to obtain for the serosal membranes. This selectivity pattern resembles that described

Whittembury, G. (1967) Sobre los mecanismos de absorción en el tubo proximal del rinon. Acta cientif. venezolana 18, (3, suppl.) pp. 71–83.

Giebisch, G., Boulpaep, E. L. & Whittembury, G. (1971) Electrolyte transport in kidney tubule cells. Proc. roy. Soc. B 262, 175–196.

for the frog skin and would suggest that the cell interior should be electrically positive with respect to the mucosal solution, an orientation opposite to that which is observed experimentally. On the other hand, when the frog skin is short-circuited or when the shunt conductance is increased by making the outer solution hypertonic with urea (Ussing & Windhager 1964) the cell interior of the frog skin becomes electrically negative with respect to the outer solution. Thus, it is quite possible that because of the high conductance shunt pathway acros rabbit ileum which permits coupling between the electromotive forces across the serosal and mucosal membranes, the electrical potential difference is opposite to the electromotive force across that membrane. In short, in the case of a tissue that is to a large extent "self short-circuited", the finding that the cell interior is electrically negative with respect to the mucosal solution does not rule out the possibility that the electromotive force across that membrane is oriented in the opposite direction. As for the question concerning the dimensions of the shunt pathway: Some years ago, Dr. Bjarne Munck and I (Munck & Schultz 1969) demonstrated that lysine movements across rabbit ileum were not significantly affected by the transepithelial potential difference. We, therefore, estimate that the dimensions of the shunt pathway restricts the diffusion of molecules with a radius of 5–6 A.

KEYNES: Why don't you like the term electrogenic that you have been partly responsible for?

SCHULTZ: Becauce even neutral pumps are electrogenic. If we inhibit a neutral pump transmembrane electrical potential differences will be dissipated. And the reason I don't like the term in fact stems from one of the early pages in professor Using's monograph of some years ago where he pointed out the fact that were it not for pumps all the electrical potential differences other than Donnan potentials would be dissipated. So that I think the term electrogenic does not distinguish between a pump that is actually generating current and a pump that is electrically neutral but at the same time results in an electrical potential difference across the membrane.

Ussing, H. H. & Windhager, E. E. (1964) Nature of shunt path and active sodium transport path through frog skin epithelium. *Acta physiol. scand. 61*, 484–504.
Munck, B. G. & Schultz, S. G. (1969) Lysine transport across isolated rabbit ileum. *J. Gen. Physiol. 53*, 157–182.

KEYNES: What does your rheogenic mean?

SCHULTZ: Rheogenic is generating current. Electrogenic implies the "genesis" of an electrical potential difference. Actually, the term "rheogenic" was suggested by Schwartz (1971) so that while I strongly adhere to this terminology, I claim no credit for introducing it.

Schwartz, T. L. (1971) Direct effects on the membrane potential due to pumps that transfer no net charge. *Biophys. J.* *11*, 944–960.

Amino Acid Transport in Intestines

Peter F. Curran

The classical concept of active transport across an epithelial membrane system was proposed by Koefoed-Johnsen & Ussing (1958) in their model for Na transport in frog skin. This model incorporated the now rather obvious concept that, if a substance is actively transported across an intact cell layer, the membranes at the two sides of the cells making up the epithelium must have different transport properties. If the membranes were entirely symmetrical, the cells could clearly maintain an intracellular environment different from the external one but they could not bring about net active transcellular transport. Koefoed-Johnsen & Ussing also provided direct evidence for the presence of asymmetrical barriers in frog skin with respect to the response of electrical potential difference to ionic composition of the external solutions.

For some years now, my laboratory has been interested in trying to obtain further information on transport properties of the individual barriers in epithelia and to understand how these properties determine the overall transport characteristics of the tissue. Such information has not been easy to acquire because of difficulties in making direct and unequivocal measurements on the individual barriers. In fact, it is fair to say, we do not yet understand any epithelial system in detail in these terms. However, I believe that we are now beginning to approach this degree of understanding for transport of neutral amino acids across the intestinal mucosa, and I would like to summarize the current position.

Neutral amino acids are actively transported from mucosa to serosa across *in vitro* preparations of small intestine from a variety of animals. During the process of transcellular transport, the mucosal cells also accumu-

Department of Physiology. Yale University School of Medicine. New Haven, Connecticut 06510 U.S.A.

»Transport Mechanisms in Epithelia«, Munksgaard, Copenhagen.

Table I. Alanine Transport in Intestine

	$[A]_o$ (mM)	$[A]_c$	J_{ms} (μmoles/hr cm^2)	J_{sm}
a) Na Present				
Rabbit	5	42.5	1.30	0.13
Turtle	5	32.5	0.60	0.12
b) Na Absent				
Rabbit	5	4.9	0.16	0.13
Turtle	5	6.7	0.10	0.07

$[A]_o$ and $[A]_c$ indicate external and cellular concentrations respectively and J_{ms} and J_{sm} indicate mucosal-to-serosal and serosal-to-mucosal fluxes respectively. Fluxes were measured with 5 mM alanine in both mucosal and serosal solutions. Data on rabbit from Field *et al.* (1967) and Schultz *et al.* (1966), on turtle from Hajjar *et al.* (1972).

late the amino acid to concentrations considerably in excess of those in the bathing media. Table Ia shows some typical data on these points for alanine transport in rabbit and turtle intestine, the two species to be considered. In both species, the mucosal-to-serosal alanine flux is several times larger than flux in the opposite direction when the two bathing solutions contain identical alanine concentrations, and the cells accumulate the amino acid to a level 5–10 times that in the external solution. We have shown in rabbit ileum that most of this cellular alanine is in free, osmotically active form since alanine accumulation is accompanied by the quantitatively expected increase in cell water (Schultz *et al.* 1966, see also Armstrong *et al.* 1970). One simple conclusion that can be drawn from these observations is that the primary active transport step for amino acid is located in the brush border and that it "pumps" amino acid into the cell from the mucosal solution. While an active transport system on the serosal side cannot be ruled out, active transport from mucosa to serosa accompanied by cellular accumulation appears to require a "pump" in the mucosal or brush border membrane.

An additional characteristic of intestinal amino acid transport, important in our considerations, is illustrated in Table Ib. If all Na ion in the bathing solutions is replaced by choline, the ability of the tissue to bring about active transmural transport and cellular accumulation is completely abolished. Similar effects are obtained if ions other than choline are used to replace Na, indicating that *in vitro* the active transport of amino acids has an absolute requirement for Na. In addition, Schultz & Zalusky (1965) have demonstrated that the rate of active Na transport from mucosa to serosa in rabbit ileum is markedly stimulated if an actively transported amino acid, such as alanine, is added to the mucosal bathing solution. Thus, there appears to be a more or less direct coupling between Na transport and amino acid transport. An understanding of the nature of this coupling is essential in developing an adequate hypothesis regarding the overall process of amino acid transport. Therefore, our beginning picture of this process, involves an active step of entry of amino acid into the cells from the mucosal solution, an exit step of unspecified nature toward the serosal solution and a marked dependence on Na.

Several years ago, we set about to explore by direct measurements the entry of amino acid into the cells from the mucosal solution. As shown in the subsequent discussion, the only way to obtain unequivocal evidence about this step is to measure it directly since other measurements such as transmural flux or cellular accumulation depend on aspects of the system other than the primary active step. The method for making this measurement has been described in detail, (Schultz *et al.* 1967). It involves a brief exposure of the mucosal surface only to solution containing radioactive amino acid and/or Na and determination of the amount of tracer taken up by the tissue. A variety of control experiments have shown that the method provides an accurate estimate of unidirectional flux from mucosal solution to cell.

Using this technique, we have characterized the influx of several neutral amino acids across the brush border membrane of rabbit ileum and examined the dependence of this process on Na. The influx of neutral amino acids was reduced by about 75% when all Na was removed from the mucosal bathing solution, but was unaffected by depletion of cellular Na, indicating that transport into the cells requires Na in the external solution but not cellular Na. More detailed studies showed that the influx of amino acid across the mucosal membrane (J_m^i) followed Michaelis-Menten type kinetics so that

$$J_m^i = \frac{J_m^{im}\,[A]_m}{K_m + [A]_m} \tag{1}$$

in which J_m^{im} is the maximal influx, $[A]_m$ is amino acid concentration in the mucosal solution and K_m is the "apparent Michaelis constant." J_m^{im} was found to be independent of Na but K_m increased as Na concentration in the mucosal solution was reduced. Table II summarizes results obtained with several amino acids. In the absence of Na, amino acid influx still obeyed equation 1 and could be competitively inhibited by other amino acids. Thus even under these conditions in which there is no active transport, amino acid entry across the brush border involves a mediated transfer process and this process appears to be the same one that is capable of active transport in the presence of Na. We have been unable to detect a significant movement of amino acids across the brush border by free diffusion.

Table II. Characteristics of Amino Acid Influx Across The Brush Border of Rabbit Ileum.

Amino Acid	$[Na]_m$ (mM)	K_m (mM)	J_m^{im} (μmoles/hr cm^2)
Glycine	140	22.7	4.5
(Peterson et al. 1970)	70	33.4	4.9
	40	68.0	5.2
Alanine	140	9.1	6.1
(Curran et al. 1967)	20	31.2	7.8
	0	70.0	6.8
Phenylalanine	140	2.7	3.5
(Hajjar et al. 1970)	20	10.2	3.8
	0	17.6	4.8
Methionine	140	1.3	3.0
(Preston & Curran unpublished)	70	2.2	3.4
	0	9.1	2.9

Na influx across the brush border is increased by the presence of amino acids in the mucosal solution; there is approximately a linear relation between Na influx and amino acid influx. Approximately one extra Na ion enters the cell with each amino acid molecule at 140 mM Na, but the slopes of the lines relating these two fluxes decrease as Na concentration is lowered (Table III).

These and several other observations have led us to a model of the amino acid transport system in the brush border that is shown in Fig. 1. It involves a transport site, X, that can combine with both Na and amino acid and mediate the movement of both solutes across the membrane. In the absence of Na, amino acid transfer can still occur via the form XA. As discussed in

Table III. Relationship Between Na And Amino Acid Influxes Across The Brush
Border Of Rabbit Ileum

Amino Acid	$[Na]_0$ (mM)	$\Delta J^i_{Na} / \Delta J^i_A$
Alanine	140	0.96
(Curran et al. 1967)	70	0.82
	25	0.55
	5	0.17
Glycine	140	0.95
Peterson et al. 1970)	40	0.60
	10	0.21

detail by Curran *et al.* (1967), this model can explain quantitatively our
observations on influx across the brush border. In formulating the model
shown in Fig. 1, we have not indicated a direct input of metabolic energy
to account for transport of amino acid into the cell against a concentration
difference. Instead, we have implicitly followed the lead of Crane (1965) who
has proposed that the Na concentration difference between mucosal solution
and cytoplasm could provide the driving force required for cellular accumulation of sugars in intestine. The model shown could achieve this end, since the
effective affinity of the amino acid for the transport system is determined by
the Na concentration. It can be easily shown that if Na concentration in the
mucosal solution is higher than that in the cytoplasm, the model predicts

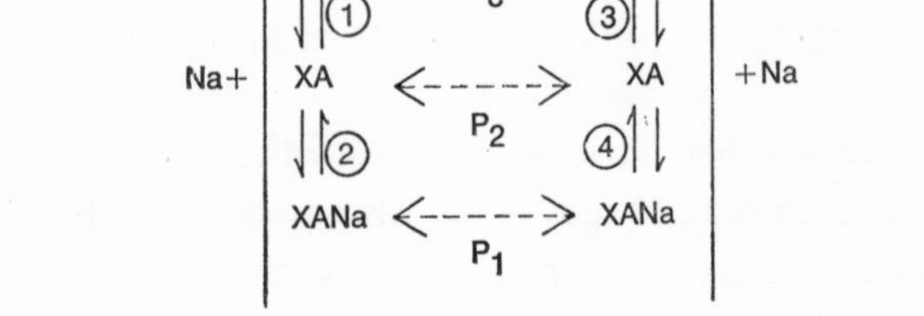

MODEL for AMINO ACID TRANSPORT

Fig. 1. Model for the amino acid transport system in the brush border of intestine.
Reproduced from Curran *et al.* (1967) by permission of Rockefeller University Press.

amino acid transport into the cell against a concentration difference. Such schemes have been discussed in some detail (Schultz & Curran, 1970; Eddy, 1967; Vidaver & Shepherd, 1968). We will not dwell on them here, or on the as yet unresolved question of whether the Na gradient is the sole energy source for active amino acid transport (Kimmach, 1970). However, the model leads to certain predictions regarding amino acid efflux across the membrane from cell to mucosal solution. This process is the second aspect of the overall transport system that we must try to evaluate.

The model suggests that the amino acid transfer system is at least to some extent reversible and that it could also mediate amino acid efflux from the cell. If the system is reversible, the direction of its operation should be determined by the relative Na and amino acid concentrations in the cell and the bathing solution. In particular, an outwardly directed Na concentration gradient (high in the cell and low outside) should be capable of driving an "active" extrusion of amino acid from the cell.

Hajjar et al. (1970) tested this possibility by incubating strips of mucosa poisoned with ouabain or cyanide in solutions containing 140 mM Na and 5 mM alanine. Under these conditions, both Na and alanine concentrations in cell water become approximately equal to those in the bathing solutions.

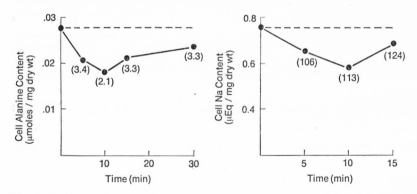

Fig. 2. (Left), alanine extrusion from mucosal cells in the presence of a Na gradient. Ouabain poisoned tissues loaded with Na and alanine were transferred at zero time to Na-free KCl medium containing 5 mM alanine. Numbers in parantheses give cellular alanine concentration in mM. Content and concentration do not change in parallel because of changes in cell volume. (Right), Na extrusion from mucosal cells in the presence of an alanine gradient. Ouabain poisoned tissues loaded with Na and alanine (40 mM) were transferred at zero time to alanine-free medium containing 140 mM Na. Numbers in parentheses are cellular Na concentrations.

The strips of tissue were then transferred to a Na-free solution containing the same concentration of alanine as the preincubation solution and cellular alanine content was measured as a function of time after transfer. As shown in Fig. 2, there was a net extrusion of alanine from the cells; cellular alanine concentration fell to a level well below the external concentration so that the extrusion occurred against a concentration difference.

Curran et al. (1970) carried out the converse experiment to examine the effect of an amino acid gradient on Na movement. Mucosal strips were poisoned with sufficient ouabain to inhibit completely the Na pump. They were loaded with Na and alanine and then transferred to a solution containing no alanine. As also illustrated in Fig. 2, there was a net extrusion of Na against a concentration difference. These net Na and alanine movements occurred mainly across the brush border membrane (Hajjar et al. 1970; Curran et al. 1970). Thus, they are quite consistent with the predictions of the model shown in Fig. 1 and suggest that efflux of amino acid from the cell across the brush border is mediated by a process that may have characteristics similar to those displayed by the influx process at this membrane.

To test this point in more detail, we have attempted to make direct measurements of efflux across the brush border by loading the cells with radioactive alanine and following washout toward the mucosal solution (Hajjar et al. 1970). From such experiments, efflux of alanine can be evaluated as functions of cellular concentrations of Na and the amino acid. There are clearly, uncertainties about such measurements, including questions regarding compartmentalization and intracellular activity, particularly for Na but the results should provide some further insight into characteristics of the efflux process.

Alanine efflux toward the mucosal solution (J_m^e) could also be described by Michaelis-Menten kinetics so that

$$J_m^c = \frac{J_m^{em} \, [A]_c}{K_m' + [A]_c} \tag{2}$$

in which $[A]_c$ is cellular concentration of amino acid. As is the case for alanine influx across this membrane, maximal efflux, J_m^{em} appeared to be independent of cellular Na concentration, but K_m' increased as cellular Na concentration decreased. The maximal flux rates for influx and efflux appeared to be similar, but under apparently comparable conditions, K_m' > K_m. For example, in normal cells having a nominal cell Na concentration of 40–50 mM (Schultz et al. 1966), K_m'=55 mM while measurements of alanine influx from a mucosal solution containing 40 mM Na yield K_m=22. This difference could be due to an inherent asymmetry in the transport system. Alternatively, it could be caused by competition for alanine exit by other amino acids present in the cellular pool. Compartmentalization of cellular Na and a low Na activity coefficient (Lee & Armstrong 1972) could also be involved (i. e. the effective Na concentration at the transport site may be considerably lower than the measured average cell Na concentration and this would tend to increase K_m'). However, these details need not concern us further for purposes of the present discussion.

The experiments summarized above provide the necessary preliminary information regarding alanine movement across the brush border membrane, but to characterize the overall transport system, we also need similar data about the serosal side of the cell layer. This information is not easy to obtain because the serosal aspect of the cells is rather less accessible for direct measurement than is the mucosal side. Fluxes across the serosal side of the cell can be estimated if the mucosal-to-serosal, serosal-to-mucosal and mucosal-to-cell unidirectional fluxes are known (Schultz et al., 1967). However, these measurements usually require the use of three separate pieces of tissue (actually four pieces if cellular concentration is also determined) and more direct measurements seem highly desirable.

Hajjar et al. (1972) have recently obtained such direct information on alanine transfer across the serosal side of the cells in turtle intestine. The turtle intestine is particularly suitable for such measurements because it can be stripped of nearly all muscle and serosal layers so that the serosal aspect of the cells is in closer contact with the bathing solution than in most other intestinal preparations. With our methods, this preparation transports alanine in a manner very similar to rabbit intestine. There is active transport from mucosa to serosa accompanied by accumulation of alanine in the epithelial cells (see Table I). Both net active transport and cellular accumulation are abolished by incubation of the tissue in Na-free medium. Thus, the behavior of the alanine transport system in this species of turtle appears to be quite similar to that in rabbit intestine and, to a first approximation, it is not unreasonable to suggest that the serosal membranes in the two species might be relatively similar. (We must note,

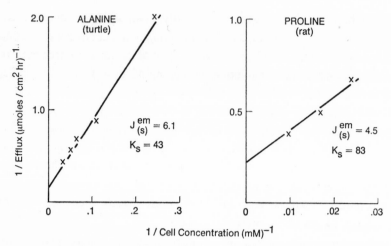

Fig. 3. Double reciprocal plot of amino acid flux from cell to serosal solution as a function of cellular concentration.

however, that Gilles-Baillien & Schoffeniels (1966, 1970) have obtained quite different results in another species of turtle for reasons that remain unclear).

We have, to date, examined two aspects of alanine movement across the serosal membrane. First, unidirectional flux from cell to serosal solution was measured by loading the cells with radioactive alanine and following washout toward the serosal solution. Different cellular concentrations of alanine were achieved by preincubating the tissue in solutions containing different alanine concentrations. The results of one series of experiments are shown in Fig. 3 in terms of a double reciprocal plot, 1/efflux vs. 1/cellular concentration. The data can be described by Michaelis-Menten kinetics

$$J_s^e = \frac{J_s^{em} [A]_c}{K_s + [A]_c} \qquad (3)$$

with a J_s^{em} of 6.1 μmoles/hr cm^2 and a K_s of 43 mM. Thus, the exit of alanine from these cells at the serosal side appears to occur by a saturable, mediated process. Munck & Rasmussen (personal communication) have recently used the indirect method referred to above to evaluate proline fluxes from the cells of rat intestine toward the mucosal and serosal solutions. Their results for efflux across the serosal barrier are shown in Fig. 3 and they are also consistent with the presence of a saturating process in this membrane. In fact, the exit processes for alanine in turtle intestine and for proline in rat intestine appear to have similar characteristics.

Hajjar *et al.* (1972) also examined the Na-dependence of the efflux process by carrying out similar experiments with cells depleted of Na by incubation in Na-free medium or loaded with Na by incubation in the presence of ouabain. Alanine efflux was unaffected by wide variations in cellular Na concentration indicating that efflux at the serosal side of the cell has different characteristics than efflux at the mucosal side (which is Na-dependent).

In a second series of experiments, we examined the ability of a Na concentration difference to cause net movement of alanine in turtle intestine. Strips of mucosa were loaded with Na and alanine and then transferred to Na-free medium containing alanine as described above for experiments on rabbit ileum. A net extrusion of alanine was observed with results nearly identical to those shown in Fig. 2. We then carried out experiments to determine whether this alanine extrusion driven by a Na concentration difference occurred across the mucosal or serosal side of the cell. Tissues loaded with Na and alanine were mounted in a chamber and only one side of the tissue was exposed to Na-free solution. Results are summarized in Fig. 4. An outwardly directed Na concentration difference across the mucosal membrane caused net alanine extrusion. A similar Na gradient across the serosal membrane caused no alanine extrusion. Thus alanine efflux across the mucosal membrane is Na-dependent (as in rabbit ileum) while efflux across the serosal membrane is Na-independent.

The one aspect of alanine movement across the intestinal mucosa for which we have no direct evidence at present is the unidirectional flux from the serosal solution into the cells (J_s^i). It seems reasonable to suggest that this process occurs via the same mediated mechanism involved in efflux across this membrane. In other words, we assume that alanine fluxes across the serosal membrane involve a facilitated transfer mechanism with net flux always in the direction of the alanine concentration difference. If we assume that these results on rabbit and turtle intestine can be combined, we can now formulate a specific description of amino acid transport in intestine. The overall similarities in alanine transport in these two species suggest that this assumption is a reasonable first approximation, and the experiments

20*

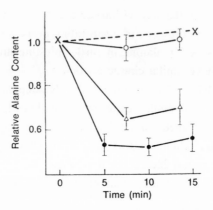

Fig. 4. Alanine extrusion from mucosal cells of turtle intestine in the presence of a Na gradient. Ouabain poisoned tissues were loaded with Na and alanine and transferred at zero time to Na-free choline medium containing 10 mM alanine. (●) both sides of tissue exposed to Na-free medium, (0) only the serosal side exposed and (△) only the mucosal side exposed.

of Munck & Rasmussen, discussed above, lend support to this concept. We can summarize the overall transport of alanine under steady state conditions as follows:

$$J_m^i - J_m^e = J_s^e - J_s^i \tag{4}$$

Equation 4 simply states that inflow of alanine into the cell is equal outflow so that cellular concentration is independent of time. If we assume that maximal fluxes across the mucosal membrane are equal for the two directions as suggested by the data of Curran *et al.* (1967) and Hajjar *et al.* (1970), Equation 4 can be written as

$$J_m^{im}\left[\frac{[A]_m}{K_m + [A]_m} - \frac{[A]_c}{K_m' + [A]_c}\right] = J_s^{em}\left[\frac{[A]_c}{K_s + [A]_c} - \frac{[A]_s}{K_s + [A]_s}\right] \tag{5}$$

in which Equations 1, 2 and 3 have been used and we have assumed that the facilitated transfer system in the serosal membrane is entirely symmetrical. Asymmetry at the mucosal side is expressed by the fact that $K_m < K_m'$.

For given values of the various parameters, Equation 5 can be solved for $[A]_c$. Using this value, unidirectional fluxes across the two membranes can be estimated as well as transepithelial unidirectional and net fluxes.

Too few experimental data are available to test this model in any detail, but a few sample calculations may be of interest. For simplicity, we assume equal amino acid concentrations in the mucosal and serosal solutions and examine the behavior of the system for four compounds that have different affinities for the transport systems. For this calculation, $J_m^{im} = J_s^{em} = 6$ μmoles/hr cm², $K_m' = K_s$ and $K_m'/K_m = 8$. This last asymmetry insures that the system is capable of "active" transport. Fig. 5 shows the results for three readily measurable properties of the system: net transepithelial flux, unidirectional flux from mucosa to serosa, and the steady state ratio of cellular to extracellular concentration. Values of K_m chosen for the calculation were 1, 2, 5, 10 mM. These are roughly equivalent to values established in rabbit ileum for methionine, phenylalanine, leucine and alanine influx across the brush border.

The results of these calculations display several points of interest. As amino acid concentration increases, net transmural flux rises to a peak and then declines. The concentration at which maximal net flux is obtained depends on K_m, the "apparent affinity" of the amino acid for the mucosal transport system. This type of behavior for net amino acid transport was observed by Mathews & Laster (1965) in studies on rat intestine. The observed decline in net flux at higher concentrations was attributed to damage to the intestine, but these calculations suggest it may be a natural concequence of the behavior of the overall transport system. A second point of interest regarding the calculated net flux is that its magnitude is not necessarily related to the "affinity" of the amino acid for the mucosal system. At low concentrations, the net flux is greater the higher the affinity, but at high concentrations, the reverse is true. At concentrations above 20 mM, the amino acid with lowest affinity displays the highest net flux.

The upper portion of Fig. 5 shows that, for the parameters chosen for this calculation, the cellular accumulation ratio is inversely related to the "affinity" of the amino acid for the mucosal transport system. That is, the amino acid with the lowest affinity has the highest accumulation ratio. Comparison of the upper and lower portions of Fig. 5 shows that there is not a necessary or consistent relationship between the ability of the cells to accumulate amino acid and net active transport across the epithelium. At low

PETER F. CURRAN

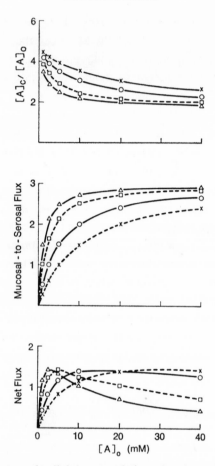

Fig. 5. Calculated fluxes and cellular accumulation of amino acids based on Equation 5. $J_m^{im} = J_s^{em} = 6$ μmoles/hr cm^2, $K_m' = K_s$, $K_m/K_m = 8$.
($\triangle$) $K_m = 1mM$, ($\square$) $K_m = 2$, (0) $K_m = 5$, (X) $K_m = 10$.

amino acid concentrations, the highest net flux occurs with the amino acid showing the lowest accumulation ratio.

Finally, there is not a simple relationship between unidirectional mucosal to serosal flux and net transport. Again, a high mucosal to serosal flux may be associated with either a relatively high or a relatively low net flux depending on external amino acid concentration. Further, for these particular conditions, the mucosal to serosal flux is inversely related to the cellular accumulation ratio. However, in this set of calculations, the mucosal to serosal flux most nearly reflects the "active" process at the mucosal border in the sense that the half maximal concentration for this flux is the same as K_m (the half maximal concentration for the active step). However, further calculations show that this is not a general relationship. If K_m' is varied, keeping all other parameters constant, the half maximal concentration for mucosal to serosal flux is appreciably different than K_m.

These few calculations are far from exhaustive. They are meant simply to show the complexities of a system such as the one proposed for intestinal amino acid transport. Such a system cannot be characterized in a satisfactory way by a single type of experiment or by studies at a single concentration. It seems clear that if we hope to understand this particular epithelial transport system, or for that matter any similar system, we must apply a variety of experimental techniques to measure a number of properties of the system. Throughout this discussion I have tacitly assumed that there is a single amino acid transport system in intestine. This is clearly incorrect in terms of the total spectrum of amino acids (Wilson 1962) and is probably incorrect, at least to some extent, even for neutral (monoamino-monocarboxylic) amino acids. In other tissues for example, Christensen (1967) has defined several transport systems for neutral amino acids that have overlapping affinities. Such a situation adds an entire level of complexity over and above that discussed here.

REFERENCES

Armstrong, W. McD., Musselman, D. L. & Reitzag, H. C. (1970) Sodium, potassium and water content of isolated bullfrog small intestinal epithelia. *Amer. J. Physiol. 219*, 1023–1026.

Christensen, H. N. (1967) Some transport lessons taught by the organic solute. *Perspect. Biol. Med. 10*, 471–494.

Crane, R. K. (1965) Na$^+$-dependent transport in the intestine and other animal tissues. *Fed. Proc. 24*, 1000–1005.

Curran, P. F., Hajjar, J. J. & Glynn, I. M. (1970) The sodium-alanine interaction in rabbit ileum. Effect of alanine on sodium fluxes. *J. gen. Physiol. 55*, 296–308.

Curran, P. F., Schultz, S. G., Chez, R. A. & Fuisz, R. E. (1967) Kinetic relations of the Na-amino acid interaction at the mucosal border of intestine. *J. gen. Physiol. 50*, 1261–1286.

Eddy, A. A., Mulcahy, M. F. & Thomson, P. J. (1967) The effects of sodium ions and potassium ions on glycine uptake by mouse ascites-tumor cells in the presence and absence of selected metabolic inhibitors. *Biochem. J. 103*, 863–876.

Field, M., Schultz, S. G. & Curran, P. F. (1967) Alanine transport across isolated rabbit ileum. *Biochim. biophys. Acta (Amst.) 135*, 236–243.

Gilles-Baillien, M. (1970) Inorganic ions and the transfer of neutral amino acids

312 PETER F. CURRAN

across the isolated intestinal epithelium of the tortoise *Testudo hermanni hermanni* Gmelin. *Arch. int. Physiol. Biochim. 78,* 119–130.

Gilles-Baillien, M. & Schoffeniels, E. (1966) Metabolic fate of L-alanine transported across the tortoise intestine. *Life Sci. 5,* 2253–2255.

Hajjar, J. J. & Curran, P. F. (1970) Characteristics of the amino acid transport in the mucosal border of rabbit ileum. *J. gen. Physiol. 56,* 673–691.

Hajjar, J. J., Khuri, R. N. & Curran, P. F. (1972) Alanine efflux across the serosal border of turtle intestine. *J. gen. Physiol. 60,* 720–734.

Hajjar, J. J., Lamont, A. S. & Curran, P. F. (1970) The sodium-alanine interaction in rabbit ileum. Effect of sodium on alanine fluxes. *J. gen. Physiol. 55,* 277–296.

Kimmach, G. A. (1970) Active sugar accumulation by isolated intestinal epithelial cells. A new model for sodium-dependent metabolite transport. *Biochemistry, 9,* 3669–3677.

Koefoed-Johnsen, V. & Ussing, H. H. (1958) The nature of the frog skin potential. *Acta physiol. scand. 42,* 298–308.

Lee, C. O. & Armstrong, W. McD. (1972) Activities of sodium and potassium ions in epithelial cells of small intestine. *Science 175,* 1261–1263.

Matthews, D. M. & Laster, L. (1965) Kinetics of intestinal active transport of five neutral amino acids. *Amer. J. Physiol. 298,* 593–600.

Munck, B. & Rasmussen, S. Private communication.

Peterson, S. C., Goldner, A. M. & Curran, P. F. (1970) Glycine transport in rabbit ileum. *Amer. J. Physiol. 219,* 1027–1032.

Preston, R. L. & Curran, P. F. Unpublished observations.

Schultz, S. G. & Curran, P. F. (1970) Coupled transport of sodium and organic solutes. *Physiol. Rev. 50,* 637–718.

Schultz, S. G. & Zalusky, R. (1965) Interactions between active sodium transport and active amino acid transport in isolated rabbit ileum. *Nature (Lond.) 204,* 292–294.

Schultz, S. G., Fuisz, R. E. & Curran, P. F. (1966) Amino acid and sugar transport in rabbit ileum. *J. gen. Physiol. 49,* 849–866.

Schultz, S. G., Curran, P. F., Chez, R. A. & Fuisz, R. E. (1967) Alanine and sodium fluxes across the mucosal border of rabbit ileum. *J. gen. Physiol. 50,* 1241–1260.

Wilson, T. H. (1962) *Intestinal Absorption* W. B. Saunders, Philadelphia.

Vidaver, G. A. & Shepherd, S. L. (1968) Transport of glycine by hemolyzed and restored pigeon red blood cells. Symmetry properties, trans effects of sodium ion and glycine, and their description by a single rate equation. *J. biol. Chem. 243,* 6140–6150.

DISCUSSION

TOSTESON: Would you like to speculate on how you think the formation of the complex with the sodium is promoting the transport and also about the specificity of that reaction for the different cations.

CURRAN: I can answer the second part more easily because there is a fairly factual answer. The specificity is really quite high for sodium although not absolute. Frizzell and Schultz (1970) found that the only ion that shows any ability to replace sodium at all is lithium, and it is a very poor substitute. The other ions that they tried were either inhibitory or had no effect at all; potassium was inhibitory, rubidium was slightly inhibitory and caesium had no effect. Thus the system is highly specific for sodium but the specificity is not quite as absolute as we once thought it was. Then you asked me to speculate about?

TOSTESON: Do you think that it is possible that the complex between amino acid and carrier is charged and that the monovalent cation renders it neutral and therefore promotes the transport?

CURRAN: That is quite possible. We have no evidence one way or another. The data we have for a variety of amino acids almost requires that we postulate that all three forms of the transport site are translocated across the membrane with identical rate constants. In that case, the effect of sodium would be simply to shift the relative concentrations of uncombined carrier to forms in which it is combined with the amino acids.

TOSTESON: The two forms, right.

CURRAN: Yes.

KIRSCHNER: But isn't the interaction between sodium and glucose virtually identical.

CURRAN: That appears to depend on the particular animal studied. The characteristics of the relationship between sodium and sugar transport determined by Crane (1965) for hamster intestine are essentially identical to the characteristics of amino acid transport that I have shown for rabbit intestine.

Frizzell, R. A. & Schultz, S. G. (1970) Effects of monovalent cations on the sodium-alanine interaction in rabbit ileum. *J. Gen. Physiol. 56*, 462–490.

However, in rabbit intestine the behavior of sugars is different in that changing sodium concentration alters maximal flux rate but causes relatively little change in the apparent Michaelis constant.

KIRSCHNER: But you do get a strong interaction between the two. What I am getting at is the suggestion from Dr. Tosteson that what you are doing really is creating a charged carrier to move: I think the similarity in the glucose system will argue against that.

CURRAN: No, not necessarily.

MOREL: What about the possibility of concentration gradients (in the steady-state condition) within the cell, along the axis of transporting microvilli for example?

CURRAN: I suppose that is possible. It is an extremely difficult question to answer. I think the only point one can make about it, is that in his radio-autographic studies Kinter (private communication) was unable to detect any concentration gradient for sugar across the cell. That is a fairly crude estimate, but Kinter felt that he would be able to detect a fairly large gradient if it existed.

Crane, R. K. (1965) Na+-dependent transport in the intestine and other animal tissues. *Fed. Proc. 24,* 1000–1006.

Coupling of Salt and Water Flow Across and Along Intestinal Epithelia

E. Skadhauge

INTRODUCTION

Water can flow across vertebrate intestinal epithelia in the absence of, or against, a transepithelial osmotic difference, along with the active transport of NaCl. This flow happens presumably because a local hypertonicity is built up in or between the cells. I shall discuss the coupling of salt and water flow across and along the two vertebrate intestinal epithelia which in nature receive fluids hyperosmotic to plasma: the bird cloaca and the intestine of the marine teleost. I have been mainly interested in understanding the regulation of salt and water absorption in these intestines, the adaptation to dehydration, and the role of the solute-linked water flow against the osmotic difference.

The strategy of the experiments has been first to study easily available species of convenient size. I chose the fowl *(Gallus domesticus)* and the European eel *(Anguilla anguilla)*. Later, more specialized species, i. e. species well adapted to dehydration were studied: the budgerygah *(Melopsittacus undulatus)*, a bird which can live without water, fed dry seeds alone, and a cyprinodont, *Aphanius dispar,* a fish that can stand very high salinity of the surrounding water, around 4000 mOsm. This approach was considered most suitable in order to elucidate the mechanisms involved.

I shall give an overview of our findings in the birds and the fish, and focus on the interaction of salt and water flow. Although water absorption against an osmotic difference originates on the cellular level, it is the interaction, as the intestinal fluid flows along the length of the gut, which permits a net water absorption from the fluids originally hyperosmotic to plasma.

Institute of Medical Physiology A, University of Copenhagen, Juliane Maries Vej 28, 2100 Copenhagen Ø, Denmark.

»Transport Mechanisms in Epithelia«, Munksgaard, Copenhagen.

BIRDS

In the dehydrated bird the urine which is hyperosmotic to plasma is regurgitated into the coprodeum and large intestine from which a resorption of salt and water may occur. An important problem is: Does the cloacal sojourn lead to a water loss counteracting the osmotic work of the kidney, or does a resorption occur?

Four types of studies were carried out: *1.* The flow rate of ureteral urine, its osmolality and ionic concentration were measured (Skadhauge & Schmidt-Nielsen 1967). *2.* The osmotic and ionic concentration of the cloacal contents were measured *in vivo* (Skadhauge 1968). *3.* The cloacal transport parameters were measured by an *in vivo* perfusion technique (Fig. 1). With this technique the maximal sodium transport rate, the apparent Km for sodium transport, the osmotic permeability coefficients, the reflexion coefficient to the solutes of plasma, and the solute-linked water flow as a

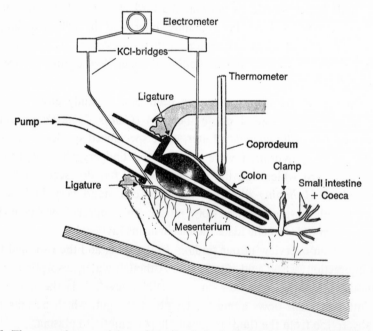

Fig. 1. The coprodeum and large intestine (colon) of the domestic fowl with a teflon insert through which perfusion fluids can be circulated. The net water flow was calculated from the change in concentration of unabsorbable water markers. The length of the coprodeum and large intestine is around 7 cm (Bindslev & Skadhauge 1971a).

function of osmolality of the luminal perfusion fluid were measured (Bindslev & Skadhauge 1971a, Bindslev & Skadhauge 1971b, Skadhauge 1967). With this knowledge at hand the cloacal modification of ureteral urine could be estimated quantitatively. *4.* The integrated salt and water absorption in the cloaca in the dehydrated state was analyzed by analog computer simulation (Skadhauge & Kristensen 1972).

Some of the experimental findings will be given in slightly more detail at this point. First the sodium concentration of the absorbate when the luminal fluid is isosmotic to plasma seemed to be higher than that of plasma, around 600 mEq/1, and the perfusion fluid will therefore eventually be diluted.

Second, when the osmolality of the perfusion fluid was sufficiently higher or lower than that of plasma (150–200 mOs) the difference between the general osmotic water flow (measured in the absence of NaCl in the perfusion fluid) and the solute-linked water flow (with NaCl in the perfusate) disappeared.

Third, in rapid perfusion experiments the osmolality at zero water flow was 60–80 mOsm higher than that of plasma. If only a limited amount of NaCl and water flows into the intestine, which is the case in the dehydrated bird, net water resorption will take place from solutions of much higher osmolality. Solutions 200 mOsm higher than plasma resulted in a zero water absorption (Bindslev & Skadhauge 1971b), and is the osmolality of ureteral urine in the dehydrated fowl (Skadhauge & Schmidt-Nielsen 1967).

Fourth, when the fowl was dehydrated, the osmotic permeability coefficient for flow in the mucosa to serosa direction went up, and the apparent Km for sodium transport was reduced. The computer calculation disclosed, however, that the total intestinal water absorption was little sensitive to the osmotic permeability coefficient. But the decreased Km resulted in an augmented sodium absorption, which caused a marked increase in water absorption.

When all the available information was used, the following picture emerged. If the bird is hydrated or salt loaded the cloacal sojourn does not quantitatively interfere with the water and salt excretion of the kidney. In the dehydrated state a small fraction of the hyperosmotic ureteral urine (water) is resorbed in the cloaca (coprodeum and large intestine) but at the expense of a pronounced NaCl absorption. Since xerophilic seedeaters often lack NaCl this may be an advantageous adaptation. With the parameters, as observed, the resorption seems maximal at the maximal urine osmolality (Skadhauge & Kristensen 1972). This is partly due to a slight reduction of the glomerular filtration rate occurring in the dehydrated state (Skadhauge & Schmidt-Nielsen 1967) presumably due to arginine vasotocin (Ames *et al.* 1971).

In the budgerygah, which concentrates better than the fowl, a computer simulation was carried out using the cloacal transport parameters of the fowl on a weight basis. The renal-cloacal interaction was, however, difficult to understand (Krag & Skadhauge 1972, Skadhauge 1972) and direct measurements in budgerygahs or similar birds must be carried out.

FISH

The marine teleost compensates for water loss to the hyperosmotic surroundings by drinking sea water. It absorbs NaCl and water in the intestine and excretes NaCl through the gills. The problem is to understand how the sea water which is hyperosmotic to plasma gets across the intestinal wall, and how the intestinal salt and water absorption is regulated in the euryhaline species. In the yellow European eel we first measured the drinking rate and the gill sodium isotope turnover as functions of the salinity of the surrounding water. Second, the intestinal transport parameters were as in the bird measured by an *in vivo* perfusion technique (Fig. 2). The following relations were observed.

The drinking rate and gill ionic turnover increased with increasing osmolality of the surrounding medium (Maetz & Skadhauge 1968). The intestinal NaCl absorption rate increased from fresh water to sea water and again in double strength sea water adapted animals. At the same time the osmolality against which the intestine could transport water went up (Skadhauge 1969, Skadhauge & Maetz 1967) (Fig. 3).

The correlation between net intestinal NaCl transport and the osmolality against which the intestine could transport water was explored further in additional experiments and a fairly good proportionality observed (Skadhauge, in preparation). A difference around 200 mOs was the highest the perfused intestine could transport against in rapid perfusion experiments. In the eels adapted to double strength sea water the incoming fluid from which net water absorption eventually occurred was 6 times more hyperosmotic to plasma. The fluid was, however, early in the intestine diluted to an osmolality which allowed net water absorption.

The maximal salinity the eel could stand was approximately double strength sea water. At this concentration the amount of NaCl being drunk seemed to match the intestinal NaCl transport capacity. Since the gills'

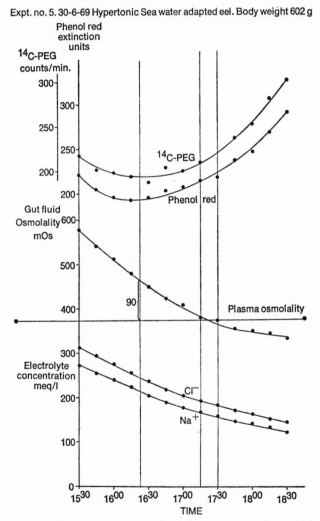

Expt. no. 5. 30-6-69 Hypertonic Sea water adapted eel. Body weight 602 g

Fig. 2. Change in concentrations of unabsobable water markers, and NaCl, and osmolality in the perfusion fluid, when the anterior and the posterior intestine of the yellow European eel is perfused with dilute sea water (Skadhauge, in preparation).

capacity for NaCl excretion seemed much larger (Mayer & Nibelle 1970), the gut NaCl transport may be an important factor limiting the ability to withstand high salinities (Maetz & Skadhauge 1968).

The eel data were also treated by a computer simulation of the flow along the intestine (Fig. 4) and the net water absorption as a function of the various parameters of the system calculated (Kristensen & Skadhauge, in preparation). The net water absorption was investigated in particular as a function of the drinking rate. The interesting result was that the drinking rate in the sea water adapted eel was exactly as large, as to give a maximal water absorption.

Since water absorption only occurs when NaCl also is absorbed, it is quite natural that drinking rate, net intestinal NaCl absorption, and net intestinal water absorption are correlated to the osmolality of the surrounding water.

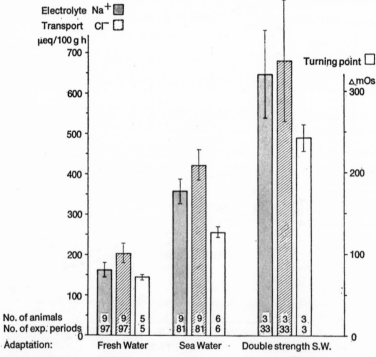

Fig 3. Net NaCl absorption and osmolality against which the intestine could transport water in eels adapted to fresh water, to sea water and to double strength sea water (Skadhauge 1969).

The water loss from the fish, presumably, goes up in waters of higher salinity since plasma osmolality is fairly stable.

In preliminary studies (Skadhauge, in preparation) in *Aphanius dispar,* the picture seemed more complicated. Drinking rate did not seem to be augmented proportional to the osmotic difference from plasma to the surrounding medium. It increased much less (Lotan 1969). This is understandable, however, when it is taken into account that double and triple drinking of fluids of double and triple osmolality will result in a NaCl intake four times and nine times higher, respectively. This might present too large a NaCl load to the intestine, which could not be absorbed. *In vitro* an increase in intestinal Na transport capacity was not found when intestines from *Aphanii* adapted to sea water and to triple strength sea water were used (Lotan & Skadhauge 1972). It seems necessary therefore to conclude that the effective osmotic permeability coefficient, presumably largely that of the gills, goes down in waters of higher sainity.

This problem deserves further study. A likely possibility is that the

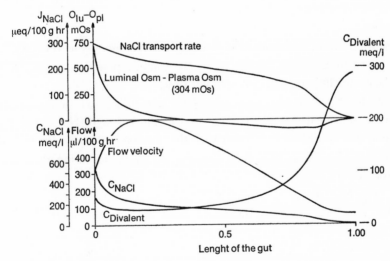

Fig. 4. A computer calculation of the concentration and flow along the length of the gut in the sea water adapted eel. Average values for drinking rate, plasma osmolality, osmotic permeability coefficients, sodium absorption rate, and solute-linked water flow as determined by the author have been used. The divalent ions were assumed not to be absorbed. 100 g body weight corresponds to a simple intestinal surface area of 10 cm². The intestine was assumed to be a rigid tube. (Kristensen & Skadhauge, in preparation).

reflexion coefficient of the gills goes down. Since this requires an energetically expensive pump-leak mechanism, it may be noted, that the gills
seem capable of excreting much more NaCl than the gut can absorb (Mayer
& Nibelle 1970). There is one possible explanation which, most likely, can
be ruled out. It was found in preliminary experiments in the author's
laboratory that the rate of metabolism was not reduced in waters of higher
salinity in spite of the smaller content of oxygen in the media. The oxygen
uptake was not significantly lower in fish adapted to 3500 mOsm than in
fish adapted to ordinary sea water.

COMPARISON OF BIRDS AND FISH

When the coupling of salt and water flow across and along these two
intestinal epithelia is compared, the following picture emerges. In both cases
a hyperosmotic fluid flows along the wall of a tube. It is diluted and some
absorption of salt and water takes place eventually. Both epithelia are very
impermeable to NaCl so the transmural flow of NaCl is almost entirely
due to active transport. The reflexion coefficient is unity. The water flow
is caused by a general osmotic flow in the two directions and by a solute-
linked water flow. Because of self compensation along the tube, the net
water absorption is little influenced by the osmotic permeability coefficients
and the osmolality of solute-linked water flow.

What is critical is the total net NaCl absorption, which is the most
important regulated parameter, compared to the inflow of NaCl. The NaCl
transport will accordingly, together with the osmolality (largely due to NaCl)
of the incoming fluid, determine the fraction of an incoming fluid which
is absorbed.

In both epithelia a net water absorption occurs against a much higher
osmotic difference between incoming fluid and plasma than the single
intestinal epithelial cell can transport against. Thus, as far as the apparently
uphill water flow is involved, the organ is more than the sum of its cells.

REFERENCES

Ames, E., Steven, K. & Skadhauge, E. (1971) Effects of arginine vasotocin on renal
    excretion of Na$^+$, K$^+$, Cl$^-$, and urea in the hydrated chicken. *Amer. J. Physiol.* 221,
    1223–1228.

Bindslev, N. & Skadhauge, E. (1971a) Salt and water permeability of the epithelium of the coprodeum and large intestine in the normal and dehydrated fowl (*Gallus domesticus*). *In vivo* perfusion studies. *J. Physiol. (Lond.) 216*, 735–751.

Bindslev, N. & Skadhauge, E. (1971b) Sodium chloride absorption and solute-linked water flow across the epithelium of the coprodeum and large intestine in the normal and dehydrated fowl (*Gallus domesticus*). *In vivo* perfusion studies. *J. Physiol. (Lond.) 216*, 753–768.

Krag, B. & Skadhauge, E. (1972) Renal salt and water excretion in the budgerygah (*Melopsittacus undulatus*). *Comp. Biochem. Physiol. 41 A*, 667–683.

Kristensen, K. & Skadhauge, E. Unpublished Observations.

Lotan, R. (1969) Sodium, chloride and water balance in the Euryhaline teleost *Aphanius dispar* (Rüppell) (Cyprinodontidae) *Z. vergl. Physiologie 65*, 455–462.

Lotan, R. & Skadhauge, E. (1972) Intestinal salt and water transport in a euryhaline teleost, *Aphanius dispar* (Cyprinodontidae). *Comp. Biochem. Physiol. 42A*, 303–310.

Maetz, J. & Skadhauge, E. (1968) Drinking rates and gill ionic turnover in relation to external salinities in the eel. *Nature (Lond.) 217*, 371–373.

Mayer, N. & Nibelle, J. (1970) Kinetics of the mineral balance in the eel *Anguilla anguilla* in relation to external salinity changes and intravascular saline infusions. *Comp. Biochem. Physicol. 35*, 553–566.

Skadhauge, E. (1967) *In vivo* perfusion studies of the cloacal water and electrolyte resorption in the fowl (*Gallus domesticus*). *Comp. Biochem. Physiol. 23*, 483–501.

Skadhauge, E. (1968) The cloacal storage of urine in the rooster. *Comp. Biochem. Physiol. 24*, 7–18.

Skadhauge, E. (1969) The mechanism of salt and water absorption in the intestine of the eel (*Anguilla anguilla*) adapted to waters os various salinities. *J. Physiol. (Lond.) 204*, 135–158.

Skadhauge, E. (1972) Salt and water excretion in xerophilic birds. *Symp. zool. Soc. Lond. 31*, 113–131.

Skadhauge, E. & Kristensen, K. (1972) An analogue computer simulation of cloacal resorption of salt and water from ureteral urine in birds. *J. theor. Biol. 35*, 473–487.

Skadhauge, E. & Maetz, J. (1967) Étude *in vivo* de l'absorption intestinal d'eau et d'électrolytes chez *Anguilla anguilla* adapté à des milieux de salinités diverses. *C. R. Acad. Sc. Paris (Ser. D) 265*, 347–350 and 923.

Skadhauge, E. & Schmidt-Nielsen, B. (1967) Renal function in domestic fowl. *Amer. J. Physiol. 212*, 793–798.

ACKNOWLEDGEMENT

Major financial support came from the Danish Research Councils and the NOVO Foundation.

DISCUSSION

SCHULTZ: Your finding that adapting the fish to salt water did enhance the absorption of sodium by the intestine is in contrast to that recently reported by Michael Smith and his collaborators (Ellory et al. 1973).

SKADHAUGE: Yes. As the eels are adapted to waters of higher salinity, the maximal net sodium transport rate goes up. But eels are euryhaline teleosts, which live in sea water or even higher salinities. Michael Smith is working on a fresh water fish, the goldfish, which only goes to ⅓ sea water. Teleologically the difference makes sense, since the gills of the eel can excrete the extra load of sodium chloride taken in orally, the gills of the goldfish cannot.

WINDHAGER: I would like to ask a question to Dr. Schultz: I think everybody who works with leaky epithelia would like to tighten up the open tight junctions. Do you know any way? Have you tried hypertonic solutions in the lumen?

SCHULTZ: Yes, we have tried to tighten the tight junctions by making the mucosal solution hypertonic but the result was only a 20 per cent increase in tissue resistance.

WHITTEMBURY: Does the resistance of the intestine change at different salinities?

SKADHAUGE: As far as it has been measured, there is little difference – the transmural PD is small, a few millivolts in the intestine of marine teleosts. This has also been observed by House & Green (1965). In the bird cloaca it is different. It has all the characteristics of mammalian colon: a fairly large PD (40–50 mV, lumen negative) and potassium secretion. In the fish the net potassium transport is almost zero.

Ellory, J. C., Nibelle, J. & Smith, M. W. (1973) The effect of salt adaptation on the permeability and cation selectivity of the goldfish intestinal eptithelium J. Physiol. (Lond.) 231,105–115.

House, C. R. & Green, K. (1965) Ion and water transport in isolated intestine of the marine teleost, Cottus scorpius. J. exp. Biol. 42, 177–189.

GENERAL DISCUSSION

HOSHIKO: Dr. Curran, have you studied the effect of intracellular alanine on alanine uptake at the mucosal border?

CURRAN: Yes, there is no effect.

HOSHIKO: Would you interpret that to mean that there is no exchange diffusion type process going on?

CURRAN: No, not really. The observation that loading the cells with amino acid does not affect the influx of amino acid into the cells across the brush border led us to postulate for our model that the coefficients for translocation of all three forms of the carrier were identical. If they are not identical, trans-stimulation or trans-inhibition should be observed.

LEYSSAC: This is a problem in relation to the low resistance epithelia. I think it is a very nice idea that you could view them as a frog skin with open shunts. But there is a problem also that these epithelia transport a very large volume of water isotonic with the mucosal solution. And I would think this large conductivity and these large passive fluxes in the extracellular space seem to be a little difficult to reconcile with a high sodium chloride concentration to drive the water from the cell to the extracellular space. There is some evidence from our laboratory using gall bladder *in vitro* that when we use ethacrynic acid, after 30–40 minutes we have a sudden drop in the total resistance and a very large increase in the passive back-flux without any effect on the net water transport. Similar observations have been obtained by Cremaschi *et al.* (1971a, b) using amphotericin B, where they observe a rapid change in the ohmic resistance with a $T_{\frac{1}{2}}$ of about 3 minutes with a very slow drop in the net fluid transport with a $T_{\frac{1}{2}}$ about 30–40 minutes. Isn't that a problem?

SCHULTZ: The shunt does not present a major problem in terms of isotonic water absorption. If one is to accept Diamond's modification of Curran's

Cremaschi, D., Montanari, C., Simonic, T. & Lippe, C. (1971a) Cholesterol in plasma membranes of rabbit gall bladder epithelium tested with amphotericin B. *Arch. int. physiol. biochim. 79*, 33–43.

Cremaschi, D., Henin, S. & Calvi, M. (1971b) Transepithelial potential difference induced by amphotericin B and NaCl–NaHCO₃ pump localization in gall bladder. *Arch int. physiol. biochim. 79*, 889–901.

»double-membrane model«, all one has done is to put a leak at the top of a cul-de-sac that was originally believed to be »tight«. If an osmotic gradient is established at the top of this quasi-cul-de-sac, it will draw water and presumably ions through the junction but osmotic equilibrium can still take place as this solution travels down the lateral intercellular space. In a certain sense, the presence of a »hole« at the end of the cul-de-sac may ease the problem of generating an isotonic absorbate.

LEYSSAC: A sudden change in this conductivity, shouldn't it be reflected somehow in the net water transport?

SCHULTZ: That's going to depend upon the rate of net solute transport and since these tissues are virtually short circuited, there is very little net ion transport under normal circumstances through the shunt pathway due to electrochemical potential differences. Thus, the rate of active ion transport will still be the major determinant of the rate of net water transport.

LINDEMANN: I think the point which Dr. Leyssac was making is that when you take a frog skin and make its shunt very leaky it will not give us isotonic absorption; there still remains a difference between these two tissues.

SCHULTZ: I agree entirely with Dr. Lindemann; indeed, the frog would do poorly if it were coated by intestine rather than its normal skin. My major question does not concern itself with water transport but rather with the properties that determine the orientation of the electromotive forces across the outer and inner membranes of the frog skin and the mucosal and basolateral membranes of the small intestine. My admittedly speculative argument is that these properties may be very similar in spite of the fact that in small intestine the cell interior is electrically negative with respect to the mucosal solution, in contrast with findings on frog skin and toad bladder, and that the difference in the observed electrical potential differences *may* be the result of the high conductance shunt pathway that effectively short-circuits small intestine.

CURRAN: You are just saying that maybe the intestine is a little more like a frog skin than one might have thought at first glance.

SCHULTZ: Yes. It may be a frog skin that has been short-circuited.

Permeability Properties of the Intestinal Epithelium of the Greek Tortoise

E. Schoffeniels

The intestinal epithelium of the Greek tortoise is certainly a material well suited for the study of intestinal transport. It is easily stripped of the underlying muscle layers and can be kept *in vitro* for many hours in excellent conditions of survival, a challenge that cannot easily be met by other biological preparations (Baillien α Schoffeniels 1961a). Also the fact that we are dealing with a poikilothermic animal makes things easier with respect to the temperature control of the incubating medium.

I shall now present the picture we arrive at when studying the relations

Table I. Values for short-circuit current and various ion net fluxes in the small intestine of the Greek tortoise. Millicoulombs cm^{-2} h^{-1} (Gilles-Baillien α Schoffeniels 1967b).

Short-circuit current (SCC)	39.8
Na	70.4
K	2.3
SO$_4$	21.8
Cl	–66.4*

* The minus sign means that the net flux is directed outwards

that exist between the inorganic ions and the flux of some amino acids. Let me first recall that when using the double tracer technique, it is easily demonstrated that there is a net flux of sodium ions from the mucosal side (i. e. the lumen of the intestine) to the serosal side. Influx and outflux values are rather high.

It is found that the short circuit current (Table I) is always smaller than

From Department of General and Comparative Biochemistry, University of Liège, Belgium and Laboratory of Marine Membrane Physiology, Duke University Marine Laboratory, Beaufort, N. C. 28516, U.S.A.

»Transport Mechanisms in Epithelia«, Munksgaard, Copenhagen.

the net flux of Na thus indicating that an anion is transported actively from the lumen to the serosal side or that a cation is transported in the reverse direction. Our studies with chloride ions indicate that there is net flux of chloride from the serosal to the mucosal side (Gilles-Baillien α Schoffeniels 1967a). But it cannot explain the discrepancy we observe between sodium net flux and short-circuit current. The net flux of sulphate directed towards the serosal saline is far from being sufficient to compensate for the chloride net flux in the opposite direction since it amounts to about 22 mcb cm^{-2}h^{-1}. Therefore other ionic species must be involved.

It has been shown that an active transport mechanism for phosphate ions in the mucosal-serosal direction exists in the rat intestine (Helbock et al. 1966) but neither this aspect nor that of calcium ions has yet been investigated in the tortoise intestine.

The unidirectional fluxes of potassium have been determined by means of ^{42}K on symmetrical portions of epithelium. The results show important variations between the different periods of the same experiment, e. g. influx varies between 7.06 and 10.04 mcb cm^{-2}h^{-1} and outflux between 4.44 and 8.14. It is thus rather difficult to conclude if there is a net flux of potassium, since the flux values are small and variable. These results show also that the contribution of this ionic species to the short-circuit current required to abolish the potential difference is rather negligible.

Interesting observations were made by using compounds known to affect ion movements. For instance, 2,4-dinitrophenol inhibits both influx and outflux of sodium (Baillien α Schoffeniels 1961b); Diamox inhibits also both chloride influx and outflux (Gilles-Baillien α Schoffeniels 1967b). These results lead us to suggest that an active transport mechanism could be localized at both borders of the jejunum epithelium. But further experimental evidence is necessary before ascertaining this fact since the two compounds could well affect the passive permeability to sodium and to chloride ions. However if one accepts the idea that an ATPase is involved in the active transport of sodium, Quigley α Gotterer (1969) have demonstrated the presence of a Na-K-stimulated ATPase at the serosal and mucosal faces of the rat intestinal epithelium.

Bioelectric potentials of the intestinal epithelium may be interpreted as resulting from the distribution of permeability characteristics to inorganic ions at the two poles of the intestinal cells (Baillien α Schoffeniels 1961b). The low potential (around 1 mV, serosal side +) measured in the small intestine can be explained by the fact that both mucosal and serosal sides of the epithelium present nearly the same permeability characteristics to inorganic ions, i. e. the mucosal and the serosal borders are permeable to sodium and chloride ions. As for the colon, the mucosal border is essentially permeable to sodium and chloride ions while the serosal border is permeable to both sodium and potassium ions thus explaining the higher potential measured (10–50 mV).

Experiments performed with microelectrodes demonstrate that the trans-epithelial potential is the sum of two potentials in series, opposite in sign but differing in magnitude; the intracellular fluid being always negative (Gilles-Baillien α Schoffeniels 1967c). The intracellular location of the microelectrode, after the first potential jump has been observed, is ascertained by the fact that a voltage pulse sensed by the microelectrode shows the capacitative effect of the membrane component.

In the small intestine, chloride and sodium ions contribute to the establishment of both transmucosal and transserosal potentials. Both sides of the epithelium are also permeable to potassium, but this permeability characteristic is only apparent if one lowers the sodium concentration. In the colon, sodium and chloride ions contribute to generating the transmucosal potential while at the serosal border, besides the constribution of sodium and potassium ions, another mechanism has to be postulated. At this level, chloride ions act by shorting the potential otherwise established.

Amino Acids

As far as the amino acids are concerned, it can be shown that in the case of glycine and L-alanine, the influx values are of an order of magnitude higher than those of the outflux in the small intestine. In the colon, the flux values are very similar and always smaller than the values found with the small intestine. By applying the generally-accepted criteria to define the force responsible for the movement of ions and molecules, one comes to the conclusion that glycine and alanine are actively transported across the epithelium of the small intestine. With glutamate, the flux ratio is very close to 1 thus suggesting a passive behavior for this amino acid. L-glutamate and L-arginine are without effect on the influx of glycine across the small intestine.

On the contrary, L-alanine reduces the influx of glycine. This result may be interpreted as showing that a common step is involved in the transfer of L-alanine and glycine (Baillien α Schoffeniels 1961a).

Now if one adds to the mucosal solution bathing a fragment of intestinal epithelium an amino acid said to be actively transported (alanine for instance), an analysis of the intracellular pool of free amino acids shows that most of the amino acids have a concentration either equal or slightly lower than that of the control, except in the case of the added amino acid which has the *same* concentration in the intracellular fluid as in the mucosal solution.

With radioactive alanine the results show also that among the different amino acids, only radioactive L-alanine is found in measurable amounts in the intracellular fluid. The specific activity of alanine is the same in both mucosal saline and intracellular fluid (Gilles-Baillien α Schoffeniels 1966). The radioactivity appearing in the serosal solution is due to L-alanine for 80%. The remaining 20% belong to other substances that have not yet been identified.

Another important observation is that the specific activity of L-alanine in the serosal saline is nearly half that measured in both intracellular fluid and mucosal saline. An interesting speculation stemming from the above results, is that the active transport of L-alanine could be in close connection with the synthesis of this amino acid at the serosal border.

This view is supported by our observation that the addition of pyruvate and NH_4Cl in the mucosal saline results in an increase in the transserosal potential identical to that obtained with alanine. Furthermore, L-alanine appears in the serosal saline.

The present results invite us to be cautious in interpreting the transcellular fluxes of molecules that can be subjected to metabolic transformation. It is indeed obvious that the concept of active transport could not be applied to characterize the influx of alanine if the results of further investigations demonstrate that the transserosal flux is solely attributable to a synthesis mechanism.

On the other hand, if this synthesis mechanism is responsible only for part of the observed flux, a measurement of radioactivity alone would still lead to an erroneous estimation of the exact quantities of alanine really subjected to the active transport.

Since the Greek tortoise is a hibernating species, the seasonal variations of the transport systems in the intestinal epithelium have been studied (Gilles-Baillien 1970). The net transfer of alanine decreases markedly in the torpid animal and the discrepancies we notice in the summer animal, between the specific activities of serosal and intracellular alanine, no longer exist. Moreover the Na fluxes are no longer affected by 2, 4-dinitrophenol; both influx and outflux, while rather high (around 300 mcb cm^{-2}h^{-1}), are equal. Cl fluxes though are still affected by Diamox ® and the net fluxes are outwardly directed.

As mentioned above when added to the saline bathing the mucosal face of an isolated intestinal epithelium alanine induces an increase in the transepithelial potential. The same effect is observed with glucose and further studies of the electrical potential profiles have indicated that the potential change is solely attributable to an increase of the potential at the serosal border.

As indicated by the results of Table II, the amino acids that are not actively transported e. g. aspartate or glutamate, affect both mucosal and serosal potentials. As a consequence the transepithelial potential is not affected since in the same conditions, the short-circuit current is increased, one has to conclude that the overall conductance of the epithelium also increases.

The problem we are now facing is to explain the origin of the change in transserosal potential, when alanine is added to the mucosal side or on both sides of the isolated epithelium. In Ca-free medium, the potential variation is prevented at least partially. Addition of calcium to the medium allows the development of the potential change.

In the presence of EDTA, the change in potential induced by alanine is completely inhibited. One has however to be very careful in interpreting this results as EDTA seems to loosen the junctions between the cell by solubilizing the intercellular cement or whatever the process may be.

Addition of potassium in the serosal saline brings about a decrease of the potential while the influx of alanine is not affected. This result is also observed when veratrine (0.4 mM) is added. Cocaine (4 mM) on the contrary enhances the action of L-alanine on the potential difference (Baillien a Schoffeniels 1963).

Table II. Action of D-glucose and some amino acids on the electrical potential profile recorded in the isolated small intestine of the tortoise (Gilles-Baillien α Schoffeniels 1965, 1968).

The potentials are given in mV and the sign refers to the extracellular fluid

L-glutamate	PD	PD_M	PD_S	L-aspartate	PD	PD_M	PD_S
0	2	10	12	0	0.5	10.5	11
20	2	18	20	20	1	15	16

L-alanine mM/1	PD	PD_M	PD_S	D-glucose mM/1	PD	PD_M	PD_S
0	3	20	23	0	2	19	21
5	13	20	33	5	8	19	27.5
20	12.5	20	33	20	7	19	25

PD = transepithelial potential; PD_M = transmucosal potential; PD_S = transserosal potential.

The passive permeability to sodium of the serosal border is certainly not implicated in the increase in transserosal potential observed after addition of L-alanine, since a large increase of the transepithelial potential still occurs when the sodium concentration of the serosal fluid has been decreased tenfold. However a tenfold dilution of the sodium chloride in the mucosal saline, reduces or even abolishes the increase of the transepithelial potential usually observed in the presence of L-alanine. This reduction may be explained by the fact that under the same experimental conditions the influx of L-alanine is very much decreased (Gilles-Baillien α Schoffeniels 1966; Gilles-Baillien 1970).

The increase of the transepithelial potential due to L-alanine exists in chloride saline as well as in sulphate saline, which eliminates the possibility that chloride is implicated in the potential change. Furthermore, since L-alanine is actively transported in sulphate saline as well as in chloride saline (Gilles-Baillien 1970) it means that chloride and sulphate are not specifically required for the active transport of L-alanine or that sulphate is a good substitute for chloride.

The net flux of chloride, which is usually directed towards the lumen of the intestinal epithelium, is reversed in the presence of alanine (Table III). This is due to an increase in influx as well as a decrease in outflux. If this result suggests a certain relationship between chloride and alanine trans-

Table III. Action of L-alanine on ion fluxes (Gilles-Baillien α Schoffeniels 1967a)

	Conditions	M_{in}	M_{out}	M_{in}-M_{out}
Na	C	467.65	409.6	58.05
	E	494.2	462.1	32.1
K	C	16.85	11.95	4.9
	E	25.2	13.7	11.5
	C	13.05	9.95	3.1
	E	17.35	12.4	5.05
Cl	C	133.65	307.95	−174.3
	E	193.8	184.15	9.65
	short-circuit			
	C	153.05	218.85	−65.8
	E	200.05	174.7	23.35

M_{in} = influx (mucosal to serosal); M_{out} = outflux.
L-alanine (20 mM except in the case of Na where it is 10 mM) added to the mucosal
   saline.
C and E = control and experimental periods of 1 hr.
Results expressed in millicoulombs cm^{-2} h^{-1}.

ports, it rules out, however, the participation of the active transport of chloride in the variation of transserosal potential induced by alanine (Gilles-Baillien α Schoffeniels 1968).

It may also be shown that the fluxes of sodium and potassium seem to be modified when L-alanine is actively transported. The net flux of sodium decreases and this is mainly due to an increase in outflux. The potassium fluxes increase and as a consequence the inwardly-directed net flux is almost doubled. These results however should be interpreted with caution because of the large variations observed even in the control periods. Nevertheless the results of Table III indicate a trend that is generally reproducible.

As suggested by many authors, in the small intestine of mammals, the rate of sugar and amino-acid transport is directly related to the sodium concentration gradient existing at the mucosal barrier of the epithelium. Sugar or amino acid, together with sodium, would combine with a carrier to form a ternary complex which crosses the mucosal cellular barrier towards the intracellular fluid because of the sodium concentration gradient (Schultz 1969).

In contrast Csaky (1963) has proposed that only the intracellular sodium concentration is important for coupling the metabolic energy to the transport mechanisms of non-electrolytes. According to this hypothesis the deficit in the sodium concentration of the mucosal saline would inhibit the transport of non-electrolytes through the intermediary of an intracellular sodium depletion.

As mentioned above the influx of alanine is reduced if the sodium concentration of the mucosal saline is diluted tenfold. At first sight this result could be interpreted in the light of one of the hypotheses proposed for the mammalian intestine. However in the case of the Greek tortoise, it can be said that the concentrative mechanism for alanine is located at the serosal barrier, since there is always an equality between the concentration of alanine in the mucosal saline and in the intracellular solution. Moreover the osmotic deficit in sodium chloride, when using a tenfold dilution, is generally counterbalanced by adding sucrose.

As already noted, we observe a decrease in alanine influx, a situation analogous to that found with mammals. But if we replace NaCl by Tris-Cl (Tris (hydroxymethyl) aminomethane), the influx of alanine is no longer inhibited. Since chloride ions may be replaced by sulphate ions without affecting the alanine flux, it may safely be concluded that neither sodium nor chloride ions are specifically or directly involved in the inhibition of the alanine influx. But an additional, intermediary step has to be postulated.

Table IV shows the ionic composition of the epithelial cells of the small intestine. An extracellular space of 10% has been taken into account to compute the results presented (Gilles-Baillien 1968).

It is certainly interesting to note that an intracellular sodium depletion is only observed when the NaCl is diluted tenfold in the serosal saline. In the same situation the influx of alanine is unaffected (Table V). The intracellular concentration in potassium increases when the sodium is diluted on the mucosal side, a situation that brings about a decrease in alanine influx. If Tris ions are used as substitute for sodium ions, we do not observe a change in the intracellular concentration of potassium; as mentioned above, the influx of alanine is not affected.

From these results it seems reasonable to assume that the influx of alanine is related to the intracellular potassium concentration rather than to the concentration of Na in the mucosal saline. The results indicate that a high intracellular potassium concentration inhibits the active transport of alanine. This interpretation finds an experimental support in the observation that

Table IV. Changes in the inorganic ion content of the jejunum mucosa when sodium chloride concentration or only sodium concentration is modified in either the mucosal or the serosal saline (Gilles-Baillien & Schoffeniels 1970).

	Na	K	Cl		Na	K	Cl
	mEq/Kg wet weight				mEq/Kg wet weight		
Na Cl/10 Sucrose M	67.0	65.3	58.1	Na/10 Tris M	49.0	47.6	61.6
Control	68.1	50.8	65.3	Control	50.2	49.3	58.3
Na Cl/10 Sucrose S	36.4	65.8	37.6	Na/10 Tris S	27.9	42.0	65.6
Control	66.4	59.1	73.1	Control	56.1	49.9	60.7
Na Cl/10 Sucrose				Na/10 Tris			
M and S	39.3	63.1	37.2	M and S	20.4	49.6	66.2
Control	71.2	55.2	60.9	Control	69.7	47.3	68.7

Na Cl/10 Sucrose = nine-tenths of sodium chloride are replaced isosmotically by sucrose.

Na/10 Tris = nine-tenths of sodium ions are replaced by Tris ions, the chloride concentration of the saline remaining unchanged.

M and S refer respectively to the mucosal and serosal saline.

4 hours of incubation.

in torpid animals during hibernation, the intracellular content in potassium goes up while the influx of alanine is depressed (Gilles-Baillien, 1966, 1969).

It is also of interest to note that when the epithelium is bathed with sulphate or chloride saline, the intracellular concentrations in sodium or in potassium are unaffected. The chloride concentration goes down from 66.4 mE/Kg wet weight to 17.3 mE. Once more it is worth emphasizing that the alanine influx is not modified under these circumstances.

As discussed, it appears that the modification of the transserosal potential induced by alanine is potassium sensitive, insofar as a slight increase in the potassium concentration of the serosal saline brings the potential almost to its original value. This finding is strongly suggestive of a change of permeability of the serosal barrier to potassium ions.

One may thus wonder if alanine *per se* could not affect the ionic content of the epithelium. It is interesting to notice that the only concentration affected is that of potassium ions. And again the conclusion to be drawn is that a net influx of alanine, i. e. from mucosal to serosal side, is always paralleled with a low intracellular concentration of potassium ions.

Table V. Influx of alanine in various conditions of ionic composition of the bathing saline (Gilles-Baillien 1970).

Mucosal	C μMcm^{-2} h^{-1}	E	Serosal	C μMcm^{-2} h^{-1}	E
Na Cl/10					
sucrose	1.470	0.873	Na Cl/10	0.875	0.836
	0.819	0.215	sucrose		
	0.532	0.097			
Na/10 Tris	0.931	0.963			
	0.725	0.727			
KCl × 10	1.220	1.276	KCl × 10	1.106	1.143
Ca Cl$_2$ × 10	0.997	1.021	Ca Cl$_2$ × 10	0.678	0.663

L-alanine (20 mM) in both salines bathing the mucosa.

Na Cl/10 sucrose = the sodium chloride concentration is diluted tenfold and sucrose balances the osmotic deficit.

Na/10 Tris = nine-tenths of sodium ions are replaced by Tris ions.

KCl × 10 = the potassium chloride concentration is increased ten times.

Ca Cl$_2$ × 10 = the calcium chloride concentration is increased ten times.

Each experiment includes one control period (C) and one experimental period (E) each of one hour.

Results expressed in μMoles per square centimetre and per hour.

It may be appropriate to speculate as to the relationship existing between inorganic ions and the transfer of alanine from mucosal to serosal side. At this point it is certainly adequate to emphasize the fact that the kind of interaction between sodium ions and amino acid transport that is proposed in the case of mammalian intestine does not apply to our case. Rather we have to conclude that sodium ions in the mucosal solution act indirectly by holding the intracellular potassium concentration below a certain level, which enables the transfer of alanine to be activated. Tris acts in the same way.

The change in the transserosal potential brought about by alanine is a consequence of the flux of alanine which is independent of this electrical potential, since its abolition by raising the potassium concentration in the serosal saline is without effect on the alanine flux. It is also clear that the reversal of the net flux of chloride ions is a consequence of the activation of the transfer mechanism for alanine.

The only constant relation we found is that of intracellular potassium concentration. It is a reciprocal relation: the flux of alanine lowers the intracellular potassium concentration and a flux of alanine can only take place if the intracellular concentration of potassium can be lowered. In the situation of a tenfold dilution of the mucosal NaCl in sucrose, the potassium concentration in the cell goes up and is locked at a level that renders impossible the activation of the transport of alanine.

It is thus safe to conclude that in the Greek tortoise intestine, the metabolically-dependent transfer of alanine is directly related to the intracellular potassium concentration. The effects observed on transserosal potential, Na and Cl fluxes are side effects expressing the metabolic changes brought about in the cell by the activation of the transfer mechanism for the amino acid.

It is likely that potassium acts intracellularly by controlling the activity of some key structure directly involved in the transfer of alanine. But it is certainly unwise to propose models and to speculate further as to the type of interaction we are dealing with, since at this stage of our knowledge in the field of transport we are in the dark as to the molecular mechanism involved.

REFERENCES

Baillien, M. & Schoffeniels, E. (1961a) Le transport actif d'acides aminés au niveau de l'épithélium intestinal isolé de la tortue grecque. *Biochim. biophys. Acta (Amst.)* 53, 521–536.

Baillien, M. α Schoffeniels, E. (1961b) Origine des potentiels bioélectriques de l'épithélium intestinal de la tortue grecque. *Biochim. biophys. Acta (Amst.)* 53, 537–548.

Baillien, M. α Schoffeniels, E. (1963) Différence de potentiel, acides aminés et calcium au niveau de l'intestin grêle de la tortue greque. *Arch. int. Physiol· Biochim.* 71, 286–288.

Csaky, T. Z. (1963) A possible link between active transport of electrolytes and nonelectrolytes. *Fed. Proc.* 22, 3–7.

Gilles-Baillien, M. (1966) L'hibernation de la tortue grecque. *Arch. int. Physiol. Biochim.* 74, 328–329.

Gilles-Baillien, M. (1968) The extracellular space of the isolated intestinal epithelium of the Greek tortoise. *Arch. int. Physiol. Biochim.* 76, 731–739.

Gilles-Baillien, M. (1969) Seasonal changes in the inorganic ion content of various tissues in the tortoise *Testudo hermanni hermanni* Gmelin. *Life Sci.* 8 (part II), 763–766.

Gilles-Baillien, M. (1970) Inorganic ions and transfer of neutral amino acids across the isolated intestinal epithelium of the tortoise *Testudo hermanni hermanni* Gmelin. *Arch. int. Physiol. Biochim.* 78, 119–130.

Gilles-Baillien, M. α Schoffeniels, E. (1965) Site of action of L-alanine and D-glucose on the potential difference across the intestine. *Arch. int. Physiol. Biochim. 73,* 355–357.

Gilles-Baillien, M. α Schoffeniels, E. (1966) Metabolic fate of L-alanine actively transported across the tortoise intestine. *Life Sci. 5,* 2253–2255.

Gilles-Baillien, M. α Schoffeniels, E. (1967a) Action of L-alanine on the fluxes of inorganic ions across the intestinal epithelium of the Greek tortoise. *Life Sci. 6,* 1257–1262.

Gilles-Baillien, M. α Schoffeniels, E. (1967b) Fluxes of inorganic ions across the isolated intestinal epithelium of the Greek tortoise. *Arch. int. Physiol. Biochim. 75,* 754–762.

Gilles-Baillien, M. α Schoffeniels, E. (1967c) Bioelectric potentials in the intestinal epithelium of the Greek tortoise. *Comp. Biochem. Physiol. 23,* 95–104.

Gilles-Baillien, M. α Schoffeniels, E. (1968) Amino acids and bioelectric potentials in the small intestine of the Greek tortoise. *Life Sci. 7,* 53–63.

Helbock, H. J., Forte, J. G. α Saltman, P. (1966) The mechanism of calcium transport by rat intestine. *Biochim. biophys. Acta (Amst.) 126,* 81–93.

Quigley, J. P. α Gotterer, G. S. (1969) Distribution of (Na^+-K^+)-stimulated ATPase activity in rat intestinal mucosa. *Biochim. biophys. Acta (Amst.) 173,* 456–468.

Schultz, S. G. (1969) The interaction between sodium and amino acid transport across the brush border of rabbit ileum: a plausible molecular model. In *The molecular basis of membrane function,* ed. Tosteson, D. C., p.p. 401–420. Prentice Hall, Inc. New Jersey.

DISCUSSION

SCHULTZ: I am concerned by the failure to find a change in the electrical potential difference across the mucosal membrane following the addition of alanine to the mucosal solution. In the paper by Wright (1966) he reported a value of −7 mV for the electrical potential difference across the mucosal membrane, a value considerably smaller than those you have reported. Fig. 1 illustrates the relation between the initial potential difference across the mucosal membrane of rabbit ileum and the change elicited by the addition of either alanine or glucose to the mucosal solution. Clearly, if the initial value is less than −20 mV and particularly if it is less than −10 mV, the change is very small and could easily be missed. Further, if the low initial value is due to damage to the punctured membrane or incomplete sealing of the membrane around the microelectrode tip, one would get little or no change across the mucosal membrane and ascribe the change in trans-epithelial electrical potential difference to a change in the potential difference across the serosal membrane.

SCHOFFENIELS: If it is a change, as you say, of a couple of mV, well maybe we miss it. But it has nothing to do with the change of 10 mV, that we

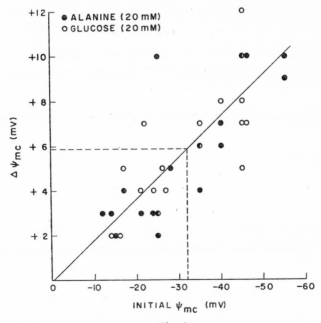

Fig. 1.

observe at the serosal side. Also, I am very confident in this type of result, because we observe an effect with glutamate. Before doing the microelectrode studies we only measured the transepithelial potential. We did at the same time a measurement of short circuit current. With alanine an increase in the short circuit current obtains. With arginine we wouldn't get any change in potential or short circuit current. With glutamate while there is no change in potential difference we do get an increase in the short circuit current. Micro-electrode studies have revealed that glutamate induces a change of potential at both the serosal and mucosal sides. Also I may add one point, that White & Armstrong (1971) get a change with the toad on both sides.

SCHULTZ: Yes, but they ascribed the change elicited by sugars or amino acids primarily to a depolarization of the potential difference across the mucosal membrane and his results are in excellent agreement with ours. I might add that similar results have recently been reported by Maruyama & Hoshi (1972) for the luminal membrane of Newt proximal tubule.

Why is it, if alanine reverses the chloride flux, that you get an increase in short circuit current?

SCHOFFENIELS: You see, if you add at first all those values, you cannot ac-count for all the short circuit current, something is still missing. The inter-esting thing is that you can use sulfate instead of chloride, and you still get an influx of alanine. So, I don't think there is any direct relation between the transfer of alanine and the chloride transport It just happens as an epipheno-menon of the transport that the chloride fluxes are reversed. But it is not necessary for the alanine transport to have chloride.

SCHULTZ: Do you also get an increase of short circuit current in sulfate solution?

SCHOFFENIELS: Yes.

CURRAN: I don't want to get involved in a long discussion at this point. I am sorry that Dr. Schoffeniels was not here yesterday, because I did present

White, J. F. & Armstrong, W. McD. (1971) Effect of transported solutes on membrane potentials in bullfrog small intestine. *Amer. J. Physiol. 221,* 194–201.
Maruyama, T. & Hoshi, T. (1972) The effect of D-glucose on the electrical potential profile across the proximal tubule of Newt kidney. *Biochim. Biophys. acta 282,* 214–225.

some data on a different species of turtle in which we found that its alanine transport appears to be quite similar to the transport that we have reported earlier in mammalian intestine. I don't really know what the difference is between our results and yours.

USSING: A very short remark to Dr. Schoffeniels: You said that you could not decide whether you were dealing with active transport or not, because you have formation of alanine in the system. This is not correct, though, because the flux ratio treatment gives the same result whether or not the substance in question is being consumed or formed in the cell. As long as you are sure that you are measuring only that species it is easy to demonstrate that the flux ratio is not dependent on whether or not the substance is being consumed or produced.

SCHOFFENIELS: Maybe I should specify what I meant. When you are measuring the influx of alanine using a radioactive alanine, the totality of the radioactive flux couldn't be used to measure the total active transport. Of course I agree with your comment on the flux ratio. We have a synthesis of alanine. We have demonstrated it by putting pyruvate and ammonium on the mucosal side, we then get alanine on the other side, and we observe an increase in the potential difference. This type of mechanism should *not* be called active transport.

The Route of Cation Transport Across the Silkworm Midgut

William R. Harvey and John L. Wood

1. INTRODUCTION

1.1 Transport route in epithelia

The route followed by sodium as it is transported across the frog skin and other sodium transporting epithelia has yet to be established. The ions might follow a mixing route through the cells or a non-mixing route through or between the cells. Even if a route were to be established unequivocally for sodium systems one could not conclude immediately that the same route is followed in all epithelial systems since these differ widely in structure and transport properties (Keynes 1969). The large silkworm midgut has a relatively simple structure; it has uniquely simple ionic requirements for transport; and it has a rapid transport rate (Harvey & Zerahn 1972). This combination of properties makes possible a relatively simple analysis of the transport route through this tissue.

1.2. Midgut transport properties

Potassium is actively transported from the blood-side to the lumen-side across the isolated midgut by an oxygen-dependent pump which requires no counter-ions, i.e., by an electrogenic pump. Rubidium is transported almost as well as potassium and can be used both as a substitute for potassium and as an additional label for potassium transport. The isolated midgut is composed of just columnar cells and goblet cells arranged in a one-cell-thick epithelium which is directly in contact with the bathing solution on the lumen-side and which is separated from that on the blood-side only by a continuous basement membrane and by discontinuous muscle layers and tracheal cells. There is no serosa as in the gut of chordates.

Department of Biology, Temple University, Philadelphia, Pennsylvania, 19122.

»Transport Mechanisms in Epithelia«, Munksgaard, Copenhagen.

1.3. Route studies in midgut

Harvey & Zerahn (1969) initiated kinetic studies on the midgut, showed that the lag-time delay for tracer influx is small and constant with variation in flux rate, and concluded that the K-pool causing the delay must likewise be small. They argued that the transport must proceed either by a non-mixing route through or between the epithelial cells or possibly by a mixing route through just the less numerous goblet cells. A route through just the goblet cells was eliminated by Wood *et al.* (1969) on electrical grounds. Moreover, by making allowance for the decay in the short-circuit current, Harvey & Wood (1972) showed that the mixing-time for tracer influx is long and corresponds to a large K-pool when the flux is large. They accounted for the constancy of the mixing-time by demonstrating that the size of the K-pool varied directly with the flux. They argued that only the epithelium itself has sufficient volume to contain such a large pool and proposed a mixing route through both columnar and goblet cells. Harvey & Zerahn (1972) then pointed out that the large pool might be after the pump in the extracellular space on the lumen-side of the tissue. In this case, potassium would be actively transported into a pool at this location and therefore the [K] in such a pool could be very high. Moreover, the [K] in such a pool would be expected to vary directly as the flux which would explain the constancy of the mixing-time with varying flux. It is our present purpose to show that the pool is not in the extracellular space on the lumen-side of the midgut and that it is not likely to be in the space on the blood-side, but that it is most likely to be in the epithelial cells themselves.

1.4. Rationale for experimental choice between transport and transported pools

There are several arrangements in which a pool could be located before a pump (transport pool) and several in which it could be located after a pump (transported pool). In the case of the midgut there are only two serious possibilities and those are illustrated in Fig. 1. If the pool is before the pump, (upper left, Fig. 1), then ions will be actively transported from it to the lumen-side compartment. The washout from the pool to the lumen-side compartment will depend on the pumping rate. On the other hand, if the pool is after the pump, (lower left, Fig. 1), then the ions simply will be

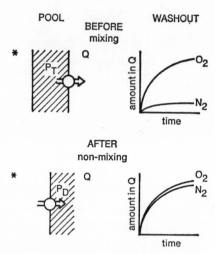

Fig. 1. Theoretical effects of pool location on tracer washout.

diffusing from it to the lumen-side compartment. The washout from the pool to the lumen-side will be governed simply by Fick's law and will be independent of the pumping rate. Therefore, the location of the pool relative to the pump can be determined by comparing the tracer washout from the loaded tissue to the lumen-side when the pump is operating to the washout when the pump is inhibited.

The midgut K-pump can be inhibited rapidly and reversibly by anoxia. Therefore, in the case of the midgut the washout to the lumen-side from any pool before the pump will be oxygen dependent (upper right, Fig. 1). By contrast, the washout from any pool after the pump will not be oxygen dependent (lower right, Fig. 1).

1.5. Prologue

We will show that the washout to the lumen-side is oxygen dependent and deduce that the pool must be before the pump. We will show that the extra-cellular space on the lumen-side, and probably that on the blood-side as well, is too small to contain the pool and conclude that the pool must be in the cells. We will support this conclusion by confirming our earlier report (Harvey & Wood 1972) that the pool size calculated from the loading of the tissue with ^{42}K (or ^{86}Rb) is the same as the pool size calculated from the influx kinetics. Combining this information with other properties of the midgut which are already published, we will be able to eliminate all but two of the theoretical models for the location of the pool relative to the pump in the midgut. One of these two remaining models is improbable since it requires that a transport pool in the blood-side space supply a pump on the lumen-side membrane. In the remaining model a large transport pool is located in the cells and potassium is transported actively from it out of the cells to the lumen-side by a pump located in the apical plasma membranes of the epithelial cells. Since this arrangement was the key to the model for the midgut K-transport route presented in Gargnano in 1971 (Harvey & Wood 1972), that model is supported by these new results.

2. METHODS

Experiments were performed on midguts (ca 100 mg wet wt.) isolated from mature fifth-instar larvae of *Hyalophora cecropia (L.)* The larvae were reared either on artificial diet (Riddiford 1968) or on cherry leaves. The midgut was isolated, mounted in a closed chamber, and short-circuited by techniques reviewed by Harvey& Zerahn (1972). It was equilibrated for 30 minutes in an oxygenated solution composed of 32 mM/l KCl or RbCl, 5 mM/l CaCl$_2$, 5mM/l MgCl$_2$, 5 mM/l Tris Cl, 1 mM/l Na$_2$SO$_4$, and 166 mM/l sucrose. Then the tissue was loaded from the blood-side with ^{42}K or ^{86}Rb for one hour while influx samples were taken from the lumen-side compartment. Finally both chambers were washed rapidly with isotope-free solution and washout samples were collected from both sides for one hour. The continued operation of the K-pump in the oxygenated solution was indicated by the normal slow decay of the short-circuit current. By contrast, the inhibition of the pump in nitrogen-containing solutions was indicated by a rapid drop of the short-circuit current to values near zero. The extracellular space was marked with ^{35}SO$_4$ by adding this marker to one or both compartments five minutes prior to the start of the washout period. ^{42}K og ^{86}Rb was distinguished from^{35}S by their energy levels using a Packard liquid scintillation counting system. The fluxes and extracellular space corrections were calculated on a computer by comparing the radioactivity in the samples with the specific activity of the isotope in standards removed from the bathing solutions.

3. RESULTS

3.1. Amount of washout to blood-side

The time-course of the amount of ^{86}Rb which washed out from the tissue to the blood-side solution is plotted (left, Fig. 2). The time-course is the same whether the tissue was equilibrated with oxygen and the K-pump was operating or was equilibrated with nitrogen and the pump was inhibited. This result is consistent with the report by Wood *et al.* (1969) that entry of K into the midgut cells is a passive process, but it is of no further interest at present.

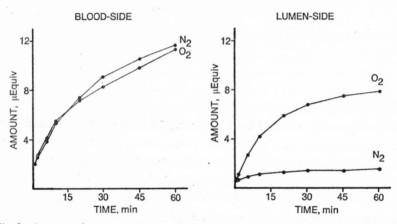

Fig. 2. Amount of tracer washout to blood-side and lumen-side in oxygen and nitrogen.

3.2. Amount of washout to lumen-side

The time-course of the amount of ^{86}Rb which washed out to the lumen-side is plotted (right, Fig. 2). By contrast with the blood-side washout just described, the amount of ^{86}Rb which washed out to the lumen-side was much greater when the tissue was equilibrated with oxygen (upper curve) and the pump was running, than when the tissue was equilibrated with nitrogen (lower curve) and the pump was inhibited.

3.3. Rate of washout to lumen-side

In the experiments just described the simultaneous washout to blood-side and lumen-side in nitrogen was carried out on a different preparation from the washout to blood-side and lumen-side in oxygen. In order to control as many variables as possible, the washout to the lumen-side in both oxygen and nitrogen was carried out on the same preparation and at as nearly the same time as possible. For this determination the midgut was loaded with ^{86}Rb exactly as described above. Then the washout to the lumen-side was started in nitrogen, while the short-circuit current was monitored to assure that the K-pump was strongly inhibited. As soon as the nitrogen washout curve was defined (after 4 min.) oxygen was introduced abruptly into the bathing solutions. The K-pump was rapidly turned on as monitored by a rapid increase in the short-circuit current. At the left, Fig. 3, the *rate* of washout to the lumen-side in oxygen (upper curve) and that in nitrogen (lower curve) are plotted. These data are from experiments similar to those used to plot the *amount* function in Fig. 2, but are corrected for extracellular spaces. The rate of washout in oxygen (upper curve) was much greater than the rate in nitrogen (lower curve). These washout data are redrawn as broken line curves (right, Fig. 3). In addition the data from the present experiment in which the stirring gas was changed during the washout are plotted and connected with a solid line (right, Fig. 3). Clearly the initial portion of the washout, which was in nitrogen, follows the typical nitrogen function (lower broken line curve). Then, at the time indicated by the arrow, when oxygen was admitted, the washout immediately shifted to the typical oxygen function

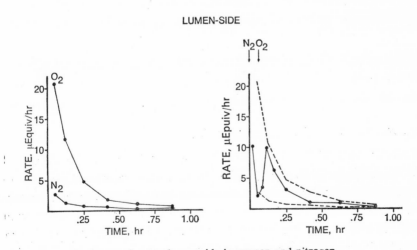

Fig. 3. Rate of tracer washout to lumen-side in oxygen and nitrogen.

(upper broken line curve). No difference was found between Rb and K in these experiments; the ^{42}K-measured K-washouts were found to be the same as the ^{86}Rb-measured Rb-washouts in both oxygen and nitrogen.

3.4. Size of extracellular spaces

The volume of the extracellular space on the blood-side and that on the lumen-side of the midgut were determined from the washout of $^{35}SO_4$ from the tissue to the two sides over a period of one hour after the tissue was pulse labeled with $^{35}SO_4$ for five minutes. As summarized in Table I, the volume of the lumen-side extracellular space was found to be 5.6 $\pm$ 0.4 μ1 in 5 determinations. The volume of the blood-side extracellular space was found to be 1.2 $\pm$ 0.8 μ lin 4 determinations. These are maximal values for the size of the extracellular spaces (for five min. of pulse-labeling), since the method used to measure the volume includes all unstirred volumes in the chamber. Therefore, the sizes of the extracellular spaces on the two sides of the midgut are very small.

Table I. Size of extracellular spaces (μ1).

Lumen-side		Blood-side	
date	volume	date	volume
20 jul 72	6	4 aug 72	0
26 jul 72	8	23 oct 72	0
28 jul 72	6	31 oct 72	3
8 aug 72	5	9 nov 72	2
9 aug 72	3		
Mean $\pm$ S. E. M.	5.6 $\pm$ 0.4		1.2 $\pm$ 0.8

3.5. Specific activity of midgut after loading

The results reported thus far lead us to the conclusion that the large pool is in the epithelial cells (see § 4.5.). In that case we should be able to demonstrate the pool directly by measuring the radioactivity in a midgut when it is removed from the chamber immediately after it is loaded with isotope. Moreover, the amount of loading should be equal to the size of the pool calculated from the influx kinetics on the same preparation. Finally, we expect that any fraction of K in the tissue which is not involved in the K-transport should not be labeled.

For this demonstration we equilibrated a short-circuited midgut in 32 mM/l Rb solution in the closed chamber for 60 min., then added ^{86}Rb to the blood-side and loaded the tissue for 90 min. while taking samples from which the influx kinetics could be calculated. Then the midgut was ashed in hot concentrated H_2SO_4 followed by hot 30 % H_2O_2, and the specific activity of Rb and the amount of K were determined. In one experiment (7 aug 72, Table II), the specific activity of Rb was 52% and the amount of labeled Rb (the loading level) in the cells was 2.6 μEquiv. There was 0.4 μEquiv. of K in the cells which had not exchanged with the labelled or unlabeled Rb even after 150 min. of exposure to Rb and 90 min. of exposure to ^{86}Rb. The size of the pool calculated from the influx kinetics on this preparation was 3.6 μEquiv. which is nearly the same as the size of the loading level of 2.6 μEquiv. Similar results were obtained when ^{42}K was used to label a midgut equilibrated in bathing solutions containing 32 mM/l K (see Table II). The size of the influx pool agreed with that of the loading pool within 99.8 $\pm$ 10.2% in 7 determinations. These results confirm the report by Harvey & Wood (1972) that there is a nonexchangeable fraction of K in pieces of midgut tissue incubated with isotope in a beaker. Moreover, these results confirm their report that there is a loading pool in the midgut with the same size as the influx pool.

Table II. Comparison of influx pool with loading pool (μ Equiv).

Date	Bathing solution	Influx pool	Loading pool	*Influx* Loading	Total cell K	Total cell Rb	Non-exchange-able amount
7 Aug 72	32 – Rb– S – Tris	3.6	2.6	138	0.4	5.0	2.8
11 Oct 72	–	2.3	2.6	88	3.8	3.2	4.4
12 Oct 72	–	2.7	3.0	90	2.6	4.9	4.5
8 Aug 72	32 – K– S – Tris	7.3	7.5	97	12.1	0.0	4.6
17 Oct 72	–	..5.2	4.0	130	8.7	0.0	4.7
18 Oct 72	–	8.6	8.8	98	12.0	0.0	3.2
19 Oct 72	–	2.1	3.6	58	8.0	0.0	4.4

Mean ± S. E. M.   99.8 ± 10.2

4. DISCUSSION

4.1. Number and size of pools

Analysis of influx kinetics reveals that there is just one pool involved in the transport route and that this pool amounts to as much as two-thirds of the midgut tissue K at moderately high transport rates (Harvey & Wood 1972). Analysis of our preliminary determinations of the kinetics of simultaneous washout to both sides shows that the half-time of washout to the blood-side and lumen-side is approximately the same and corresponds to a single pool with about the same size as that deduced from the influx kinetics. With confidence, then, we can say that there is a single pool involved in active K transport across the midgut and that the pool size is large.

4.2. Location of pool with respect to pump

Washout to the lumen-side was found to be oxygen-dependent (Figs. 2 and 3). This result demonstrates that the large pool is located before the pump. However, a small fraction of the washout to the lumen-side was found to be oxygen-independent. This result indicates that a small pool exists after the pump. The existence of this small pool after the pump is partially confirmed by the presence of a small but significant extracellular space on the lumen-side of the midgut (Table I). However, the finding which must be stressed is that the large pool is located before the pump.

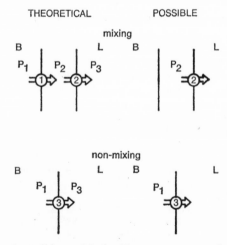

Fig. 4. Theoretical and possible models for K-transport across the midgut; B – blood-side, L – lumen-side, P – pool, and double arrow – pump.

4.3. Theoretical models

The midgut behaves electrically either like a two-barrier, three-compartment system or possibly like a one-barrier, two-compartment system (Wood *et al.* 1969); this behavior is consistent with its structure, a one-cell-thick epithelium communicating directly via an extracellular space on either side with the respective bathing solutions (Anderson & Harvey 1966). As diagrammed (upper left, Fig. 4), for a two-barrier, three-compartment system, the pump could be located on either barrier and the pool could be in any one of the three compartments. The barriers are most likely the basal plasma membrane, marked "1", and the apical plasma membrane, marked "2"; the compartments are the extracellular space on the blood-side marked "P_1", the cells, marked "P_2", and the extracellular space on the lumen-side, marked "P_3". For the one-barrier, two-compartment system the barrier is most likely the "tight junctions" between cells, marked "3", whereas the two compartments are the extracellular spaces on blood-side and lumen-side, marked P_1 and P_3 as before.

4.4. Possible models

4.4.1. *Elimination of basal pump (1)* – A pump located between the blood-side space and the cells has already been eliminated on electrical grounds (Wood *et al.* 1969). Moreover, Harvey & Zerahn (1969) showed that the passive movement of potassium between lumen-side and cells is negligible. If ions were pumped from the blood-side space into the cells, they could not pass out of the cells to the lumen-side. Therefore, we can eliminate a pump which is located between the blood-side compartment and the cells.

4.4.2. *Rejection of extracellular blood-side pool (P_1)* – A pool located in the extracellular space on the blood-side in the mixing model can be rejected, because a basal pump location has been eliminated (§ 4.4.1. above). However, in the non-mixing model, a pool could be located in the extracellular space on the blood-side and supply a pump located on the apical barrier by a non-mixing route through or between the cells. Such a pool location, which would be before the pump, would be consistent with the oxygen-dependent washout to the lumen-side. But, the measured sulfate space on the blood-side is negligible (Table I). Moreover, such a pool would be exhausted by the pump and could have a [K] no greater than that in the blood-side bathing solution. Thus, a pool located in the measured extracellular space on the blood-side could not contain more than about 0.1 μEquiv., whereas the size of the transport pool, whether calculated from influx kinetics, washout kinetics, or loading levels, can be greater than 8.0 μEquiv. (Table II). Therefore, the measured extracellular space on the blood-side is too small to contain the pool. However, we cannot eliminate outright an extracellular blood-side pool, due to the well-known difficulties in measuring and interpreting sizes of extracellular spaces. On these grounds, we can only state that an extracellular blood-side pool is rendered highly improable.

4.4.3. *Elimination of extracellular lumen-side pool (P_3)* – A pool located in the extracellular space on the lumen-side in both the mixing and non-mixing model can be eliminated outright. The strongest evidence is the oxygen dependency of the lumen-side washout (§ 3,2., § 3.3. and § 4.2.). Moreover, the size of the extracellular space on the lumen-side is only 5.7 ± 0.4 μl (Table I). For a pool of about 3 μEquiv, the mean concentration of Rb in a pool located in this small volume would have to be over 500 mM/l which is possible, but highly unlikely. The rejection of a pool in the extracellular space on the lumen-side is likewise supported by electron micrographs which reveal no structure on the lumen-side of the plasma membrane which might contain a large pool. The only space visible is that between microvilli of the columnar cells and that in the cavity of the goblet cells. Otherwise the plasma membrane on the lumen-side is in direct contact with the bathing solution (Anderson & Harvey 1966). However, it is possible that the small oxygen-independent fraction of the washout to the lumen-side (§ 4.2.) could be located in the goblet cavities. In summary, an extracellular pool on the lumen-side must be eliminated on three grounds: (1) it should wash out independently of the pumping rate, whereas the actual washout is dependent on the pumping rate, (2) the volume of the extracellular (sulfate) space on the lumen-side is not large enough to contain it, and (3) there is no structure on the lumen-side of the tissue large enough to restrict it.

4.4.4. *Summary of possible models* – From the above consideration of theoretical models, we are left with two possible models for the midgut. The first possible model (upper right, Fig. 4) postulates a mixing pathway in which the pool is located in the cells (P_2) and the pump is located on the apical membrane (2). The second possible model (lower right, Fig. 4) postulates a non-mixing pathway in which the pool is located in the blood-side extra-cellular space (P_1) and the pump is located between the cells (3). However, the non-mixing model is rendered unlikely because the measured extracellular space on the blood-side is far too small to contain the pool calculated from the influx kinetics (§ 3.5.). Therefore, although we cannot eliminate it rigorously, a non-mixing route is highly unlikely and we are left with the mixing model.

4.5. Probable model for K-transport across the midgut

The most probable model for K-transport across the midgut is the mixing model (upper right, Fig. 4), but this model has some difficulties. We will first describe this model, then point out its difficulties, and finally present evidence which enables us to overcome the difficulties.

4.5.1. *Description of the mixing model* – The mixing model, as proposed by Harvey & Wood (1972), is summarized in Fig. 5. The pump (2) is located on the apical plasma membrane and has an s-shaped velocity-con-

PROBABLE

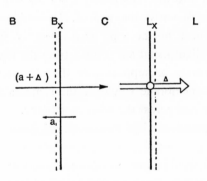

Fig. 5. Probable model for K-transport across the midgut; B – blood-side, C – cellular pool, L – lumen-side, B_x – unstirrer layer on blood-side, L_x – unstirred layer on lumen-side, a – passive exchange across blood-side barrier, $\triangle$ active transepithelial flux.

centration dependency. The K concentration, which determines the pumping velocity is that in the pool, $[K]_p$, and is given by the amount of K in the cellular pool, S_p, divided by the volume of the pool, V_p. The amount of K in the pool, S_p, is determined by the rate at which K enters the pool from the blood-side, $a + \triangle$, relative to that at which K leaves the pool to the blood-side, a, and is transported to the lumen-side, $\triangle$. Parenthetically, this facet of the model is appreciated only when the system is not in a steady state and the influx to the pool and the efflux from it are not equal. Finally, the rate of entry of K into the pool, $a + \triangle$, is determined by the electrochemical gradient for potassium across the blood-side barrier, which is primarily determined by the K concentration in the blood-side bathing solution, $[K]_b$. To summarize the mixing model, potassium passively diffuses from the blood-side bathing solution into the cellular pool across the basal plasma membranes, mixes with cell K, and is then actively transported out of the cells to the lumen-side bathing solution by the pump located on the apical plasma membranes.

4.5.2. *Evidence apparently contradictory to the mixing model* – Two lines of evidence would be explained more easily by a non-mixing model than by the presently proposed mixing model. First, Zerahn (1971; see also Harvey & Zerahn 1972 and Zerahn, 1973) has shown that the sequence for loading the alkali metal ions into the cells differs from the sequence for the overall transport of these ions across the midgut. This difference in loading and transport sequences need not be a problem, however, if the true affinity of the K-pump for the alkali metal ions differs from the loading sequence. That is, there is no reason to presuppose that the affinity sequence for active transport across one membrane would be the same as the sequence for the passive entry of these ions across the other membrane.

Second, the proposed mixing model would have to account for the measured variations in pool size through variations in the cell K, whereas a non-mixing model would have changes in pool size occurring in a pool outside the cells. However, we have demonstrated that both the total tissue K and the tracer loading level do vary directly with the external [K] (Harvey & Wood 1972, and unpublished results); therefore, while it might be considered unreasonable that the tissue K should vary, the fact is that it does (see § 4.5.3. below). In summary, there are two lines of evidence which could be explained more easily by a non-mixing model, but these two lines of evidence are not necessarily incompatible with the presently proposed mixing model.

4.5.3. *Evidence compatible with the mixing model* – First, the blood-side barrier has a sufficiently large passive permeability to potassium to maintain the tissue K level. This passive permeability has been demonstrated in wash-out experiments (Fig. 2) and also in loading experiments conducted in beakers in which the amount and rate of loading were the same in oxygen as in nitrogen (Harvey & Wood, unpublished results). Moreover, the blood-side barrier has a sufficiently large, oxygen independent, electrochemical gradient to allow the passive entry of potassium (Wood *et al.* 1969). Finally, the passive permeability of the blood-side barrier is affected by calcium in such a way that sodium cannot readily enter the tissue when calcium is present. Incidently, the best explanation (Harvey & Wood 1972) for how calcium inhibits sodium transport (Harvey & Zerahn 1971) remains that calcium inhibitis the entry of sodium into a cellular pool.

Second, the lumen-side barrier has the properties expected of a transporting membrane. The passive permeability to potassium of this barrier approaches zero (Harvey & Zerahn 1969). This important property has been confirmed in the present investigation in that even after 60 min. the cells were less than 6 % labeled from the lumen-side (one determination). Moreover, the oxygen-dependent electrogenic step corresponding to the K-pump has been localized on the lumen-side barrier (Wood *et al.* 1969). Locating the pump on the lumen-side barrier would account for the oxygen dependent washout to the lumen-side (Figs. 2, 3). In fact, there is no doubt that the midgut K-pump is located on the lumen-side barrier.

Third, the K-pump has an s-shaped velocity-concentration curve. The midgut ceases to transport K when it becomes equilibrated after an hour with bathing solutions containing as much as 8 mM/l Rb (one determination). On the other extreme, the K-pump saturates when the midgut is equilibrated with bathing solutions containing approximately 100 mM/l of K. These statements do indicate that the midgut K-pump has an s-shaped velocity-concentration dependency, but this dependency would be equally compatible with both a mixing and a non-mixing model.

And fourth, the transport pool is shown to be a cellular pool by several lines of evidence. The kinetic influx pool contains the same amount of K as the amount of labeled K in the tissue (Table II), whether the latter is deter-

23*

mined from the loading pool or from the washout pool (preliminary experiments). The similarity in size of these three types of pool argues that they are one and the same pool and that this one pool is located in the cells forming the pool and the loading pool have been shown to vary in synchrony with external [K] (Harvey & Wood, unpublished results). It should be noted that the loading pool is not the same as the total tissue K because there is a sizeable fraction of tissue K which does not exchange with tracer even after an hour (Table II). Moreover, the rate of K-transport appears to be determined by the concentration of K in some compartment in addition to the blood-side bathing solution because that concentration remains constant whereas the rapid decay in the short-circuit current is highly correlated with the loss of tissue K (Wood 1972).

Five lines of evidence are more easily explained by the presently proposed mixing model than by a non-mixing model: (1) that calcium decreases the sodium permeability of the blood-side barrier in concentrations which inhibit sodium transport, (2) that the passive permeability of the lumen-side barrier to potassium is near zero, (3) that the K-loading pool, the K-washout pool, and the K-influx pool are all the same size, (4) that both the K-influx pool and the K-loading pool vary directly as the external [K], and (5) that the rate of decay of K-transport varies directly with the loss of tissue K, while the [K] in the bathing solutions remains constant. These five lines of evidence could be made compatible with a non-mixing model, but this would require that we ascribe additional properties to the midgut, whereas, these five lines of evidence are made compatible with the mixing model simply by stating that there is a cellular pool.

5. SUMMARY

Our previous work has shown that there is a large pool in the route followed by potassium as it is transported across the midgut. In the present paper we have shown that this large pool must be before the pump, because the washout to the lumen-side is oxygen-dependent. Furthermore, from kinetic measurements of the extracellular spaces, the pool cannot be located in the extracellular space on the lumen-side and it is not likely to be located in the extracellular space on the blood-side, because the measured volumes are too small. Moreover, the loading pool, as determined from the labeling of the tissue, is confirmed to be almost identical in size and half-time with the kinetic influx pool, as determined from the mixing-time of the tracer influx. These three lines of evidence argue that the transport pool is located in the cells, that potassium follows a mixing route through the cells, and that K is transported out of the cells. This new evidence strongly supports the mixing model for K-transport across the midgut presented previously and reviewed here.

This resarch was supported in part by a research grant (AI-05903) from the National Institute of Allergy and Infectious Diseases, U. S. Public Health Service. We thank Mr. Jerome J. Jordan for competent technical assistance.

REFERENCES

Anderson, E. & Harvey, W. R. (1966) Active transport by the Cecropia midgut II. Fine structure of the midgut epithelium. *J. Cell. Biol. 31*, 107–134.

Harvey, W. R. & Wood, J L. (1972) Cellular pool involved in active K-transport across the isolated Cecropia midgut. Ii Role of *membranes in secretory processes,* eds. Bolis, L. Keynes, R. D. and Wilbrandt, W. North-Holland Publ., Amsterdam. 310–331.

Harvey, W. R. & Zerahn, K. (1969) Kinetics and route of active K-transport in the isolated midgut of *Hyalophora cecropia. J. exp. Biol. 50*, 297–306.

Harvey, W. R. & Zerahn, K (1971) Active transport of sodium by the isolated midgut of *Hyalophora cecropia. J. exp. Biol 54*, 269–274.

Harvey, W. R. & Zerahn, K. (1972) Active transport of potassium and other alkali metals by the isolated midgut of the silkworm. *Current Topics in Membranes and Transport 3,* eds. Bronner, F. and Kleinzeller, A. Academic Press, N. Y. and London, 367–409.

Keynes, R. D. (1969) From frog skin to sheep rumen. *Qart. Rev. Biophys. 2*, 177–281.

Riddiford, L. (1968) Artificial diet for Cecropia and other Saturniid silkworms. *Science 160*, 1461–1462.

Wood, J. L. (1972) Some aspects of active potassium transport by the midgut of the silkworm, *Antheraea pernyi.* Ph. D. Thesis, Cambridge University.

Wood, J. L., Farrand, P. S. & Harvey, W. R. (1969) Active transport of potassium by the Cecropia midgut VI. Microelectrode potential profile. *J. exp. Biol. 50*, 169–178.

Zerahn, K. (1971) Active transport of the alkali metals by the isolated mid-gut of *Hyalophora cecropia. Phil. Trans. Roy. Soc. London. B 262*, 315–321.

Zerahn, K. (1973) Properties of the cation pump in the midgut of Hyalophora cecropia, in Ussing, H. H. & Thorn, N. A. (ed.). Transport mechanisms in Epithelia. Alfred Benzon Symposium V. Copenhagen 1973, pp. 360–367.

DISCUSSION

KEYNES: I entirely accept your electrical arguments, but it seems very extra-ordinary that if the energy for driving the pump is coming from the mito-chondria, the cells should carefully place the mitochondria so that they are as far away as possible from the membrane, where the work is being done.

HARVEY: You are thinking, I presume, of the mitochondria in the columnar cells which indeed are located in infoldings of the basal membrane many microns away from our postulated pump location in the apical membrane. However, the goblet cavity is lined with apical projections each of which contains a large elongated mitochondrion. These mitochondria are only 100 Ångstrom units away from the postulated pump locations. Moreover, on the cytoplasmic surface of the plasma membrane lining these projections there are small elementary particle-like components. These components are not found elsewhere on the plasma membrane of the goblet cell nor on any part of the columnar cells. The common occurrence of such particles on insect membranes with Type V pumps has led Smith, Oschman and others to implicate them in K-transport.

KEYNES: Then the mitochondria in columnar cells just sit there looking pretty?

ZERAHN: About the long lag time you have found, I will treat that in my lecture later. I would like to see your Table II again — about the exchange. You usually find very low exchanges, even if you have an hour for exchange. The amount of K you have exchanged with Rb in the first line in Table II, Exp. 7 August is 5.0 of a total of 5.4, but later you find less than 50 % with radio potassium. These results do not agree.

HARVEY: No, because what we have called the "non-exchangable amount" does exchange at a very slow rate. Since the tissue was exposed to Rb longer than to ^{86}Rb, it is not surprising that there is a greater exchange of Rb with K than ^{86}Rb with Rb or K.

ZERAHN: I still find it surprising and I cannot find the slow exchange of K in guts bathed in 32 mM K solutions and we did not find this in the earlier experiments. I think the disagreement may be caused by different thickness of the midguts or other differences in the experimental conditions.

SACHS: Due to the location of the mitochondria in these apical regions of the goblet cells, has anybody looked at the effect of valinomycin on potassium transport in the species?

HARVEY: We have tried valinomycin but found no inhibition yet. Gramicidin has an interesting effect first observed by J. Bialawski in my laboratory in 1966. In low concentrations (10 μg/ml) it inhibits the I_{sc} and influx almost completely but does not affect the oxygen uptake. We are experimenting with the effects of these and other cyclic peptides at the present.

Properties of the Cation Pump in the Midgut of Hyalophora Cecropia

Karl Zerahn

INTRODUCTION

When the midgut of the cecropia larva is bathed in a suitable solution containing an alkali metal ion and some sucrose the cation will be actively transported. (Harvey & Zerahn 1971) When more than one of the alkali metals are present they will compete for the transport mechanism, so it is concluded that the ions share the same ion pump.

It is essential for the transport of Na and Li ions that the solutions are Ca free; the presence of 5 mM Ca will also influence the transport of Cs, but not change the transport of K and Rb in the usual concentrations used of 32 mM. The competition for the transport mechanism depends mainly on the concentration ratio of the ions, but also on the absolute concentration. The experiments for competition were therefore performed in solutions containing 16 mM of each of the competing ions. The sequence found for competition for transport was K>Rb>Cs>Na>Li.

When bathed in Ca free solution containing 32 mM Na, the midgut will take up Na ions and lose K rapidly; in 5 minutes the Na concentration in midgut cells may be 100 mM Na. When the midgut is bathed in solutions containing competing ions it will take up both the competing ions, and a sequence for uptake was also determined. This sequence, different from the sequence for transport, was found to be K>Rb>Na>Li>Cs.

The conclusion drawn, that the midgut content of K is not directly taking part in the active transport is in agreement with the results of Harvey and Zerahn (1969). These results have been doubted recently by Harvey and Wood (1972), but evidence is given here that their criticism is not valid.

Institute of Biological Chemistry A. University of Copenhagen, 13 Universitetsparken, DK-2100 Copenhagen Ø - Denmark.

»Transport Mechanisms in Epithelia«, Munksgaard, Copenhagen.

METHODS

The larvae were grown either on willow or on artificial diet as described by Riddiford (1968). The midgut was put up in the apparatus described by Harvey and Zerahn (1972), which makes it possible to distend the midgut in a controlled way and to keep the shape and thickness constant

Table I. Composition of solutions used.

Abbreviation	alkali metal ion	Cl⁻	HCO₃⁻	sucrose
32-K	32 mM K	30 mM	2 mM	166 mM
32-Na	32 mM Na	30 mM	2 mM	166 mM
16-K,16-Na	16 mM K	30 mM	2 mM	166 mM
	16 mM Na			
3.2-K,28.8-Na	3.2 mM K	30 mM	2 mM	166 mM
	28.8 mM Na			

during an experiment. The bathing solutions were either 32–K or solutions where some or all of the K was substituted by another alkali metal ion. The composition can be seen in Table I. The flux of the ions are measured by labelling the ions in the blood-side solution with radio-isotopes and measuring their appearance in the lumen solution.

Uptake of the alkali metal ions

When the midgut is bathed in a solution in which the K is replaced by another alkali metal ion, the gut will take up the other ion rapidly and lose K. The concentration of the cations in the midgut cells was determined as follows. The midgut was removed from the apparatus, blotted very gently on a piece of soft absorbent paper and weighed. Some bathing solution will adhere to the gut and some will be in the extracellular space. Allowance for this was made by determining the amount of sucrose in the midgut and assuming that the concentration of sucrose and ions in the extracellular space was the same as the concentration in the bathing solution.

The midgut was mashed together with one ml of perchloric acid and put in the coldroom for later determination of sucrose and alkali metal ions. The cations were determined by flame photometry; the sucrose by the glucose oxidase method after hydrolysis. The amounts of reducing substances in the midgut are small compared to the large amount of glucose derived from the

362 KARL ZERAHN

sucrose and may be neglected. By subtracting the weight of adhering solution and extracellular space and the content of the cations, the intracellular concentration of the cations could be determined as μEq. per gram cells.

RESULTS

Sequence of the competition for the transport mechanism

The extent to which the different alkali metal ions compete for the pump is described by Zerahn (1971). The potassium and rubidium are by far the preferred ions; when Na and K are present in the same concentration of 16 mM, Na transport accounts for only about 5 % of the total active flux. If the relative concentration is increased to 90 % Na (28.8 mM) and K is only 10 % (3.2 mM), the two ions will each contribute about half of the total flux. This indicates that the ratio of the concentration of the competing ions in the bathing solution is the determining factor. When equal concentrations of K and Na are used the estimation of the relative fluxes can not be very accurate, but enough to get the sequence. The accuracy for the comparison between the ions Na, Li, Cs is much better because they compete more equally. The sequence is given in Table II; it can be seen that the sequence is the sequence IV given by Diamond and Wright (1969) for the sequence of active transport processes.

Sequence for uptake of the competing ions by the gut cells

When the midgut is bathed in solutions with 16 mM of each of the competing ions, the ions will be taken up by the tissue cells and after 10 to 20 min. they will be close to a stable value. Simultaneously the K will be decreasing within the cells. Corrections were made for the adhering and extracellular space solution and the ions in it. The sequence for the uptake of the ions is given in Table II.

When the sequence for the uptake is compared with the sequence for the active transport, it is seen that the two sequences are different. Further it can be seen that the quantitative differences are very pronounced. The differences for uptake are much more moderate than for transport.

These differences indicate that the concentration of the ions in the midgut has no close connection with the active transport of the ions through the gut, the important concentrations are those in the bathing solution.

This so far would agree with the results given by Harvey and Zerahn

Table II. Competition sequence of the active transport of the alkali metal ions in the upper line. The figures below are the ratios between the transport of each ion and the transport of K. The lower line shows the sequence for the uptake of the ions by the gut cells, and again the figures below are the ratios between the uptake of each ion and the uptake of K.

Active transport									
K	>	Rb	>	Cs	>	Na	>	Li	
1.00		0.90		0.08		0.05		0.05	
Uptake									
K	>	Rb	>	Na	>	Li	>	Cs	
1.00		0.90		0.41		0.30		0.20	

in 1969, where it was found that the K passing the midgut by active transport does not mix with the main pool of K in the midgut tissue.

Lag time

When the blood-side solution is suddenly labelled by injecting ^{42}K there will be a delay before the radioisotope appears in the lumen solution and it will take even longer before the steady state is obtained. When the flux is constant with time this delay can be pictured by the lag time defined in Fig. 1. This lag time was determined to be 2 to 4 min.

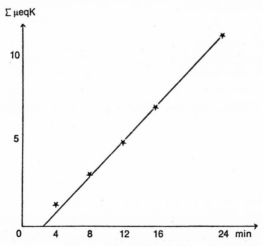

Fig. 1. Time course of ^{42}K movement from blood-side to lumen of an isolated midgut is plotted for a representative experiment in which the gut is equilibrated with 32-K. The lag time is estimated by extrapolating the steady state line to the abscissa. The ordinate is the amount of ^{42}K which at a certain time has appeared in the lumen.

This work has recently been criticized by Harvey and Wood (1972). They found that the lag time in a different preparation of the midgut was much longer, when they used certain corrections and even without corrections when the midgut was several hours old. The midgut was in these determinations put up as a flat sheet and not, as in the earlier experiments, as a tube distended to a sphere, they concluded that the earlier results were artifacts. Considering this before the meeting the question arose; what really influences the lag time?

When the old data were reconsidered it was clear that the short lag times were found in the thin midgut preparations, so it was natural to test the influence of the thickness of the midgut on the lag time. It seemed obvious to measure the lag time of the same midgut in a thick and later in a thin state.

When the midgut is put up in the closed chamber as a tube it is as thick as it can be in that arrangement and it is easy to distend it later to a large sphere by increasing the lumen side volume. The lag time was determined first when the midgut was thick, then the gut was distended and both compartments and the midgut washed free of radioactivity with 32-K solution. When the gut was without tracer a new lag time was determined on the distended gut.

Table III. Determination of lag time on the same midgut when the thickness is varied. The larvae were between 6.5 and 10 grams grown on willow (salix babylonica).

Date	lag time given in minutes cylinder (thick)	sphere (thin)
22 June	4.8	2.2
28 June	7.5	3.8
3 July	6.6	1.9
5 July	7.2	2.6
Mean value	6.5	2.6

The gut was kept in the open state in respect to the potential difference because we know that this will not change the lag time. Table III shows the results. It is obvious that the thickness of the midgut has a large influence on the lag time. The longest lag time found of 7.5 min. is of the same order as found by Harvey and Wood (1972), and the shortest of 1.9 min. is close to some of the very short lag times found in the earlier experiments. The midgut in the flat sheet preparations are not distended and are of a "thickness" of around 60 mg/cm^2, while the spherical preparations are 15 to 30 mg/cm^2. So the difference in lag time between the two preparations are what could be expected from the table. Thus a direct comparison between the lag times for the two preparations as done by Harvey and Wood is not justified.

The importance of the thickness of the midgut for the determination of the lag time may be explained by folds in the membrane. The thickness of the gut is measured by weighing the midgut and knowing the actual area used. But when the midgut is weighed it is first put on a blotting paper and a fraction of the solution in the folds may leak out and this may give a wrong value for the weight of the gut *in situ*. The folds may contain solution which may be a part of the pool ions in front or after the transport mechanism, which would appear in the kinetic determination but not elsewhere.

Cs pool

When the midgut is bathed in the Ca free solutions containing Cs the midgut will take up Cs. Fig. 2 shows how the concentration of K and Cs in the gut cells will vary with time when the bathing solution is changed from 32-K to 32-Cs at time zero. Already after 5 min. the concentration of K in the midgut cells is lower than the concentration of Cs, but it takes 2 to 3 min. before the concentration ratio Cs/K exceeds the ratio found in the midguts bathed in 16-Cs, 16-K and this ratio shows only 8 % Cs flux of the total flux. Furthermore it takes much longer time before the ratio for the steady state is obtained with respect to the concentration ratio Cs/K.

When the midgut is bathed in a solution containing two competing ions, it will take 5 to 10 min. before the gut is in a steady state. At that time we may label the competing ions in the blood-side solution and follow their appearance in the lumen. The delay will be due at least partly to a simple diffusion, so the ion with the lowest mobility will be delayed most and show the longest lag time. Preliminary experiments with Na and K showed that the lag time for these two ions will be nearly inversely proportional to their mobility; a mean of 4 experiments showed a ratio of 1.8 instead of 1.5 for the ratio of the mobilities. When the mobilities for the two competing ions are equal, the two ions should appear at the same time and show about the same lag time in the steady state experiments.

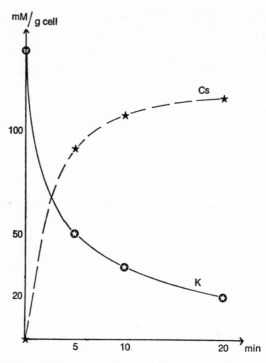

Fig. 2. Change in Cs and K concentration in midgut cells with time, after changing to a bathing solution of 32-Cs at zero time.

But if the midgut is bathed in 32-K at the start and suddenly the solution is replaced with solution 6.4-K,25.6-Cs and both the competing ions are labelled on the blood-side, the picture may be different. Now the labelled K will go to the transport pool and at once be ready for a preferred K transport because the K is there in high concentration. The Cs will have to wait until the concentration ratio between Cs and K is high enough. If we assume that the gut K is the transport pool, it is obvious from Fig. 2 that this will take several minutes and 5 to 10 min. before the steady state is really approximately reached. So we should find very little Cs in the first and second 4 min. period in this kind of experiments. Table IV shows the results. There is no change in the ratio not even at the first 4 min. period; the change in the later 4 min. periods is of no importance and is expected because the midgut may change its properties by changing the ion contents in the cells.

From Fig. 2 it is seen that the ratio for Cs/K is 1.8 at 5 min., 3.1 at

Table IV. Ratio of Cs flux to K flux at successive time periods after a sudden change of the solution on the blood-side from 32-K to 6.4-K,25.6-Cs. Each value is the mean of 5 experiments.

min.	ratio μEqCs/hr-μEqK/hr
4	2.3
8	2.4
12	2.3
16	1.9
24	1.9

10 min., and 5.8 at 20 min. These ratios are for replacing the bathing solution with 32-Cs not with the solution 6.4-K,25.6 Cs which would undoubtedly change the curve to some degree. However the first periods will keep higher K concentrations in the cells and so favor K transport, and the steady state is not obtained significantly faster with the higher K concentration in the midgut.

The explanation for these results is concluded to be that the pool located in front of the mechanism is a small one and not of the size of the midgut K. It is not possible at the moment to make any estimate of how large the pool is. Later experiments may however make it possible. The results do however confirm the earlier results of Harvey and Zerahn from 1969.

REFERENCES

Diamond, J. M. & Wright, E. M. (1969) Biological Membranes: The physical basis of ion and non-electrolyte selectivity. *Ann. rev. Physiol. 31*, 581.

Harvey, W. R. & Wood, J. L. (1972) Cellular pools involved in active K-transport across the isolated Cecropia midgut. In *Role of Membranes in Secretory Processes*, ed. Bolis, L. North-Holland Publ., Amsterdam, p. 310–331.

Harvey, W. R. & Zerahn, K. (1969) Kinetics and route of active K-transport in the isolated midgut of *Hyalophora cecropia. J. exp. Biol. 50*, 297–306.

Harvey, W. R. & Zerahn, K. (1971) Active transport of sodium by the isolated midgut of *Hyalophora cecropia. J. exp. Biol. 54*, 269–274.

Harvey, W. R. & Zerahn, K. (1972) Active transport of potassium and other alkali metals by the isolated midgut of the silkworm. *Current Topics in Membranes and Transport 3*. Academic Press. New York, London.

Riddiford, L. (1968) Artificial diet for Cecropia and other Saturniid silkworms. *Science 160*, 1461–1462.

Zerahn, K. (1971) Active transport of the alkali metals by the isolated mid-gut of *Hyalophora Cecropia*. Phil. Trans. Roy. Soc. Lond. B. *262*, 315–321.

DISCUSSION

HOSHIKO: I would like to comment that in washout of frog skin we found that sodium washout in the whole skin was at the order of 2–4 minutes, as you know, but the isolated epidermis gives a washout time that we can attribute to the cells of about 20 seconds. So this, I think, corresponds with your experience that thinning that tissue decreases the time.

ZERAHN: Yes.

TOSTESON: Could you tell us exactly how those numbers for the selectivity ratios were obtained. I missed that.

ZERAHN: The bathing solutions are containing: sucrose 166 mM, potassium chloride: 16 mM, and cesium chloride: 16 mM, besides some bicarbonate and chloride. You equilibrate and then you add the isotope to measure the flux through the midgut, or measure the short circuit current, or do both. In the same gut you can determine how large the flux is of for instance K and the flux of cesium and then later determine how high the content is of K, cesium or sodium and all the different alkali metals.

TOSTESON: So the numbers then are ratios of the transport rates, either all the way across the epithelium or into the cells.

ZERAHN: All the way across the tissue.

TOSTESON: With equal molar concentrations of the ions in the bathing solution?

ZERAHN: Yes.

TOSTESON: You do one pair, and then another pair of metal ions?

ZERAHN: You compare for instance sodium and cesium. But you can directly measure the flux of sodium and cesium through the gut and you can directly measure the content of sodium and cesium in the gut.

SACHS: In the absence of added alkali metal, for instance in choline chloride solution, does *Cecropia* transport acid?

ZERAHN: The choline chloride has been tried for the midgut, but the tissue is damaged. I don't know if there will be any secretion of hydrogen under these conditions. But I do know that there is no secretion of hydrogen ions during transport of the alkali metals.

MEARES: I wonder if there is any evidence about the power of the thickness which is related to the lag time. For instance does the lag time appear to follow inversely with the first power or with the second power of the thickness? Also when you were showing replacement of, was it potassium by cesium or cesium by potassium you had this plotted against the first power of time? I wonder if you had plotted it against the square root of time or, better still the square root of time divided by the thickness, whether you could get any slopes and kinetic constants from this which could be correlated.

ZERAHN: I would like to have the figures, but it is really not so easy, because in the experiments we are weighing the midgut and we don't know how much solution is in between the folds. This makes it very difficult to estimate the real thickness. And if we have the sphere and distend it, we have to distend much to get a uniform thickness. Most of the experiments we have done are not quite uniform midguts. The part tied to the glass tube is thicker than the rest. But I could of course plot the figures in the way you suggested.

KEYNES: It seems to me that if you are asking where the pool is in this tissue and you take the drawing that Dr. Harvey put up on the board, one's immediate first suggestion would be that the pool is in the middle of those goblet cells. You have presumably ruled this out, but when Dr. Zerahn stretches the tissue, might that not open up the hole, the communication with the outside world of the apex of the goblet cells? Have you worked out what the diffusion times are in the system?

HARVEY: No, we have not worked out these diffusion times but we should do so. From a rough estimate of the combined volume of all the goblet cavities based on measurements of their size and numbers in electron micrographs, the volume is not nearly great enough to contain the transport pool at any reasonable concentration. A stronger argument is that the washout experiments which we have reported here demonstrate that the pool is not after the pump. As you mentioned these goblet cavities are on the lumen side of the apical membrane which of course is after the pump. They might contain the small oxygen-independent component of the washout pool but there is insufficient volume for the large oxygen-dependent component.

CRABBÉ: Along the same line of reasoning, I wonder whether one could document changes in cell volume when one manipulates the rate of potassium transport by for instance withdrawing oxygen from the incubation medium.

ZERAHN: I don't know. It hasn't been done. Of course you could do it, but then you have to be working very fast. The moment you change the oxygen concentration, you are really changing the whole system. And it goes fast.

HARVEY: Dr. Zerahn contends that our argument for the long mixing time and correspondingly large pool size is invalid. He shows in his Table III that the lag time is shorter in stretched midguts than in unstretched ones. However, from Zerahn's own experiments (Table X of Harvey and Zerahn 1972) the pool size determined in spherical preparations showing the short lag times and fast fluxes amounts to an average 4 μEq of K for midguts which contain about 6 μEq of K.

Dr. Zerahns second criticism is that we should not compare kinetic data from flat sheet preparations with that from spherical preparations. Dr. Wood and I have now used our steady-state treatment including current corrections on spherical preparations and found the same long lag times and large pool sizes which we found earlier in the flat sheet preparations. When the data are expressed on a wet weight basis corrections for volume in adhering and extracellular fluid, the short-circuit currents and pool sizes from sphere and flat sheet in the steady-state are almost identical and are large (Wood & Harvey 1973).

ZERAHN: Your first arguments are for a short lag time. If you increase the lag time in Table X three times but keep the flux constant, the pool will have to increase three times and becomes larger than the K in the whole midgut (Andersen & Zerahn 1963). I will return to your second comment later.

Harvey, W. R. & Zerahn, K. (1972) Active transport of potassium and other alkali metals by the isolated midgut of the silkworm. In: *Current Topics in Membranes and Transport*, pp. 367–410. Academic Press, New York.

Wood, J. L. & Harvey, W. R. (1973) Active transport of potassium by the Cecropia midgut VII. Tracer kinetic theory and transport pool size. (In preparation).

Andersen, B. & Zerahn, K. (1963) Method for non-destructive determination of the sodium transport pool in frog skin with radiosodium. *Acta physiol. scand. 59*, 319–329.

WALL: When you stretch out the midgut epithelium, could there still be folds in the midgut?

ZERAHN: I would have to ask the people who are making pictures about it.

WALL: Well, I think that there are still folds, even when you stretch the midgut, and I don't think you can get rid of the folding of the epithelial cell layer unless you cut the muscles. When there is less stretching and the midgut is thicker, you are just compounding the folds. So you have lots of extracellular space, more than you would when you stretch the midgut out fully.

ZERAHN: Yes. I think that may be true.

WALL: Another point about the goblet cavity. There is a lot of extracellular material there, which may in some way bind a lot of ions.

MAETZ: Some of the discrepancies in your experiments concerning potassium transport may result from the heterogeneity of the epithelial potassium pool. While all the cells would rapidly exchange their potassium from the blood side, only the small pool related to the goblet cells would be transported across the membrane.

ZERAHN: I have not been able to measure the pools, which only make up a small fraction of the midgut K.

Transport of Amino Acids in Cecropia Midgut

Signe Nedergaard

INTRODUCTION

Uptake of amino acids from the vertebrate intestine seems to be associated mainly with the active sodium transport across the tissue. Sodium is actively transported from the lumen of the intestine to the blood, which is in the same direction as the transport of amino acids. The results of research on amino acid transport have been reviewed by Schultz & Curran (1970).

The midgut of the larva of the American silkmoth, *Hyalophora cecropia,* does not transport sodium but has an active transport of potassium from the blood into the lumen (Harvey & Nedergaard 1964). The present paper describes the measurement of amino acid fluxes in the two directions across the tissue. A possible relation to the active transport of potassium is investigated.

MATERIALS AND METHODS

The animal used is the fifth instar larva of *Hyalophora cecropia,* most of which were fed an artificial diet. The midgut is dissected out and a piece of about 100 mg is mounted in a modified Ussing chamber. The details of the experimental set-up is described by Nedergaard & Harvey (1968), and the short-circuiting arrangement by Harvey *et al.* (1967).

All the bathing solutions contained 5 mM $CaCl_2$ and 5 mM $MgCl_2$ in addition to 2 mM $KHCO_3$ and either 30 or 62 mM KCl. The osmolarity was in all cases made up to 260 mOsm by sucrose. The pH of the bathing solutions was about 8. In some experiments all potassium was substituted by equal amounts of sodium. The test amino acid was α-aminoisobutyric acid and in some experiments lysine. The bathing solutions contained, of the

---

University of Copenhagen, Denmark, Institute of Biological Chemistry A.

»Transport Mechanisms in Epithelia«, Munksgaard, Copenhagen.

Table I. α-Aminoisobutyric acid fluxes across the isolated *Cecropia* midgut

| | Lumen → blood | | Blood → lumen | | |
Exp.	Flux µmole/h.	PD mV	Exp.	Flux µmole/h.	PD mV
29/7	22.2	109	1/8	0.20	105
30/7–I	17.7	115	4/4†	0.23	105
30/7–II	11.9	110	11/4†	0.50	73
31/7	22.9	113	13/4†	0.30	78
6/8	11.8	102	14/4†	0.16	101
7/8	22.2	95			
23/1†	9.0	86			
Mean flux	16.8±5.9		Mean flux 0.28±0.13		

Table I: The amino acid fluxes are mean values of 2 to 4 consecutive periods. The potential differences are mean values for the same periods. 10 min. periods for the lumen to blood fluxes, and 15 min. periods for the blood to lumen fluxes. † Midguts from animals fed on artificial diet.

following carbon-14 labelled test substances, either α-aminoisobutyric acid (10 mM), lysine (10 mM), isobutyric acid (1 mM), or methylamine (10 mM). The volume of the bathing solution on the bloodside was 50 ml and on the lumen side 5 ml. Both sides were aerated and stirred by bubbling oxygen through the bathing solutions.

RESULTS AND DISCUSSION

Active transport of an amino acid

The fluxes of α-aminoisobutyric acid were measured in the two directions across the isolated midgut with identical solutions bathing both sides. The results are shown in Table I. It is seen that the flux from lumen to blood is much larger than the flux in the opposite direction; the average flux ratio for the experiments of table I is about 50.

Fig. 1 shows the effect on the two fluxes when the aerobic metabolism is inhibited by aerating the bathing solutions with nitrogen instead of oxygen. Fig. 1 shows a decrease in the amino acid flux from lumen to blood compared to the control value, while the flux in the opposite direction is increasing, so the flux ratio is falling and becomes close to 1. The decrease in potential difference produced by the anoxia is shown by the dotted line in Fig. 1.

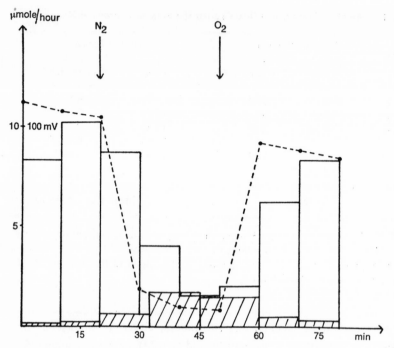

Fig. 1. Effect of anoxia on α-aminoisobutyric acid fluxes across the Cecropia midgut. The white columns are α-aminoisobutyric acid fluxes from lumen to blood, the hatched columns are fluxes from blood to lumen. The dotted line is the potential difference across the midgut. The two fluxes are not measured on the same midgut. The potential difference shown belongs to the lumen to blood experiment, but the variations in the potential difference of the other experiment were very similar. The arrows indicate the time for the change in aerating gas.

The observations that α-aminoisobutyric acid is transported 50 times faster from lumen to blood than in the opposite direction, when the midgut is bathed with identical solutions on both sides, and that the flux ratio for the amino acid approaches 1, when the aerobic metabolism is inhibited, suggest that α-aminoisobutyric acid is actively transported from lumen to blood of the *Cecropia* midgut. A more detailed report on this subject has been published (Nedergaard 1972).

Dependence on active potassium transport

The midgut transports potassium from blood to lumen, which is in the opposite direction to the active transport of amino acid. The active transport of

Table II. Effect of short-circuiting the midgut potential on α-aminoisobutyric acid flux across the *Cecropia* midgut

	Open-circuit		Potential short-circuited	
	Flux	PD	Flux	SCC
Exp.	μmole/h.	mV	μmole/h.	μAmp.
		Lumen → blood		
1/2	7.1	101	2.7	2000
2/2–I	9.1	94	4.1	2310
2/2–II	13.9	71	6.1	1700
14/4–II	6.5	86	2.1	1330
15/4–I	5.8	63	3.6	1470
17/4–I	6.6	98	2.0	1965
17/4–II	8.5	109	2.0	2240
17/4–III	10.2	82	3.9	1840
Mean flux	8.5±2.5		Mean flux 3.4±1.3	
		Blood → lumen		
3/2-I	0.37	71	1.02	1060
5/2	0.37	76	1.02	1100
6/2	0.23	104	0.58	1510
4/4	0.23	105	1.19	1820
11/4	0.50	73	1.77	1440
13/4	0.29	78	0.76	1690
14/4-I	0.16	101	0.52	1820
Mean flux	0.31±0.11		Mean flux 0.98±0.40	

Table II: Each column is the mean value of at least three 10 min. periods in the lumen to blood experiments and three 15 min. periods in the blood to lumen experiments. SCC: short-circuit current. All larvae reared on artificial food.

potassium is independent of the presence of sodium in the bathing solutions. In order to study if the amino acid transport is related to or influenced by transport of potassium, the active transport of potassium was altered and the effect on the amino acid fluxes was measured. The active potassium transport was increased 100 percent by short-circuiting the midgut potential. The results are shown in Table II. There is a considerable fall in the amino acid transport from lumen to blood compared with the control value, while the amino acid flux in the opposite direction is increased. The flux ratio falls from 30 in the control periods to about 3 in the experimental periods. The overall effect of the increase in active potassium transport seems to be a fall in the active amino acid transport.

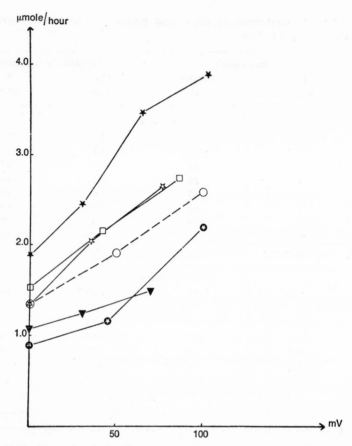

Fig. 2. Effect of the midgut potential on α-aminoisobutyric acid flux from lumen to blood.

    Abscissa: potential difference in mV. Ordinate: μmole α-aminoisobutyric acid flux from lumen to blood per hour. Each symbol is one midgut, i.e. one experiment. The large open circles are the mean values of all experiments of the figure, see text.

Dependence on the midgut potential difference

It seems surprising that the active transport of potassium in the midgut produces an inhibition of amino acid transport, as the two functions have to operate at the same time in the living animal. The inhibition could, however, be due to elimination of the potential difference across the tissue, so the effect might instead be attributed to the missing potential difference during short-circuiting. Fig. 2 shows the amino acid flux from lumen to blood as a function of the potential difference across the midgut. Each symbol in

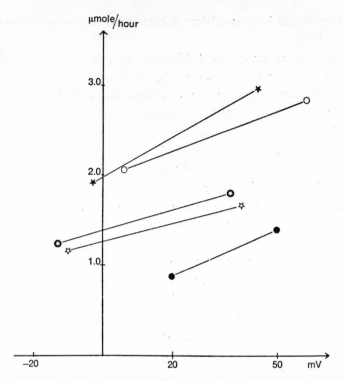

Fig. 3. Effect of an applied midgut potential on α-aminoisobutyric acid flux from lumen to blood in the absence of potassium.

Abscissa: potential difference in mV. Ordinate: μmole α-aminoisobutyric acid flux from lumen to blood per hour. Each symbol is one midgut, i.e. one experiment.

the figure represents one midgut, the highest point is the amino acid flux at the natural spontaneous potential difference and the point on the ordinate is the flux from lumen to blood when the midgut potential is short-circuited. From the results of Fig. 2, it may be concluded that there is a relation between the potential difference and the active amino acid flux.

Effect of an applied potential difference

In all experiments so far mentioned the bathing solutions contained potassium and potassium is essential for maintenance of the potential difference, when calcium is present in the bathing solutions (Harvey & Nedergaard 1964). When all potassium is removed from the bathing solutions and substituted by sodium, the electrical potential difference across the midgut will be small or even negative, as long as calcium is kept in the bathing solutions (Harvey & Zerahn 1971).

The inhibition of the potential difference by removal of potassium from the bathing solutions is irreversible. Fig. 3 shows the effect of an applied electrical difference across the tissue on the amino acid flux from lumen to blood in such potassium depleted midguts. Each symbol represents one midgut. The point to the left is the amino acid flux from lumen to blood in the control periods, and the point to the right is the flux with an applied potential difference. In all cases there is an increase in amino acid flux from lumen to blood when the artificial potential difference is applied. This shows that the amino acid can be transported without potassium present in the bathing solutions and that an electrical potential difference increases the amino acid transport.

In order to be able to compare Fig. 2 and 3, the mean value of the experiments of Fig. 2 is shown as a dotted line in the figure. The numbers are arrived at by taking the mean of the fluxes at 100% open condition and when half of the potential difference is short-circuited (50%), and finally when the midgut is short-circuited (0%). It is seen by comparing Fig. 2 and 3 that the average slopes of the two sets of experiments are of the same order of magnitude.

The importance of the overall charge on the amino acid

As pH of the bathing solutions is about 8, α-aminoisobutyric acid will carry an overall negative charge. The midgut potential is positive on the lumen side, so the amino acid, due to its negative charge, is moving against the electrochemical gradient. In order to test if a negative charge is essential for the transport mechanism, the fluxes in the two directions of lysine were measured at pH 7.5, where the overall charge of lysine is positive. Table III shows the results. It is seen that the ratio between the two fluxes is about 10 indicating that lysine is transported from lumen to blood. When the pH is changed to about 10 on the lumen side, lysine is still transported into the blood, although it is now virtually uncharged. The overall charge of the amino acid seems therefore not to be involved in the transport mechanism.

Table III also shows that the flux ratio of lysine transport is smaller at the high pH, while both the flux from lumen to blood and the flux in the opposite direction increase. This might suggest a higher degree of leakiness

Table III. Effect of luminal pH on lysine fluxes across the isolated *Cecropia* midgut

Exp.	Normal pH			High pH		
	pH	Flux µmole/h.	PD mV	pH	Flux µmole/h.	PD mV
		Lumen → blood				
2/7	7.5	1.07	100	10.1	3.33	85
5/7	7.5	1.46	75	10.1	3.95	69
6/7-I	7.5	1.22	52	10.0	2.56	43
6/7-II	7.5	1.86	110	10.1	4.26	104
	Mean flux 1.52			Mean flux 3.52		
		Blood → lumen				
8/7	7.6	0.06	123	9.9	0.38	99
9/7	7.5	0.10	105	9.9	0.29	113
1/11	7.7	0.22	98	9.4	1.26	71
2/11	7.7	0.26	86	9.8	1.46	57
3/11	7.6	0.04	98	9.9	0.14	111
	Mean flux 0.14			Mean flux 0.71		

Table III: The amino acid fluxes are mean values of 2 to 4 consecutive periods. The potential differences are mean values for the same periods, as are the pH values.

of the midgut at the high luminal pH. The effect of the increase in pH is that the net transport into the blood of lysine is about twice as high at the high luminal pH as at pH 7.5. It should be mentioned that the pH of the midgut lumen is about 10 in the living animal.

The behaviour of the two reactive groups of an amino acid

Methylamine is transported from lumen to blood of the midgut with a flux ratio of about 20, Table IV, and when the midgut potential is short-circuited so there is no electrical potential gradient with a flux ratio of at least 10. The transport of methylamine is in the opposite direction to the active potassium transport, but in the same direction as the active amino acid transport. Methylamine was used, as labelled isopropylamine was not commercially available.

Table IV also shows that isobutyric acid is transported by the midgut, but in the opposite direction of methylamine – from blood to lumen, and also when the potential difference is short-circuited.

Table IV. Methylamine and isobutyric acid fluxes across the isolated *Cecropia* midgut

Exp.	Lumen → blood Flux μmole/h	PD mV	Exp.	Blood → lumen Flux μmole/h	PD mV
			Methylamine		
16/2-I	16.9	90	16/2-II	0.96	52
18/2-I	22.3	65	15/5-I	0.51	77
18/2-II	15.7	95	15/5-II	0.62	58
10/5	7.8	98			
Mean flux	15.6		Mean flux	0.70	
			Isobutyric acid		
20/1-I	0.68	24	20/1-II	3.7	42
21/1-I	0.52	73	21/1-II	7.8	60
16/5-I	0.54	92	16/5-II	6.2	67
Mean flux	0.58		Mean flux	5.9	

Table IV: The fluxes are mean values of 2 to 4 consecutive periods. The potential differences are mean values for the same periods. The concentration of methylamine in the bathing solutions is 10 mM and of isobutyric acid 1 mM. All larvae grown on artificial food.

CONCLUSION

It can be concluded that α-aminoisobutyric acid is transported actively by the *Cecropia* midgut from lumen to blood. Futhermore, this active transport is independent of the presence of potassium in the bathing solutions, and is dependent mainly upon the electrical potential difference across the midgut. The transport mechanism does not seem to be affected by the overall charge of the amino acid.

REFERENCES

Harvey, W. R., Haskell, J. A. & Zerahn, K. (1967) Active transport of potassium and oxygen consumption in the isolated midgut of *Hyalophora cecropia*. *J. exp. Biol. 46*, 235–248.

Harvey, W. R. & Nedergaard, S. (1964) Sodium-independent active transport of potassium in the isolated midgut of the Cecropia silkworm. *Proc. nat. Acad. Sci. (Wash.) 51*, 757–765.

Harvey, W. R. & Zerahn, K. (1971) Active transport of sodium by the isolated midgut of *Hyalophora cecropia*. *J. exp. Biol. 54*, 269–274.

Nedergaard, S. (1972) Active transport of α-aminoisobutyric acid by the isolated midgut of *Hyalophora cecropia*. *J. exp. Biol.* *56*, 167–172.

Nedergaard, S. & Harvey, W. R. (1968) Active transport by the Cecropia midgut. IV. Specificity of the transport mechanism for potassium. *J. exp. Biol.* *48*, 13–24.

Schultz, S. G. & Curran, P. F. (1970) Coupled transport of sodium and organic solutes. *Physiol. Rev.* *50*, 637–718.

DISCUSSION

HARVEY: The sizes of the amino acid fluxes are quite a lot smaller than the potassium fluxes. That is right, isn't it? So, that would be another argument against there being any kind of coupling between them. At least the coupling ratio would have to be quite far from unity.

NEDERGAARD: If we take the experiments shown in Table II and assume that the active potassium transport under open conditions is about half of the value in the short-circuited condition we get an average value of 35 μEq per hour for the potassium transport and 8.5 μEq per hour for the amino acid transport in the same 8 experiments. So there is at least not a one-to-one relationship between the two transports; moreover the two transports are in opposite directions, contrary to sodium and amino acid transport in the vertebrate intestine.

SCHULTZ: Have you investigated whether or not the amino acid transport mechanism is sodium dependent?

NEDERGAARD: It is very unlikely that the amino acid transport is dependent on sodium, as there is no sodium at all present in the bathing solutions of most of the experiments.

SCHULTZ: Is it possible that the relation between the potential difference across the gut and the rate of amino acid transport is due to change in the sodium profile across the tissue?

HARVEY: Related to that comment, in the living insect there is approximately 1 mM of sodium per kgm of tissue water in the midgut and in all other tissues that have been analyzed.

ULLRICH: What is the normal amino acid and potassium concentration in the hemolymphatic fluid of the larva?

NEDERGAARD: The blood of the living silkworm contains 27 mM potassium, 6 mM sodium, and a considerable amount of free amino acids. For another lepidoptera the free amino acids amount to about 40 per cent of the total ionic concentration of the blood. The osmolarity of the blood of the fifth instar larva of the *Cecropia* silkworm, which is the one we use, is 260 mOsm, but exactly how much of this is due to free amino acids, I do not know.

SACHS: When you short-circuit the tissue, do you happen to know whether there is any accumulation in the midgut of amino acids?

NEDERGAARD: Unfortunately, I have no good measurements of the amino acid concentrations in the tissue.

GENERAL DISCUSSION

ZADUNAISKY: With respect to *cecropia:* These animals must have a high potassium concentration in the hemolymph. What is the composition of cells of the intestine with a hemolymph of 100 mM potassium?

HARVEY: The [K] is 23 mM in the hemolymph. For midgut tissue taken freshly from the living insect and rinsed but not corrected for extracellular spaces it is 90 mM and in the lumen contents it is 285 mM.

OSCHMAN: There is a most important structural feature of the *cecropia* midgut epithelium that relates to the washout experiments and to the determinations of extracellular space. This epithelium has very extensive basal infoldings or channels that open in only a few places to the haemocoel, or blood cavity. Thus there is a relatively large fluidfilled compartmental system that communicates relatively poorly with the blood. This arrangement exaggerates the unstirred layer phenomenon, and would be expected to have considerable influence upon the results of washout studies. The goblet cavity, of course, forms a similar compartment on the apical or lumen side. For details, see Berridge (1970).

Berridge, M. J. (1970) A structural analysis of intestinal absorption. In: *Insect Ultrastructure,* ed. Neville, A. C., pp. 135–151, Blackwell, Oxford.

CURRAN: Since Drs. Harvey and Zerahn seem to be arguing about this time lag. I wonder if one of them would explain to us as clearly as possibly exactly what that time lag really means.

ZERAHN: It is a delay of the flux.

CURRAN: A delay in reaching the steady-state flux-level?

ZERAHN: Yes.

CURRAN: What does that mean precisely. It means different things in different systems. Dr. Meares was asking a question about this with reference to the behaviour that one might observe in an artificial membrane, weren't you, Dr. Meares?

MEARES: That's right.

CURRAN: The lag obviously has an entirely different meaning in such a simple system than it has in the complex systems you are dealing with.

ZERAHN: It is derived in the work on the frog skin, and it is used here mostly because it is one of the data you can measure representing the delay.

CURRAN: But I think it is not clear exactly what is being measured. It is measuring the delay in reaching the steady-state, but what does that mean?

KIRSCHNER: I think that the presumption is that you are estimating pool size.

HARVEY: Wood & Harvey (1973*) have derived equations describing influx kinetics across epithelial membranes. They assumed (1) that the tissue is in a steady-state with respect to pumping rate (2) that influx sampling is continued until the tracer steady-state is reached (3) that the entire I_{sc} is carried by the ion whose influx is being measured and (4) that the efflux of that ion is constant with time. Since the I_{sc} of epithelial membranes decays with time, it is necessary to correct for this decay in order to satisfy the first assumption (constant pumping rate). The correction equation is:

$$\triangle^n = \triangle^n_u + (I^o_{sc} - I^n_{sc}) \left[\frac{\triangle^n_u}{(I^n_{sc} + \text{efflux})} \right] \qquad (1)$$

where, $\triangle$, is the corrected influx, $\triangle_u$, is the uncorrected influx, I_{sc}, is the short-circuit current, and the superscripts refer to the time elapsed after isotope is added. When the midgut is equilibrated in an adequate physiological solution for 120 minutes before the addition of isotope, the corrections are small and the other three assumptions are satisfied.

When the current-corrected influx values from the midgut are plotted by the method of Hoshiko & Ussing (1960), which we will refer to as the "influx logarithmic method", the result is a straight line whose equation is:

$$\ln \left(1 - \frac{\triangle^n}{\triangle^\infty} \right) = -\alpha t + \gamma \qquad (2)$$

where the slope, α, is given by:

$$\alpha = \frac{\ln \left[\dfrac{\triangle^\infty - \triangle^1}{\triangle^\infty - \triangle^2} \right]}{(t^2 - t^1)} \qquad (3)$$

and the y-intercept, λ, is given by:

$$\lambda = \ln \left(1 - \frac{\triangle^o}{\triangle^\infty} \right) \qquad (4)$$

Solving for $\triangle^n$, the equation for the results plotted by the "influx time-course method" is:

$$\triangle^n = \triangle^\infty (1 - e^{-\alpha t + \lambda}) \qquad (5)$$

Integrating the rate, $\triangle^n$, to determine the amount of labelled ion accumulating in the cold-side copartment, S_C, with time yields the equation for

---

Hoshiko, T. & Ussing, H. H. (1960) The kinetics of Na[24] flux across amphibian skin and bladder. *Acta physiol. scand.* 49, 74–81.

* see p. 370.

the results plotted by the method introduced by Andersen & Zerahn (1963) which we will refer to as the "amount time-course method".

$$S_C = \triangle^\infty \left[t + \frac{1}{\alpha} (e^{-\alpha t + \lambda} - e^{\lambda}) \right] \tag{6}$$

This amount time-course reaches an asymptote whose equation is:

$$S_{C_{as}} = \triangle^\infty t - \frac{\triangle^\infty}{\alpha} e^{\lambda} \tag{7}$$

The intercept of this line on the x-axis is the parameter designated by Andersen and Zerahn as the lag-time. Its relationship to the 75 percent mixing time, t_{75}, is given by letting $\triangle^n = 0.75 \triangle\infty$ and setting $S_C = O$. Then the lag-time, X, is:

$$X = \frac{t_{75}}{(\ln 4)\ e^{-\lambda}} \tag{8}$$

Equations (7) and (8) define the lag-time mathematically as requested by Dr. Curran. Wood & Harvey (1973) have also shown that (λ) is small for the midgut. Therefore, Equation (8) says that, if the *midgut is in a steady-state* so that the influx logarithmic plot follows a single exponential curve and provided that suitable corrections for current decay are applied to satisfy the constant pumping rate assumption, then the lag-time is 1/ln4 times the 75 percent mixing-time determined from the influx logarithmic plot.

ZERAHN: You want to study if the pools in the midgut are mixed with the transported K. You use the equations derived by Hoshiko & Ussing to obtain your results, but these corrections are based on the assumptions of a model with a completely mixed layer for the transported ions through the pools.

SCHULTZ: At first you said that there were at least two elements involved in the lag. One is the effect of diffusion delays such as those found in unstirred

Andersen, B. & Zerahn, K. (1963) Method for non-destructive determination of the sodium transport pool in frog skin with radiosodium. *Acta physiol. scand. 59,* 319–329.

layers and which you would find across any artificial membrane and the other is the time required to obtain uniform specific activity in the transporting cells.

HOSHIKO: I think that the nature of that curve as it is analyzed this way will fail to reveal the presence of another exponential component. In a sense you may fail to see whether there is more than one compartment involved. So the actual interpretation of the meaning of this, of course you people know this very well, depends upon the particular model one chooses, and in some way you are prejudicing the answer by taking this approach.

Can I ask one more question: How fast do you sample in your washout experiments? We found that washing out the skin we have to sample at least one per second in order to resolve the components.

HARVEY: The components of the influx kinetics are not revealed by the lag-time method, and the interpretation of the data by this method would be prejudiced by the particular model one chooses. However, the components of the influx kinetics can be analyzed by the influx logarithmic method. Moreover, the pool size can be calculated from the influx data with no assumptions regarding model so that the answer is not prejudiced. The pool size is given by the area between the influx time-course curve described by Equation 5 and the theoretical instantaneous steady-state time-course curve. This is the same approach used by Harvey & Zerahn (1972, fig. 7) in their geometric approximation of the pool size. The resulting equation for the size of the influx pool, S , after integration is:

$$S_I = \frac{\triangle^\infty}{\alpha} (1 + \lambda) \tag{9}$$

The pool size is large whether calculated from this equation, from the y-intercept of the amount time-course asymptote, or approximated by the geometric method of Harvey & Zerahn. As mentioned on page 350 the size of the pool, even when calculated from the uncorrected amount time-course data by Harvey & Zerahn (1972; Table X) is about 2/3rds of the midgut tissue K. There are components which would not be expected to participate in an influx pool such as the K in muscle cells and possibly that in compartments within the epithelial cells such as the mitochondria. Finally, from our Table

II it can be seen that the loading pool is very nearly the same size as the influx pool. So the simplest hypothesis is that the large transport pool is located in the epithelial cells.

We are aware that our sampling technique is not perfect for determining all of the washout components, but we were interested only in the relative effects of oxygen on the lumen-side washout. Moreover, the washout calculations are complicated by the fact that the washout to the blood-side is independent of oxygen, and this affects the relative specific activity of the ions washing out to the lumen-side. Finally, the half-time of mixing within the bulk of the bathing solutions in the chamber is about 15 seconds, rendering intervals of sampling shorter than about one minute rather meaningless. The main point is that although our sampling technique was imperfect, we could demonstrate a large relative effect of oxygen on the lumen-side washout.

ZERAHN: In my lecture, I have given an explanation why the tissue K may be determined too low by weighing the gut and determining the K chemically.

KEYNES: I am not sure whether this is sensible or not, but in some of the curves that Dr. Zerahn showed for washing out, I had the impression that although the size of the pool was obviously much smaller in the absence of oxygen, the time constant for getting to the steady-state was much the same in the two curves. Is that right?

ZERAHN: Yes, that is right.

KEYNES: Is that not inconsistent with the argument about nitrogen?

ZERAHN: No, that is not quite true. You have to have a certain amount of oxygen to measure from blood-side to lumen and to get the time delay, and in these washout curves I think the oxygen is completely removed by nitrogen, so I wouldn't like to imply that the delay is the same under these circumstances because that I don't know.

TOSTESON: I would like to return to the other experiment that Dr. Zerahn showed. I am most impressed with the observation that with the cesium substitution the ratio of K transport to cesium transport seems to be independent of the intracellular concentration. That can only mean that the pool through which the transport is occurring isn't the pool that you are measuring when you measure the tissue concentration of K and cesium.

HARVEY: Not necessarily. The midgut ion pump may have a very high affinity for cesium. There is no reason to expect that the sequence for alkali ion uptake across the basal membrane of the epithelial cells should be in any way related to the sequence of alkali ion affinity for the pumping sites if they are located on the apical membrane as postulated by Dr. Wood and me. I believe that Dr. Curran expressed a similar view yesterday.

ZERAHN: A high affinity of the pump for Cs should show up in the experiments with 16-Cs, 16-K. The gut contains much Cs but a very small fraction of the active transport of ions is Cs. K is still the ion preferred by the pump.

TOSTESON: You change over a range of an order of magnitude both concentrations, both for the K and the cesium and you don't change the ratio of the fluxes. That is astounding. How do you interpret it, Dr. Zerahn?

ZERAHN: I of course interpret it that the amount of K in the midgut is not the K which is transported. The K is going into the bathing solution through some unknown pathway, as far as I can see. I can't say about the amount of K^+ in the goblet cells, because I don't know about that. But it is not the K in the whole midgut, because this I can measure.

TOSTESON: I understand that when you take away the K outside and replace it with cesium, don't you change the ratio of K^+ to cesium transport, even though both were going extracellularly. There must be a pool.

ZERAHN: There must be a pool, that is quite certain, but it is a small one.

HARVEY: At present we cannot explain why Dr. Zerahn's cesium/potassium transport ratios are constant from the start (see his table IV) while the cesium/potassium concentration ratio for the tissue is increasing rapidly (see his fig. 2). Obviously the midgut is far from the steady-state.

Under the experimental conditions used by Dr. Zerahn, the I_{sc} decays very rapidly (Zerahn 1970). Moreover, after exposure to the low K concentrations used in this experiment the midgut does not recover (Harvey & Zerahn 1972). How can these experiments involving a non-steady state

Zerahn, K. (1970) Active transport of caesium by the isolated and short-circuited midgut of *Hyalophora cecropia*. *J. exp. Biol.* 53, 641–649.

midgut undergoing an irreversible decrease in pumping rate be used to argue about the size of the K-influx pool in a steady-state midgut?

ZERAHN: There is a misunderstanding here. The bathing solution is not the one Dr. Harvey has referred to, it is a solution free of Ca and Mg. In this solution the reference is Harvey & Zerahn (1971). The curve shown in fig. 1, 16-Na, 16-K is almost representative for the curve of transport for K and Cs as measured by the short-circuit current in the solution 6.4–K,25.6-Cs used.

SCHMIDT-NIELSEN: I would like to change the discussion to a question that Dr. Wall raised. In the cavity of the goblet cells there is mucopolysaccharide. This is being found, too, in the cavity of cells in the fish gills which secrete chloride. Dr. Wall said that maybe it binds ions. I would like to hear what people here think about that.

ZADUNAISKY: In the corneal stroma of the rabbit, for instance, there is an excess of sodium ion of about 30 mM when the concentration is compared to the one in the aqueous humor. Since it is connective tissue, it is assumed that sodium and other ions should be at the same concentration as in the most immediate extracellular fluid, the aqueous humor.

HOGBEN: Obviously, with a gel like mucus you can change the pH to the extent that it is a buffer. You have a Donnan equilibrium within the gel, but you are not going to affect importantly the diffusion constants of all but very large colloids. Unless you are worrying about measuring rates of transfer of hydrogen or bicarbonate ion, mucus may be inconsequential. It is not a diffusion barrier, and it is just sitting here. It does constitute an unstirred layer excluding convection and hence may diminish radical concentration changes at the face of the plasma membrane.

SKADHAUGE: When the mucopolysaccharide "slime" appears in macroscopic quantity on the inner surface of the epithelium in vertebrate guts, you can take this material out and analyze it, and it has the same sodium-chloride-potassium concentration as the bulk fluid. This suggests free diffusion of these ions.

LOEWENSTEIN: The mucopolysaccharides on some cell surfaces at least can bind a lot of divalent cations. One can certainly not neglect them especially since the mucopolysaccharides are constantly made.

SCHULTZ: Dr. Nedergaard, the only other sodium-free system that I am aware of where the amino acid transport has been fairly carefully studied, apart from terrestrial bacterial, is yeast. In this instance it appears as if hydrogen ion acts very much the way sodium ion acts in mammalia and in a variety of invertebrate species. Have you measured any pH changes in the lumen during the course of amino acid transport?

NEDERGAARD: No there is very little change in pH during an experiment, but in the living animal the pH of the lumen where the amino acids are absorbed from, is from 9.5 to 10.5

Binding of Calcium to Cell Membranes

James L. Oschman and Betty J. Wall

INTRODUCTION

Many cellular processes involve or are regulated by the binding of ions such as Na^+, K^+, or Ca^{++} to surface or intracellular membranes. The morphologist can contribute to an understanding of various aspects of physiology by identifying the anatomic sites of ion binding. We would expect the different surfaces of cells to differ in their affinities for ions, and for these affinities to correspond to those predicted from physiological data.

We recently reported (Oschman & Wall 1972) that when calcium was added to the solutions used for preparation of tissues for electron microscopy, electron-opaque deposits appeared along the inner surfaces of the plasma membranes in several transporting epithelia. Although the evidence is not yet conclusive, we suspect that these deposits may localize binding sites that could be involved in cellular regulation or transport. Here we describe how we found these deposits and our present ideas of their significance.

BACKGROUND

Our interest in ion binding to membranes arose during some studies we were doing in Professor Ussing's laboratory. We were using a variety of extracellular tracers that could be observed with the electron microscope, using the method devised by Ussing (1970). As an extra precaution in our studies we omitted the heavy metal stains that electron microscopists routinely use, as we did not want to confuse our tracers with binding of the stains. We simply fixed our tissues in glutaraldehyde, dehydrated and embedded them, and examined sections without staining. When we examined intestinal epithelia (from rats, toads, and insects) we literally began to see spots before our eyes, and some of them were in the wrong places – in addition to the extra-

From the Department of Biological Sciences, Northwestern University, Evanston, Illinois, USA.

»Transport Mechanisms in Epithelia«, Munksgaard, Copenhagen.

cellular tracers, we observed extensive deposits along the intracellular surfaces of both mucosal and lateral membranes. The deposits were less abundant toward the basal (serosal) surface and were absent from the basal plasma membrane. These deposits were somewhat disturbing until we examined our control preparations and found that the intracellular "spots" along the membranes were still present, and thus had nothing to do with the tracers we were using. The deposits also occurred in tissues that were quickly removed from the animal and fixed with glutaraldehyde. While this restored completely our confidence in the tracer experiments, the question remained as to the origin of the dense material along the inside of the membranes.

SOURCE OF THE DEPOSITS

At the outset we suspected that the deposits might be related to the biologically important divalent cations, calcium or magnesium, which are components of a variety of biological membranes (eg. see Manery 1966). Thus we did a preliminary study on bullfrog intestine in which we treated the tissue with EDTA *in vitro,* fixed it as for electron microscopy, and then checked the calcium content of the tissue by flame photometry. As anticipated, intestines incubated in calcium free medium containing EDTA had a much lower calcium content (71 mg/kg. dry weight) than controls (203 mg/kg. dry weight) incubated in the presence of calcium. The surprise came when we analyzed pieces of the tissues that had been incubated in EDTA and then fixed. These contained considerably more calcium (202 mg/kg. dry weight) than before fixation. Thus large amounts of calcium must have been taken up from the fixative. When we analyzed the components of the fixative, we found that calcium was present both in the glutaraldehyde (ca. 4 mM/1) and in phosphate buffer (ca. 1 mM/1).

The next step was to examine glutaraldehyde and buffers from various sources to find a combination that would be low in calcium. We finally settled on distilled glutaraldehyde and distilled collidine buffer.[1] From these reagents, we prepared a fixative that produced no reading on the flame photometer, which is sensitive to calcium concentrations as low as 10^{-6}M. This fixative also had very low concentrations of sodium and potassium. The deposits were very difficult to find in tissues fixed in this mixture. When

---

[1] Both available from Polysciences, Inc., Paul Valley Industrial Park, Warrington, Pennsylvania, U.S.A.

we added calcium to our "ion free" fixative, we obtained abundant deposits as before, even with calcium concentrations as low as $5 \times 10^{-5}M$. At $5 \times 10^{-6}M$. it was still possible to observe deposits, but they were faint and difficult to photograph.

These observations show that calcium has something to do with formation of the deposits, but we still do not know what the deposits are composed of. To some readers the deposits may seem to be ubiquitous and obnoxious artifacts of fixation that can be prevented by using calcium free fixatives. This may indeed prove to be the case. We can think of several spurious means for the formulation of the deposits. For example, the fixative may alter the configuration of certain tissue components that do not normally bind calcium in such a way that negative groups are exposed. Another possibility is that one of the many intracellular components that can form ligands with calcium diffuses to the cell surface at the time of fixation. Examples of groups that complex with calcium include citrate, carbonate, phosphate, ATP, polyglutamic acid, and various phospholipids and proteins. In any case, the deposits occur at the same intracellular sites regardless of the calcium concentration in the fixative, over a 1000-fold range of concentration, and, as we shall see below, the deposits occur only along certain portions of the plasma membranes.

Calcium concentrations in the physiological range seem adequate to saturate the supposed sites, and increases in calcium concentration above $5 \times 10^{-5}M$. produce little alteration in the number, size, and location of the deposits. It has recently been possible to find very faint deposits in tissues fixed in "ion free" fixatives by examining them at very high magnifications. These may be due to traces of calcium in our solutions or they may represent the residual calcium bound in the living tissues. Further study should clarify this point.

We find that if tissues are fixed in the presence of calcium and then washed with EDTA the deposits do not appear. Extensive washing in calcium free buffer fails to remove the deposits, indicating a low solubility for the complex. We also find that tissues fixed in ion-free fixatives, washed, and then soaked in buffer containing calcium have the deposits. This would argue against non-specific precipitation with small intracellular anions, which should be removed during the washing. The deposits are still present after glutaradehyde fixation followed by osmium post-fixation, provided calcium is added to all of the solutions used in processing the tissue. The image obtained with osmium post-fixation differs from that obtained with glutaraldehyde alone, as the opaque regions appear to be more intimately associated with the membranes, which are much better preserved with osmium post-fixation.

Images were often obtained in which the opaque regions were patches on the inner leaflets of the bilayer membranes, on the side facing the cytoplasm (Fig. 1)*. This latter finding indicated that the hypothetical calcium binding substance might be a component of the plasma membrane. Should this prove to be the case, the findings obtained thus far could be relevant to several physiological processes in which calcium has a key role. It seems that when cells are fixed in the presence of calcium, this ion enters and attaches to all of the available sites that have a high affinity for calcium. An ion-exhange type of cation adsorption is presumed to form a relatively insoluble complex that is retained within the tissue during preparation for the electron microscope. It is, of course, a large step from finding these deposits in fixed tissue to any conclusion that the binding is the same as that involved in the living system.

We are not sure if the opacity of the deposits is due only to calcium atoms or if it is due to some opaque component of the tissue that is rendered insoluble by calcium ions. For example, phosphoric acid groups on proteins and lipids can have a high affinity for calcium. Calcium phosphate has a low solubility in water and alcohol, and could, if sufficiently concentrated, be electron opaque. For example, under appropriate conditions calcium phosphate plus an organic matrix can form within mitochondria, yielding opaque deposits (Lehninger *et al.* 1963, Peachey 1964). Indeed, it is possible that the deposits we observe along the membranes may be similar in composition to the mitochondrial deposits. We do not know if the deposits form during fixation or during dehydration of the specimen. However, by using the assumption that the deposits localize calcium binding sites as a working hypothesis, a number of additional studies are suggested. For the remainder of this paper we will suggest some physiological aspects of calcium binding that can lead to further study.

CELL COUPLING

Recent studies, summarized by Professor Loewenstein at the beginning of this symposium, have indicated that cellular calcium levels regulate cell to

---

* see figure insert opposite p. 400.

cell communication. Oliveira-Castro and Loewenstein (1971), studying *Chironomus* salivary glands, have demonstrated that when intracellular levels of Ca^{++}, Mg^{++}, Sr^{++}, or Ba^{++} are raised, junctional conductance is greatly depressed. The suggested explanation for this was that the presence of elevated divalent cation concentrations at or near junctional membranes bring about uncoupling.

These findings from Professor Loewenstein's group could be related to the observations we have made on intestines, since we observe opaque deposits along the junctional membranes. We tested the other divalent cations that uncouple cells and found that they produced deposits similar to those produced by calcium. We have some reservations about these results, however, because calcium occurs as an impurity in the salts of divalent cations from which we prepare the solutions, and very low concentrations of calcium can produce the opaque deposits. In these studies it has been necessary to use the usual precautions required for quantitative analysis of ions, including use of carefully cleaned glassware or plastic containers, and constant monitoring of reagents for calcium content. To determine precisely which divalent cations are effective in producing the deposits will require use of highly purified reagents. However, keeping these reservations in mind, it seems that the divalent cations that inhibit coupling also produce opaque deposits along the junctional membranes. The method may thus localize binding sites that have a regulatory function in controlling cell-to-cell communication.

A key question is the anatomical site of the low resistance pathway from cell to cell. In insect systems such as *Chironomus* salivary glands and cockroach intestine there are several possibilities. For the reader unfamiliar with these tissues, Fig. 2 summarizes the morphology of the region of cell contact. Proceeding from the serosal surface one encounters first an intercellular space of variable width, and then a junctional region consisting of both gap and septate junctions, as well as desmosomes (macula adherens). The proportion of the junctional area devoted to either gap or septate junction and their arrangement varies from organism to organism. Since gap junctions appear to be sites of electrical coupling in other epithelia (eg. Payton *et al.* 1969, Gilula *et al.* 1972) it is possible that they have a similar role in insect epithelia. However, we have seen from Professor Loewenstein's presentation that there is good evidence for the movement of a tracer, horse-radish peroxidase, through the septa of the septate junction (see also Larsen 1972).

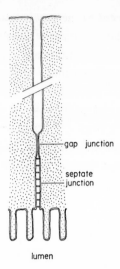

Fig. 2. Junctions that may be involved in cell to cell coupling in invertebrate epithelia.

These results do not rule out the gap junctions as sites of coupling, but do provide strong evidence that the septa are channels connecting adjacent cells.

At this point the reader must be warned not to confuse the insect salivary gland, the subject of extensive study by Professor Loewenstein and his colleagues, and the insect intestine which we have found convenient for the studies described here. *Chironomus* salivary glands are ideally suited for coupling studies because they are comprised of very large cells that can be readily penetrated by one or more microelectrodes. The intestinal epithelium we study is not well suited for coupling experiments because the cells are small. However, the small size is an advantage when one wishes to examine the morphology and binding properties of the junctions, as junctional membranes are abundant and easily located, whereas junctions are few and far apart in salivary glands.

We wished to know if *Chironomus* salivary gland junctions would also show the deposit along the membranes, since this is the system that has been studied most extensively. Dr. Larsen of the University of Miami kindly provided us with some *Chironomus* salivary glands, which we fixed in the presence of calcium, and we found the deposits along the junctional membranes. The deposits were observed most readily in tissue that had been treated with osmium with calcium present. An interesting preliminary observation, however, was that the deposits seemed to be confined to the septate junctions, and did not appear adjacent to the gap junctions. Returning to our intestinal preparations, with their large sample of junctional membranes, we were able to confirm this, ie. the gap junctions are consistently free of deposits while adjacent septate junctions have many deposits associated with them (Fig. 3–5)*.

These findings, combined with those of Oliveira-Castro & Loewenstein (1971) and Larsen (1972) focus our attention upon the septate junction as a possible site of cell to cell movement of large molecules, as well as a site of calcium interactions that regulate coupling. A problem that remains is the role of the gap junction in these tissues.

CALCIUM BINDING TO APICAL MEMBRANES

Recent studies have revealed that vertebrate intestinal epithelia have a calcium-binding protein (CBP) that, from evidence of various sorts, seems to be involved in calcium uptake (Taylor & Wasserman 1969). We suspect that the deposits along the apical membrane in the insect intestine might localize such a protein. However, no physiological data are available on calcium uptake in insect intestine. We therefore examined rat intestine, since this is a tissue in which the CBP has been well described. We found a distribution of opaque deposits similar to that in insect intestine (Fig. 6)*. The location of the deposits corresponds to the location of CBP achieved by Taylor & Wasserman (1969) by means of immuno-fluorescence. The next step will be to determine if the deposits are depleted in animals that are deprived of vitamin D, and that will thus have less CBP.

PYROANTIMONATE LOCALIZATION

Komnick (1962) introduced a method intended for the electron microscopic localization of sodium ions. The method involves precipitation of sodium ions with electron-opaque antimonate. While anatomists have made extensive use of this method, physiologists generally take a skeptical view of the findings, mainly because of the problems of diffusion. For quantitative and accurate localization of sodium, precipitation must be completed before sodium diffuses from one region within the cell to another and before extracellular sodium has entered the cell. While the fixative makes the cell

---

* See figure insert opposite p. 400.

membrane permeable to pyroantimonate, the fixative also makes the membrane permeable to sodium and the location of the reaction product thus depends on a variety of parameters that are difficult to control. In addition, it is now clear that pyroantimonate also precipitates calcium and magnesium (Bulger 1969). In fact, the antimonate reaction is about 3 to 4 orders of magnitude more sensitive to calcium and magnesium than to sodium (Klein et al. 1972). An inescapable conclusion is that in many cases the dense deposits obtained with pyroantimonate may be calcium or magnesium, and an independent analytical procedure is required to determine which cation is localized.

The deposits along membranes that can be produced simply by adding calcium ions to the fixative are strikingly similar to those obtained in a number of studies in which pyroantimonate has been used in an attempt to localize sodium. In particular I refer to studies by Bulger (1969), Ochi (1968), Kaye et al. (1965), Satir & Gilula (1970), and Lane & Martin (1969). Only in the latter study was the precipitate identified as sodium by microprobe analysis. In the other examples we do not know what the precipitates are, although we suspect some may prove to contain a significant amount of calcium.

DEPOSITS IN OTHER EPITHELIA

In addition to intestinal epithelial cells from rat, toad, and insect, we have observed deposits adjacent to membranes in neurons, muscle, liver, and erythrocytes. We include a micrograph illustrating the deposits in rat erythrocytes (Fig. 7)*. Harrison & Long (1968) have shown that the small amount of calcium present in erythrocytes is retained by ghosts obtained by hemolysis, so the calcium must be bound to the membrane. Only by EDTA treatment could most of the calcium be removed. Further study by Long & Movat (1971) revealed that calcium added to the medium was bound to the cells, and other ions compete for the binding sites. A survey of various species yielded a good correlation between sialic acid content and calcium binding capacity. Ghosts washed in $CaCl_2$ bound 2–3 times as much calcium, indicating that many of the binding sites are present on the inner surface of the membrane. These studies on erythrocytes encourage our view that the deposits we observe in tissues fixed with calcium may reflect binding sites within tissues.

---

* See figure insert opposite p. 400.

CONCLUSIONS

Deposits thought to contain calcium have been observed adjacent to plasma membranes in a variety of cells. To observe the deposits one needs only to add calcium to the solutions used in processing tissue for electron microscopy. It is clear from many other studies that the inner surface of the plasma membrane should have ion-exchange sites that can adsorb calcium and other ions. Further study will be required to determine if the deposits accurately reflect the distribution of binding sites within the living organism. Should this prove to be the case, the technique will be useful in testing the many theories about the role of membrane-associated calcium in controlling a variety of processes, such as neuro-transmitter release, excitation-contraction coupling, adhesion, secretion, pinocytosis, resting and action potentials, and the like.

ACKNOWLEDGEMENTS

The research has been supported by National Institutes of Health Grants FR-7028 and AM-14993. It is a pleasure to acknowledge the assistance of Mr. Bill Heddon and JEOL, Inc. for use of the electron microscope at Woods Hole.

SUMMARY

When traces of calcium are added to the various solutions used in preparing tissues for electron microscopy, opaque deposits are observed along the inner surfaces of certain parts of the plasma membranes. The phenomenon has been observed in a variety of cells from both vertebrates and invertebrates, including intestine, nerve, muscle, and erythrocytes. In insect intestine similar deposits are observed with other divalent cations (Mg^{++}, Sr^{++}, Ba^{++}, Mn^{++}) but deposits are very difficult to detect if ion-free solutions are used. While the deposits may be artifacts, our working hypothesis is that the deposits form because of divalent cation adsorption onto the inner surfaces of the membranes, and that the method thus localizes ion binding sites. Rat and insect intestinal cells are polarized with regard to the location of the deposits as they are abundant along apical (mucosal) surfaces and nearby lateral (junctional) membranes, and sparse toward basal (serosal) surfaces. Deposits along lateral membranes may localize binding sites predicted by the finding that calcium regulates cell-to-cell communication. In insect intestine deposits

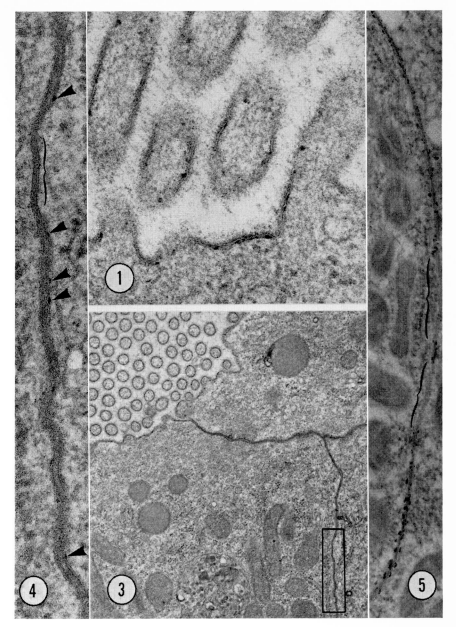

Fig. 1. Deposits on inner leaflet of apical membrane, insect intestine. Fixed in glutaraldehyde, post-fixed in osmium, section stained with lead and uranium, with 5 mM/l CaCl₂ in all solutions used for processing tissue. X98,000.

Fig. 3. Low magnification of junctional region near apical (lumen) surface, insect intestine. Preparation as in Fig. 1. Area within rectangle shown at higher magnification in Fig. 4.X20,000.

Fig. 4. Detail of region within rectangle in Fig. 3. Deposits (arrows) occur along septate junctions but not gap junction (bracketed). Many other pictures of this sort have been obtained, but deposits have never been observed adjacent to gap junctions. X115,000.

Fig. 5. Lateral membranes, insect intestine. Glutaraldehyde fixation only, section lightly stained with uranium. Deposits occur along septate junctions but are absent adjacent to gap junctions (brackets). X51,000.

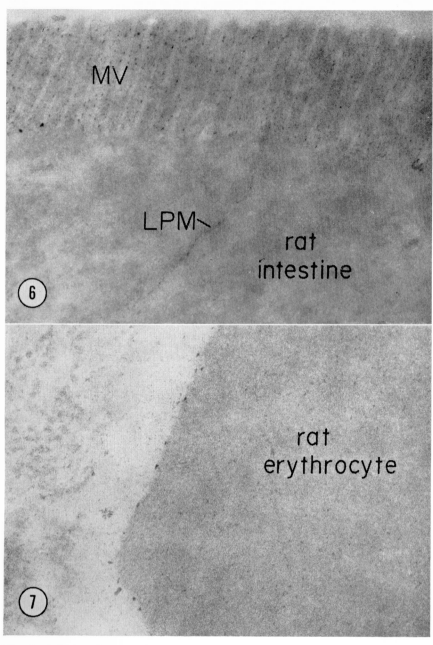

Fig. 6. Apical surface, rat duodenum. Fixed in glutaraldehyde plus 5 mM/l CaCl₂. Section unstained. Deposits along microvilli (MV) and lateral plasma membranes (LPM). X42,000.

Fig. 7. Rat erythrocyte, preparation as in Fig. 6 except that section lightly stained with uranium. X119,000.

occur adjacent to septate but not gap junctions, supporting other evidence that the septate junction may be responsible for coupling. Deposits along mucosal membranes may form at sites involved in control of membrane permeability or in calcium uptake, perhaps localizing the calcium binding protein described by other workers. The results also permit interpretation of earlier studies in which pyroantimonate has been used in an effort to localize sodium ions. In some cases pyroantimonate deposits along membranes may localize calcium rather than sodium.

REFERENCES

Bulger, R. E. (1969) Use of potassium pyroantimonate in the localization of sodium ions in rat kidney tissue. *J. Cell Biol. 40*, 79–94.

Gilula, N. B., Reeves, O. R. & Steinbach, A., (1972) Metabolic coupling, ionic coupling and cell contacts. *Nature (Lond.) 235*, 262–265.

Harrison, D. G. & Long, C. (1968) The calcium content of human erythrocytes. *J. Physiol. (Lond.) 199*, 367–381.

Kaye, G. I., Cole, J. D. & Donn, A. (1965) Electron microscopy: sodium localization in normal and ouabain-treated transporting cells. *Science 150*, 1167–1168.

Klein, R. L., Yen, S. & Thureson-Klein, A. (1972) Critique on the K-pyroantimonate method for semiquantitative estimation of cations in conjunction with electron microscopy. *J. Histochem. Cytochem. 20*, 65–78.

Komnick, H. (1962) Elektronen Mikroscopie Lokalisation von Na^+ and Cl^- in Zellen und Geweben. *Protoplasma 55*, 414–418.

Lane, B. P. & Martin, E. (1960) Electronprobe analysis of cationic species in pyroantimonate precipitates in Epon-embedded tissue. *J. Histochem. Cytochem. 17*, 102–106.

Larsen, W. J. (1973) Contributed comments in Symposium on Membrane-membrane interactions in differentiation. *Fed. Proc.* In press.

Lehninger, A. L., Rossi, C. S. & Greenawalt, J. W. (1963) Respiration dependent accumulation of inorganic phosphate and Ca^{++} by rat liver mitochondria. *Biochem. biophys. Res. Commun. 10*, 444–448.

Long, C. & Movat, B. (1971) The binding of calcium ions by erythrocytes and "ghost" cell membranes. *Biochem. J. 123*, 829–836.

Manery, J. F. (1966) Effects of Ca ions on membranes, *Fed. Proc. 25*, 1804–1810.

Ochi, J. (1968) Electron microscopical detection of sodium ions in the sweat glands of rat foot pads. *Histochemie 14*, 300–307.

Oliveira-Castro, G. M. & Loewenstein, W. R. (1971) Junctional membrane permeability. Effects of divalent cations. *J. Membr. Biol. 5*, 51–77.

Oschman, J. L. & Wall, B. J. (1972) Calcium binding to intestinal membranes. *J. Cell Biol. 55*, 58–73.

Payton, B. W., Bennett, M. V. L. & Pappas, G. D. (1969) Permeability and structure of junctional membranes at an electrotonic synapse. *Science 166*, 1641–1643.

Peachey, L. D. (1964) Electron microscope observations on the accumulation of divalent cations in intramitochondrial granules. *J. Cell Biol. 20,* 95–110.

Satir, R. & Gilula, N. B. (1970) The cell junction in a lamellibranch gill ciliated epithelium. Localization of pyroantimonate precipitate. *J. Cell Biol 47,* 468–487.

Taylor, A. N. & Wasserman, R. H. (1969) Correlations between the vitamin D-induced calcium-binding protein and intestinal absorption of calcium. *Fed. Proc. 28,* 1834–1838.

Ussing, H. H. (1970) Tracer studies and membrane structure. *In Capillary Permeability,* Crone, C. & Lassen, N. A., ed. Alfred Benzon Symposium II, Copenhagen, Munksgaard, pp. 654–656.

DISCUSSION

ZADUNAISKY: Did you do an electrondiffraction.

OSCHMAN: No.

ZADUNAISKY: The main point is whether these are calcium deposits or not. In mitochondria it was easy to know because you can separate them and the granules are there. If you do electron diffraction in the same section that you are doing transmission, I think you will be able to tell if it is calcium or not.

OSCHMAN: The approach that we have been interested in is microprobe analysis.

TOSTESON: Have you looked in freeze fractured preparations, and have you tried to square the quantitative estimates of Long, of the calcium content of the membrane of red cells with how much calcium you would have to have there in order to be able to see it in the microscope.

OSCHMAN: We have not yet examined freeze-fracture specimens, although this approach could prove quite useful. With regard to the amount of calcium (or of any other atom, for that matter) that is needed to produce an electron-opaque spot, I can say this is an aspect of electron microscopic technique that is not well understood. We do find from tissue analysis that fixation of toad intestines in the presence of calcium (5 mM/l) results in about a 20-fold increase in the tissue calcium content, compared to fresh tissue.

TOSTESON: To see it, I think you would have to have quite a bit. I mean you are not seeing individual calcium but small precipitates.

OSCHMAN: This is correct. We think the deposits are clumps or aggregates of calcium ions along with the anionic substance with which they have precipitated.

26*

Transport Across Rumen Epithelium

C. E. Stevens

The ruminant forestomach is a large, compartmentalized outgrowth of the conventional mammalian stomach. This evolutionary development allows for microbial digestion of cellulose and synthesis of protein prior to those phases of gastrointestinal digestion common to all mammals. The ecological distribution of ruminants to almost all climates and altitudes testifies to its success. The forestomach is lined with a stratified squamous, nonglandular epithelium. Its structural similarity to esophageal epithelium originally suggested that it was derived from the latter and it was presumed that it served as a relatively impermeable barrier between lumen contents and blood. However, embryological studies have indicated that the forestomach is derived from tissue distal to the cardiac sling of gastric oblique muscle (Warner, 1958) and functional studies have since demonstrated a wide range of substances which are effectively absorbed from this organ.

The glandular mucosa of the mammalian stomach can be divided into three morphologically distinct regions: the cardiac, proper gastric and pyloric mucosa. Of these, the proper gastric mucosa has attracted the greatest interest due to its secretion of HCl and pepsinogen. Fig. 1 shows the general distribution of these mucosal regions in the stomachs of six species. It can be seen that although the stomachs of man and the dog show only a relatively narrow band of cardiac mucosa, this region occupies almost one-half the stomach of the pig. The pig stomach also contains a fourth, nonglandular region, consisting of stratified squamous epithelium. In the horse stomach the cardiac mucosa is again a narrow band of tissue while the nonglandular, stratified squamous epithelium occupies the cranial one-half of the stomach. In *Camelidae,* such as the llama, cardiac mucosa lines most of the third compartment while much of the first and part of the second compartment are lined with stratified squamous epithelium (Cummings *et al.* 1972, Valle-

Department of Physiology, Biochemistry and Pharmacology, New York State Veterinary College, Cornell University, New York, 148500.

»Transport Mechanisms in Epithelia«, Munksgaard, Copenhagen.

nas *et al.* 1971). However, the latter two compartments also contain glandular pouches lined with tissue morphologically indistinguishable from cardiac mucosa. Finally, we have the ruminant stomach with a small region of cardiac mucosa at the junction between the proper gastric mucosa and a large, compartmentalized, forestomach lined entirely with stratified squamous epithelium.

The gastric invasion by stratified squamous epithelium is seen in species which belong to a wide range of mammalian orders. In addition to the representatives of *Artiodactyla* (pig, llama, cow) and *Perissodactyla* (horse) described above, similar developments can be seen in orders such as *Edentata* (sloth), *Rodentia* (rat, hamster, lemming, vole, muskrat), *Marsupialia* (Kangaroo) and *Primates (Colobus, Semnopithecus)*. This demonstration of evolutionary convergence was noted by Oppel (1897) and Bensley (1902–3), whose

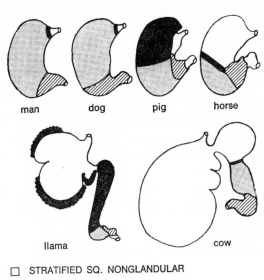

STRATIFIED SQ. NONGLANDULAR
CARDIAC
PROPER GASTRIC
PYLORIC

Fig. 1. Some variations in the distribution of gastric mucosa. The stomach of each species demonstrates regions of cardiac, proper gastric and pyloric mucosa. The pig stomach contains a relatively large region of cardiac glandular mucosa as does the llama. In the latter, this tissue is also present as islands of secretory saccules in the first two compartments. The pig, horse, llama and cow also show increasing areas of stratified squamous nonglandular epithelium. (Stomachs are not represented on same scale, e. g. the volume capacity of adult bovine stomach is approximately 70 times that of man's stomach, or 14 times the capacity per Kg body weight).

studies also concluded that the cardiac mucosa may represent an intermediate stage in the replacement of proper gastric mucosa with nonglandular, stratified squamous epithelium. Although this theory has not found wide acceptance it has some interesting implications when considered in the light of recent functional studies, discussed below.

As an introduction to a discussion of the transport of substances across rumen epithelium, it is useful to consider briefly the composition of rumen contents. Very great volumes of saliva enter the rumen. The volume of saliva secreted daily by adult sheep and cattle exceeds that of either their fore-stomach or their extracellular fluid space. The contribution of this to the rumen contents has been the subject of recent reviews (Dobson & Phillipson 1968, Schneyer & Schneyer 1967).

The parotid glands of the sheep secrete an isotonic saliva, even at low rates of flow. The Na/K ratio is normally high and the major anions are PO_4 (40 mM), HCO_3 (80–130 mM), and Cl (20–30 mM). Concentrations of the latter two can reach their higher values at increased rates of flow. It has been estimated that an adult sheep can secrete 1.2 to 1.5 moles of Na/day (Kay 1960) and that approximately one-half of this amount is reabsorbed from the rumen (Dobson 1959, Hydén 1961). Chloride can be absorbed from rumen pouches containing solutions with Cl concentrations as low as 6–7 mM (Hydén 1961).

The major byproducts from the microbial digestion of cellulose and other carbohydrates are the volatile fatty acids (primarily acetate, propionate and butyrate), and the gases CO_2 and CH_4. As a result, glucose is normally present in extremely small amounts and the VFA's provide the major source of anions following feeding. It has been estimated that as much as 3.0 moles of VFA can be absorbed daily from the rumen of sheep (Dobson & Phillipson 1968) along with as much as one-third of the CO_2 produced (Hungate 1966). The maintenance of rumen contents at a pH above 5.5, even at the height of fermentation, has been attributed to the rapid absorption of VFA as well as to the buffering capacity of the saliva.

TRANSPORT OF UREA, AMMONIA AND WATER

Relatively little amino acid has been found in the bacterial-free filtrate of rumen contents. The microbial protein is derived from both dietary protein and non-protein N_2. Endogenous urea, which enters the rumen via saliva and passage across the rumen wall, can also act as an important substrate. Houpt (1970) reviewed the evidence which indicates that in sheep, fed a low protein diet, endogenous urea is utilized in the net synthesis of protein by rumen microbes. While relatively large amounts of urea enter the rumen by diffusion through the epithelium, he concludes that the transfer of urea N into the rumen could be fully explained by passive diffusion and the presence of bacterial urease in the cornified layer of epithelium as well as in the rumen itself. The addition of CO_2 to solutions bathing an isolated rumen compartment of unanesthetized cows resulted in a delayed but marked increase in their rate of urea accumulation (Thorlacius *et al.* 1971). It was concluded that this effect of CO_2, which was localized to the compartment studied and independent of changes in rumen blood flow, was due to the release of a substance which increased epithelial permeability to urea.

Engelhardt (1970) has recently reviewed the studies of water movement across rumen epithelium. No net flux of H_2O was observed when solutions within the range of 265 to 325 mosmoles/l were added to the rumen pouch of the goat. An epithelial tissue model was proposed for the explanation of these results. Similar studies in cattle indicated that although H_2O followed its osmotic gradient in the absence of CO_2, it was absorbed against an apparent osmotic gradient when solutions were gassed with CO_2 (Dobson *et al.* 1972). These workers concluded that this could be explained by the additional osmotic activity of CO_2 absorbed into the blood perfusing the rumen. Although there is some circumstantial evidence that ADH may increase the permeability of rumen epithelium, the administration of ADH to intact animals did not result in an increase in the permeability to HTO (Engelhardt 1970) or urea (Dobson *et al.,* 1972, Houpt 1970).

TRANSPORT OF INORGANIC IONS

Following the demonstration that Na was absorbed against an electrochemical potential gradient from the rumen of the anesthetized sheep (Dobson 1959), *in vitro* studies of the isolated, short-circuited rumen epithelium of the cow and goat (Stevens 1964) and the sheep (Ferreira *et al.* 1964, Harrison *et al.* 1968) showed that this tissue can actively transport both Na and Cl in the direction of lumen to blood. Active transport of much smaller amounts of K in the direction of blood to lumen was also demonstrated in studies of sheep rumen epithelium (Ferreira *et al.* 1964).

Table I. Simultaneous measurements of Na and Cl transport across the isolated, short-circuited rumen epithelium.

	μEq/cm² x hr		
	Influx L → B	Efflux B → L	z x Net Flux L → B
Na	2.2±0.2	0.9±0.1	1.3±0.2
Cl	8.8±1.3	5.3±0.7	−3.6±1.0
Σ Na+Cl			−2.3±1.1
SCC			0.55±0.04

Bidirectional fluxes of Na and Cl were simultaneously measured by a quadruple labelling procedure. Values listed in column 4 are expressed as z x Net Flux where z indicates the valence of Na ($+1$) and Cl (-1) to allow direct comparison with the SCC. All values are expressed in μEq/cm² x hr. Experiments included in this and all subsequent tables were conducted on short-circuited tissues bathed on both surfaces with an identical solution, unless otherwise indicated (Chien & Stevens 1972, Courtesy of American Journal of Physiology).

Although rumen epithelium shows a number of histological (Steven & Marshall 1970) and histochemical (Hendrickson 1971) similarities to frog skin, it demonstrates a wider spectrum of ion transport. Furthermore, the net transport of Na, Cl and K still do not appear to account for the short-circuit current (SCC) and it was concluded that this discrepancy could best be explained by the additional active transport of H toward the blood or the active transport of either HCO_3 or fatty acid anions in the direction of blood to lumen (Ferreira et al. 1972).

Studies of bovine rumen epithelium (Chien & Stevens 1972) have provided similar results. Table I gives the results of simultaneous measurements of the SCC and the bidirectional Na (Na[22], Na[24]) and Cl (Cl[36], Cl[38]) fluxes across isolated, bovine, rumen epithelium. It can be seen that neither the net Na flux nor the net fluxes of both Na and Cl accounted for the SCC measured in these tissues, indicating the active transport of additional ion(s) at quite a substantial rate. Isosmotic replacement of either the Mg or the PO_4 in the Ringer solution bathing both surfaces of the tissue had little or no effect on the SCC. Although the tissue actively transported Ca in the direction of lumen to blood, this accounted for less than 1 % of the SCC. However, isosmotic replacement of the Na in the bathing solutions with K resulted in

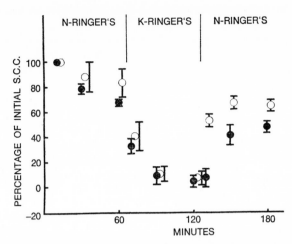

Fig. 2. Changes in SCC when the Na in the Ringer solution bathing both tissue surfaces is replaced with K. Graph shows changes in SCC measured with tissues bathed in tricine (o) or bicarbonate (●) buffered Ringer solution. Tissues were bathed first with normal Ringer, then with a Ringer solution in which all of the Na was replaced with K and finally with the original normal Ringer. Replacement of Na with K resulted in a complete, but reversible, inhibition of SCC. (Chien & Stevens 1972, courtesy of American Journal of Physiology).

essentially complete inhibition of SCC, which was reversible (Fig. 2). The substitution of choline for Na also resulted in a large (90 %), reversible reduction in this current, suggesting that the SCC was primarily dependent upon Na transport.

Table II shows the effect of replacing various ions in the bathing solutions on the tissue's SCC and ability for actively transporting Na and Cl. Tricine (tris [hydroxymethyl] methylglycine) was used as the only replacement for HCO_3 buffer since earlier studies had indicated that its use resulted in SCC values similar to those obtained with HCO_3 buffer. Bidirectional fluxes of Na and Cl were measured by the double-labelling technique (Levi & Ussing, 1949). All bathing solutions were buffered at a pH of 7.40 ± 0.03 (SE) and adjusted to an osmolality of 295.6 ± 2.3 (SE) mOsm/kg. The normal and Ac-bicarbonate Ringer contained 143 mEq of Na/l. This was reduced by about 10 % in the preparation of normal tricine Ringer solution and increased by approximately 15 and 25 % in the preparation of SO_4-tricine and SO_4-bicarbonate solutions respectively. It was assumed that

Table II. Effect of various bathing solutions on net transport of Na and Cl by short-circuited rumen epithelium.

RINGER'S BATHING SOLUTION	Ion(s) Removed	Na Transport Exps μEq/cm² x hr		Cl Transport Exps μEq/cm² x hr	
		z×Net Flux L → B	SCC L → B	z×Net Flux L → B	SCC L → B
Bicarbonate	– – –	1.7 ± 0.2	0.48 ± 0.04	–2.8 ± 0.2	0.50 ± 0.02
Tricine	HCO$_3$	1.0 ± 0.1	0.50 ± 0.03	–1.4 ± 0.2	0.37 ± 0.03
Acetate Bicarb.	Cl	0.8 ± 0.1	0.56 ± 0.08	– – –	– – –
SO$_4$ Bicarb.	Cl	0.7 ± 0.1	0.80 ± 0.05	– – –	– – –
SO$_4$ Tricine	Cl, HCO$_3$	0.6 ± 0.04	0.64 ± 0.02	– – –	– – –
Tricine +					
1.1 mM Acetazolamide	HCO$_3$	– – –	– – –	–1.0 ± 0.1	0.34 ± 0.04
4.5 mM Acetazolamide	HCO$_3$	– – –	– – –	–1.2 ± 0.3	0.41 ± 0.07
Choline Bicarb.	Na	– – –	– – –	–0.7 ± 0.2	0.13 ± 0.03
Choline Tricine	Na, HCO$_3$	– – –	– – –	–0.4 ± 0.2	0.11 ± 0.03

Net transport of Na and Cl were each determined by separate double labelling experiments and compared with SCC measured in the same tissue. The first column lists the solutions used to bathe both surfaces of the short-circuited tissue and the second column lists the ion removed from these solutions. Subsequent columns list the mean (±SE) values obtained during 2.5 hr experimental periods (Modified from Chien & Stevens 1972).

Na transport would not be concentration-dependent at these high levels. The concentration of all other ions, with the exception of those intentionally removed, remained unaltered.

Tissues bathed with normal bicarbonate Ringer actively transported Na at a rate approximately 3.5 times that accountable to SCC. The replacement of bicarbonate buffer with tricine resulted in a decrease in the net transport of Na, although it was still significantly greater ($p < 0.01$) than the SCC. This effect may have been due to the lower concentrations of Na in tricine Ringer. Replacement of the Cl in the bathing solution with either acetate or SO$_4$ resulted in even a greater reduction in net Na transport, to levels which could account for the SCC. The SCC itself was relatively unaffected by these changes in bathing solution, except for a small but significant increase noted with SO$_4$-bicarbonate Ringer's. This effect, also reported in studies of sheep rumen epithelium (Keynes 1969), may have been due to the higher concentrations of Na in these solutions. Therefore, it appeared that the active transport of Na was at least partly dependent upon the presence of Cl ion. However the SCC did not appear to be dependent upon either the active transport of Cl or the active transport of most of the Na.

Net Cl transport showed an even greater reduction when bicarbonate buffer was replaced with tricine (Table II). Acetazolamide was added to the tricine-buffered solutions in an attempt to inhibit tissue production of HCO_3. This tissue has been previously shown to contain relatively high concentrations of carbonic anhydrase (Aafjes 1967, Carter 1971, Stevens & Stettler 1967). Two levels of acetazolamide, 1.1 and 4.5 mmoles/l, were tested. Neither dosage level produced a significant inhibition in the net transport of Cl. Inhibition of net Cl transport by sheep rumen epithelium has been reported with the use of 10 mM acetazolamide in Ringer solution buffered with HCO_3 (Harrison 1971).

The small decrease in SCC, observed when tricine Ringer was used in the study of Cl transport (Table II), was not observed when these solutions were used to study Na transport (Table II) or the effect of K replacement (Fig. 2). Replacement of the Na in the bathing solution with choline resulted in a very marked reduction in the net transport of Cl and a marked drop in SCC, similar to that previously noted.

The above studies suggested that there was interaction between the active transport of Na and Cl and that the active transport of Cl, and possibly Na, was inhibited by the replacement of the bicarbonate buffer system with tricine. They also indicated that while the SCC may be entirely dependent upon the presence of Na, most of the Na was actively transported by a Cl-dependent, nonelectrogenic mechanism.

Fig. 3 compares the results of the above studies of bovine rumen epithelium with those obtained in studies of ion transport across sheep rumen epithelium, in which undirectional fluxes were measured in paired tissues (Ferriera et al. 1972). The two studies give almost identical values for net Na transport and SCC. The tissue from sheep showed a lower net Cl transport and a relatively small net transport of K in the direction of blood to lumen, with a discrepancy between the sum of the measured net ion fluxes and the SCC of approximately 0.9 $\mu Eq/cm^2 \times hr$.

It was concluded that this result might be explained by either the additional net transport of HCO_3 or fatty acid anions in the direction of blood to lumen, or a net transport of H in the opposite direction. The bovine tissue showed an even greater difference between net ion transport and SCC (1.6

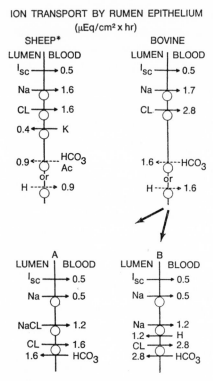

ION TRANSPORT BY RUMEN EPITHELIUM
(μEq/cm² x hr)

Fig. 3. Diagramatic representations of net ion transport and SCC observed in rumen epithelium. All values are expressed in μEq/cm² x hr. The tissue is represented as a single membrane bathed on both surfaces with identical solutions of HCO_3^- Ringer. The first two diagrams (top) summarize results obtained from studies of sheep (Ferreira et al. 1972) and bovine (Chien & Stevens 1972) tissue. The dotted lines indicate alternative explanations given for the discrepancy between net ion transport and SCC seen in the sheep studies. The possibility of fatty acid transport is indicated by the Ac symbol. Diagrams A and B (bottom) represent two alternatives which might account for the effects of ion substitution, observed with bovine rumen epithelium.

μEq/cm² × hr). Although K transport was not measured in the latter tissue, a net transport similar to that of sheep rumen would have increased this difference. The discrepancy observed with bovine tissue could be similarly explained by an additional net transport of HCO_3 or H (fatty acids were not included in the bathing solutions), however, this in itself would not account for the results obtained in the ion substitution experiments.

The lower half of Fig. 3 suggests alternative mechanisms which might account for the ion transport interdependency observed with bovine rumen epithelium. Both arrangements assume that the SCC is entirely dependent on the active transport of Na by a conventional Na pump and that the remaining Na and all of the Cl are actively transported by electrically neutral mechanisms. In tissue A these consist of a neutral NaCl pump, such as that postulated for the roach bladder (Diamond 1962), and a mechanism which exchanges the remaining Cl for HCO_3. In tissue B these consist of Na-H and Cl-HCO_3 exchange mechanisms. Both arrangements could account for an interpendency between Cl and HCO_3, the NaCl pump (tissue A) would account for a further interdependence between Na and Cl transport. While diagram B provides no explanation for an interpendence between Na and Cl transport, and neither diagram would account for the apparent inter-dependence between Na and HCO_3, these interactions might be explained if the respective transport systems were serially arranged in the opposing membranes of a tissue compartment.

Further evidence for Cl-HCO_3 exchange by bovine rumen epithelium was given by the results of recent *in vivo* studies (Breukink *et al.* personal communication). Isotonic solutions of NaCl, Na_2SO_4 or glucose containing 5 mmoles $NaHCO_3/l$ were added to the rinsed, isolated, ventral sac of the rumen in an unanesthetized cow and gassed with 5 % CO_2 – 95 % N_2. Serial samples of these solutions were then collected and analyzed for HCO_3 by acidification and titration back to their initial pH. The NaCl solutions were found to accumulate HCO_3 at a relatively constant rate, of about 0.6 mmoles/min. However, this was reduced by two-thirds when the Cl of the lumen solution was replaced with either SO_4 or glucose. While the appearance of HCO_3 in these solutions could represent either a secretion of HCO_3 or a preferential absorption of H, the lower rate of HCO_3 appearance in Cl-free solutions would support the argument for an exchange of HCO_3 for Cl.

In consideration of this evidence for Cl-HCO_3 exchange by stratified squamous epithelium, it is interesting to note that recent studies of cardiac glandular mucosa indicated that it also appears to secrete HCO_3 into the lumen in exchange for Cl. Solutions containing NaCl and HCl, placed in gastric pouches formed from the cardiac glandular region of the pig stomach were shown to accumulate HCO_3 and lose Cl (Höller 1970). From studies of diurnal variation and the effects of sham-feeding and pentagastrin injection, it was concluded that HCO_3 secretion by the cardiac mucosa of pigs was inhibited by gastrin but might serve to buffer excess HCl as plasma gastrin levels declined following a meal. Studies of secretions collected from

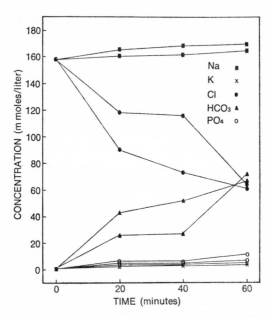

Fig. 4. Changes in the composition of an isotonic NaCl solution placed in the isolated second compartment of the llama stomach (2 experiments). Results indicate the addition of small amounts of K and PO_4 and a slight change in Na concentration. These changes were accompanied by a marked increase in the concentration of HCO_3 and a reciprocal decrease in Cl concentration. (Eckerlin & Stevens 1973, Courtesy of Cornell Veterinarian).

the glandular pouches of the first compartment of the llama stomach (Eckerlin & Stevens 1973) indicated that these also secrete HCO_3, and experiments similar to those described above in the cow showed that this was associated with a reciprocal absorption of Cl (Fig. 4). Therefore, it would appear that cardiac mucosa and stratified squamous epithelium may contain similar mechanisms for the exchange of HCO_3 for Cl.

TRANSPORT OF VOLATILE FATTY ACIDS

The absorption, metabolism and transport of volatile fatty acids (VFA) by rumen epithelium has been the subject of recent reviews (Dobson & Phillipson 1968, Stevens 1970). Acetate, propionate and butyrate are rapidly absorbed, even at the relatively neutral pH of rumen contents. Rumen epithelium is also capable of metabolizing substantial amounts of each of these fatty acids and this appears to be the major factor which governs their relative rates of absorption and transport.

It was originally presumed that VFA was absorbed entirely as the undissociated acid. This was supported by early studies showing an increased rate of absorption with a decrease in the pH of rumen contents (Danielli *et al.* 1945). However, their relatively rapid rate of absorption from rumen contents at a pH well above their pKa (4.8) was difficult to explain. Furthermore, it was found that with rumen contents at a pH near neutrality the absorption of two equivalents of VFA was associated with the appearance of one equivalent of HCO_3 within the rumen (Masson & Phillipson 1951).

Additional studies of the relationship between acetate absorption and rumen HCO_3 and P_{CO_2} levels suggested that only one-half of the acetate was absorbed as anion (Ash & Dobson 1963). Anion absorption was also suggested by *in vitro* studies showing that acetate absorption and transport were increased to a greater degree when the acetate concentration of the lumen bath was increased three-fold, at a pH of 7.4, than when the pH of the lumen bath was lowered by one unit to provide a ten-fold increase in undissociated acid (Stevens & Stettler 1966a). Yet *in vitro* studies, comparing the effects of transepithelial chemical and electrical potential gradients of equal magnitude, indicated that the tissue contained a barrier impermeable to the passive diffusion of acetate anion (Stevens & Stettler 1966b).

Although tissue metabolism of fatty acid makes it difficult to design a test for their active transport, a steady state can be reached by preincubation of the tissue in the solution to be tested. When the tissue was preincubated in 10 mM Ac-Ringer's and then bathed on both its surfaces with this same solution, short-circuited tissues bathed with bicarbonate-buffered solutions showed a net uptake of acetate from the blood side and no significant exchange with the lumen bath (Table III). However, the substitution of either imidazole or tricine for the bicarbonate buffer resulted in a net transport of acetate in the direction of blood to lumen. Under steady state conditions the tissues maintained a 2:1 concentration ratio of acetate between the solutions bathing its lumen and blood surfaces. Similar studies, conducted with two other weak organic acids and two weak organic bases (Table IV) indicated that the tissue also transported these against a transepithelial electrochemical gradient; acids from blood to lumen and bases from lumen to blood.

From the above studies it was concluded that if the tissue contained a

Table III. Net transport of acetate by short-circuited bovine rumen epithelium.

Buffer	Net Change of Acetate in Bath. Solutions μmoles/cm^2 x hr	
	Lumen Side	Blood Side
Bicarbonate	$+0.1 \pm 0.2$	-0.7 ± 0.1
Imidazole	$+0.6 \pm 0.1$	-1.1 ± 0.2
Tricine	$+0.7 \pm 0.2$	-1.1 ± 0.1

Tissues were preincubated in 10 mM Ac-Ringer for 0.5 hr. They were then placed between chambers containing this same solution and short-circuited. Mean values ($\pm$SD) are those obtained over the subsequent 2.0 hr period. (Modified from Stevens & Stettler 1967).

compartment separated from its opposing bathing solutions by two membranes which differed in their relative permeability to the dissociated and undissociated forms of the weak organic electrolyte, the observed transport of acid and base could be driven by an electrochemical gradient of H ion between compartment and bathing solutions. For example, if it is assumed that the blood-facing membrane is permeable only to the undissociated acid and the lumen-facing membrane is permeable only to the anion, an electrical

Table IV. Net transport of organic acids and bases by short-circuited bovine rumen epithelium.

Acid or Base Conc. of Initial Bath. Solution	Net Change in Bath. Solutions μmoles/cm^2 x hr	
	Lumen Side	Blood Side
Acids		
10mM Acetate	$+0.7 \pm 0.2$	-1.1 ± 0.1
10mM 3,3-dimethyl-butyrate	$+0.5 \pm 0.1$	0.6 ± 0.2
10mM DMO	$+0.38 \pm 0.07$	-0.32 ± 0.02
Bases		
2.0mM Morphine	-0.09 ± 0.02	$+0.11 \pm 0.02$
0.5mM Strychnine	-0.05 ± 0.01	$+0.03 \pm 0.01$

Results obtained with acetate (TABLE III) are compared with those obtained by the same procedure conducted with other organic acids and bases (Modified from Stevens *et. al.* 1969).

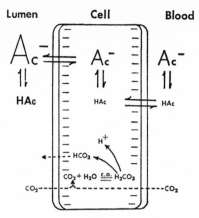

Fig. 5. A model depicting the short-circuited bovine rumen epithelium as a single cellular compartment capable of maintaining a 2:1 concentration gradient of acetate between lumen and blood. This model assumes the alternative that the blood-facing membrane is permeable only to the undissociated form and the lumen-facing membrane is permeable to only the anion. Mathematical analysis showed that this extreme a difference in permeability characteristics is not required. Under these conditions a negative (cell contents to bathing solution) electrical potential gradient would favor the observed transport of organic acid in the direction of blood to lumen and organic base in the opposite direction. A positive H ion gradient between cell compartment and bathing solutions (not indicated) would favor transport in the opposite direction, a condition which could explain the effects of HCO_3^- CO_2 buffer and acetazolamide.

gradient between cell contents and lumen would tend to transport acetate toward the lumen (Fig. 5).

Conversely, a H ion gradient would favor the transport of acetate in the opposite direction. The effect of such a H ion gradient could explain the inhibitory action of bicarbonate buffers on net blood-to-lumen transport of acetate, since hydration of CO_2 within the compartment could serve to increase its H ion concentration. Studies of acetate transport down a concentration gradient between lumen and blood lend support to this interpretation. These show that the transport from lumen to blood was decreased to an equal degree by either the addition of acetazolamide to bicarbonate-buffered solutions or the replacement of bicarbonate buffer with tricine (Table V).

Although the rumen and abomasum (glandular stomach) of the goat demonstrate similar ratios in their relative rate of HTO and antipyrine absorption, the abomasum normally absorbed VFA at much higher rates

Table V. Absorption and transport of acetate by short-circuited rumen epithelium.

Buffer	Acetazolamide 1.1 mM	Acetate Transport μmoles/cm² x hr	
		Loss Lumen Side	Gain Blood Side
Bicarbonate		2.8 ± 0.2	1.2 ± 0.1
Bicarbonate	+	2.3 ± 0.4	0.7 ± 0.1
Tricine		2.3 ± 0.4	0.6 ± 0.1
Tricine	+	1.9 ± 0.6	0.6 ± 0.1

A 30mM Ac-Ringer solution was added to the lumen side of the tissue and normal (Ac-free) Ringers was added to the blood side. Results represent the mean (±SE) rates of acetate loss from the lumen bath and gain on the blood side of the tissue. Note that both the absorption and transport of acetate were inhibited to an equivalent degree by the addition of acetazolamide to bicarbonate-buffered solutions or substitution of bicarbonate buffer with tricine. All tissues contained similar amounts of acetate at the end of the 2.5 hr experiment. These results, indicating that use of bicarbonate buffer increases the rate of acetate transport from lumen to blood agree with those showing that it decreases the rate of acetate transport in the opposite direction (Table III).

(Engelhardt *et al.* 1968). However VFA absorption was greatly reduced when the pH of abomasal contents was elevated. VFA transport was studied in the isolated, short-circuited cardiac and proper gastric mucosa for comparison to results obtained with rumen epithelium (Argenzio, personal communication). An equimolar mixture of VFA was included in the lumen bath, at a pH of 7.4. The cardiac mucosa of the pig transported VFA to its blood side at approximately one-half the rate seen with rumen epithelium, while the proper gastric mucosa of the pig and horse transported VFA at only 12 and 4% of this rate, respectively. This would suggest that the proper gastric mucosa is considerably less permeable to VFA when it is predominately in its dissociated form.

CONCLUSION

It appears that the stratified squamous epithelium of the rumen is capable of transporting a relatively wide range of strong and weak electrolytes against their transepithelial electrochemical gradients. This would include the transport of Na, Cl, Ca and weak organic bases in the direction of lumen to blood, and the transport of HCO_3, K and weak organic acids in the opposite direction. The physiological advantage of mechanisms which would aid in the absorption of Na and Cl and the exclusion of K are obvious when one considers the amounts of Na and Cl in the saliva and the high K, low Na content of plant roughage. The release of HCO_3 into the rumen could serve as an additional means of buffering the VFA in rumen digesta.

The purpose of a mechanism which favors the transport of small amounts of weak organic acids into the rumen is not obvious since this is normally overwhelmed by the large concentration gradients which favor diffusion in the opposite direction. However, it is interesting to note that the net transport of Na, Cl, HCO_3, VFA, urea and H_2O all appear to be increased when solutions bathing the lumen surface of the epithelium are gassed with CO_2. This occurs under conditions which cannot be explained by increased blood flow or a change in lumen pH. while these effects of CO_2 are not necessarily due to a common mechanism, they could be largely explained by an effect on the electrochemical gradient of H ion between a tissue compartment and its bathing solutions.

The mechanism by which various regions of the gastrointestinal tract transport Cl, HCO_3 and VFA would seem to deserve further comparative study. It appears that the stratified squamous epithelium of the rumen and the cardiac mucosa both transport VFA and Cl into the blood and HCO_3 into the lumen. The proper gastric mucosa, from which these may have derived, seems less capable of transporting VFA and transports its Cl and HCO_3 in the opposite direction. In many mammals the large intestine is faced with problems of Cl retrieval and VFA production similar to those of the rumen. Comparisons of transport in these morphologically dissimilar tissues may lead to better understanding of the underlying mechanisms.

REFERENCES

Aafjes, J. H. (1967) Carbonic anhydrase in the wall of the forestomach of cows. *Brit. Vet. J. 123*, 252–258.

Argenzio, R. A. Personal communcation.

Ash, R. W. & Dobson A. (1963) The effect of absorption on the acidity of rumen contents. *J. Physiol. (Lond.) 169*, 39–61.

Bensley, R. R. (1902–3) The cardiac glands of mammals. *Amer. J. Anat. 2*, 105–156.

Breukink, H. G., Dobson A. & Sellers A. F. (personal communcation).

Carter, M. J. (1971) The carbonic anhydrase in the rumen epithelial tissue of the ox. *Biochem. biophys. Acta (Amst.) 235*, 22–236.

Chien, Wan-Ju & Stevens, C. E. (1972) Coupled active transport of Na and Cl across forestomach epithelium. *Amer. J. Physiol. 223,* 997–1003.

Cummings, J. F., Munnel, J. F. & Vallenas, A. (1972) The mucigenous glandular mucosa in the complex stomach of two new-world camelids, the llama and guanaco. *J. Morph. 137,* 71–109.

Danielli, J. R., Hitchcock, M. W. S., Marshall, R. A. & Phillipson, A. T. (1945) The mechanism of absorption from the rumen as exemplified by the behavior of acetic, proprionic and butyric acids. *J. exp. Biol. 22,* 75–84.

Diamond, J. M. (1962) The mechanism of solute transport by the gallbladder. *J. Physiol. (Lond.) 161,* 474–502.

Dobson, A. (1959) Active transport through the epithelium of the reticulorumen sac. *J. Physiol. (Lond.) 146,* 235–251.

Dobson, A. & Phillipson, A. T. (1968) Absorption from the ruminant forestomach. In *Handbook of Physiology,* Alimentary Canal, Washington, D. C. Amer. Physiol. Soc., Sect. 6, Vol. V., pp. 2761–2774.

Dobson, A., Harrop, J. F. & Phillipson, A. T. (1972) Osmotic effects of carbon dioxide absorption from the rumen. *Fed. Proc. 31,* 260.

Eckerlin, R. H. & Stevens, C. E. (1973) Bicarbonate secretion by the glandular saccules of the llama stomach. *Cornell Vet. 63,* In press.

Engelhardt, W. v. (1970) Movement of water across the rumen epithelium. In *Physiology of Digestion and Metabolism in the Ruminant,* ed. Phillipson A. T. Oriel Press, Newcastle, pp. 132–146.

Engelhardt, W. v., Ehrlein, H. J. & Hörnicke, H. (1968) Absorption of volatile fatty acids, tritiated water and antipyrine from the abomasum of goats. *Quart. J. exp. Physiol. 53,* 282–295.

Ferreira, H. G., Harrison, F. A. & Keynes, R. D. (1964) Studies with isolated rumen epithelium of the sheep. *J. Physiol. (Lond.) 175,* 28–29.

Ferreira, H. G., Harrison, F. A., Keynes, R. D. & Zurich, L. (1972) Ion transport across an isolated preparation of sheep rumen epithelium. *J. Physiol. (Lond.) 222,* 77–93.

Harrison, F. A. (1971) Ion transport across rumen and omasum epithelium. *Phil. Trans. 262,* 301–305.

Harrison, F. A., Keynes, F. D. & Zurich, L. (1968) The active transport of chloride across the rumen epithelium of the sheep. *J. Physiol. (Lond.) 194,* 48–49.

Hendrickson, R. C. (1971) Mechanism of sodium transport across ruminal epithelium and histochemical localization of ATPase. *Exp. Cell Res. 68,* 456–458.

Höller, H. (1970) Untersuchungen über Sekret und Sekretion der Cardiadrüsen zone im Magen des Schweines. II. Versuche zur Beeinflussung der Spontansekretion der isolierten Cardiadrüsenzone, Flüssigkeits-und Elektrolytsekretion in den mit verschiedenen Flüssigkeiten gefüllten isolierten kleinen Magen. *Zbl. Vet.-Med. 17A* (10), 857–873.

Houpt, T. R. (1970) Transfer of urea and ammonia to the rumen. In *Physiology of Digestion and Metabolism in the Ruminant,* ed. Phillipson, A. T., Oriel Press, Newcastle, pp. 119–131.

Hungate, R. E. (1966) *The Rumen and Its Microbes.* Academic Press, New York, p. 181.

Hydén, S. (1961) Observations on the absorption of inorganic ions from the reticulorumen of the sheep. *Kgl. Lantbrukshögskol. Ann. 27,* 273–285.

Kay, R. N. B. (1960) The rate of flow and composition of various salivary secretions in sheep and calves. *J. Physiol. (Lond.) 150,* 515–537.

Keynes, R. D. (1969). From frog skin to sheep rumen: a survey of transport of salts and water across multicellular structures. *Quart. Rev. Biophys. 2,* 177–281.

Levi, H. & Ussing, H. H. (1949) Resting potential and ionic movements in the frog skin. *Nature (Lond.) 164,* 928–929.

Masson, M. J. & Phillipson, A. T. (1951) The absorption of acetate, propionate and butyrate from the rumen of sheep. *J. Physiol. (London) 113,* 189–206.

Oppel, A. (1897) *Lehrbuch der Vergleichenden Mikroskopischen Anatomie der Wirbeltiere,* Zweiter Teil. Schlund und Darm, Jena. pp. 35–1353.

Schneyer, L. H. and Schneyer, C. A. (1967) Inorganic composition of saliva. In *Handbook of Physiology,* Alimentary Canal, Washington, D. C. Amer. Physiol. Soc., Sect. 6 Vol V, pp 497–530.

Steven, D. H. & Marshall, A. B. (1970) *Physiology of Digestion and Metabolism in the Ruminant,* ed. Phillipson, A. T., Oriel Press, Newcastle, pp. 80–100.

Stevens, C. E. (1964) Transport of sodium and chloride by the isolated rumen epithelium. *Amer. J. Physiol. 206,* 1099–1105.

Stevens, C. E. (1970) Fatty acid transport through rumen epithelium. In *Physiology of Digestion and Metabolism in the Ruminant,* ed. Phillipson A. T., Oriel Press, Newcastle, pp. 101–112.

Stevens, C. E., Dobson, A. & Mammano, J. H. (1969) A transepithelial pump for weak electrolytes. *Amer. J. Physiol. 216,* 983–987.

Stevens, C E. & Stettler, B. K. (1966) Factors affecting the transport of volatile fatty acids across rumen epithelium. *Amer. J. Physiol. 210,* 365–372.

Stevens, C. E. & Stettler, B. K. (1966b) Transport of fatty acid mixtures across rumen epithelium. *Amer. J. Physiol. 211,* R 264–271.

Stevens, C. E. & Stettler, B. K. (1967) Evidence for active transport of acetate across bovine rumen epithelium. *Amer. J. Physiol. 213,* 1335–1339.

Thorlácius, S. O., Dobson, A. & Sellers, A. F. (1971) Effect of carbon dioxide on urea diffusion through ruminal epithelium. *Amer. J. Physiol. 220,* 162–169.

Vallenas, A., Cummings, J. F. & Munnel, J. F. (1971) A gross study of the compartmentalized stomach of two new-world camelids, the llama and guanaco. *J. Morph. 134,* 399–423.

Warner, E. D. (1958) The organogenesis and early histogenesis of the bovine stomach. *Amer. J. Anat. 102,* 33–63.

DISCUSSION

MAETZ: I was wondering whether you have made any progress in solving the problem concerning the mechanism of acidification by the rumen. By following the suggestions made by Brodsky & Schilb (1967) you should be able to decide between H^+ secretion or HCO^-_3 absorption.

STEVENS: No, we have not attempted to measure pCO_2 and, as mentioned in reference to the in vivo studies of Breukink et al. (personal communication) this should provide the answer. Ash & Dobson (1963) have made similar measurements in the rumen of sheep but under conditions which would not answer this particular question. We plan to examine this next.

KEYNES: Can I put something on the board? I would just like to present another way of looking at the rather complicated reactions that Dr. Stevens has been describing, because I think our data agree with his perfectly. But I would like to suggest that it is quite instructive to look at the equivalent electrical circuit. If you take the Ussing + Zerahn model reduced to its simplest terms, it just incorporates a sodium resistance, a sodium potential, and a passive leak parallel pathway for chloride. Now this model predicts that if you short circuit the tissue the sodium flux should increase, which is what actually happens; and it also predicts correctly that if you take away the chloride from the system it won't make any difference to the flux of sodium or to the short circuit current. In order to explain the facts we had to produce a much more complicated equivalent circuit in which you include several sources of potential with series resistances, covering potassium and chloride as well as sodium, together with another that we put down as X, which fits very nicely with what Dr. Stevens has as bicarbonate. Then if you also put in a series resistance R_S, you can in fact quite easily achieve reasonable figures for the batteries and for the resistances, and you can explain the quite different behaviour of the rumen epithelium, which is that if you take away the chloride on either side, the short circuit current increases a bit and the sodium flux decreases. On the other hand if you short circuit this system and reduce the potential to O, it

Brodsky, W. A. & Shilb, T. P. (1967) Mechanism of acidification in turtle bladder. Fed. Proc. 26, 1314–1321.

Breukink, H. G., Dobson, A. & Sellers, A. F. (personal communication).

Ash, R. W. & Dobson, A. (1963) The effect of absorption on the acidity of rumen contents. J. Physiol. (Lond.) 169, 39–61.

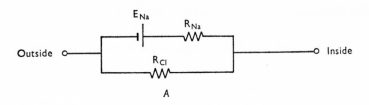

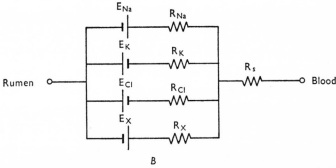

Equivalent electrical circuits for frog skin (A) and rumen epithelium (B). The batteries represent the potentials generated by the ion pumps. The unidentified component of the active flux in rumen epithelium is shown as if it were a cation X+. In order to account for the observed increase in short-circuit current and decrease in net sodium flux in sulphate Ringer, the common series resistance R_s in B would need to be about five times greater than R_{Na}.*

has no effect at all on the sodium flux, in considerable contrast to what happens in frog skin. This is quite a helpful way of looking at the problem, and I don't think there is any disagreement between us.

STEVENS: No, we have no disagreement here.

KEYNES: The trouble with such an equivalent electrical circuit is that it is not an unambiquous way of describing the system. I think you can put some of the batteries in a different place and still expect the same behaviour. So I don't want you to take me too literally in putting all the transport systems in parallel in one membrane, and then a sort of series coupling resistance on the opposite side. I think that Dr. Rehm produced a very similar arrangement

---

* From: Ferreira, H. G., Harrison, F. A., Keynes, R. D. & Zurich, L. (1972) Ion transport across an isolated preparation of sheep rumen epithelium. *J. Physiol. (Lond.)* 222, 77–93.

Rehm, W. S., Davis, T. L., Chandler, C., Gohmann, E. & Bashirelahi, A. (1963) Frog gastric muccosae bathed in chloride-free solutions. *Amer. J. Physiol.* 204, 233–242.

to this (Rehm *et al.* 1963) to explain what happens to the coupling between chloride movements and bicarbonate movements and hydrogen ion movements in gastric mucosa. I am rather surprised, by the way, that more people haven't measured the effect of short circuiting on the flux in various preparations. When I last searched the literature it seemed to me that I couldn't find anyone who had actually measured the flux in one and the same frog skin, first during a period of short circuiting and then on open circuit.

STEVENS: The equivalent circuit which you have proposed could account for our results, although I would think that the series resistance would have to be either relatively low or vary with available ion species. The distribution of these pumps in serial membranes would also complicate the picture, e. g. requiring an additional in-series resistance.

TOSTESON: It might be pertinent to mention that in the past year or so evidence in a number of laboratories, particularly that of Dr. Wieth and his colleagues in Copenhagen, has demonstrated that more than 99 % of the chloride transport across the red cell membrane occurs by a process of anion exchange which *in vivo* is with the main exchange partner being bicarbonate. So I think we have clear evidence now in at least one membrane for a very high ion exchange, a high capacity anion exchange system. I think that will be particularly necessary in trying to get a system that would do something like what you want in what you showed in your last slide, where you want to vary the ratio of permeability of the ionic and non-ionic forms. I think one is going to require some kind of facilitated anion transport system in order to accomplish changes in that ratio.

STEVENS: The transport of Cl and HCO_3 across rumen epithelium could be explained by the presence of a membrane similar to that of the red cell at the lumen surface of the tissue. Even a passive exchange of Cl for HCO_3 produced within the epithelial cells could account for the net transepithelial transport of Cl, if the opposing blood-facing membrane was much more permeable to Cl than HCO_3. The system could then be driven by the intracellular hydration of CO_2 if electrical neutrality could be maintained by an equivalent loss of intracellular H. While a more general mechanism for facilitated anion transport is attractive, its apparent nonspecificity (e. g. bicar-

Stevens, C. E., Dobson, A. & Mammano, J. H. (1969) A transepithelial pump for weak electrolytes. *Amer. J. Physiol. 216,* 983–987.

bonate, acetate, 3,3-dimethylbutyrate, DMO) makes this seem unlikely. Furthermore, mathematical analysis (Stevens *et al.* 1969) indicates that the observed net transport of the weak organic acids and bases could be accomplished with a reasonably small difference in the relative permeability of the opposing membranes to the dissociated and undissociated forms.

SCHULTZ: I might just add that your model A seems to match what we have recently found in the distal ileum.

STEVENS: Yes, the foregut of herbivores with a more highly developed stomach does appear to perform many of the functions characteristic of the lower intestine. The foregut absorbs substantial amounts of the Na and Cl secreted in the saliva while releasing HCO₃ which aids in the buffering of VFA. The lower intestinal tract must also help retrieve the substantial amounts of Na and Cl secreted into the upper gastrointestinal tract (Soergel & Hofmann 1971). The secretion of HCO₃ by the ileum and colon may also aid in the buffering of VFA since the latter provide the major anions in the feces of a variety of mammals, including man. Our results indicate that the mucosa of the equine large intestine also transports VFA in a manner qualitatively and quantitatively similar to that of bovine rumen epithelium and the cardiac glandular mucosa of the pig. In fact, even the proper gastric glandular mucosa of the pig and horse appears to take up VFA from the lumen bath as readily as the above tissues, it simply does not seem to allow as rapid a release to the blood side.

HOGBEN: That has been analyzed in the frog and the lactate goes out from the cell to the serosa solution (Hogben 1962, 1965). In the rat and dog lactate also appears in the serosal compartment.

STEVENS: How about acetate?

HOGBEN: We haven't looked at that.

Soergel, K. H. & Hofmann, A. F. (1971) Intestinal absorption. In: *Pathophysiology: Altered Regulatory Mechanisms in Disease*, Frohlich, E. D., ed., Lippencott, Philadelphia.

Hogben, C. A. M. (1962) Ultrastructure and transport across epithelial membranes. *Circulation 26*, 1179–1188.

Hogben, C. A. M. (1965) Introduction: the natural history of the isolated bullfrog gastric mucosa. *Fed. Proc. 24*, 1353–1359.

STEVENS: When bathed with solutions containing either glucose or propionate, rumen epithelium produces substantial amounts of lactate and this is also preferentially released to the serosal side. The model which we have proposed would predict a preferential release of weak organic acid to the lumen side and while lactic acid has a pK_a lower than that of the other acids tested this would not account for the reversal of direction. Therefore it would seem that lactate is handled differently. It is even conceivable that its production and release may help drive the transport of other weak electrolytes.

Transport Mechanisms in Sea-Water Adapted Fish Gills

J. Maetz

The gills of seawater teleosts transport sodium chloride towards the exterior against a chemical gradient of about 4:1. The question arises whether this is accomplished by means of separate cation and anion pumps, or whether one ion passively follows the active transport of the other, or whether separate ion carriers are linked in some fashion. Several observations made in our laboratory suggest separate branchial Na^+ and Cl^- exchange mechanisms (Motais *et al.* 1966). To get acquainted with the gill transport phenomena, the reader is referred to recent reviews (Motais 1967, Maetz 1971, Motais & Garcia Romeu 1972). The present paper should be considered as a progress report presenting recent unpublished observations on Na^+ transport, Cl^- transport and the linkage between the two ionic pumps.

Na^+ TRANSPORT AND BRANCHIAL Na-K DEPENDENT ATP-ase ACTIVITY

A Na-K activated ATP-ase has been implicated in the branchial sodium extrusion mechanism (Epstein *et al.* 1967, Kamiya & Utida 1968, Motais 1970). Kamiya (1972) very recently showed by cell fractionation techniques that the enzyme is located in the so-called 'chloride cells' or mitochondria-rich cells which almost certainly are the site of active NaCl extrusion. Electron-microscopic studies of the localization of the enzyme in eel gills (Mizuhira, *et al.* 1970, Utida, Pers. comm.) show the enzyme on the apical cell border and along the endoplasmic microtubular system which is an extension of the basal and intercellular cell border.

Experiments reported below strongly suggest that the Na-K dependent ATP-ase located on the apical cell membrane plays an essential role in Na extrusion and that this is accomplished by an exchange with K^+ derived

---

* Groupe de Biologie Marine, Département de Biologie du Commissariat à l'Energie Atomique, Station Zoologique, Villefranche-sur-Mer 06230, France.

»Transport Mechanisms in Epithelia«, Munksgaard, Copenhagen.

from the external medium (concentration of K^+ in sea water: 10–11 milli-moles per liter, about 3 times that of the plasma).

Perturbance of the sodium balance in relation to mucosal ATP-ase activity

Two types of experiment support the proposed model.

a) Flounders kept for 24 hrs in renewed K-free sea water show a progressive increase of the internal sodium content, amounting to about 100 micromoles hr^{-1}. $(100 \ g)^{-1}$, a value which would be expected if the Na^+ extrusion mechanism had failed. After return to sea water for 48–72 hrs, the internal Na^+ level declined significantly as if the Na^+ pump had been 'turned-on' upon addition of external K^+ (Maetz 1969).

b) Eels exposed for 24 hrs to $10^{-4}M$ ouabain in sea water, show an increased internal sodium level indicating an impairment of the Na^+ excreting pump (Motais & Isaia 1972).

Isotopic kinetic studies of the branchial sodium exchange in relation to mucosal Na-K ATP-ase activity

Isotopic studies showed that the unidirectional exchange fluxes across the gill represent 20 to 50% of the internal sodium per hr in sea water teleosts. Thus the sodium efflux is at least 10 times greater than the net sodium extrusion rate expected from the drinking rate and balancing intestinal absorption (Motais & Maetz 1965).

Sodium efflux components

The sodium efflux may be divided into 3 to 4 components (Motais *et al.* 1966, Maetz 1969, Bornancin *et al.* 1972, Motais & Isaia 1972).

(1) a small net extrusion component balancing the intestinal absorption of salt swallowed by the fish. Maetz (1969) suggested that this is the active component mediated by a one-for-one exchange with external K, amounting to about 100 micromoles $hr^{-1}(100 \ g)^{-1}$ in the eel and the flounder.

(2) a large fraction dependent of the external Na concentration linked with the branchial Na influx in a manner suggesting Michaelis-Menten kinetics. Various mechanisms have been proposed for this Na-Na exchange. Motais *et al.* (1966) suggested the presence of an exchange-diffusion carrier.

Maetz (1969) proposed that since sea water contains 50 times more Na than K, external Na competes with K on the K-site of the Na-K exchange pump and thus at least part of the exchange diffusion fluxes are the result of a competitive process similar to that observed in human red cells by Garrahan & Glynn (1967). Smith (1969) demonstrated on theoretical grounds that passive sodium exchange along electro-chemical gradients may well show a flux-concentration relationship analogous to Michaelis-Menten saturation processes.

(3) a residual efflux observed upon transfer to fresh water, independent of external cations which may be considered as a passive leak component. This leak is relatively small, about 10 to 30% of the total efflux, in various euryhaline teleosts such as the eel or the flounder, and large, 50 to 80 %, in the euryhaline *Fundulus heteroclitus* and stenohaline sea perch *Serranus* (Motais *et al.* 1966).

Correlation between K influx and Na efflux via the Na-K dependent ATP-ase system.

This correlation is evidenced by rapid transfer experiments and by pharmacological studies.

(1) Transfer of fish from sea water (SW) to K-free SW is accompanied by a reduction of the sodium efflux. This reduction is relatively small in the flounder (3 $\pm$ 0.44%), large in the sea perch (20 $\pm$ 5.6%) and in the seawater adapted trout (Maetz 1969, Greenwald & Kirschner 1971, Motais & Isaia 1972). On the other hand, no reduction is observed in *Fundulus kansae* (Potts & Fleming 1971) and in *Anguilla anguilla* (Motais & Isaia 1972). According to Motais and Isaia, this indicates that in some species the net excreting pump exchanging internal Na for external K switches to exchanging Na for Na after removal of external K. Thus in K-free SW, the Na-Na exchange mechanisms is operative alone as shown by the high Na turnover rate observed in the fish kept in this medium. But obviously it is not the vital mechanism permitting the maintenance of the Na balance of the fish.

(2) Comparison of the Na efflux in SW, in FW and in FW containing KCl concentrations ranging from 2 to 50 mM/l distinctly demonstrated the K-dependent Na efflux component. Figure 1 illustrates typical experiments on *Anguilla* and *Serranus,* two fish which differ widely in their respective Na-free effect observed upon transfer to FW. Addition of K to FW increases

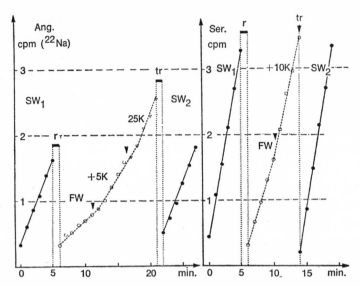

Fig. 1. ^{22}Na appearance into the external medium. Effect of external K. On the left, experiment on *Anguilla*. Transfer from sea water (SW) to fresh water (FW). Effect of the addition of KCl into the external medium: conc. 5 mM at first arrow, and 25 mM at second arrow. Return to SW (SW$_2$). Note the important reduction of the efflux upon transfer into FW. On the right, experiment on *Serranus*. Transfer from SW to FW. At arrow, addition of KCl, final conc.: 10 mM. The reduction of the flux upon transfer into FW is only slight in this species. r: rinse in FW to remove external Na; tr: transfer without rinse. Abscissa: External radioactivity in 10^3 cpm. Ordinate: Time in min. Printout of ext. ^{22}Na conc., every minute.

the Na efflux to a value which is nearly identical to the level observed in SW. This augmentation is clearly related to the external K concentration as shown in the experiment on the eel (on the left).

In Figure 2, the K-depending Na efflux of the flounder has been plotted as a function of the external K concentration. It may be seen that this flux obeys saturation kinetics and that the apparent affinity of the Na carrier for K corresponds to a K$_m$ of 10 mM/l. The maximal rate attained in this fish is about one-half of the Na efflux depending on external cation.

In the same figure, the K influx measured with the help of a total body counter has been plotted comparatively in the same group of flounders. It may be seen that for the K influx also saturation kinetics prevail and that the apparent affinity of the hypothetical K carrier corresponds to a K$_m$ of 10 millimolar, the same as that of the Na efflux component. This identity of the

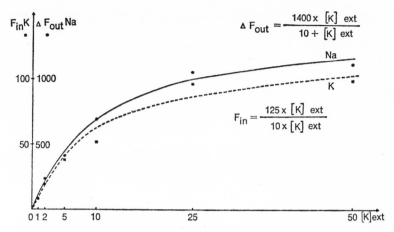

Fig. 2. Relationship between K influx and K dependent Na efflux and external K concentration in the flounder. Ordinates: fluxes in μmoles hr^{-1} (100 g)$^{-1}$. Abscissa: concentration in millimoles litre^{-1}. The equations of the two curves are given. $\triangle f_{out}$ or K dependent Na efflux is the difference between the fluxes measured in FW and in the various KCl concentrations, immediately following transfer from SW. Each point represents the mean of at least 4 measurements. The same group of 15 fish were used for influx and efflux determinations.

apparent affinities is a strong argument in favour of a common Na-K exchange carrier. The maximal rate of K influx is nearly 10 times less than the Na efflux component. It does not mean that the stoechiometry of the Na/K exchange is 10:1, because preliminary experiments suggest that the limiting factor for the K influx across the branchial epithelium is the internal facing membrane and that the fluxes across the external border are much faster than the K – transflux would indicate. New experiments are planned to assess the Na-K stoechiometry, and to explain why different techniques for estimating the K influx yield widely different values.

Competition between external K and Na for the Na-K exchange carrier

A direct demonstration of the inhibitory effect of external Na on K influx was obtained by Dr. J. C. Rankin in my laboratory by means of an eel gill perfusion experiment (Maetz 1971). With the help of the total body counter technique, I was able to confirm that K influx is 2 to 4 times less in SW than in a 10 millimolar KCl solution in the flounder, the eel and in the sea perch (Table I).

Table I. Branchial K exchanges and K dependent Na efflux in sea water adapted teleosts.

1. K influx measurements (total body counter technique)

Ext. Medium	Anguilla	Platichthys	Serranus
SW	18 ± 5.5	24 ± 4.9	34.5
	(6)	(6)	(2)
FWK	77 ± 23.9	56 ± 9.5	96.5
	(6)	(6)	(2)

2. K-dependent Na efflux

	Anguilla	Platichthys	Serranus
FWK	496 ± 102	693 ± 139	364 ± 182
	(6)	(9)	(4)

Fluxes in μ moles hr^{-1}. (100 g)$^{-1}$. Number of animals in brackets.

The depressing effect of external Na on K dependent Na efflux observed by Maetz (1969) in the flounder indirectly confirms the inhibitory effect of external Na on K influx linked to Na efflux.

Experiments are in progress on the effects of various Na/K concentration ratios on the K influx and K-dependent Na efflux. They suggest that it is the apparent affinity of the carrier for K rather than the maximal rate of flux which is altered by external Na. This altered affinity would indicate a true competitive process.

Action of ouabain added to the external medium on Na/K and Na/Na exchanges

According to Motais and Isaia (1972), the addition of 10^{-4}M ouabain to the external medium results in an inhibition of the Na efflux in the eel.

a. If the external medium is SW, the reduction is relatively small (16.1 $\pm$ 2.3%) but sufficient to account for a complete inhibition of the net extrusion rate suggested by the Na balance experiments reported above.

b. If the external medium is 10 millimolar KCl, the reduction attains 27 $\pm$ 6.0%. This inhibition is however far from attaining 100% as one would expect if the totality of the exchanges passed by a Na-K pump. Motais and Isaia suggest that this is due to an antagonism between K and the inhibitor similar to that observed by Glynn (1957) and Hoffman (1969) on the human red cell. The possibility of such an antagonism is confirmed by the fact that in 2 millimolar KCl solution, the K dependent Na efflux is virtually completely inhibited by $10^{-4}M$ ouabain.

Relationship between Na/Na and Na/K exchanges

Indirect evidence suggests that there must be a close correlation between the net extrusion rate of sodium, the exchange-diffusion component and branchial Na-K dependent ATPase. All these three parameters are depressed after hypophysectomy, adrenalectomy or actinomycin D treatment in the seawater adapted eel (Maetz et al. 1969a, Maetz et al. 1969b, Motais 1970, Maetz 1971). More recent observations, however, strongly argue against a close relationship between Na/K and Na/Na exchange. This evidence will be summarized below.

The Na/Na exchange is ouabain insensitive

An unexpected observation by Motais & Isaia (1972) was that there is no significant reduction of the Na efflux by ouabain in the eel transferred to K-free SW, while in SW there is a distinct inhibition. The Na/Na exchange component which in the human red cell results from a competition between external Na and K for a common carrier is definitely ouabain sensitive according to Garrahan & Glynn (1967). Two hypotheses have been advanced by Motais and Isaia to explain the lack of sensitivity of the branchial Na/Na exchange to cardiac glucosides: either interaction between ouabain and pump is prevented by an antagonism resulting from a steric interference between Na and inhibitor, or by a change in membrane conformation when the Na-K pump turns into a Na-Na pump. The first hypothesis seems rather improbable since at 500 millimolar NaCl concentration ouabain can have either an inhibitory effect (in SW) or none (in K-free SW).

Differential effects of Ca^{++} removal from the external medium on Na/Na and Na/K exchanges

Bornancin et al. (1972) observed that Ca^{++} removal from SW produces a depressing effect on the Na/K exchange component of the eel gill reducing it by half. Simultaneously the Na dependent Na efflux and the Na influx are increased by 100%, while the leak component is tripled. The inhibition of the Na extrusion pump by Ca^{++} lack is confirmed by the augmentation of the internal Na level and space observed in Ca-free SW. Eels do not survive more than 24 hrs in this medium.

Differential effects of temperature changes

Maetz and Evans (1972) observed that the branchial Na/K exchange of the flounder is far more temperature-sensitive than the Na/Na exchange. For instance transfer of eels from 17°C (adaptation temperature) to 7°C produces only a 50% reduction of the Na/Na exchange flux, while the Na/K component is depressed 6 fold. In cold water, the internal Na content of the fish slowly augments as a result of the impairment of the Na extrusion pump.

Differential effects of insulin treatment

In unpublished experiments on the eel, Dr. R. Langford in my laboratory showed that insulin treatment results in a complete inhibition of the Na/K exchange pump while the Na/Na exchange component remains unchanged. Failure of the Na pump is also illustrated by the significant increase of the internal Na level. More experiments are necessary to ascertain the mechanism of insulin action.

In conclusion, there is now ample proof that a mucosal Na/K dependent ATP-ase is closely connected to the branchial Na extrusion pump in seawater adapted teleosts. The very fast Na/Na exchange which accompanies the Na/K exchange in some species seems to be only loosely connected to the pump. It is probable that external Na interferes with the K-site of the pump by effecting conformational changes. External agents (temperature, Ca^{++}, inhibitors) or endogenous hormones may thus affect both components in a widely different manner.

CHLORIDE TRANSPORT AND INTERACTION BETWEEN Na AND Cl PUMPS

Little is known of the mechanism responsible for the extrusion of Cl^- across the gill, though since the exterior of the gill electrically negative to the interior of the fish, there is ample reason to postulate an active transport requiring the expenditure of energy. The results reported in this section were obtained in collaboration with professor F. H. Epstein in my laboratory.

Effect of Na SCN on Cl⁻ efflux of the eel in SW

As thiocyanate inhibits active halide transport in other tissues (thyroid, stomach, cornea) its effect on Cl⁻ efflux across the gill was examined in *Anguilla*. NaSCN was injected intraperitoneally as a 0.7 molar solution, 200μl per 100 g body weight, corresponding to an internal plasma concentration of 7 millimoles litre⁻¹.

On the right, Fig. 3, is a typical experiment, showing that within 2 minutes (up to 6 min. in other fish) there is a marked fall of the ³⁶Cl efflux, while control injection of 0.7 mM (left) produces no change.

Fig. 4 (left, stippled columns) summarizes the effect obtained on 6 fish. The 65% reduction is highly significant ($p < 0.001$). The magnitude of this inhibition is much greater than would be expected if only the net Cl⁻ extrusion rate (about 100 μmoles hr⁻¹ . (100 g)⁻¹) component balancing intestinal absorption were affected. In fact, the flux observed after inhibition is nearly equal to the Cl⁻ leak observed after transfer to FW. This is illustrated

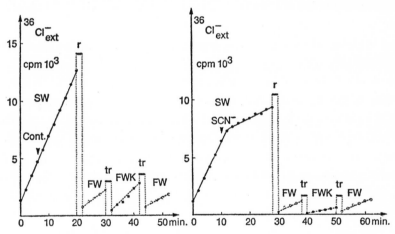

Fig. 3. ³⁶Cl appearance rate in seawater adapted *Anguilla*. On the left, control experiment. Efflux measured successively in SW, after a 2 min. rinse (r), in FW, in FWK (10 milliequivalent K as K₂SO₄) and in FW. In SW, intraperitoneal injection of NaCl. On the right, effect of Na SCN injection on SW efflux and on FW and FWK fluxes. Ordinate: external ³⁶Cl conc. in 10³ cpm. Abscissa: time in minutes. The printout of the ext. ³⁶Cl was every two minutes. Note, the reduction of efflux upon transfer into FW and the increased efflux produced by external K in the control fish. Note the depressing effect of SCN⁻ on Cl efflux in SW and the absence of K effect in this fish. The flux in SW following SCN⁻ injection is similar to the FW flux.

28*

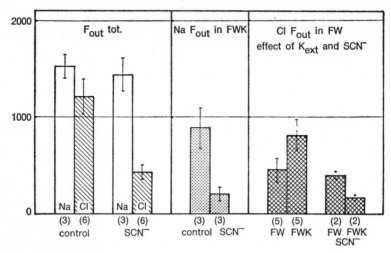

Fig. 4. Summary of the experiments concerning SCN⁻ inhibition. On the left: comparison of the effects of SCN⁻ on Na and on Cl effluxes in SW. In the middle: effect of SCN⁻ on the Na efflux observed in FWK (10 millimolar as K_2SO_4). On the right: Cl efflux upon transfer from SW into FW: effect of the addition of K into the external medium. Effect of SCN⁻ injection. Ordinates: fluxes in μmoles hr⁻¹. (100 g)⁻¹. Number of fish in brackets.

in Fig. 3 by the comparison of the slopes of the ^{36}Cl appearances in FW and in SW after SCN⁻ injection (see also absolute values of the fluxes in Fig. 4).

It may be concluded that the exchange-diffusion component which characterizes the Cl⁻ exchange is completely inhibited. Thiocyanate is known to inhibit carbonic anhydrase, but the effect on Cl⁻ flux cannot be ascribed to carbonic anhydrase inhibition. In two experiments, acetazolamide in amounts (10 mg/100 g) sufficient to produce marked inhibition of this enzyme (Maetz 1956) did not alter the efflux of Cl⁻, at least not during the 20 min. following injection.

Effect of Na SCN on plasma Na⁺ and Cl⁻

Table II summarizes experiments on eels in which Na SCN was injected intraperitoneally and also placed in the external bath in the same concentration as that estimated to be in the plasma after injection in order to retard SCN⁻ loss from the body and its rapid dissipation into the bath. Thiocyanate

Table II. Effect of NaSCN injection on plasma Na$^+$ and Cl$^-$ (in μmoles ml^{-1})

Eel body weight	Conc. SCN in	Conc. SCN out	Time (hrs)	Na	Cl
62	–	–	0	160	132
	7	14	4	250	235
42	–	–	0	160	145
	2	2	4	173	150
40	–	–	0	161	136
	2	2	4	180	150
330	–	–	0	161.5	136
	5	5	4	178	154
	5	5	24	198	163
240	–	–	0	159	137.5
	5	5	4	172	154
	5	5	24	198	162.5
450	–	–	0	160	137
	5	5	4	165	145
	5	5	24	185	146

The bigger eels having slower sodium and chloride turnover rates, the effect of SCN$^-$ injection is faster in smaller animals.

at concentrations ranging from 2 to 7 mM produced over several hours a sustained rise in plasma Cl$^-$ and Na$^+$, which was reversible when the fish was returned to running SW. Thus SCN$^-$ produces a marked impairment of the Cl$^-$ balance which confirms its inhibitory effect on the Cl$^-$ pump. But Na$^+$ balance is also perturbed, a point which was investigated further.

Effect of SCN$^-$ on Na$^+$ efflux and its components

The above mentioned observation suggests that SCN$^-$ also acts on the Na pump. It can be seen on the left of Fig. 4 that in contrast to its pronounced action on Cl$^-$ efflux, SCN$^-$ produced only a slight and insignificant effect on the total Na efflux. But as discussed in the preceding section, Na efflux comprises several components. The Na/K exchange which may be tested after transfer of the fish into FW with 10 millimolar KCl is an excellent indicator of the Na pump efficiency. To avoid any interference with the Cl exchange diffusion mechanism, the 'K-tests' were made by transferring the eel into K$_2$SO$_4$ solutions. Preliminary experiments showed that K solutions have similar effects on the Na efflux irrespective of the accompanying

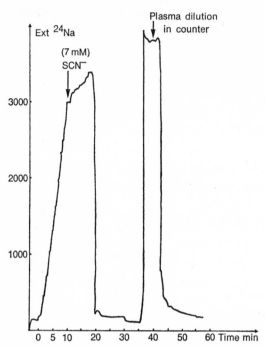

Fig 5. Recording of the ²⁴Na appearance observed in the eel after transfer into FWK (10 millimolar as K₂SO₄). Effect of SCN⁻ injection. The recording of a dilution of a plasma sample taken from the fish at the end of the experiment permitted the determination of the internal ²⁴Na specific activity and the flux calculation. Ordinate, ²⁴Na in cpm/ml.

anion. Accordingly the effect of SCN⁻ injection on the Na efflux in FWK medium was tested.

Fig. 5 reproduces a typical recording of the ²⁴Na appearance into the external medium in FWK. After injection of SCN⁻, there is a considerable decrease of the Na efflux. The results of 3 such experiments are summarized in the middle of Fig. 4. The flux attained after SCN⁻ inhibition (about 200 μmoles hr⁻¹ 100g⁻¹) is similar to the leak-flux observed after transfer of the fish into FW. Thus while total Na efflux remains virtually unchanged after SCN⁻ treatment, the Na/K exchange is totally inhibited and also the Na pump, as shown by the increase of internal Na level discussed above. Thus SCN⁻ inhibition offers one more instance of independence between Na/Na and Na/K exchanges. SCN⁻ interferes with the Na extrusion of the gill by blocking Na-K exchange. This suggests a link between the transport of

Na against K across the gill and the transport of Cl⁻. In order to elucidate this linkage further, more experiments were performed.

Stimulation of Cl⁻ efflux by K_2SO_4 in FW and its blockade by SCN⁻

Fig. 3 (left) and Fig. 4 (right) show that the addition of 10 millimolar K_2SO_4 to FW nearly doubled the Cl⁻ efflux. This stimulation was completely abolished, even reversed after injection of SCN⁻ into the fish. For this reversed effect, we have as yet no clear explanation. It may result from changes of diffusion potential induced by external K on the passive movement of Cl⁻.

Thus these results illustrate the intimate connection between the active extrusion of anions and cations across the gill. The nature of the linkage is not clear, but the simplest assumption is that it is an electrical one. Stimulation of Na⁺ efflux by external K⁺ also increases Cl⁻ efflux. Blockage of the Cl⁻ carrier by SCN⁻ not only inhibits Cl⁻ efflux but also eliminates the K dependent Na efflux and inhibits Na extrusion from the body. A general biological inference is that the transport of Na Cl across an epithelium may be accomplished by two pumps *in tandem* and that transport can be blocked if either pump is inhibited.

SUMMARY

1. A model suggesting that an apical Na-K dependent ATPase is associated with the Na extrusion mechanism across the gill of various seawater teleosts (eel, flounder and sea perch) is proposed:
   (a) Rapid readjustments of the Na efflux are observed with changes of the external K concentration in SW and after transfer to FW. A linkage between K influx and the K dependent component of the Na efflux is suggested by the identity of their apparent affinity constant for K.
   (b) Addition of ouabain to the external medium either SW or FW with added K, produces an inhibition of the K dependent component of the Na efflux.
   (c) The relationship between Na/K exchange and Na exchange diffusion possibly as a result of competition between external Na and K for the K-site, remains in doubt. Some observations suggest an independence between the two exchange systems.
2. Cl extrusion across the gill is effected by an active pump and is linked to an exchange diffusion mechanism.
   (a) Injection of SCN⁻ blocks both pump and exchange diffusion.
   (b) SCN⁻ also inhibits the Na/K exchange pump, while the Na/Na exchange and total Na efflux remain unchanged.

3. A linkage between Cl and Na pumps is suggested by:
   (a) Addition of K to the external medium following transfer to FW produces a simultaneous increase of Cl and Na effluxes.
   (b) Injection of SCN blocks this effect of K.

REFERENCES

Bornancin, M., Cuthbert, A. C. & Maetz, J. (1972) The effects of calcium on branchial sodium fluxes in the sea-water adapted eel, *Anguilla anguilla, J. Physiol. (Lond.) 222*, 487–496.

Epstein, F. H., Katz, A. I. & Pickford, G. E. (1967) Sodium and potassium activated adenosine triphosphatase of gills: role in adaptation of teleosts to salt water. *Science 156*, 1245–1247.

Garrahan, P. J. & Glynn, I. M. (1967) The behaviour of the sodium pump in red cells in the absence of external potassium. *J. Physiol. (Lond.) 192*, 159–174.

Glynn, I. M. (1957) The action of cardiac glucosides on sodium and potassium movements in human red cells. *J. Physiol. (Lond.) 136*, 148–173.

Greenwald, L. & Kirschner, L. B. (1971) Sodium extrusion across the gills of rainbow trout in seawater. *Am. Zool. 11*, 664. (abstract 234).

Hoffman, J. F. (1969) The interaction between tritiated ouabain and the sodium – potassium pump in red blood cells. *J. gen. Physiol. 54*, 343–353.

Kamiya, M. (1972) Sodium-potassium-activated adenosinetriphosphatase in isolated chloride cells from eel gills. *Comp. Biochem. Physiol. 43B*, 611–617.

Kamiya, M. & Utida, S. (1968) Changes in activity of Na-K activated adenosine tri-phosphatase in gills during adaptation of the Japanese eel to sea water. *Comp. Biochem. Physiol. 26*, 675–685.

Maetz, J. (1956) Les échanges de sodium chez le poisson *Carassius auratus* L. Action d'un inhibiteur de l'anhydrase carbonique. *J. Physiol. (Paris) 48*, 1085–1099.

Maetz, J. (1969) Sea-water teleosts: evidence for a sodium-potassium exchange in the branchial sodium-excreting pump. *Science 166*, 613–615.

Maetz, J. (1971) Fish gills: mechanisms of salt transfer in freshwater and in sea water. *Phil. Trans. B 262*, 209–249.

Maetz, J. & Evans, D. H. (1972) Effects of temperature on branchial sodium-exhange and extrusion mechanisms in the seawater-adapted flounder *Platichthys flesus* L. *J. exp. Biol. 56*, 565–585.

Maetz, J., Motais, R. & Mayer, N. (1969) Isotopic kinetic studies on the endocrine control of teleostean iono-regulation. *Excerpta Med. 184*, 225–232.

Maetz, J., Nibelle, J., Bornancin, M. & Motais, R. (1969b) Action sur l'osmorégulation de l'anguille de divers antibiotiques inhibiteurs de la synthèse des protéines ou du renouvellement cellulaire. *Comp. Biochem. Physiol. 30*, 1125–1151.

Motais, R. (1967) Les mécanismes d'échanges ioniques branchiaux chez les téléostéens *Ann. Inst. Oceanog. Monaco 45*, 1–84.

Motais, R. (1970) Effect of actinomycin D on the branchial Na-K dependent ATP ase activity in relation to sodium balance of the eel. *Comp. Biochem. Physiol. 34*, 497–501.

Motais, R. & Garcia Romeu, F. (1972) Transport mechanisms in the teleostean gill and amphibian skin. *Ann. Rev. Physiol. 34*, 141–176.

Motais, R., Garcia Romeu, F. & Maetz, J. (1966) Exchange diffusion effect and eury-halinity in Teleosts. *J. gen. Physiol. 50,* 391–422.

Motais, R. & Isaia, J. (1972) Evidence for an effect of ouabain on the branchial sodium-excreting pump of marine teleosts: interaction between the inhibitor and external Na and K. *J. exp. Biol. 57,* 367–373.

Motais, R. & Maetz, J. (1965) Comparison des échanges de sodium chez un Téléostéen euryhalin (le Flet) et un Téléostéen sténohalin (le Serran) en eau de mer. Importance relative du tube digestif et de la branchie dans ces échanges. *C. R. Acad. Sci. (Paris) 261,* 532–535.

Mizuhira, V., Amakawa, T., Yamashina, S., Shirai, N. & Utida, S. (1970) Electron microscopic studies on the localization of sodium ions and sodium-potassium-activated adenosine triphosphatase in chloride cells of eel gills. *Exp. Cell Res. 59,* 346–348.

Potts, W. T. W. & Fleming, W. R. (1971) The effects of environmental calcium and prolactin on sodium balance in *Fundulus kansae. J. exp. Biol. 55,* 317–327.

Smith, P. G. (1969) The ionic relations of *Artemia salina* L. II. Fluxes of sodium chloride and water. *J. exp. Biol. 51,* 739–757.

ACKNOWLEDGEMENTS

The author wishes to thank Dr. Walshe-Maetz for correcting the manuscript, MM. Tanguy and Lahitette for technical help, Prof. Epstein for his outstanding contribution to this project and for allowing the release of our unpublished data and to Prof. Motais and Dr. Isaia for permitting the use of their results still in press.

DISCUSSION

HOGBEN: For the gastric mucosa, it may be pedagogically convenient though possibly conceptually dangerous to speak of two chloride pumps: an "electrogenic" or electromotive Cl pump and a "non-electrogenic" HCl pump. Thiocyanate specifically inhibits the "HCl" pump but not the "electromotive Cl" pump. An excessive concentration of SCN does lead to irreversible deterioration of the mucosa.

MAETZ: Yes, this is a possibility. Under thiocyanate inhibition, however, the Cl flux ratio across the gill epithelium follows more or less Ussing's criterion for passive exchange, while the potential across the gill changes only slightly (experiments done in collaboration with Dr. C. R. House, unpublished). Thus the potential across the gill originates elsewhere.

KIRSCHNER: I agree with some elements in the model proposed by Dr. Maetz. There is no question about a rise in ATPase, although I was not convinced of this in 1967. There is also no question about a sodium-potassium exchange at the apical border. We can reproduce these data in the trout. But I don't believe the sodium-potassium exchange has been demonstrated to be ATP-linked. We are unable to show convincing ouabain inhibition of Na extrusion in sea water adapted trout, even if we delete potassium completely from the medium. I can't explain the discrepancy between our results and Dr Maetz's. But our observation is consistent with others. For example, Kamiya has shown (Kamiya &Utida 1968) that extrusion of sodium by isolated, sea water eel gill was inhibited by oubain on the nutrient (blood)side. And this is supported by Zaugg's demonstration that injection of ouabain into the blood almost completely blocks ATPase activity in the sea water salmon gill (Zaugg & McLain 1971). Some recent work on ATPase localization in salt gland is also relevant (Ernst 1972). It is the first convincing histochemical demonstra-

Kamiya, M. & Utida, S. (1968) Changes in activity of Na-K activated adenosine triphosphatase in gills during adaptation of the Japanese eel to sea water. *Comp. Biochem. Physiol. 26,* 675–685.

Zaugg, W. S. & McLain, L. R. (1971) Gill sampling as a method of following biochemical changes: ATPase activities altered by ouabain injection and salt water adaptation. *Comp. Biochem. Physiol. 38B,* 501–506.

Ernst, S. A. (1972) Transport adenosine triphosphate cytochemistry. II. Cytochemical localization of ouabain-sensitive potassium-dependent phosphatase activity in the secretory epithelium of the avian salt gland. *J. Histochem. Cytochem. 29,* 39–55.

tion of a K-dependent, ouabain-sensitive enzyme in a transport epithelium, and I commend it to your attention. The work is unfortunately not on gill, but in ultrastructure the salt gland cell resembles the gill "chloride cell", and the organs have the same job to do: excreting salt from blood into a hypertonic solution. There is enough resemblance to suppose that they function similarly.

MAETZ: There is no potassium outside, however.

KIRSCHNER: (cont'd) There probably is K^+ in the lumen since the fluid secreted contains it. Dr. Ernst was kind enough to provide pictures of the two poles of a secretory cell stained for ATPase activity. In the picture one can see the basal infoldings on the nutrient side. The infoldings are open to the blood. The system of intracellular membranes is heavily stained with products of the enzyme reaction. In the presence of ouabain or if K^+ is deleted the deposits are absent. The second figure shows that this internal membrane system ends blindly short of the apical border; this also true in the gill chloride cell. One notes absence of deposits on the apical membrane. These pictures are consistent with ready access of ouabain to the enzyme when added to the blood rather than to the apical side. If the cellular ATPase distribution is the same in trout gills, it explains why we can't demonstrate ouabain inhibition of Na extrusion, but not why Dr. Maetz can in eels. An alternative is that gills, unlike salt gland, have some ATPase on the apical membrane. This would explain his results though not ours. That is the substance of our "disagreement".

One other point is worth noting. Both salt gland and gill chloride cells are characterized by the profuse system of membrane infoldings that stop short of the cell apical border. The pump ATPase is located on these membranes, at least in the salt gland. This suggests the ned for rethinking the role of this enzyme in salt throughput. Is it to transport Na^+ into channels which are already open to blood Na^+, or to protect the cytoplasm from Na^+ loading in the face of an increasing chemical gradient in the channels?

MAETZ: May I comment on that. It is unfortunate that you did not bring any slides concerning the chloride cell. In their preliminary report concerning

Ernst, S. A. & Philpott, C. W. (1970) Preservation of Na-K-activated and Mg-activated adenosine triphosphatase activities of avian salt gland. *J. Histochem. Cytochem. 18,* 251–263.

the technique for Na-K dependent ATPase preservation with formaldehyde as fixative Ernst & Philpott (1970) do not give any clear localization for Na-K ATPase for the chloride cell. Very recently, I had the visit of Dr. S. Utida from Tokyo. He showed slides with very clear cut ATPase reaction in the apical face of the chloride cell taken from the gill of the sea water adapted eel. According to his coworker specialist of electronmicroscopy (Dr. Mizuhira) this reaction is ouabain sensitive.

KIRSCHNER: Then he has to explain his old experiment showing that ouabain inhibits from the inside and is ineffective from the outside.

MAETZ: Dr. Mizuhira and Dr. Utida also show the ATPase reaction along the infoldings of the chloride cell, a localization very similar to that given by Dr. Ernst for the salt gland of birds. I entirely agree that this second ATPase also plays a role in transepithelial sodium transport.

KIRSCHNER: Zaugg injects ouabain into intact fish, and it completely inhibits the ATPase from the inside.

MAETZ: Another point I wanted to raise about apical Na-K dependent ATPase is that it should offer the possibility to measure directly the number of transport sites with ^{3}H-ouabain. In collaboration with Dr. M. Bornancin, I have recently measured ouabain-binding comparing sea water and fresh water eel gills. This was done by shaking excised gills at room temperature in 10^{-7} M tagged ouabain either in a choline chloride solution or in a potassium chloride solution. After 2 min contact, the gills were rinsed in the same media without ouabain and a near-zero temperature for 5 min. As K competes for the binding site, the gills kept in KCl solution served as controls for the appreciation of the non specific binding. Only the seawater adapted gill showed significant binding with ouabain. Admitting a gill surface of 200 cm² per 100 g body weight the number of sites would be about $3000/\mu^2$. About Motais and Isaia's experiments (1972) two points should be stressed. Ouabain inhibition from the outside is extremely rapid, within 2 min of addition of the inhibitor. This points has been verified by me in subsequent experiments. It seems that K as well as Na present in sea water interfere with ouabain binding.

Motais, R. & Isaia, J. (1972) Evidence for an effect of ouabain on the branchial sodium-excreting pump of marine teleosts: interation between the inhibitor and external Na and K. *J. exp. Biol.* 57, 367–373.

GENERAL DISCUSSION

LOEWENSTEIN: First of all, I would like to congratulate Drs. Oschman & Wall on their results. These findings open an entirely new appraoch to study membrane-bound calcium and functions of calcium, if you can confirm that it is really calcium that is sticking to the membrane by other methods, say by electron probe. There are a number of interesting experiments here one can immediately do even at the present stage. Some of these we have discussed privately. Let me point out a few other ones. Oschman & Wall could now examine the junction under conditions of uncoupling in absence of extracellular calcium, such as the uncoupling produced by metabolic inhibition. The uncoupling produced by the substitution of lithium for sodium would be another situation; and what could perhaps be even more interesting is to see whether recoupling the cell system – and this can be done by simply raising the resting potential in one of the cells – whether under these conditions you switch back to the non-stained situation. It would also be worthwhile to look into situations of competition between calcium and hydrogen ions. Hydrogen ions and calcium ions are known to compete for binding to membrane, and biological membranes usually seem to have a higher affinity for the hydrogen ion.

HOGBEN: Why not try frozen dried sections where there is no artifact at all?

OSCHMAN: I think we have seen from Dr. Voutes presentation that there are artifacts.

HOGBEN: We know there are small crystals and undoubtedly water shifts, but you could see if there is really calcium there or not.

OSCHMAN: This is a good idea, as is the freeze etching which Dr. Tosteson suggested.

ZADUNAISKY: A point about the fixation. Gluteraldehyde fixation will permit diffusible ions to move out into the solution, but for something that is bound maybe it is not so bad.

OSCHMAN: I agree. Conventional fixation methods seem suitable for study of ions that are mainly in bound form, such as calcium and magnesium, while one must resort to cryo-methods such as those described by Dr. Voûte, for mobile ions such as sodium and potassium.

WHITTEMBURY: In squid axons Gloria Villegas (1969) has observed at the inside face of the double layer membrane structure thickenings which might correspond to what you are observing. Whether that is calcium or not is a different story. I don't know how feasible it might be, you might try autoradiographs to see whether this is really calcium. I think calcium has a low enough activity, so you might get the dots at the right spots.

OSCHMAN: Sampson *et al.* (1970) have done an autoradiographic study of rat intestine. They found that the location of the electron-opaque deposits corresponds with that of autoradiographic label. They fed rats with water containing ^{45}Ca, waited 15 to 30 minutes, removed the intestines, and prepared the tissues for electron microscopic autoradiography. Silver grains were present over microvilli and mitochondria. Since the tissues were fixed for 2 hours, it is clear that any unbound calcium would be washed out, so the label must indicate the location of bound calcium.

TOSTESON: I would make a comment to the general problem of the binding of calcium to membranes. Dr. Butler in our laboratory worked on this problem several years ago and was able to demonstrate calcium and magnesium binding to red cell membranes. However, this process is completely nonspecific and depends entirely on the ionic strength of the medium. As you know, any negative surface will bind cations. It is important to try to distinguish between specific binding of calcium as compared with this nonspecific process.

SCHULTZ: If one wanted to establish a physiological role for the calcium or apparent calcium binding to the brush border, one might suggest using rachitic animals and vitamin D treated animals and seeing whether there are any differences.

OSCHMAN: We actually began such a study, but then found out that Sampson *et al.* (1970) had already shown that there was no great difference in the number of granules present within the microvilli of rachitic and normal rat intestines. Instead, rachitic animals lacked mitochondrial granules, whereas vitamin-D treated rats had numerous mitochondrial granules. These findings, and their implications for intestinal calcium transport, are most interesting.

Villegas, G. M. (1969) Electron microscopic study of the giant nerve fiber of the giant squid Dosidicus gigas. *J. Ultrastructure Res. 26,* 501–514.

Sampson, H. W., Matthews, J. L., Martin, J. H. & Kunin, A. S. (1970) An electron microscopic localization of calcium in the small intestine of normal, rachitic, and vitamin-D-treated rats. *Calc. Tiss. Res. 5,* 305–316.

Electrolyte Transport across the Body Surface of Fresh Water Fish and Amphibia

Leonard B. Kirschner

INTRODUCTION

Much of the modern research on ion transport has been done on isolated preparations such as the frog skin and toad bladder. The amount of information, especially about sodium transport across such epithelia, is enormous and has provided most of the elements that shape our current models. Recent work on the same types of transport epithelia *in vivo* has also provided some useful information about the mechanisms involved. Some of these data can easily be correlated with information on the isolated preparations, but others cannot.

In this communication I propose to describe some studies on active salt transport across the body surface of intact fresh water animals. The emphasis will be on three aspects. One is to deduce something about the transport mechanisms from data obtained in circumstances actually experienced by fresh water animals, that is under open-circuit conditions and with dilute external solutions. Another is to point out some apparent differences between the function of similar systems *in vivo* and *in vitro*. And last, I would like to convey an impression of the remarkable similarity displayed by transport systems from totally unrelated animals drawn from several phyla.

The methods employed to study active NaCl transport in intact animals have been described recently (Kirschner 1970) and will not be discussed in any detail here. A recent review (Maetz 1971) covers many of the questions treated below. Although its emphasis is on fish, the discussion is applicable to NaCl transport in other forms.

Washington State University, Department of Zoology, Pullman, Washington 99163 USA

»Transport Mechanisms in Epithelia«, Munksgaard, Copenhagen.

RESULTS

Active Transport and Electrogenesis

Clear evidence for active salt transport was first obtained in fresh water animals by Krogh (1937, 1938). He showed that when they were salt depleted and then placed in pond water (NaCl = 1 mM) they absorbed salt into the more concentrated body fluids. With the development of tracer methodology and the flux ratio criterion (Ussing 1949a,b) more sophisticated analysis became possible. Table I shows the results of flux ratio analysis for a number of fresh water animals. It is clear that both Na^+ and Cl^- are actively transported from very dilute solutions. With the recent demonstration that chloride can be actively transported across isolated frog skin (Kristensen 1972) information from *in vitro* and *in vivo* preparations is in qualitative agreement on this point.

Table I. Flux Ratio Analysis of NaCl Transport in Some Fresh Water Animals

| Animal | Ion | Concentration (mM) | | $c_o/_1$ | E* (mV) | J_{in}/J_{out} | |
		out	in			expected	found
Amblystoma	Na	1.2	100	0.012	+14	0.007	2.6
Rana	Na	3	100	0.030	+85	0.001	0.6
Salmo	Na	1.0	150	0.007	+15	0.004	1.4
Astacus	Na	2	205	0.010	+ 5	0.008	1.1
Amblystoma	Cl	1.2	80	0.015	+14	0.026	1
Astacus	Cl	0.3	184	0.016	÷28	0.021	1.0
Salmo	Cl	1.0	130	0.008	+15	0.014	1.8

*Sign is that of the body fluids (Adapted from Kirschner 1970)

Transport epithelia also generate potential differences *in vivo,* and Table I shows some typical values reported for animals in dilute solution. The transepithelial potential difference (TEP) is a function of $[Na]_{out}$ in the larval salamander (Dietz *et al.* 1967); at higher external concentrations it is larger in the presence of a non-penetrating anion (sulfate) than with chloride. Fig. 1 shows the same phenomena in the intact frog (Kirschner 1970). This is very similar to the behavior of isolated skins.

However, there have been a number of reports that some animals in dilute media develop TEPs of the opposite polarity; that is, body fluids

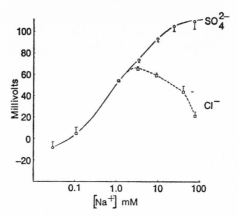

Fig. 1. External [Na⁺] and TEP in intact *Rana pipiens*. Concentrations were changed by adding NaCl (-☐-) or Na_2SO_4 (-0-) to the external medium (KCl or K_2SO_4, 10 meq/1). TEP was independent of anion below 10 mM but not at higher concentrations. The frogs had been salt-depleted for 10 days, and these TEPs are larger than those in pond water-adapted animals.

negative. These include several fresh or brackish water fish (House 1963, Maetz & Campanini 1966, Evans 1969) and the crayfish (Chaisemartin 1966). As will be seen later, the transport systems in these animals appear to be very similar to those in amphibia, hence polarity differences in TEPs are hard to understand.

Interpretation is further complicated by the observation that the polarity of the TEP may depend on the presence or absence of Ca^{2+} in the external medium (Kerstetter *et al.* 1970). The discrepancy here appears to be between amphibian epithelia and those from other animals rather than a difference between the behavior of *in vitro* and *in vivo* preparations. The role of active ion transport in epithelial electrogenesis warrants further investigation in the light of these differences.

Mechanisms

Most of what follows concerns the entry of Na⁺ and Cl⁺ into the epithelial cells from the external medium. When sodium influx (J_{in}^{Na}) is studied at different [Na]$_{out}$ many fresh water epithelia appear to show saturation kinetics. Data for the rainbow trout gill are shown in Fig. 2, and comparable observations have been made on the crayfish (Shaw 1959) and frog (Greenwald 1971). K_m values (external concentration giving a halfmaximum

Table II. K_m Values for Sodium Influx Systems of Different Animals.

Animal	Class	K_m (mM)	Habitat Preference
Carcinus maenas	Crustacea	20	Marine to brackish estuaries
Marinogammarus finmarchius	Crustacea	6–10	Marine to brackish estuaries
Fundulus heteroclitus	Osteichthyes	2.0	Brackish, does not survive well in fresh water
Gammarus duebeni (brackish race)	Crustacea	1.5–2.0	Brackish water
G. tigrinus	Crustacea	1–1.5	Brackish water
G. zaddachi	Crustacea	1–1.5	Brackish water
Lumbricus terrestris	Oligochaeta	1.3	Moist earth
Bufo americanus	Amphibia	>1	Terrestrial
Eriochier sinensis	Crustacea	1.0	Breeds in seawater, matures in fresh water
Platichthys flesus	Osteichthyes	0.8	Marine to brackish, can live in fresh water
Salmo gairdneri (gills)	Osteichthyes	0.5	Fresh water, can live in seawater
G. duebeni (fresh water race)	Crustacea	0.4–0.5	Fresh water
Rana cancrivora	Amphibia	0.4	Fresh water to seawater
Amblystoma gracile	Amphibia	0.3–0.5	Fresh water larvae of terrestrial form
Limnea stagnalis	Gastropoda	0.25	Fresh water
Actacus pallipes	Crustacea	0.25	Fresh water
R. pipiens	Amphibia	0.2	Fresh water to semi-terrestrial
Amhiuma means	Amphibia	~0.2	Fresh water
G. pulex	Crustacea	0.15	Fresh water
G. lacustris	Crustacea	0.1–0.15	Fresh water
Lampetra planeri	Agnatha	0.13–0.26	Fresh water; after metamorphosis will migrate to the ocean
Carassius auratus	Osteichthyes	<0.05	Fresh water
Xenopus laevis	Amphibia	<0.05	Fresh water

(Adapted from Greenwald 1972).

influx) for many animals are shown in Table II (from Greenwald 1972). Except for a few animals they lie between 0.1–1.0 mM. This is much lower than values reported for isolated, short-circuited frog skin (Kirschner 1955, Snell & Leeman 1957). Recent measurements on sodium entry into the

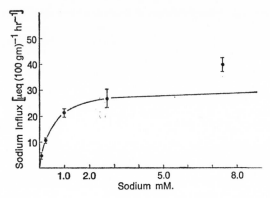

Fig. 2. External [Na$^+$] and J_{in}^{Na} in rainbow trout. Each point is the mean for a group of animals. Vertical bars show ±1 s.e.m. The high value at 7.5 mM is reproducible and has also been seen in larval salamanders. It may be due to a permeability change at higher concentrations.

transport pool in isolated, short-circuited skin also appear to give a high K_m (14.3 mM; Biber & Curran 1970).

The difference may be due to some uncharacterized change in the skin when it is excised. It could be due to the use of high ionic strength external media (using choline or K$^+$ salts) with isolated preparations. Or it may be due to the fact that short-circuiting, used in the *in vitro* work, stabilizes the PD across the outer membrane (Cereijido & Curran 1965), while the studies *in vivo* are under open-circuit conditions. In the latter situation the PD increases logarithmically with external [Na] (Biber *et al.* 1966). Smith (1969) showed that a concentration-dependent PD can give rise to apparent saturation kinetics even in a diffusive system without transport. At present we can only say that the difference in K_m needs further investigation.

Chloride transport also exhibits saturation kinetics with very low K_m values. For the rainbow trout the $K_m = 0.2$ mM (Kerstetter & Kirschner 1972) and for the crayfish about 0.1 mM (Shaw 1960c). Krogh's experiments (1937, 1938) showed that Na$^+$ and Cl$^-$ influxes are independent of each other. Both ions could be absorbed from solutions with non- or poorly-penetrating counterions. This observation has been extended to include a number of fresh water animals.

Table III shows representative data (rainbow trout) for J_{in}^{Cl} from 1 mM solutions of NaCl, MgCl$_2$ and choline chloride as well as J_{in}^{Na} from NaCl, Na$_2$SO$_4$ and Na$_2$HPO$_2$/NaH$_2$PO$_4$. The counterions were all either non-or

29*

Table III. J_{in}^{Na+} and J_{in}^{Cl-} from Solutions of Different Counterions (rainbow trout gill).

| | Influx, μeq (100 gm)$^{-1}$ hr^{-1} | |
Counterion	Sodium	Chloride
Chloride	21.7 ± 1.4 (18)	– – –
Sulfate	14.9 ± 1.7 (7)	– – –
Phosphate	20.1 ± 1.4 (11)	– – –
Sodium	– – –	33.1 ± 4.6 (12)
Choline	– – –	36.9 ± 3.5 (5)
Magnesium	– – –	30.9 ± 3.5 (5)
Potassium	– – –	19.9 ± 3.8 (11)

poorly-penetrating. It is obvious that J_{in}^{Cl} is the same with choline or Mg^{2+} as with Na$^+$ and only slightly reduced in the presence of K$^+$. Similarly, J_{in}^{Na} was only a little lower from SO$_4^{2-}$ than from Cl$^-$ and was not reduced at all in phosphate solutions. Under open-circuit conditions entry of either ion into an epithelial cell without its counterion demands simultaneous extrusion of an ion of like sign from the cell if electrical neutrality is to be preserved.

Krogh (1938) suggested that either NH$_4^+$ or H$^+$ might be the exchange partner for Na$^+$ while bicarbonate was an obvious candidate for the Cl$^-$ system. Some evidence was developed favoring a coupled Na$^+$/NH$_4^+$ exchange. When animals are in 1 mM solutions J_{out}^{NH3} across the body surface is roughly equivalent to J_{in}^{Na} in crayfish (Shaw 1960a) and salamander (Dietz et al. 1967). Injection of NH$_4^+$ stimulates J_{in}^{Na} and high concentrations in the external medium inhibits it (Shaw 1960b, Maetz & Garcia Romeu 1964). On the other hand, stoichiometry was often poor; carp (deVooys 1968) and crayfish (Shaw 1960a) excreted as much NH$_3$ in distilled water ($J_{in}^{Na} = 0$) as when J_{in}^{Na} was rapid.

The lack of correspondence between NH$_3$ efflux and Na$^+$ influx was convincingly demonstrated in a simple but elegant study on the frog *Calyptocephalella gayi*. This animal excretes far too little NH$_3$ across the skin to account for J_{in}^{Na} under any conditions (Garcia Romeu et al. 1969). Instead there is nearly exact equivalence between sodium uptake and H$^+$ excretion by the skin.

This question was examined more thoroughly in our laboratory. Most of the work was done on the rainbow trout *(Salmo gairdneri)*, but some experiments were conducted on the frog *(Rana pipiens)* and crayfish *(Procam-*

Table IV. The Effect of Transport Inhibitors on Fluxes of Na^+, NH_3 and H^+ in Various Fresh Water Animals

Animal Inhibitor Period			Ion Flux $[\mu eq\ (100\ gm)^{-1}\ hr^{-1}]$			
			J_{out}^{Na}	J_{in}^{Na}	$J_{out}^{NH_3}$	J_{out}^{H}
Trout	Diamox	Control	15.7 ± 3.6	18.0 ± 1.6	15.4 ± 1.6	23.6 ± 1.3
		Exp't'l	10.2 ± 2.6	2.6 ± 0.9	12.6 ± 1.2	9.4 ± 3.4
Trout	Amiloride	Control	15.7 ± 1.2	29.9 ± 2.1	27.1 ± 2.4	22.4 ± 2.1
		Exp't'l	13.9 ± 1.4	6.2 ± 0.7	19.0 ± 1.6	10.0 ± 1.5
Frog	Amiloride	Control	10.7 ± 3.9	26.6 ± 6.0	<4	26.5 ± 3.9
		Exp't'l	9.0 ± 1.8	14.4 ± 7.0	<4	16.0 ± 2.4
Crayfish	Amiloride	Control	20.4 ± 7.9	88.7 ± 9.4	33.8 ± 4.2	69.1 ± 6.2
		Exp't'l	22.6 ± 2.1	<10	15.5 ± 3.5	<4

barus spp.). The general plan was to place the animal in a standard medium containing no penetrating anions (usually 0.5 mM Na_2SO_4 in 1 mM K_2HPO_4/KH_2PO_4, pH 7.1). J_{in}^{Na} was measured and compared with $J_{out}^{NH_3}$, and J_{out}^{H} was estimated simultaneously. Then J_{in}^{Na} was varied, either by changing [Na] $_{out}$ (in the trout), or by using the inhibitors Diamox® (trout) or amiloride (trout, frog, crayfish). J_{in}^{Na} was measured again as were $J_{out}^{NH_3}$ and J_{out}^{H}. The results are compiled in Tables IV and V and may be summarized as follows:

1. In control animals Table IV shows that either NH_3 or H^+ efflux might balance Na^+ influx in the trout, but in the normal frog and salt depleted crayfish NH_3 efflux is far too small. In contrast, there is good correlation between J_{in}^{Na} and J_{out}^{H}.

2. When Diamox® is injected into the trout J_{in}^{Na} is almost completely inhibited. This is accompanied by an equivalent reduction in J_{out}^{H} but virtually no change in $J_{out}^{NH_3}$.

3. Amiloride inhibited Na^+ influx in the trout, the frog and crayfish. Some reduction in $J_{out}^{NH_3}$ occurred, but it was insufficient to balance the reduction in J_{in}^{Na}. The largest change was in J_{out}^{H}, suppression of which in the frog and crayfish almost exactly equaled inhibition of Na^+ uptake.

4. Table V shows that when Na influx varies at different [Na]$_{out}$ there is no corresponding change in $J_{out}^{NH_3}$ This was also observed in the goldfish (Maetz 1972). J_{out}^{H} was not calculated for these experiments, but pH changes in the external medium were measured. They indicate that the more rapid the Na^+ uptake the faster H^+ is excreted.

Table V. Effect of External Na^+ Concentration on J_{in}^{Na}, $J_{out}^{NH_3}$ and pH of the Medium*.

Sodium conc. mM	J_{in}^{Na}	$J_{out}^{NH_3}$	$\triangle$pH
7.0	33.4 ± 6.3	15.4 ± 2.4	$\div 0.31 \pm 0.08$
1.0	17.9 ± 1.1	17.4 ± 0.6	$\div 0.07 \pm 0.04$
0.1	6.2 ± 1.5	12.4 ± 1.8	$+0.07 \pm 0.08$

*in perfused trout gill.

Na^+ efflux was also measured, and the data in Table IV indicate that it is little modified by these treaments. This suggests that neither inhibitor modified the permeability of the epithelia.

Such flux comparisons show that there is little relationship between NH_3 efflux and Na^+ transport. The approximate equivalence reported previously for fish and crayfish in 1 mM solutions is, in the light of this evidence, fortuitous. There is much closer correlation between J_{out}^{H} and J_{in}^{Na} under all conditions tested. The evidence suggests that most, if not all, of the obligatory exchange is between these two ions.

There is less information about the anion excreted when Cl^- is taken up in the absence of Na^+. In the goldfish injection of HCO_3^- stimulates J_{in}^{Cl}, while Diamox® inhibits it, probably by reducing the supply of HCO_3^- ions within the transport cells (Maetz 1956, Maetz & Garcia Romeu 1964). Cl^- uptake in the frog is accompanied by excretion of base, the titration curve of which was identical with that for HCO_3^- (Garcia Romeu et al. 1969). These observations suggest that bicarbonate is the exchange partner.

The Action of Inhibitors

Several compounds have been found which appear to act on Na^+ uptake across the outer membrane of epithelial cells. Injection of acetazolamide (Diamox®) into the goldfish was shown to have marked inhibitory action on both Na^+ and Cl^- transports in the goldfish (Maetz 1956, Maetz & Garcia Romeu 1964). Given the evidence summarized above that H^+ and HCO_3^- are required as exchange partners, these data suggest that their source is blood CO_2 diffusing into the transport cells where it is hydrated and ionized. Apparently carbonic anhydrase is necessary to insure that hydration occurs at a rate commensurate with the normal fluxes of Na^+ and Cl^- from the external medium. They also indicate that the exchange systems are necessary even when both Na and Cl are present together in the environment.

Diamox® also inhibits Na^+ transport in the trout (Kerstetter *et al.* 1970), but Cl^- movement is unaffected (Kerstetter & Kirschner 1972). This may be because the serosal (nutrient) membrane in the trout is sufficiently permeable to permit plasma HCO_3^- to enter the cells for exchange with Cl^- at the outer membrane. Cl^- transport has also recently been shown to be Diamox®-sensitive in the isolated frog skin (Kristensen 1972).

Amiloride (N-amidino-3,5-diamino-6-chloropyrazine carboxamide) has been shown to inhibit active sodium transport in a number of epithelia including isolated frog skin (Salako & Smith 1970a,b). The compound is active when added to the outside medium and appears to prevent Na^+ entry into the epithelial cells (Salako & Smith 1970b). Fig. 3 shows that it inhibits J_{in}^{Na} in the trout; inhibition is just perceptible at 10^{-5} M but nearly complete at 10^{-4} M; it is completely and rapidly reversible. Sodium efflux is unchanged (cf. Table IV) with the result that net uptake during the control period is converted to net loss at the higher concentration. Chloride fluxes are unaffected.

The same observations (inhibition of J_{in}^{Na} with no change in J_{out}^{Na}, J_{in}^{Cl} and J_{out}^{Cl}) have been made on the frog and crayfish and provide the basis for our use of amiloride in the experiments shown in Table IV. The locus of action is the same *in vivo* as *in vitro*, and presumably the same mechanism is involved. One major difference, however, is in the concentrations required. In isolated skins J_{in}^{Na} is nearly completely inhibited at 10^{-6} M while in intact frogs 10^{-4} M suppresses influx less than 50% (Table IV). It is only a little more effective in the trout and crayfish. The basis of this difference is not yet known.

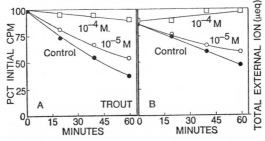

Fig. 3. Inhibition of J_{in}^{Na} by amiloride in a rainbow trout. A. Isotope (^{22}Na) uptake from the external medium. B. Total Na^+ in the external medium. The control data and those with amiloride were obtained from the same fish.

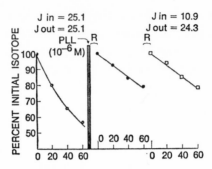

Fig. 4. Inhibition of J_{in}^{Na} by poly-L-lysine in a rainbow trout. Isotope (^{22}Na) uptake is shown for the control period (first panel). The animal was exposed to PLL for 2 minutes (vertical bar) then to PLL-free solution for 8 minutes (R). A second uptake measurement was then made (middle panel) followed by 10 minute rinse period (R) and a third uptake measurement (last panel).

Influx and efflux in μeq (100 gm)$^{-1}$ hr^{-1} are shown above the figure for the first and third periods. J_{in}^{Na} was suppressed by 60%; J_{out}^{Na} was unaffected.

Poly-L-lysine (PLL) is also a potent inhibitor of sodium influx across the trout gill as has also been suggested for proximal tubular reabsorption in the mammalian nephron (Sato & Ullrich 1972). Fig. 4 shows the effect on isotopic Na$^+$ uptake when the gill is exposed to 10^{-6} M PLL (MW = 10^5) for 2 min. followed by 8 min. of washing in PLL-free artificial pond water. Influx during the next hour was reduced to about 40% of the control value, and there was no sign of reversal in a subsequent measurement.

Fig. 5 shows data for net Na$^+$ movement in the same experiment. The animal maintained a sodium steady state during the control period (hence $J_{in}^{Na} = J_{out}^{Na}$). During the first few minutes of the next flux measurement (after adding and removing PLL) there was a large loss of Na$^+$ to the medium. This amounted to 50–500 μeq in a number of experiments (76 in the experiment shown), but the kinetics and magnitude are still to be characterized completely because no measurements were made during the 2 min. period of PLL application or the wash that followed it. The rapid loss was followed by a persistent, slower net efflux such as always ensues when influx is inhibited and efflux remains unchanged (cf. Fig. 3 for a similar action of amiloride).

Inhibition was induced by polymers of 30, 70 and 100 × 10^3 M.W. but not by lysine even at high (10 mM) concentrations. However, the methyl ester of lysine was inhibitory at 10^{-3} M which is about the same concentration as the lysine residues in PLL at 10^{-6} M. In contrast to the action of PLL,

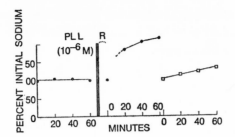

Fig. 5. Net Na⁺ movement before and after PLL. Same animal as in Fig. 4. Note the transient loss of a large quantity of Na⁺ from the animal (middle panel, first 15 minutes).

inhibition by lysine methyl ester did not induce a rapid net Na⁺ loss, and its action was completely reversible. In these regards it resembled amiloride more than PLL.

DISCUSSION AND SUMMARY

In intact fresh water animals both Na⁺ and Cl⁻ are actively transported inward. A TEP develops in amphibia and the open-circuit electrical behavior resembles that of skins *in vitro*. However, in other animals transporting NaCl at about the same rates (fish and crayfish) TEPs of the opposite polarities have been reported. The relationship between active ion transport and epithelial potentials is still not clear. Transport of each ion is independent of the other, and both can be absorbed from solutions of nonpenetrating counterions. Both J_{in}^{Na} and J_{in}^{Cl} show saturation kinetics with very low K_m values. Ion exchange is an obligatory part of the entry step; H⁺ efflux is coupled with Na⁺ uptake and HCO_3^- with Cl⁻. Intact animals do not lend themselves to study of the second step in translocation, transfer across the serosal surface into the blood. Studies on isolated preparations show that this step is active for Na⁺ and might involve the Na⁺ + K⁺-activated ATPase system (Skou 1969, Bonting 1970). Chloride transfer is probably passive at the inner membrane since there is a large increase in electrical potential across it.

Some of the inhibitor studies fill in details; others are suggestive of possible models. If Diamox® acts by inhibiting carbonic anhydrase, then catalytic hydration of metabolic CO_2 appears to provide H⁺ and HCO_3^- at rates commensurate with NaCl influx. PLL, lysine methyl ester and amiloride all inhibit J_{in}^{Na} when added to the external medium. All have more than one amino group, presumably ionized around pH 7, and this might indicate that they have a common mechanism.

One possibility is that a necessary first step in Na^+ transfer requires combination with anionic sites at the outer membrane. Such a step accounts for the saturation kinetics described above. Binding to anionic sites would explain the inhibition of sodium transport by polyamino compounds, and also by high NH_4^+ concentrations in the external medium (Shaw 1960b, Maetz & Garcia-Romeu 1964). It must be kept in mind that some characteristics of these inhibitors (e.g. reversibility) differ, and the correlation between polyamino structure and inhibitory action might be fortuitous. Nevertheless, the possibility of a common mechanism of action is worth exploring further.

Finally, the mechanism of sodium uptake in the animals studied appears to be remarkably uniform. All of them show saturation behavior with K_ms in the same order of magnitude. Those described here excrete H^+ in exchange for the Na^+ absorbed and transport is inhibited by amiloride.

ACKNOWLEDGEMENT

The research in my laboratory has been supported by grants GM 04254 and GM 01276 from the National Institute of General Medical Sciences.

REFERENCES

Biber, T. U. L. & Curran, P. F. (1970) Direct measurement of uptake of sodium at the outer surface of the frog skin. *J. gen. Physiol. 56*, 83–99.

Biber, T. U. L., Chez, R. A. & Curran, P. F. (1966) Na transport across frog skin at low external Na concentrations. *J. gen. Physiol. 49*, 1161–1176.

Bonting, S. (1970) In *Membranes and Ion Transport*, ed. Bittar, E. E., Vol. I. Wiley-Interscience, London.

Cereijido, M. & Curran, P. F. (1965) Intracellular electrical potentials in frog skin. *J. gen. Physiol. 48*, 543–557.

Chaisemartin, C. (1966) Gradients ioniques, potentiels bioélectriques et rôle de l'épithelium branchial chez *Austropotamobius pallipes* en état d'équilibre calcique (intermue). *C. R. Acad. Sci. (Paris) 160*, 1305.

deVooys, C. G. N. (1968) Formation and excretion of ammonia in Teleostei. I. Excretion of ammonia through the gills. *Arch. int. Physiol. Biochim. 76*, 268–273.

Dietz, T., Kirschner, L. B. & Porter, D. (1967) The roles of sodium transport and anion permeability in generating transepithelial potential differences in larval salamanders. *J. exp. Biol. 46*, 85–96·

Evans, D. H. (1969) Sodium, chloride and water balance of the intertidal teleost, *Pholis gunnellus. J. exp. Biol. 50*, 179–190.

Garcia Romeu, F., Salibian, A. & Pezzani-Hernandez, S. (1969) The nature of the *in vivo* sodium and chloride uptake mechanisms through the epithelium of the chilean frog, *Calyptocephalella gayi. J. gen. Physiol. 53*, 816–835.

Greenwald, L. (1971) Sodium balance in the leopard frog *(Rana pipiens). Physiol. Zool. 44*, 149–161.

Greenwald, L. (1972) Sodium balance in amphibians from different habitats. *Physiol. Zool. 45*, 229–237.

House, C. R. (1963) Osmotic regulation in the brackish water teleost, *Blennius pholis. J. exp. Biol. 40*, 87–104.

Kerstetter, T. H. & Kirschner, L. B. (1972) Active chloride transport by the gills of rainbow trout *(Salmo gairdneri). J. exp. Biol. 56*, 263–272.

Kerstetter, T. H., Kirschner, L. B. & Rafuse, D. D. (1970) On the mechanisms of sodium ion transport by the irrigated gills of rainbow trout *(Salmo gairdneri). J. gen. Physiol. 56*, 342–359.

Kirschner, L. B. (1955) On the mechanism of active sodium transport across the frog skin. *J. cell. comp. Physiol. 45*, 61–87.

Kirschner, L. B. (1970) The study of NaCl transport in aquatic animals. *Amer. Zool. 10*, 365–376.

Kristensen, P. (1972) Chloride transport across isolated frog skin. *Acta physiol. scand. 84*, 338–346.

Krogh, A. (1937) Osmotic regulation in fresh water fishes by active absorption of chloride ions. *Z. vergl. Physiol. 24*, 656–666.

Krogh, A·)1938) The active absorption of ions in some fresh water animals. *Z. vergl. Physiol. 25*, 335–350.

Maetz, I. Les échange de sodium chez le poisson *Carassius auratus*. Action d'un inhibiteur de l'anhydrase carbonique. *J. Physiol. (Paris)* 48, 1085–1099.

Maetz, J. (1971) Fish gills: mechanisms of salt transfer in fresh water and sea water. *Trans. roy. Soc. (Lond.) 262*, 209–249.

Maetz, J. (1972) Branchial sodium exchange and ammonia excretion in the goldfish Carassius auratus. Effects of ammonia-loading and temperature changes. *J. exp. Biol. 56*, 601–620.

Maetz, J. & Campanini, G. (1966) Potentiels transépithéliaux de la branchie d'Anguille *in vivo* en eau douce et en eau de mer. *J. Physiol. (Paris) 58*, 248.

Maetz, J. & Garcia Romeu, F. (1964) The mechanism of sodium and chloride uptake by the gills of a fresh water fish, *Carrassius auratus.* II. Evidence for NH_4^+/Na^+ and HCO_3^-/Cl^- exchanges. *J. gen. Physiol. 47*, 1209–1227.

Salako, L. A. & Smith, A. J. (1970a) Effects of amiloride on active sodium transport by the isolated frog skin: evidence concerning site of action. *Brit. J. Pharmacol. 38*, 702–718.

Salako, L. A. & Smith, A. J. (1970b) Changes in sodium pool and kinetics of sodium transport in frog skin produced by amiloride. *Brit. J. Pharmacol. 39*, 99–109.

Sato, K. & Ullrich, K. J. (1972) Effects of acid and basic polyelectrolytes on the rat proximal tubular isotonic reabsorption. *Pflügers Arch. ges Pysiol. 332*, R30.

Shaw, J. (1959) The absorption of sodium ions by the crayfish *Astacus pallipes*. I. The effect of external and internal sodium concentrations. *J. exp. Biol. 36*, 126–144.

Shaw, J. (1960a) The absorption of sodium ions by the crayfish *Actacus pallipes*. II. The effect of the external anion. *J. exp. Biol. 37*, 534–547.

Shaw, J. (1960b) The absorption of sodium ions by the crayfish *Astacus pallipes*. III. The effect of other cations in the external solution. *J. exp. Biol. 37*, 548–556.

Shaw, J. (1960c) The absorption of chloride ions by the crayfish *Astacus pallipes*. *J. exp. Biol. 37*, 557–572.

Skou, J. C. (1969) In *The Molecular Basis of Membrane Function*, ed. Tosteson, D. C. Prentice Hall, Englewood Cliffs, New Jersey.

Smith, P. G. (1969) The ionic relations of *Artemia salina*. II. Fluxes of sodium, chloride and water. *J. exp. Biol. 51*, 739–757.

Snell, F. M. & Leeman, C. P. (1957) Temperature coefficients of the sodium transport system of isolated frog skin. *Biochim. biophys. Acta (Amst.) 25*, 311–320.

Ussing, H. H. (1949a) The distinction by means of tracers between active transport and diffusion. *Acta physiol. scand. 19*, 43–56.

Ussing, H. H. (1949b) The active transport through the isolated frog skin in the light of tracer studies. *Acta physiol. scand. 17*, 1–37.

DISCUSSION

CURRAN: If sodium influx is completely balanced by an exchange for hydrogen ion efflux what balances the sodium efflux?

KIRSCHNER: Chloride, which is coming out of the animal. And in fact it is a complication for us, because then there is a back flux with a very low K_m. We have to work relatively rapidly because of the back flux of chloride if we let it build up.

SCHMIDT-NIELSEN: Were all the animals catheterized?

KIRSCHNER: The fish is in air and never sees water behind the gills which are irrigated by a closed recirculating system. Any urine produced drops out of the chamber without contacting the circulating medium.

SCHMIDT-NIELSEN: What about the frog?

KIRSCHNER: It is catherized. The crayfish has got a plug of dental cement in each nephropore.

SCHMIDT-NIELSEN: Were they salt-depleted before distilled water?

KIRSCHNER: Only the crayfish.

LEAF: If the sodium efflux is balanced by chloride what causes the chloride influx?

KIRSCHNER: Probably bicarbonate, though I don't think we can exclude hydroxyl ion on the basis of present data.

LINDEMANN: Would you like to elaborate on your idea concerning the K_m values in frog skin.

KIRSCHNER: The membrane potential at the mucosal border is very concentration-dependent in the range $1/10$ to 1 mM (Biber *et al.* 1966). I think the change was about 40 mV per 10-fold increase in $[Na^+]$, with the inside becoming more positive. Dr. Peter Smith has done a constant field analysis of a system in which both the flux and the membrane potential are concentration dependent. And under these conditions, because the membrane poten-

Biber, T. U. L., Chez, R. A. & Curran, P. F. (1966) Na transport across frog skin at low external Na concentrations. *J. Gen. Physiol. 49,* 1161–1176.

tial is more and more opposing entry, the system appears to saturate. I don't think we can exclude that possibility here (Smith 1969).

TOSTESON: One might expect that the effect of poly-L-lysine would be to reduce the sodium concentration at the membrane surface. If that were the case you might expect that there would be a change in the effective K_m, but not in the maximum sodium transport rate. Did you measure the K_m for sodium in the poly-L-lysine experiments?

KIRSCHNER: We have done just the experiments I showed you, plus a few with lower molecular weight polymers and with monomers. Neither Dr. Ullrich nor we could demonstrate inhibition by lysine up to very high concentrations. But it occurred to me that possibly the carboxyl group was interacting with the epsilon-amino group of another acid. It turns out that methyllysine, the methyl ester, is a good inhibitor. But we have done no other experiments.

TOSTESON: But were these experiments done at a sodium concentration high enough to saturate the system, or at a lower concentration?

KIRSCHNER: 1 mM, it is well over half maximum, say, 80 percent saturated.

KEYNES: You mentioned I think that choline doesn't substitute for sodium, what about lithium?

KIRSCHNER: We haven't used lithium on these animals.

KEYNES: It would be extremely interesting to know whether in this system lithium will substitute for sodium or will not.

KIRSCHNER: My guess is that it would. But poorly as in other fresh water preparations.

SACHS: Would you care to assign a role to the ATPases found in the gills in these transport systems?

KIRSCHNER: No. In fact I may be the only person in the world who is not convinced that the NaK ATPase plays any role other than to maintain cell sodium low and cell potassium high even in these preparations. I am not convinced that a direct role in Na transfer is dictated by available data.

Smith, P. G. (1969) The ionic relations of *Artemia salina* L. II. Fluxes of sodium chloride and water. *J. exp. Biol.* 51, 739–757.

SACHS: Have you thought through models on the bicarbonate ATPase in the gill?

KIRSCHNER: We have found the bicarbonate ATPase in the gill, and of course any imaginative person can make all kinds of models, starting with parallels to the NaK enzyme. I don't think it is worth spending time with them here.

MAETZ: I would like to state briefly that in recent experiments I was able to measure H^+ ion excretion by the gill of Carassius (1972). Both H^+ and NH^+_4 ions are used by the fish in exchange of Na^+, depending on the prevailing external conditions. At low external Na ammonia excretion occurs in the unionized NH_3 form. At high Na, in an alkaline medium, both H^+ and NH^+_4 are used for the exchange. At acidic pH (6.0 for example) part of the ammonia is excreted in the NH_3 form, while the remainder is used in the NH^+_4 form in exchange of Na^+ taken up at a much slower rate than in alkaline media.

ULLRICH: Did you measure the excretion of hydrogen ions during application of poly-L-lysine?

KIRSCHNER: No.

ULLRICH: Mamelak et al. (1969) have published a paper that said that poly-L-lysine when applied for a longer period of time to the apical surface of the toad bladder causes electronmicroscopic changes and lysis of many cells.

KIRSCHNER: Also in plant cells, so we have been careful to stay dilute and brief.

SCHULTZ: Are you saying that there is a forced sodium-potassium exchange, a hydrogen exchange and a forced chloride-bicarbonate exchange?

KIRSCHNER: Well, if you are ambiguous enough, as I insist on being about what we mean by 'forced' then yes. Otherwise you develop blue sparks.

J. Maetz (1972) Branchial sodium exchange and ammonia excretion in the goldfish, Carassius auratus. Effects of ammonia-loading and temperature changes. *J. Exp. Biol. 56,* 601–620.

Mamelak, M., Wissig, L. L., Bogoroch, R. & Edelman, I. S. (1969) Physiological and morphological effects af poly-L-lysine on the toad bladder. *J. Membrane Biol. 1,* 144–176.

Ion Transport and Corneal Transparency

J. A. Zadunaisky

The presence of ionic transport in the cornea is related to the preservation of hydration and transparency of this ocular tissue. The cornea consists of a thick layer of transparent connective tissue covered outside by a stratified epithelium bathed by the tear film. In the inside, in contact with the aqueous humor, the cornea is covered by a one layer thick cellular endothelium. Well developed junctional complexes and desmosomes are found in the epithelium, while the endothelial cells are separated by sinuous channels where discrete passage of colloidal particles has been observed. The stroma of the cornea is composed of layers of collagen fibers, embedded in a matrix of mucopolysaccharides. The stroma is always at a minimum of hydration and the mucopolysaccharides have a strong tendency to take up water. The cornea has the unusual property of being transparent to visible light. This is accomplished by the geometrical arrangement of the collagen fibers which has been compared to a perfect lattice. Its transparency has been explained in terms of the optical properties of such a lattice structure (Maurice 1957, Benedek 1971). Any disruption of the organization of the corneal stroma, due to the accumulation of fluid in the stroma, produces opacification of the cornea with a subsequent partial or total loss of sight.

The epithelium and the endothelium of the cornea exhibit transport functions which control the hydration of the stroma. In this paper are described the localization and characterization of ionic pumps and their role in the control of corneal hydration and transparency.

Ion pumps in the cornea

Before we can consider the pumps present in the epithelium and endothelium of the cornea, it is important to mention that we will refer to two species:

From the Departments of Ophthalmology & Visual Science and Physiology, Yale University, School of Medicine, New Haven, CT 06510 U.S.A.

»Transport Mechanisms in Epithelia«, Munksgaard, Copenhagen.

mammalian, the albino rabbit; and amphibian, the bullfrog. The general properties of both types of corneas are very similar from the point of view of the structure to the production of a potential difference and the capacity to show the temperature reversal phenomenon. However, there are differences between them. For instance, the dominant pump in the frog corneal epithelium is a chloride transport, and in the rabbit, a sodium pump. Therefore, we are going to consider the frog cornea and the rabbit cornea separately and a comparison between the two is provided throughout this paper.

a) The frog cornea (Rana catesbeiana)

The cornea of the bullfrog is a very stable and durable tissue amenable to the performance of Ussing's type experiments (Zadunaisky 1966). These corneas are thinner than the mammalian corneas; they measure about 100 μ in thickness. The electrical properties are fully manifested in the presence of both chloride and sodium in the bathing solutions, but the short circuit current can be completely accounted for by an active transport of chloride from the aqueous to the tear side (Zadunaisky 1966, Ploth & Hogben 1967). Sodium ions, however, were found to be essential for the transport of chloride ions (Zadunaisky 1966, 1969, 1972). A kinetic analysis of the activation by sodium of the chloride transport indicated that the effect was not competitive (Zadunaisky 1972). A small sodium transport in the opposite direction to the chloride pump has been reported (Candia & Askew 1968); however, this sodium pump is extremely difficult to evidence or activate.

Bromide can substitute for chloride efficiently in the maintenance of the short circuit current, while iodide cannot. Potassium ions in the endothelial side have a tendency to reduce the resting potential of the cornea. In the outside, if care is taken to utilize an impermeant anion, changes in potassium concentration have no definite effects on the electrical properties of the frog cornea (Davis & Zadunaisky 1967). The intraocular pressure apparently induces passage of solution across the cornea by bulk flow. Measurements of both sodium and chloride fluxes under the influence of a hydrostatic pressure, similar to the intraocular pressure, produced identical increase in the flux of sodium and chloride in the direction of the gradient (Zadunaisky et al. 1971).

The chloride transport is only partially dependent on respiration. Anoxia, or inhibition of the transport with blockers of the respiratory chain such as antimycin or amytal induced a reduction of 40% in the chloride transport, the remaining fraction of the chloride current was dependent on glycolytic processes. The study of the two main pathways for the oxidation of glucose indicated that the frog cornea metabolizes glucose through the pentose shunt at a significant rate. However, the Embden-Meyerhoff pathways were sensitive to changes in chloride concentration while the pentose-shunt was not affected (Zadunaisky et al. 1971).

During the study of the effect of metabolic inhibitors it was found that arsenite (10^{-4} to 10^{-3} N) induced a transient stimulation of the short circuit current before complete inhibition. The stimulation was due to an increase in the net chloride transport at the expense of the unidirectional flux from endothelium to epithelium. This activation occurred during anoxia and induced further accumulation of lactic acid than during anoxia alone, suggesting stimulation of a glycolytic source of energy (Zadunaisky et al. 1971, Zadunaisky & Lande 1971).

In order to determine if the chloride pump could control the hydration and scatter of light by the stroma, experiments were devised where the light transmitted through the cornea was measured simultaneously with the short circuit current or the radioisotopic flux. The device utilized consisted of a modified flux chamber with two windows at the ends through which light of a preset wavelength was collimated onto the corneal epithelium and detected at the endothelial side of the cornea. The preparation was bypassed with fiber optics in order to check the stability of the electronic unit utilized for the optical measurements. Changes in optical density of the corneas as they became hydrated were compared to their water content. In this manner it was found that after a gain of 28% in water the stroma of the frog cornea became rapidly opaque. In this region of the hydration-opacification curve the chloride transport was stimulated or inhibited with several agents.

These optical experiments demonstrated that stimulation of the chloride pump induces dehydration and reduces the light scattered by partially opaque corneas. Furthermore, the removal of chloride ions from the bathing solutions did not permit recovery of transparency and dehydration of partially opaque corneas. The chloride pump is located in the epithelium, and a cation such as sodium must accompany the movement of the anion. Then this transport of salt and water towards the tears can effectively oppose the tendency for hydration of the corneal stroma and also compensate for leaks occurring at the bordering membranes (Zadunaisky & Lande 1971).

The electrical profile of the frog cornea and especially the one of the epithelium has been studied (Candia *et al.* 1968, Akaike 1971). The main findings indicate that the stroma is positive with respect to the outside and that the potential of the basal layer of cells of the epithelium is greater than the one of the superficial cells. However, the actual electrical relationships between the epithelial cells and the most likely model for the location of resistances and pumps has not yet been completed. A more extensive revision of this point can be found elsewhere (Zadunaisky 1971, Klyce 1972).

The structural basis for transparency have been studied in the frog cornea and compared to the semitransparent nictitating membrane of the frog. It was found that fiber size and interspaces between the fibers were of critical importance for transparency, with a good lattice type arrangement of the fibers in the stroma of the nictitating membrane (Lande & Zadunaisky 1970). In the frog cornea, as much as in the rabbit and human cornea, the phenomenon of opacification is the consequence of the accumulation of fluid in the cornea. Two mechanisms can induce opacification. On one hand there is the disruption of the order of the optical structure provided by the collagen fibers of identical size which became scattering units as soon as the lattice was distorted (Maurice 1957). On the other hand, the small lakes or drops of water which have been observed in the electron microscope (Goldman *et al.* 1968) can reach a big enough size to become scatter centers themselves.

In any of these two processes the removal of water by ion transport by the epithelium apparently is the more rational explanation for the control of the transparency of the frog cornea. The function performed by the frog corneal endothelium in this process is not known.

b) The rabbit cornea

The rabbit cornea has been used very extensively as an experimental object (Maurice 1969). Rabbit corneas mounted in Ussing's type chambers produce a potential difference (outside negative) short circuit current (Donn *et al.* 1959, Green 1966, Klyce 1972). Most of this short circuit current is carried by a net transport of sodium from outside to inside, residing in the epithelium (Donn *et al.* 1959). There is, however, a small net chloride transport from aqueous to tear side that also resides in the epithelium (Klyce *et al.* 1972), as will be discussed later on.

The endothelium of the cornea, however, has been the subject of many studies, from measurements of permeability (Hedbys & Mishima 1969) to the detection of small potential differences (Fischbarg 1972, Barfort & Maurice 1972). The endothelium is one layer of flat cells with an intracellular compartment complex enough to suggest a very active cell. It is electrically silent, or it produces a potential difference in the order of a few hundred microvolts (Fischbarg 1972, Barfort & Maurice 1972). However, when isolated rabbit corneas are permitted to swell and become partially opaque, the endothelium is able to dehydrate the stroma in the absence of epithelium. This is true as long as the bare surface of the stroma on the epithelial side is not bathed by salt solution (Mishima *et al.* 1969, Dickstein & Maurice 1972).

The actual nature of the pump residing in the rabbit corneal endothelium is not precisely known. However, there is evidence to suggest that sodium ions might be involved, with pumping towards the anterior chamber from the stroma (Mishima *et al.* 1969, Lande & Zadunaisky unpublished). The absence of a sizable potential difference permits the assumption that a neutral ionic pump could well be the source of the osmotic work performed by the endothelium.

The active transport of the epithelium of the rabbit cornea has been studied again in the recent past (Green 1966) and the idea has been put forward that fixed charges in the corneal stroma are replenished by the inward sodium pump and that control of stromal hydration could be exerted by this mechanism (Green 1969). However, it is possible to think that in fact, the chloride pump activated by epinephrine (Klyce *et al.* 1972) could well explain the dehydration of partially opaque rabbit corneas by the epithelium alone.

The electrical profile of the epithelium of the rabbit cornea has received more attention than in the case of the frog (Kikkawa 1964, Ehlers 1970, Klyce & Zadunaisky 1970, Klyce 1972).

The findings indicate that two main levels of potential difference exist in the rabbit corneal epithelium. The superficial cells have a lower potential difference than the basal cells of the epithelium; this potential tends to be inversely proportional to the transepithelial potential. The resting potential of the germinative cells, on the other hand, is not correlated to the transepithelial potential of the epithelium (Klyce 1972). One possible interpretation of these findings is that the inward sodium transport is located in the more superficial cells, where most of the drop in electrical resistance also occurs in this epithelium.

The rabbit cornea, therefore, can control the hydration of the stroma by the fluid pump of the endothelium and apparently also by the ionic pumps located in the epithelium.

Regulation of corneal transport by epinephrine and Cyclic AMP

The actions of epinephrine on the eye are well known, especially the effect on pupillary size and the changes it can induce in intraocular pressure (Davson 1972). As much as in other tissues (Robison et al. 1971), cyclic AMP and adenylcyclase are involved in the action of epinephrine in the eye (Neufeld et al. 1972). Recent evidence also points to a function of cyclic AMP in photoreception (Miller et al. 1971).

The addition of epinephrine to the cornea of both frogs and rabbits produced a rapid and characteristic increase in short circuit current with little change in potential difference (Chalfie et al. 1972, Klyce et al. 1972). In both types of corneas the response was observed when epinephrine, theophylline or the dibutyryl derivative of cyclic AMP were added to the epithelial or the endothelial side. The rabbit cornea was more sensitive to these agents on the epithelial side than the endothelial side.

The stimulatory effect of epinephrine was diphasic, that is, the current tended to decrease slowly with time after stimulation. With the inhibitors of adenyl cyclase (aminophylline for instance) in both the frog and the rabbit cornea, the stimulation was more sustained and a smaller tendency to decrease was observed in the current. Cyclic AMP had a smaller effect most probably because of lack of penetration into the corneas. Dibutyryl cyclic AMP instead which is more penetrating (Robison et al. 1971) produced a maximal effect on the rabbit cornea and a good response in the frog.

Ionic fluxes were measured during stimulation with aminophylline in the frog cornea. Aminophylline was used because of the sustained activation it produces on the current of the frog cornea (Chalfie et al. 1972). Fluxes were also measured during activation by dibutyryl cyclic AMP in the rabbit cornea (Klyce et al. 1972). The results proved to be of interest and will be described separately for the frog and the rabbit corneas.

In the frog corneas, it must be recalled that a transport of chloride ions is present normally from aqueous to tear side. Under stimulation with aminophylline, the chloride transport increased proportionally to the increase in the short circuit current. In the frog cornea, then, the activation by compounds that tend to increase the tissue levels of cyclic AMP consists in a stimulation of the pre-existing chloride transport (Chalfie *et al.* 1972).

In the rabbit cornea, the short circuit current is carried mostly by sodium ions during resting conditions. During activation by dibutyryl cyclic AMP the increase in current was due to a net chloride transport from aqueous to tear side, in a direction opposite to the pre-existing sodium transport. The increase in current was similar to the increase in the net chloride transport towards the tears (Klyce *et al.* 1972).

It has to be pointed out that due to the levels of current measured in the isolated rabbit cornea during resting conditions, most of it appears to be carried by sodium ions alone. However, there is a definite but small chloride pump operating in the direction aqueous to tears side also during resting conditions. Therefore, agents that increase the levels of cyclic AMP in the rabbit cornea activate a chloride pump. The sodium flux in the direction tears to aqueous was not modified appreciably by the cyclic AMP related agents (Klyce *et al.* 1972).

The existence of a chloride pump in the mammalian cornea is of great interest to the understanding of the control of hydration by the epithelium. The epithelium alone is able to dehydrate partially opaque corneas in the absence of endothelium as long as salt solution does not become in contact with the bare endothelial surface of the stroma (Zadunaisky unpublished). It would also explain why the epithelium can dehydrate rabbit corneas and at the same time, changes in sodium concentration on the epithelial side do not affect the hydration of the stroma (Riley 1971).

In the diagram of Fig. 1 is shown the present number and location of ionic pumps in the frog and rabbit corneas. The epithelium is the site of a chloride pump outwards and a sodium pump inwards in both the amphibian and the mammalian. The level of performance of these pumps is different in the two species, and in the mammalian the chloride pump is well evidenced under stimulation with epinephrine. In the rabbit the endothelium induces water movement most probably by the transport of an ion pair. The endothelium of the frog cornea is not well known with respect to its function in the control of hydration of the stroma.

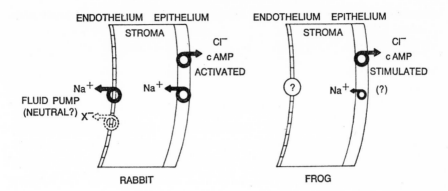

In summary, the types of ion transport in the cornea of the amphibian and the mammalian has been described and their relevance to the control of hydration and transparency of the cornea discussed at the light of recent evidence.

The work of the author and co-workers quoted here was supported by the National Eye Institute, Research Grant ≠00382, Fight For Sight, Inc. and The Seeing Eye Incorporated, U.S.A..

REFERENCES

Akaike, N. (1971) The origin of the basal cell potential frog corneal epithelium. *J. Physiol. 219,* 57–75.

Barfort, P., Maurice, D. M. (1972) Transport and electrical potential across the corneal endothelium. *Fed. Proc.* 298Abs.

Benedek, G. B. (1971) Theory of transparency of the eye. *Appl. Optics 10,* 459–473.

Candia, O. A., Askew, W. A. (1968) Active sodium transport in the isolated bullfrog cornea. *Biochim. biophys. Acta* (Amst.) *163,* 262–265.

Candia, O. A., Zadunaisky, J. A., Bajandas, F. (1968) Electrical profile of the isolated frog cornea. *Invest. Opthal. 7,* 405–415.

Chalfie, M., Neufled, A. H., Zadunaisky, J. A. (1972) Action of epinephrine and other cyclic AMP mediated agents on the chloride transport of the frog cornea. *Invest. Opthal. 11,* 644–650.

Davis, T. L., Jackson, J. W., Day, B. E., Shoemaker, R. L., Rehm, W. S. (1970) Potentials in the frog cornea and microelectrode artifact. *Amer. J. Physiol. 219,* 178–183.

Davis, T. L., Zadunaisky, J. A. (1967) Potassium permeability of the frog cornea. *Fed. Proc. 25,* 632.

Davson, H. (1972) *The physiology of the eye,* p.p. 391–394 & 42–52. Academic Press, New York.

Dickstein, S., Maurice, D. M. (1972) The metabolic basis to the fluid pump in the cornea. *J. Physiol. 221*, 29–41.

Donn, A., Maurice, D. M., Mills, N. L. (1959) Studies in the living cornea *in vitro* II: The active transport of sodium across the epithelium. *Arch. Ophtal. 62*, 741–747.

Ehlers, N. (1970) Intracellular of the corneal epithelium. *Acta. physiol. scand. 78.* 741–477.

Fischbarg, J. (1972) Electrically neutral fluid transport across the corneal endothelium. *Biophysical Soc.* 201 A.

Goldman, J. M., Benedek, G. B., Dohlman, C. H., Kravitt, B. (1968) Structural alterations affecting transparency in swollen human cornea. *Invest. Ophthal. 7*, 501–519.

Green, K. (1966) Ion transport across the isolated rabbit cornea. *Exp. Eye Res. 5*, 106–110.

Green, K. (1969) Relationship of ion and water transport to corneal swelling. In *The Cornea: Macromolecular organization of connective tissue*, ed. Langham, M., Johns Hopkins Press, Baltimore. pp. 35–67.

Hedbys, B. O., Mishima, S. (1969) The flow water across the corneal layers. In *The Cornea: Macromolecular organization of a connective tissue*, ed. Langham, M., Johns Hopkins Press, Baltimore. pp. 69–77.

Kikkawa, Y. (1964) The intracellular potential of the corneal epithelium. *Exp. Eye Res. 3*, 132–140.

Klyce, S. D., (1972) Electrical profiles in the corneal epithelium. *J. Physiol. 225*, In press.

Klyce, S. D., Zadunaisky, J. A. (1970) Microelectrode profile of rabbit corneal epithelium. *Biophys. J. 10*, 199a.

Klyce, S. D., Neufeld, A. H., Zadunaisky, J. A. (1973) The activation of chloride transport by epinephrine and Db cyclic-AMP in the cornea of the rabbit. *Invest. Opthal. 12*, 127–139.

Lande, M. A., Zadunaisky (1970) The structure and membrane properties of the frog's nictitans. *Invest. Ophthal. 9*, 477–491.

Maurice, D. M. (1957) The structure and transparency of the cornea. *J. Physiol. 136*, 263–286.

Maurice, D. M. (1969) The cornea and sclera. In *The Eye Vol. 1.*, ed. Davson, H., Academic Press, New York, London. pp. 489–600.

Maurice, D. M. (1972) The location of the fluid pump in the cornea. *J. Physiol. 221*, 43–54.

Miller, W. H., Gorman, R. E., Bitensky, M. W. (1971) Cyclic Adenosine monophosphate: Function in photoreceptors. *Science 174*, 295–297.

Mishima, S. G. K., Takahashi, G. H., Kudo, T., Trenbergh, S. M. (1969) The function of the corneal endothelium in the regulation of corneal hydration. In *The Cornea: Macromolecular organization of a connective tissue*, ed. Langham, M., pp. 207–235. John Hopkins Press, Baltimore.

Neufeld, A. H., Jampol, L. M., Sears, M. L. (1972) Cyclic AMP in the aqueous: The effects of adrenergic agents. *Exp. Eye Res.* 14, 242–250.

Ploth, D. W., Hogben, C. A. M. (1967) Ion transport by the isolated frog cornea. *Invest. Ophtal. 6*, 340–347.

Riley, M. V. (1971) The Role of the Epithelium and Control of Corneal Hydration. *Exp. Eye Res. 12*, 128–137.

Robison, A. G., Butcher, R. W., Sutherland, E. W. (1971) *Cyclic AMP*. Academic Press, New York.

Zadunaisky, J. A. (1966) Active transport of chloride in frog cornea. *Amer. J. Physiol. 221*, 506–512.

Zadunaisky, J. A. (1969) The active chloride transport of the frog cornea. *The Cornea: Macromolecular organization of a connective tissue,* ed. Langham, M., pp. 3–34. Johns Hopkins Press, Baltimore.

Zadunaisky, J. A. (1971) Electrophysiology and transparency of the cornea. In *Electrophysiology of Epithelial Cells,* ed. Giebisch, G., pp. 224–250. F. K. Schattauer Verlag, Stuttgart, New York.

Zadunaisky, J. A. (1972) Sodium activation of chloride transport in the frog cornea. *Biochim. biophys. Acta (Amst.) 292*, 255–257.

Zadunaisky, J. A., Lande, M. A. (1971) Active chloride transport and the control of corneal transparency. *Amer. J. Physiol. 221*, 1837–1844.

Zadunaisky, J. A., Lande, M. A. Hafner, J. (1971) Further studies of chloride transport in frog cornea. *Amer. J. Physiol. 211*, 1832–1836.

DISCUSSION

HOGBEN: For those of you who have not worked with this particular preparation, I would like you to appreciate that this is a very fragile tissue, and that it took great skill to conduct the studies that Dr. Zadunaisky has accomplished.

ZADUNAISKY: About a year after we published the paper describing the chloride pump in the frog cornea, Dr. Hogben also published some interesting data which confirmed our findings.

WHITTEMBURY: I am interested in what is the effect of temperature on these rates of pumping. Also have you tried whether ADH does something to this?

ZADUNAISKY: Temperature effects are very clear cut. In the classical "temperature reversal" experiment low temperature induces corneal swelling which is reversible on returning the temperature to normal. The ion transport of the epithelium is affected in a similar manner by temperature.

WHITTEMBURY: Did the epithelium swell also?

ZADUNAISKY: Not very much, in proportion to the swelling of the stroma. Most of the thickness increase, due to water accumulation, occurs in the stroma and not in the epithelium. Your other question referred to ADH. It does not increase the current of the isolated frog cornea. ADH tends to produce a very small reduction in the current, with an increase in the passive chloride flux from tears to aqueous, and on this basis one could interpret that it produces some change in the permeability of the epithelium.

SCHOFFENIELS: We have been doing quite a number of experiments on the rabbit cornea. We found that the hydrostatic pressure inside is very important. With equal pressure on both sides of the cornea, the potential difference is low, and as we increase the pressure up to the physiological value which is 28 cm of water, then we get the maximum potential difference. Now I am wondering if you get the same thing on the frog cornea.

ZADUNAISKY: First of all we do keep a small gradient of pressure because once the cornea wrinkles the potential tends to drop. So I think what is happening here is mostly not the pressure *per se*, but the unfolding of the epithelium that gives better potentials. We have measured the electrical properties, the current and the fluxes, with or without a hydrostatic pressure of the same value as the intraocular pressure. The pressure produced an increase in both the sodium and the chloride flux in the direction of the gradient. I think if you do some experiments measuring both things in the rabbit you will find exactly the same. It is just ultrafiltration.

SCHOFFENIELS: Another difference also is that the opacity of the cornea seems to be related to the metabolic integrity of the endothelium rather than that of the epithelium. If we realize a keratectomy, for instance, then we don't see any change in the opacity of the cornea, but if we scratch the endothelium we see that the opacity is spreading from the site of scratch.

ZADUNAISKY: This is very well known. The surgeons know that.

SCHOFFENIELS: But don't you get the same thing with the frog?

ZADUNAISKY: You do. Exactly the same. If you scrape the endothelium of the frog, the stroma swells, if it is in contact with a solution. But if you scrape the epithelium off the cornea, the stroma also swells.

SCHOFFENIELS: It does not in the rabbit.

ZADUNAISKY: It does if you incubate it with Ringer's solution. You have to wait.

SCHOFFENIELS: And what is the sodium content of the stroma in the frog. Is it as high as in the rabbit?

ZADUNAISKY: It is as high. The stroma is connective tissue and is equilibrated with the fluid in the anterior chamber. So it has about the same concentration of sodium. Except that when you take the stroma alone and analyze it carefully, there is an excess of sodium with respect to the aqueous humor.

ORLOFF: To which side do you add the theophylline and the cyclic AMP and why?

ZADUNAISKY: In both the frog and the rabbit we have put them on both sides, and they work on both sides. In the frog we have obtained the best actions from the inside and in the rabbit the best actions are from the epithelial side. I do not know why. There must be a question of penetration in the frog.

ORLOFF: That's strange. Is the hormone on the tear surface of the eye and if so how does it get there?

ZADUNAISKY: Not on the tear surface, but in the blood. The cornea contains cyclic AMP, and the blood contains epinephrine. Now the nutrition of the cornea occurs from fluid that diffuses from the anterior chamber but also from the vessels of the limbus. Around the limbus there are the vessels, and it is well known that solutes can diffuse into the cornea from the limbus. So the possibility exists that this is a regulatory mechanism for dehydration. And the fact that the rabbit has the two pumps with epinephrine makes it a more interesting story.

SACHS: How does the blood know that the cornea requires to be hydrated or dehydrated?

ZADUNAISKY: The ionic concentration of the stroma could trigger it. How does this work? One of the ideas we have in this problem is that there must be some feed-back between the stroma and the pumps. The concentration in the stroma, the hydration of the stroma, and the pumps. Of course, this is just a possibility.

SACHS: There are probably two other possibilities: that there is some communication between the two sides of the cornea which is cyclic AMP mediated or hormone mediated. Or simply, that there may be a required response of the cornea to change the oxygenation of the aqueous humor.

ZADUNAISKY: On application of epinephrine on the intact eye and after a delay that looks like diffusion through a thickness like the one of the cornea, there is a rapid increase of cyclic AMP in the anterior chamber. Now, whether this has physiological significance or not we don't know, but it changes the intraocular pressure.

CRABBÉ: When you hydrate your stroma, say, by lowering the tonicity of the fluid, do you also get opacification of your cornea? On the other hand, could you stimulate the transport of chloride when your epinephrine is incubated in hypotonic fluid the way sodium transport is stimulated in the case of amphibian epithelia?

ZADUNAISKY: I really don't know. We haven't done that. We know that it swells more rapidly in water than in Ringer's solution.

CRABBÉ: For the cornea of mammals, inhibition of your stroma also results in opacification, doesn't it?

ZADUNAISKY: This is a general phenomenon.

CRABBÉ: And you have to have a 20–30 percent increase in hydration to get this opacification?

ZADUNAISKY: Yes.

CRABBÉ: There are some clinical conditions associated with drastic falls in the tonicity of the "milieu interieur". My question is: do such patients get opaque corneas?

ZADUNAISKY: No, not really, not usually. They first get a lower intraocular pressure. And it is more complicated, because the intraocular pressure depends very much on blood pressure and osmotic regulation. People with glaucoma, when they have high pressure they see halos, and the halos are accompanied by the pain they get in an acute glaucoma. Glaucoma is an increase in the pressure inside the eye when the channel of movement of fluid is blocked or secretion is augmented, and the cornea can be opaque under those conditions. Another condition is when the endothelium is drastically touched, or when there are burns or when there is a deep scratch in the stroma.

SACHS: The cyclic AMP levels that you measured, were these done on whole cornea or were they done on separated endo- and epithelium?

ZADUNAISKY: They were done on the whole cornea, because we couldn't get enough material from the epithelium alone to do the analysis. And I think that is reason why the values are low and the stroma probably doesn't have so much cyclic AMP.

SACHS: It would seem to me very important to try to get enough tissue, because if the chloride pump is cyclic AMP dependent and is present in epithelium you would expect that probably there would be adenyl cyclase in that tissue. That would be one model you would have, so that you would get changes in the cyclic AMP there rather than in the endothelium.

ZADUNAISKY: Some of that is being done now in that direction with the epithelium, and it is just a question of getting enough epithelium from frogs and rabbits.

Electrophysiological Study of Ionic Transports in Pancreatic and Salivary Acinar Cells

O. H. Petersen

At present the most direct way of studying ionic transport processes in a single cell type from complex tissues like the salivary glands and the pancreas, is to measure membrane potentials. The acinar cells are the most easily studied in both these tissues, since they make up the bulk of the parenchyma.

The main task of the salivary acinar cells is to secrete fluid and a number of organic substances, of which the most important probably is the α-amylase (Schneyer *et al.* 1972). The pancreatic acinar cells secrete a number of enzymes (DeReuck & Cameron 1962), but are probably not playing an important role in HCO_3^- and H_2O transport (Schulz *et al.* 1969).

Both the salivary and the pancreatic acinar cells have a dual stimulatory system. The salivary cells are stimulated by acetylcholine (ACh) and noradrenaline (NA), whereas the pancreatic cells are stimulated by ACh and pancreozymin. In the salivary glands ACh is a much more powerful stimulant of fluid transport than is NA or adrenaline whereas the latter is a more powerful stimulant of amylase secretion. ACh and pancreozymin are both capable of stimulating pancreatic enzyme release (DeReuck & Cameron).

The electrophysiology of the salivary acinar cells has already been the subject of several studies (Lundberg 1958, Petersen 1971a, b), whereas the first successful results of measurement of membrane potentials in the exocrine pancreas have not been described until recently (Dean & Matthews 1972).

In the following some new data on the effect of varying the extracellular

Institute of Medical Physiology C, University of Copenhagen, Denmark.
This work was carried out during tenure of a Wellcome-Carlsberg Travelling Research Fellowship at the Department of Pharmacology, University of Cambridge. Additional support came from H. and I. Weimann's legacy.

»Transport Mechanisms in Epithelia«, Munksgaard, Copenhagen.

concentrations of some important ions on the membrane potentials in the resting and stimulated state will be presented and discussed in relation to previous findings.

MOUSE SALIVARY GLANDS

Segments of submandibular or parotid glands from young female mice were mounted in a perspex tissue bath through which a Krebs-Henseleit solution warmed to 37° C was pumped at a constant rate. Membrane potentials were measured, with the help of high resistance (100–200 MΩ) potassium citrate-filled micro-electrodes from cells lying immediately beneath an exposed and decapsulated surface area of the gland.

The resting membrane potential

The resting membrane potential in the submandibular gland ranged widely from -20 to -70 mV with a mean of -47.0 mV (n = 105) (Fig. 1). In the parotid gland the mean value was -61.0 mV (52–70). The wide range of submandibular membrane potentials might indicate that several cell types had been studied, but this is, however, probably not the case.

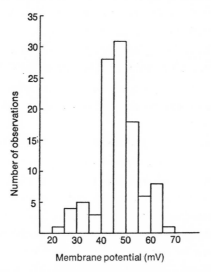

Fig. 1. Frequency distribution of membrane potentials recorded from surface cells of mouse submandibular glands (Petersen, unpublished observations).

It has generally been assumed that low resting potentials are derived from acinar cells, whereas higher potentials can be obtained from the granulated cells and still higher from the striated duct cells (Schneyer *et al.* 1972). The fact that all measurements in the present study were derived from surface cells makes it highly unlikely that striated duct cells have made any quantitatively important contribution to the material presented in Fig. 1.

In the rat submandibular gland it has been suggested that resting potentials above -35 mV are mainly derived from granulated tubule cells whereas those that are lower are from the acinar cells (Schneyer & Schneyer 1965b). Since the acinar cells are more numerous than the granulated tubule cells it is easily inferred from the results of Fig. 1 that in this case the explanation of Schneyer does not apply.

Since a large majority of the membrane potentials were above -40 mV it is clear that the mean acinar cell membrane potential is much higher than the results from previous studies on rat and cat salivary glands have indicated (Petersen 1971b). This conclusion is also in agreement with very recent results obtained from the cat submandibular gland (Nishiyama & Kagayama 1973).

The fact that the mean resting potential in the parotid gland, where there are no granulated tubules, is even higher than in the submandibular gland makes it unreasonable to argue that the high mean membrane potential from the submandibular gland could be explained by a significant contribution from the granulated tubule cells.

It has been pointed out that generally cell membranes from non-excitable tissues are less influenced by variations in extracellular potassium concentration ($[K]_0$) than those from nerve and muscle tissues (Williams 1970). This finding has been confirmed also for the salivary glands (Yoshimura & Imai 1967, Petersen 1970b).

In the mouse submandibular gland the slope of the linear curve relating membrane potential to log $[K]_0$ was 45 mV/10-fold increase in $[K]_0$ when $[K]_0$ was above 20 mM. This value is far higher than anyone previously reported for salivary gland cells and is of course consistent with the higher resting potentials obtained in the present study.

Still the dependence of the potential on $[K]_0$ is less than that expected for a K-selective electrode and it is likely that the hypothesis of Williams (1970) of a higher pNa/pK ratio in non-excitable than in excitable tissue also applies to the mouse submandibular gland cells.

The effect of ACh

Addition of a single dose of ACh directly to the tissue bath to achieve for a short period a concentration of 5.5×10^{-6} M, always evoked a change in the membrane potential. In most cells a short-lasting small depolarization or hyperpolarization of about 5 mV was followed by a long-lasting hyperpolarization (Fig. 2A). In cells with relatively low resting potentials short-lasting hyperpolarizations of a relatively large amplitude were followed by long-lasting hyperpolarizations (Fig. 2B).

When the dependence of the initial phase of the secretory potential on the resting potential (RP) was looked at more closely, a regular pattern emerged (Fig. 3), showing that the amplitude and the polarity of the initial ACh-induced potential change was very dependent on RP. The reversal potential appeared to be about -50 mV.

The delayed long-lasting hyperpolarization may depend somewhat on RP, although these results do not allow a definite conclusion. Preliminary results indicate that the initial phase of the secretory potential is very sensitive to variations in $[K]_o$. Thus at $[K]_o = 1$ mM, RP was not different from that

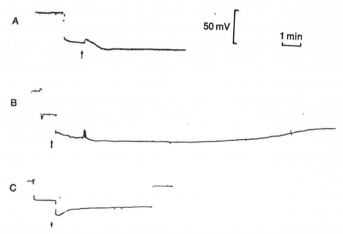

Fig. 2. Mouse submaxillary gland: Membrane potential measurement. Downward deflections represent increased negativity of the micro-electrode. The sudden jump in potential seen in the left part of all three tracings corresponds in time to the insertion of the micro-electrode into the acinus. The arrows denote additions of ACh to the tissue bath to obtain for a short period a maximum concentration of 10^{-6} g/ml (5.5×10^{-6} M). A and B were obtained during exposure of the gland to a normal Krebs-Henseleit solution while C was obtained during exposure to Strophanthin-G (10^{-3} M)-containing solution. B and C are from the same preparation. (Petersen 1973).

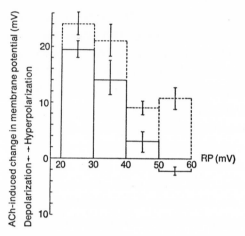

Fig. 3. Mouse submaxillary gland: Histogram showing the amplitude and polarity of the ACh-induced potential changes at different resting membrane potentials (RP). The fully drawn columns represent the first phase of the secretory potential while the columns outlined by the broken lines represent the second or delayed phase of the secretory potential. (Petersen 1973).

measured at $[K]_o = 4.7$ mM. But at this lowered $[K]_o$ large hyperpolarizations of about 20 mV were seen, rather than the small de- or hyperpolarizations usually seen at this RP; at $[K]_o = 20$ mM, where the mean value of RP was about -37 mV, small depolarizations were always seen. It is thus seen that in the range $[K]_o = 1$ to 20 mM, where RP is little influenced by variations in $[K]_o$, ACh increases markedly the sensitivity of the cell membrane to variations in $[K]_o$.

This finding is in complete accord with a previous finding of Petersen (1970b) in which it was shown that the amplitude of the hyperpolarizing secretory potentials from the cat submandibular acinar cells was very much influenced by varying $[K]_o$ between 0 and 20 mM. The findings that the initial phase of the secretory potential is very dependent on RP and also on $[K]_o$ strongly suggest that ACh increases the membrane permeability to K.

Assuming that $[K]_i$ in the mouse submandibular glands is similar to that of the rat submandibular gland, the K equilibrium potential (E_K) across the acinar cell membrane must be about -90 mV (Schneyer & Schneyer 1965a). Extrapolating the curve relating membrane potential to log $[K]_o$ to a membrane potential of O gives a value for $[K]_o$ of about 130 mEq/l; assuming that this is the true $[K]_i$, a value for E_K of -80 mV is obtained.

Anyhow, there is no doubt that the reversal potential for the initial phase of the secretory potential (−50 mV) is different from E_K, indicating that ACh must also increase the permeability to an ion other than K. As has been suggested earlier this ion is probably Na (Petersen 1970b, 1971a, b).

Thus the present data, although rather different from previous data (Petersen 1971a, b), in that RP is considerably higher and that ACh evokes biphasic potentials, confirm the hypothesis of Petersen (1971b): the main action of ACh is to increase the membrane permeability to K and Na with subsequent K efflux and Na influx.

The same conclusion based on results which are in many ways similar to those described here has recently been reached by Nishiyama & Kagayama (1973). The importance of these currents in evoking salt and water transport has been discussed several times (Petersen 1971a, b, 1972 a, b).

The delayed and long-lasting hyperpolarization has never been reported before, as previously it has not been possible to maintain the membrane potential from a salivary gland cell during intra-cellular recording for a long period after stimulation with ACh. During exposure of the tissue to Strophanthin-G (10^{-3} M), the delayed hyperpolarization was abolished (Fig. 2c); RP was lowered explaining that ACh always evoked hyperpolarization in these cases. This reaction might indicate that the delayed hyperpolarization was caused by some electrogenic ion pump.

On the other hand preliminary observations showed that the delayed hyperpolarization was rather sensitive to variations in $[K]_o$. This finding might indicate a delayed increase in K permeability causing this hyperpolarization. However, if one assumes that an electrogenic Na pump would be stimulated by the ACh induced Na-influx (Thomas 1972), it is conceivable that the strength of this inflowing Na current would depend on $[K]_o$, in such a manner that an increase in $[K]_o$ would reduce ACh-induced Na-influx and *vice versa* (Petersen 1971a). This phenomenon, therefore, might explain the dependence of the delayed hyperpolarization on $[K]_o$ assuming that it is caused by an electrogenic Na pump.

Further evidence in favour of the pump theory is that during exposure of the tissue to a Na-free tetraethylammonium solution delayed hyperpolarization was not seen. In the acinar cells of the salivary glands there are propably two kinds of Na pumps (Petersen 1971c) as is also the case in the proximal kidney tubule (Whittembury 1971): a Na-K exchange pump and a Na pump transporting Na, Cl and H_2O transcellularly. In the necturus proximal kidney tubule, Whittembury (1971) has shown that it is the latter pump that is electrogenic.

While the finding that Strophanthin-G abolishes the delayed hyperpolarization might primarily indicate an importance of the Na-K pump for this potential change, it must be noted that in the cat salivary glands both pumps can be inhibited by Strophanthin-G, although the Na-K pump is more sensitive to this drug than the solute pump (Petersen 1970a, 1971c). Since we do not yet have any information on the sensitivity of these two presumably independent pumps in the mouse submandibular acini, it is not possible to pursue this question any further at the moment.

MOUSE PANCREAS

Membrane potentials were measured in this tissue with the same methods as those described in the section on the salivary glands.

The resting membrane potential

The resting membrane potential was found to be -39.2 mV $\pm$ 0.5 (n = 66) which is very close to the value given by Dean & Matthews (1972). Above $[K]_o$ of 10 mM, there was a linear relation between membrane potential and log $[K]_o$ with a slope of 28 mV/10-fold increase in $[K]_o$. Strophantin-G (10^{-3} M) slowly depolarized the acinar cells (about 10 mV/hr) (Matthews & Petersen 1972). These results indicate that the concentration gradient for K across the cell membrane is mainly responsible for the existence of the resting potential.

The effect of ACh and pancreozymin

Fig. 4 shows examples of the depolarizing effect of both ACh and pancreozymin. Dose-response curves for the depolarizing effect of these two stimulants of protein secretion were parallel. The maximal amplitude of the stimulation-induced depolarization was in both cases about 15 mV, in close agreement with values for the depolarization caused by acetyl-β-methylcholin, electrical stimulation of the pancreatic nerves and continuous exposure to pancreozymin (Dean & Matthews 1972). The depolarizing effect of ACh

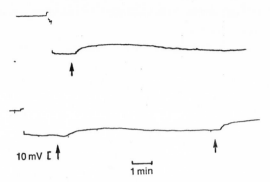

Fig. 4. Two tracings of membrane potential recordings from mouse pancreatic acinar cells. A deflection downwards represents an increased negativity of the microelectrode. The sudden jump in potential seen in the left part of both tracings corresponds in time to the insertion of the microelectrode into the acinus. The arrow in the upper tracing denotes addition of ACh to the bath to obtain a concentration of 10^{-5} g/ml (5.5×10^{-5} M), whereas in the lower tracing the arrows denote addition of CCK-Pz to obtain concentrations in the bath of 150 mU/ml (Crick-Harper-Raper units). (Petersen & Matthews 1972).

but not that of pancreozymin was abolished by atropine (Petersen & Matthews 1972).

Although salt and water secretion from the exocrine pancreas is very sensitive to reductions in extracellular HCO_3 concentration (Case *et al.* 1970; Schulz *et al.* 1971), the amplitude of the ACh-induced depolarization was found to be uninfluenced by replacing the superfusion fluid CO_2/HCO_3 buffer by a phosphate or TRIS buffer. The resting potentials were, however, somewhat reduced (5 mV) by this procedure.

Reducing the resting membrane potential from the normal −40 mV to about −20 mV by exposing the tissue to Strophantin-G (10^{-3} M), 2,4-dinitrophenol (10^{-4} M), $[K]_o = 20$ mM or amiloride (10^{-4} M) did not significantly reduce the amplitude of the ACh-induced depolarization. However reducing $[Na]_o$, which also reduced the resting potential (in a Na-free solution RP was about −24 mV), reduced the amplitude of the ACh-induced depolarization, the magnitude of the reduction being dependent on $[Na]_o$. However, even during exposure to a Na-free solution ACh still evoked small depolarizing potentials with an amplitude of about 3–4 mV (Fig. 5) (Matthews & Petersen 1972). Although these results suggest that the ACh-induced depolarization is carried mainly by Na^+ they also show that at least under some experimental conditions, some other ion can carry this current.

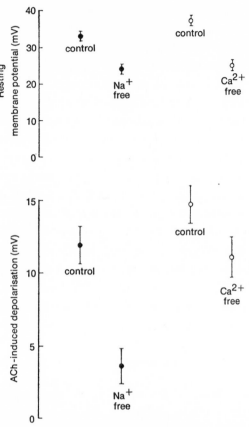

Fig. 5. Pancreatic acinar cells: Effect of omitting Ca^{2+} or Na^+ from the superfusion fluid on the resting membrane potential and the amplitude of the ACh-induced depolarization (Matthews & Petersen, unpublished observations).

As seen in Fig. 5 the amplitude of the ACh-induced depolarization is slightly reduced during superfusion of the tissue with a Ca-free solution. Furthermore such substances, which are known from other tissues to reduce or abolish Ca currents, as D-600(α-Isopropyl-α-[(N-methyl-N-homoveratryl)-γ-aminopropyl]3, 4, 5-trimethoxy-phenylacetonitril) (10^{-4} M), tetracain (10^{-3} M) or La^{3+} (5 mM) severely reduced the amplitude of the ACh-induced depolarization in the pancreatic acinar cells (Matthews & Petersen, in preparation). Therefore, although Na influx in response to ACh or pancreozymin stimulation appears to account for most of the depolarization, there may also be an important contribution from Ca.

In the salivary acinar cells, ACh evokes an increase in Na and K permeability. Since the main action of ACh on the pancreatic acinar cell membrane is to increase Na conductance, it might be of interest to determine whether an increase in K conductance also occurs.

In a series of experiments the amplitude of the ACh-induced depolarization obtained during exposure to a control solution ($[K]_o = 4.7$ mM) was compared with that seen during exposure to a solution having a K concentration of 20 mM. Although the mean value was enhanced in the latter case, the difference was not statistically significant.

Nevertheless, in individual experiments, it was clearly seen that the depolarization in response to ACh was more marked in the periods of superfusion with a solution containing a high extracellular K concentration than in the control periods (Matthews & Petersen, in preparation). It would appear, therefore, that ACh evokes an enhanced membrane permeability to commonly occurring ions, but that the increase in Na conductance is the quantitatively important process accounting for the depolarization, although the ACh-induced Ca-influx may be the crucial step in stimulus-secretion coupling (Douglas 1968, Rubin 1970).

CONCLUSION

Available evidence from the present study on mouse salivary glands and pancreas and from previous studies on the cat salivary glands (Petersen 1972b) indicates that ACh in the acinar cells of both glands evokes an increase in membrane permeability to at least Na and K. In the pancreas the increase in Na permeability dominates, explaining why the resulting potential change is always a depolarization, whereas, in the salivary glands the permeability increase for both Na and K appears to produce Na and K fluxes of about equal magnitude, which explains why the result may be a depolarization or a hyperpolarization.

REFERENCES

Case, R. M., Scratcherd, T. & Wynne, R. D'A. (1970) The origin and secretion of pancreatic juice bicarbonate. *J. Physiol. (Lond.) 210,* 1–15.

Dean, P. M. & Matthews, E. K. (1972) Pancreatic acinar cells: Measurement of membrane potential and miniature depolarization potentials· *J. Physiol. (Lond.) 225,* 1–13.

De Reuck, A. S. V. & Cameron, M. P. (eds.) (1962) *The Exocrine Pancreas,* Churchill, London.

Douglas, W. W. (1968) Stimulus-secretion coupling: the concept and clues from chromaffin and other cells. *Brit. J. Pharmacol. 34*, 451–474.

Lundberg, A. (1958) Electrophysiology of salivary glands. *Physiol. Rev. 38*, 21–39.

Matthews, E. K. & Petersen, O. H. (1972) The ionic dependence of the membrane potential and acetylcholine-induced depolarization of pancreatic acinar cells. *J. Physiol. (Lond.) 226*, 98–99 P.

Nishiyama, A. & Kagayama, M. (1973) Biphasic secretory potentials in cat and rabbit submaxilary glands. *Experientia (Basel), 29*, 161–163.

Petersen, O. H. (1970a) Some factors influencing stimulation-induced release of potassium from the cat submandibular gland to fluid perfused through the gland. *J. Physiol. (Lond.) 208*, 431–447.

Petersen, O. H. (1970b) The dependence of the transmembrane salivary secretory potential on the external potassium and sodium concentration. *J. Physiol. (Lond.) 210*, 205–215.

Petersen, O. H. (1971a) The ionic transports involved in the acetylcholine-induced change in membrane potential in acinar cells from salivary glands and their importance in the salivary secretion process. In *Electrophysiology of Epithelial Cells*, ed. Giebisch, G., Schattauer Verlag, Stuttgart, New York, pp. 207–221.

Petersen, O. H. (1971b) Initiation of salt and water transport in mammalian salivary glands by acetylcholine. *Phil. Trans. B 262*, 307–314.

Petersen, O. H. (1971c) Formation of saliva and potassium transport in the perfused cat submandibular gland. *J. Physiol. (Lond.) 216*, 129–142.

Petersen, O. H. (1972a) Electrolyte transports involved in the formation of saliva. In *Oral Physiology*, ed. Emmelin, N. & Zotterman, Y., Pergamon Press, Oxford, pp. 21–31.

Petersen, O. H. (1972b) Acetylcholine-induced ion transports involved in the formation of saliva. *Acta physiol. scand.* suppl. 381, 1–58.

Petersen, O. H. (1973) Membrane potential measurement in mouse salivary gland cells. *Experientia (Basel), 29*, 160–161.

Petersen, O. H. & Matthews, E. K. (1972) The effect of pancreozymin and acetylcholine on the membrane potential of pancreatic acinar cells. *Experientia (Basel), 28*, 1037–1038.

Rubin, R. P. (1970) The role of calcium in the release of neurotransmitter substances and hormones. *Pharmacol. Rev. 22*, 389–428.

Schneyer, L. H. & Schneyer, C. A. (1965a) Salivary secretion in the rat after ouabain. *Amer. J. Physiol. 209*, 111–118.

Schneyer, L. H. & Schneyer, C. A. (1965b) Membrane potentials of salivary gland cells of rat. *Amer. J. Physiol. 209*, 1304–1310.

Schneyer, L. H., Young, J. A. & Schneyer, C. H. (1972) Salivary secretion of electrolytes. *Physiol. Rev. 52*, 720–777.

Schulz, I., Yamagata, A. & Weske, M. (1969) Micropuncture studies on the pancreas of the rabbit. *Pflügers Arch. ges. Physiol. 308*, 277–290.

Schulz, I., Ströver, F. & Ullrich, K. J. (1971) Lipid soluble weak organic acid buffers as "substrate" for pancreatic secretion. *Pflügers Arch. ges. Physiol. 323*, 121–140.

Thomas, R. C. (1972) Electrogenic sodium pump in nerve and muscle cells. *Physiol. Rev. 52*, 563–594.

Whittembury, G. (1971) Relationship between sodium extrusion and electrical potentials in kidney cells. In *Electrophysiology of Epithelial Cells,* ed. Giebisch, G., Schattauer Verlag, Stuttgart, New York, pp. 153–178.

Williams, J. A. (1970) Origin of transmembrane potentials in non excitable cells. *J. theor. Biol . 28,* 287–296.

Yoshimura, H. & Imai, Y. (1967) Studies on the secretory potential of acinar cell of dog's submaxillary gland and the ionic dependency of it. *Jap. J. Physiol. 17,* 280–293.

DISCUSSION

KEYNES: Is there any hope of getting two electrodes into these cells?
PETERSEN: No, these cells are relatively small and we are in fact happy to get *one* electrode into the cell.

KEYNES: You could not get a double-barrel electrode in?

PETERSEN: Maybe. Of course this has been done in the cat salivary glands by Lundberg (1957a), although the results may be open to criticism (Petersen 1972). But one could try that, obviously.

SCHULTZ: Is it possible to drive the electrode through the acinar cell into the lumen and get some estimate of the transepithelial potential difference?

PETERSEN: This would obviously be of great interest. The only gland in which this has been done is the cat sublingual gland in the work of Lundberg (1957b). There the lumen is relatively large. In the cells that I have talked about today the lumen is very small. When you look at it through the dissecting microscope you cannot see the lumen at all. So you could never be really sure whether the tip of the electrode is free in the lumen or not. You would have to blow up the lumen by applying some back pressure, which would be possible with a whole gland preparation. This cannot be done in this kind of preparation that I have used here.

SCHULTZ: Yes, but what is the electrical potential difference between the duct and the blood. One would expect that that should not be very different from the luminal value.

PETERSEN: In the cat sublingual gland, again I refer to Dr. Lundberg's studies – you can compare the secretory potential measured with the microelectrode inside the cell with that measured when the microelectrode is in the acinar lumen and there is very little difference, which means that there is little change in the luminal membrane potential. This is the only case where this has been done.

Lundberg, A. (1957a) The mechanism of establishment of secretory potentials in sublinqual gland cells. *Acta physiol. scand. 40*, 21–34.
Petersen, O. H. (1972) Acetylcholine-induced ion transports involved in the formation of saliva. *Acta physiol. scand.* Suppl. 381, pp. 1–58.
Lundberg, A. (1957b) Secretory potentials in the sublinqual gland cells. *Acta physiol. scand. 40*, 35–58.

ULLRICH: Did you apply cyclic AMP, and do you think that cyclic AMP acts as the second messenger for pancreozymin?

PETERSEN: We have tried, yes. Of course there is much uncertainty about the effect of cyclic AMP on protein secretion from the acinar cells but we applied both cyclic AMP and dibutyryl-cyclic AMP in a concentration of 1 mM to the bathing solution and we saw no effect on the pancreatic membrane potentials. We have not done this in the salivary glands, only in the pancreas.

LEAF: Would it be fair to ask you to speculate on how depolarization of the cell membrane causes a secretion?

PETERSEN: I think that maybe the important thing here is the calcium influx rather than the sodium influx becauce our data indicate that both of these ions may be involved in the depolarization. There is much evidence for an involvement of calcium in exocytosis processes. And this is the sort of process that goes on in the pancreatic acinar cells. But the detailed mechanism of action of calcium is entirely unknown.

LEAF: It does seem to be a rather general phenomenon. Insulin is secreted as a result of depolarization of the islet cells.

PETERSEN: In the islet cells the notion is that it is calcium mainly, that carries the current underlying the glucose-induced action potential (Dean & Matthews 1970) and that calcium triggers off the secretion of insulin. But of course, again there the detailed mechanism of what is happening inside the cell is entirely unknown.

Dean, P. M. & Matthews, E. K. (1970) Electrical activity in pancreatic islet cells: Effect of ions. *J. Physiol. (Lond.) 210,* 265–275.

The Role of the Terminal Bars for Transepithelial Current Flow Through Necturus Gallbladder Epithelium

E. Frömter

INTRODUCTION

The terminal bars constitute one part of the permeability barrier of epithelia. They are interposed between neighbouring cells and hold the cells together so that macroscopically a uniform "tight" membrane is formed. However, whether the terminal bars are indeed tight in a microscopic sense, or whether they permit exchange of matter between the mucosal and serosal fluid compartments, was not clear. Electronmicroscopic studies (Farquhar & Palade 1963) had shown that the terminal bars consisted essentially of cell adhesions in which neighbouring cell membranes seemed to fuse. They had also shown that the terminal bars were impermeable to large tracer molecules. Both observations were generally interpreted as meaning that the terminal bars were tight and consequently the name "tight junction" was widely used as a synonym in the literature. Recently however, a number of observations was made in frog skin (Ussing & Windhager 1964), proximal kidney tubules (Windhager *et al.* 1967, Hoshi & Sakai 1967, Boulpaep 1971, Frömter *et al.* 1971), gallbladder (Barry *et al.* 1971), and intestine (Clarkson 1967, Frizzel & Schultz 1972), which suggested the presence of low conductance parallel pathways in these epithelia and it has been speculated whether such shunt pathways may lead through the terminal bars. We have tried to answer the question of the shunt paths and their location with the help of microelectrode techniques.

METHODOLOGY

Necturus gallbladder was chosen for the experiments. It was opened and exposed as a flat sheet, mucosal surface upside, in a modified Ussing chamber, providing separate fluid compartments on the mucosal and serosal surface.

Max-Planck Institut fur Biophysik, 6 Frankfurt a. M., Germany.

»Transport Mechanisms in Epithelia«, Munksgaard, Copenhagen.

Necturus gallbladder epithelium is strictly single layered and flat; it does not exhibit villi, folds or crypts. It consists of large uniform cells of cuboidal shape (Fig. 1*) which can be impaled with microelectrodes from the mucosal surface and maintain potential differences of 40 to 70 mV, cell interior negative, against the mucosal Ringer's solution.

The experimental approach was as follows: In a first series of experiments we determined the resistance of the luminal and basal cell membrane as well as the resistance of the overall epithelium to current flow in transepithelial direction. By comparing these values, we were able to calculate the resistance of the paracellular shunt path. In a second series of experiments we made an attempt to determine the location of the shunt path, by exploring the electric field that builds up under transepithelial current flow at the mucosal surface of the epithelium.

RESULTS

Determination of the cell membrane resistances

To determine the cell membrane resistances it would seem appropriate to place a microelectrode into a cell, to pass current from the cell into both external fluid compartments simultaneously and to measure the current induced voltage drop across the mucosal or serosal cell membrane. By dividing voltage through current, the input resistance of an individual cell (R_z) should be obtained. Since the current should flow across the mucosal and serosal cell membrane the respective membrane resistances R_m and R_b should be given by

$$\frac{1}{R_z} = \frac{1}{R_m} + \frac{1}{R_b} \tag{1}$$

Furthermore, if we pass current across the epithelium and measure the current induced voltage drop across the mucosal and serosal cell membrane ($\triangle V_m$ and $\triangle V_b$) separately by means of an intracellular microelectrode, the ratio of these voltages should be equal to the ratio of R_m over R_b:

$$\frac{\triangle V_m}{\triangle V_b} = \frac{R_m}{R_b} \tag{2}$$

Hence by determining R_z and $\triangle V_m/\triangle V_b$ we should be able to calculate the individual cell membrane resistances.

* see figure insert opposite p. 496.

Application of this principle to gallbladder epithelium turned out to be complicated by the fact that epithelial cells are electrically coupled (Loewenstein 1966) and form some kind of "electrical syncytium". In this case it is no longer sufficient to pass current into one cell (hereafter named current cell) and to determine the voltage drop across its membrane, since a great deal of the current will be lost into neighbouring cells so that the input resistance will be much smaller than the input resistance of an isolated cell of otherwise identical properties. To properly determine R_z in an electrically coupled cell system, we have to correct for the current loss into the neighbouring cells. This can be done by measuring the voltage attenuation from cell to cell and applying a proper analysis (Fig. 2).

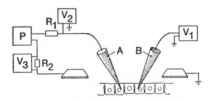

Fig. 2: Experimental set up for determination of voltage attenuation in gallbladder epithelium. Microelectrode A served for passing current pulses from the pulse generator P into a cell; microelectrode B recorded the potential change in neighbouring cells. $V_1 - V_3$ are electrometers for recording potential difference and current (across R_2).

Since voltage attenuation was practically uniform in all directions of the epithelial plane, the theoretical analysis was relatively simple. As published elsewhere (Frömter 1972) the theory predicted a zero order modified Bessel function of second kind for the voltage $\triangle V$ as a function of x, the radial distance from the current cell, under dc conditions:

$$\triangle V = A I_o \left(\frac{x}{\lambda}\right) \tag{3}$$

In this equation I_o denotes the Bessel function and A and λ are constants from which R_z can be obtained according to equation

$$R_z = \frac{2\pi A \lambda^2}{i_o} \tag{4}$$

where i_o is the total current applied.

Fig. 3 shows the result from a typical experiment. Current pulses were passed into the cell located at x = 0 and the voltage change ($\triangle V$) was measu-

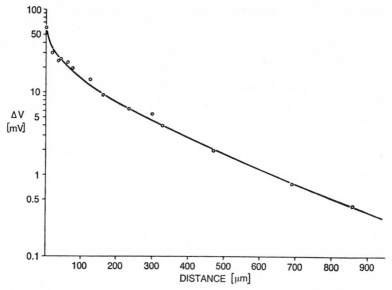

Fig. 3: Voltage attenuation in gallbladder epithelium. Abscissa: distance between current and voltage cell in μm. Ordinate: current induced voltage change in logarithnic scale in mV.

red in a number of cells located at varying distances (x) from the current cell. Current strength was 5×10^{-8}A. At this current strength the current voltage relation was still approximately linear. The pulse duration was 800 msec., thus approaching dc conditions. As tested in a series of pilot experiments, this time was sufficient for the voltage deflections to reach a constant plateau level even over cell to cell distances of > 1 mm.

The solid line in Fig. 3 shows the Bessel function, which has been fitted to the data according to the method of Shiba (1971). The fit is remarkably good. From a total of nine such experiments in 6 gallbladders mean values of the constants A = 8.9 mV and λ = 440 μm were obtained. From these data an average value of R_z = 1700 Ωcm² was calculated according to equation 4. This value may be conceived as the input resistance of a hypothetical isolated flat gallbladder cell of 1 cm² area into which the current is injected through a flat square intracellular electrode of same size.

In the same experiments the ratio $\triangle V_m / \triangle V_b$ was also determined. The mean value was $\triangle V_m / \triangle V_b$ = 1.8. Taking R_z and $\triangle V_m / \triangle V_b$ from each individual experiment, the individual values for R_m and R_b were calculated from equations (1) and (2). They yield an average of R_m = 4500 Ωcm² for the specific resistance of the mucosal cell membrane and an average of R_b = 2900 Ωcm^2 for the specific resistance of the basal cell membrane.*) These values can be used for comparison with the transepithelial resistance. If we pass current through the epithelium and *if we assume* that all current flows through the cells, the measurements would predict a transepithelial resistance of $R_m + R_b$ = 4500 + 2900 = 7400 Ωcm^2.

Comparison of cell membrane resistance and transepithelial resistance

In the same six gallbladders which had been used in the foregoing experiments, also the transepithelial resistance was measured by passing square wave current pulses through the epithelium and recording the transepithelial voltage drop. Since the voltage response to dc currents of greater than 5 μA/cm^2 exhibited already typical polarization effects, either low current densities were used in the experiments, or the resistance was calculated from the instantaneous voltage response, measured 18 msec. after onset of current, when the membrane capacitance was fully charged. Both techniques yielded identical results. The average resistance for transepithelial current flow was R_t = 310 Ωcm^2 in the six gallbladders. This value is approximately 25 times smaller than the resistance for current flow through the cells which was estimated above to be 7400 Ωcm^2. Since it is unlikely that the difference can be attributed to technical errors, we must assume that the high conductance of the epithelium is achieved by some unknown paracellular channels along which current can cross the epithelium without flowing through the cells. An equivalent diagram which illustrates this situation is shown in Fig. 4. From Kirchhoff's law the resistance of the shunt path (R_s) can be calculated according to

$$\frac{1}{R_t} = \frac{1}{R_m + R_b} + \frac{1}{R_s} \tag{5}$$

to be R_s = 330 Ωcm^2. This result means that approximately 96% of the transepithelial current through *Necturus* gallbladder flow through a paracellular shunt and only 4% flow through the cells.

---

* It should be noticed that these values do not refer to the actual surface of the mucosal or serosal cell membranes but rather to the entire area of the respective membrane surfaces contained in 1 cm^2 of epithelium.

bladder, gastric mucosa and salivary ducts, the terminal bars seem to be comparatively tight. Under physiologic conditions the latter epithelia have many characteristics in common which point to a *serial* array of two or more membrane barriers, rather than to the *parallel* array of membrane elements which has been demonstrated by the present experiments in *Necturus* gallbladder.

REFERENCES

Barry, P. H., Diamond, J. M. & Wright, E. M. (1971) The mechanisms of cation permeation in rabbit gallbladder. Dilution potentials and biionic potentials. *J. Membr. Biol. 4,* 358–394.

Boulpaep, E. (1971) Electrophysiological properties of the proximal tubule: Importance of cellular and intercellular pathways. In *Electrophysiology of Epithelial Cells,* ed. Giebish G., pp. 91–112. F. K. Schattauer Verlag, Stuttgart.

Clarkson, T. W. (1967) The transport of salt and water across isolated rat ileum. Evidence for at least two distinct pathways. *J. Gen. Physiol. 50,* 695–728.

Dobson, J. G. jr· & Kidder HI, G. W. (1968) Edge damage effect in vitro frog skin preparations. *Amer. J. Physiol. 214,* 719–724.

Farquhar, M. G. & Palade, G. E. (1963) Junctional complexes in various epithelia· *J. Cell Biol. 17;* 375–412.

Frizzel, R. A. & Schultz, S. G. (1972) Ionic conductance of extracellular shunt pathway in rabbit ileum. Influence of shunt on transmural sodium transport and electrical potential differences. *J. Gen. Physiol. 59,* 318–346.

Frömter, E. (1972) The route of passive ion movement through the epithelium of *Necturus* gallbladder. *J. Membr. Biol. 8,* 259–301.

Frömter, E. & Diamond, ,. M. (1972) Route of passive ion permeation in epithelia. *Nature New Biol. 235,* 9–13.

Frömter, E., Müller, C. W. & Wick, T. (1971) Permeability properties of the proximal tubular epithelium of the rat kidney studied with electrophysiological methods. In *Electrophysiology of Epithelial Cells,* ed. Giebisch, G., pp. 119–146. F. K. Schattauer Verlag, Stuttgart.

Hoshi, T. & Sakai, F. (1967). A comparison of the electrical resistances of the surface cell membrane and cellular wall in the proximal tubule of the Newt kidney. *Jap. J. Physiol. 17,* 627–637.

Loewenstein, W· R. (1966) Permeability of membrane junctions. *Ann. N. Y. Acad. Sci. 137,* Pt. 2, 441–372.

Pockrandt-Hemstedt, H., unpublished.

Shiba, H. (1971) Haevisides "Bessel cable" as an electric model for flat simple epithelial cells with low resistive junctional membranes. *J. Theoret. Biol. 30,* 59–68.

Ussing, H. H. & Windhager, E. E. (1964) Nature of shunt path and active sodium transport path through frog skin epithelium. *Acta physiol. scand. 61,* 484–504.

Windhager, E. E., Boulpaep, E. L. & Giebisch, G. (1967) Electrophysiological studies on single nephrons. *Proc. 3rd Int. Congr. Nephrol. Wash. 1966, 1,* 35. Karger, Basel, New York.

DISCUSSION

GIEBISCH: I wonder whether Dr. Frömter had an opportunity to study the ionic selectivity properties of the two membranes.

FRÖMTER: Which two membranes do you mean? I think we have to consider at least three.

GIEBISCH: I was referring to the luminal and the antiluminal.

FRÖMTER: In leaky epithelia like gallbladder and proximal tubule the exact determination of selectivity properties of cell membranes is rather compli-cated because we are dealing with two parallel channels. During ion substitu-tion in the external compartments for example both channels will generate diffusion potentials which will be shunted through circular current flow. Hence the potential change which can be measured across a single cell membrane does not reflect the concentration potential relation of the individual cell membrane properly but includes also at least two contributions from I–R drops. Or in other words, contrary to common practice, such measurements do not allow to calculate transference numbers directly be-cause the condition I = O is not fulfilled for the individual cell membrane. For these reasons we have thus far concentrated on resistance measurements only and have postponed all plans to measure potential changes during ion substitution which could only yield »qualitative« information at best.

KEYNES: Ever since the time when Dr. Diamond did his original experiments on gall bladder, for which I was supposed to be his research supervisor, I have been pressing him as to whether his argument that the sodium pump in gall bladder has to be tightly coupled to chloride movements at the cell membrane level, is really the only possible one. I have wondered all along whether there might not be an electrical coupling between sodium and chloride at the cell membrane level, and then some kind of anatomical explanation as to why you don't actually record that coupling potential, when you look at the potential between the bulk phases. Which hypothesis do you think that your very nice results support?

FRÖMTER: I would not say that my results necessarily support the second hypothesis and they certainly do not disprove Diamond's concept. However, they can provide an experimental basis for the second hypothesis in demon-strating the existence of two parallel pathways. In addition they suggest that the absence of a large potential difference across leaky epithelia may be first of all due to the paracellular shunt. To prove 1:1 coupling of cation and anion transport would therefore require more reliable potential measurements during symmetrical anion and cation substitutions than those which have been published by Diamond (1968) thus far. Measuring the potential difference more carefully Machen & Diamond (1969) have rather observed that gall bladder generates a significant transport potential (lumen positive) in identical Ringer's solutions and we have obtained a similar result in rat proximal tubule recently. These findings seem to contradict the electroneutral salt pump concept. On the other hand, however, Dr. Schultz told me yesterday, that his collaborators had found evidence for a 1:1 coupled entry of Na and Cl through the brush border membrane into intestinal cells.

SCHULTZ: Just to address myself to this point: Dr. Frizzell in my laboratory has been measuring the unidirectional influx of sodium and chloride across the brush border of rabbit gall bladder using the techniques that we have employed in intestine. There appears to be no doubt that there is a coupling of sodium and chloride influx across the brush border and that removing sodium from the mucosal solution inhibits chloride influx. Removing chloride from the mucosal solution inhibits sodium influx. I think even in the early papers of Dr. Diamond where non-permeating anions were used and no potential difference was developed, the possibility that a very leaky membrane was responsible for the zero PD was essentially ruled out.

SACHS: But there is a PD now, isn't there?

SCHULTZ: Yes, but it is a very small PD ranging between 0.2–0.5 mV.

FRÖMTER: The significance of a small potential difference remains questionable of course, since from the type of black-box experiments, which trans-epithelial potential measurements are, we cannot even decide whether the potential is generated across a membrane or only in some adjacent unstirred layer compartment, as for example in the lateral intercellular space. On the other hand, I want to mention that active Na-transport and electrically driven

Diamond, J. M. (1968) Transport mechanisms in the gallbladder. In: *Handbook of Physiology*, Alimentary Canal, vol. 5, pp. 2451–2482, Washington D. C.

Machen, T. E. & Diamond, J. M. (1969) An estimate of the salt concentration in the lateral intercellular spaces of rabbit gall-bladder during maximal fluid transport. *J. Membrane Biol. 1*, 194–213.

chloride absorption is not the only alternative to Diamond's tightly coupled electroneutral NaCl pump. In collaboration with Ullrich and Sauer we have recently found (Ullrich *et al.* 1972) that in rat proximal tubule only 40 percent of net sodium transport is active and that the remaining 60 percent as well as practically all of the chloride is transferred passively, mostly through solvent drag. Since the active part of sodium absorption has about the same magnitude as the entirely active HCO_3 absorption, it is conceivable that the tubule does not generate much of a potential difference in pure identical salt solutions. Such a mechanism could probably also account for most of Diamond's potential data with salt substitutions. In contrast to his concept, however, it would not require tight stoichiometric 1:1 coupling. That active sodium transport is inded independent of active bicarbonate or H^+ transport in rat proximal tubule can be demonstrated for example with glucose, which stimulates active Na transport and makes the tubular lumen negative.

KEYNES: The point that I was trying to make to Dr. Diamond was that you have got an additional factor which could attenuate a real coupling potential. If you locate the pumping at the extreme inner end of the intercellular canal, then you have an electrotonic attenuation of that potential between that point and the outside. Combined with the effect of the intercellular shunt this could reduce the recordable coupling potential to a very small value.

WINDHAGER: Is there any way of assigning a regulatory role for net transport to the shunt in the gall bladder?

FRÖMTER: I have not done any experiments on this line. According to our measurements approximately 50 percent of the shunt resistance was located in the lateral intercellular spaces. From this we can predict that any factor which changes the space width would thereby affect the membrane properties and eventually also transport. I would not dare to speculate, however, about a possible physiologic role of such a mechanism, the more so, since it is not clear whether the lateral spaces contribute a significant amount to the gall

Ullrich, K. J., Sauer, F. & Frömter, E. (1972) Transport parameters for sodium, chloride, and bicarbonate in the proximal tubules of the rat kidney. In: *Recent Advances in Renal Physiology,* ed. Wirz, H. & Spinelli, F., pp. 2–13, S. Karger, Basel.

bladder resistance *in vivo*. The data in Smulders *et al.* (1972) speak rather against this possibility. On the other hand, whether the terminal bars themselves can act as regulatory devices, I do not know.

STEN-KNUDSEN: It appears to me that the formula presented by you to describe the membrane polarisation as a function of the distance from the current injecting electrode is similar to that which can be derived if it is assumed that the internal specific resistance of the region between the two surface layers has the same value everywhere. That condition does not seem to apply to your situation. Therefore, are your calculations based upon the assumption of having a constant internal resistance or does it turn out that the inhomogenity does only give rise to second order effects.

FRÖMTER: I think I have applied essentially the same analysis which you mention, but I might have forgotten in the talk to make it clear enough that all resistance values refer only to 1 cm² of epithelium rather than to the unit area of the cell membranes proper. You will agree that the plane double membrane model can still be applied if the two membranes are folded provided that i) folding is uniform in all parts of the upper and/or lower membrane, ii) that the repetition period of the folding pattern is very small compared to the space constant of horizontal voltage spread in the sheet and iii) that we do not forget that the specific membrane resistance and core resistivity which can be calculated by applying the plane sheet model, are now lumped values which refer to units of the macroscopic system. Analogously you will agree that the plane homogeneous model can still be applied, when the interior consits of a regular pattern of identical units of heterogeneous substructure (like cells) provided that the same restrictions i to iii are again fulfilled especially that the subunit space is small compared to the space constant. This condition is obviously met in gallbladder, since the space constant was 15 to 37 times larger than the cell diameter. By the way the fact that the cell membrane resistances R_m and R_b are obtained per 1 cm² of epithelium is not disturbing since this is the dimension which we need for comparison with the transepithelial resistance, in order to determine the shunt. Furthermore, it does not matter that the application of the plane

Smulders, A. P., Tormey, J. M. & Wright, E. M. (1972) The effect of osmotically induced water flows on the permeability and ultrastructure of the rabbit gallbladder. *J. Membrane Biol. 7*, 164–197.

sheet model to gallbladder implies the use of a lumped value for the horizontal conductivity or resistivity (R_x in the paper by Frömter 1972), since
the calculations do not require to insert a figure or an estimate for the
horizontal resistivity.

A real problem, however, might have arisen, if the resistance for current
flow from the lateral spaces into the interstitium would have been of the
same order of magnitude as the cell membrane resistance. In that case the
labyrinth of lateral spaces would have contributed to horizontal current
spread and the figure for the basal cell membrane resistance would have
been obscured. This possibility could be excluded, however, by experimental
variation of the space width through application of transepithelial osmotic
gradients. Maximal dilatation of the spaces which reduces their resistance to
values of a few Ohms per cm^2 of epithelium, as can be calculated from the
dimension and the resistivity of interstitial fluid, did not change the results
significantly.

OSCHMAN: Earlier Dr. Leaf and his colleagues proposed 'limiting junction'
as a new name for the tight junction or zonula occludens. I would like to
very gently object to Dr. Frömter's suggestion that we should revert to the
term 'terminal bar' because we now know that the terminal bar of light microscopy in fact consists of three different sorts of junctions, the zonula
occludens, the zonula adherens, and the macula adherens. It is not correct
to use the terminal bar to describe a single part of the junctional complex. I
would approve of the term 'limiting junction' because this term is consistent
with our present view that these junctions can be permeable to ions while
still being impermeable to larger species, such as proteins.

Frömter, E. (1972) The route of passive ion movement through epithelium of *Necturus*
gallbladder. *J. Membrane Biol. 8*, 259–301.

Comparative Aspects of Transport through Epithelia

R. D. Keynes

Due to space limitations, I shall restrict myself to an attempt to summarise in a table what is known – and what is not known – about the different types of pump that must exist at the level of the cell membrane for the transport of monovalent ions. To avoid confusion, I shall enumerate them with the Roman numerals that I used in my review (Keynes 1969). The list in Table I is obviously very far indeed from complete. There are many other ions of biological importance, and many types of living organism whose ion transport systems have yet to be properly examined. Perhaps the widest gap, at any rate in my own knowledge, concerns the mechanisms of ion pumping in plant cells.

In view of the similarity in chemical properties between sodium and potassium, there cannot be very many alternative ways open to Nature of discriminating between the alkali metal ions; and one would surely expect to find that the plant and animal kingdoms have both exploited all of the possibilities. The situation in plants has been authoritatively summarised by MacRobbie (1970), but although it is clear from her account that the ouabain-sensitive Pump I does occur, there is insufficient evidence to identify the other Pumps with any confidence. Nevertheless, there certainly are in plants processes for K^+/H^+ exchange, $K^+ + HCO_3^-$ uptake, and $K^+ + Cl^-$ uptake that are not ATP-dependent and which could have something in common with Pumps II–V. This is an area in which further comparative studies are badly needed.

I will confine myself to commenting on only a few of the points raised in Table I:

Agricultural Research Council Institute of Animal Physiology, Babraham, Cambridge, England.

»Transport Mechanisms in Epithelia«, Munksgaard, Copenhagen.

Table I. The principal types of pump for monovalent ions.

Pump No.	Ion(s) transported	Tissue	Electro-genicity	Inhibitors	Hormonal control	Specificity	Energy source
I	[1]Na^+ (K^+)	All animal cells, many plant cells, many secretory epithelia	[1]Variable	[2]Cardiac glycosides	Aldosterone	[3]$Na \gg Li$	ATP
IIa	[4]Na^+	Outward-facing epithelia in fresh water organisms e.g. frog skin, gills of fish and crustacea, anal papillae in insect larvae	?	[5]Amiloride	Pitressin	?	[6]? concn. gradient of counterion
IIb	Cl^-			?	?	?	
III	Cl^-	[7]Oxyntic cells, chloride cells in fish gill, kidney tubule, intestinal mucosa	[8]Yes	[9]Diamox	?	$Cl^- \gg SO_4^{2-}$	? concn. gradient of HCO_3^-
IV	H^+	Oxyntic cells, *Dolium* salivary gland, mollusc skin glands	[8]Yes	Thiocyanate	Histamine	?	[11]? ATP
V	K^+	[10]Insect Malpighian tubule and rectal glands, *Cecropia* midgut, stria vascularis	Yes	? ouabain has no effect	?	$K = Rb$ $K \gg Na$	[12]Not ATP

1. There is now plenty of evidence (Thomas 1972) that Na, K-ATPase can operate either in an electrogenic mode in which Na^+ ions are extruded on their own in such a way that the pump hyperpolarizes the membrane, or in a coupled mode in which there is an electrically neutral exchange of one Na^+ for one K^+ ion. The active transport process may therefore involve Na^+ ions alone, moving outwards, or K^+ ions moving inwards as well. In frog muscle (Cross *et al.* 1965) it is clear that the tightness of coupling, and hence what I have termed the 'electrogenicity', varies with such factors as the concentration of sodium inside the cell. This variability is likely to be a quite general phenomenon.

2. It is difficult to underemphasize the importance of Schatzmann's (1953) discovery of the inhibitory action of the cardiac glycosides on the sodium pump. It is vital to almost all experiments concerned with the mechanism of Na, K-ATPase, whether at the molecular or at higher levels, to have access to a really specific inhibitor. The principal reason why so much more attention has been paid to Pump I than to the others on my list, is the lack of inhibitors known to block them with comparable specificity.

3. Perhaps the most striking difference between the selectivity of the sodium pump and that of the channels for downhill passage of Na^+ ions across nerve and muscle membranes is that the pump transports Li at least ten times more slowly than Na, whereas the electric excitability channels discriminate hardly at all between the two ions. I have the feeling that this difference in specificity for uphill and downhill transport has not been exploited as much as it might have been as a tool for investigating the nature of the mechanisms involved. It would also be interesting to have more information about the selectivity of Pump IIa.

4. As we have heard from Kirschner (1973) and others, there is good evidence that although this type of pump may normally transport Na^+ and Cl^- ions inwards in roughly equal quantities, it is possible, by selective ionic depletion, to set up a situation in which Na^+ is being transported inwards while Cl^- moves outwards, and *vice versa*. Hence the cationic and anionic sections of the pumping mechanism can operate more or less independently, and should be classified as IIa and IIb respectively.

5. In describing amiloride as the specific inhibitor for Pump IIa, and pitressin – acting via cyclic AMP – as its controlling hormone, I am tacitly assuming that the pathway for passive entry of NaCl into the epithelial cells of frog skin exposed on the outside to Ringer's solution can be identified

with the pathway for active uptake of NaCl from a low concentration. This assumption could do with some verification.

6. There seems no doubt that the operation of Pump IIa involves an exchange between Na^+ ions and H^+ or sometimes NH_4^+ ions, while in IIb Cl^- is exchanged for HCO_3^- (or OH^-). The possibility should therefore be considered that the energy source in each case is simply the concentration gradient for the counter-ion. Since the limiting external salt concentration from which NaCl can be taken up is 10 μM or even less (Keynes 1969), this would require the pH of the cell interior to be some 4 units lower than the external pH, perhaps an implausible proposition. But in the absence of alternatives it should not be ruled out.

7. The chloride pump seems likely to me to be more ubiquitous than is sometimes supposed, and many other tissues could easily be added to the list, such as the squid giant axon (Keynes 1963) and the rumen epithelium (Stevens 1973). An important question in this connexion is the extent to which, in tissues like the proximal tubule of the kidney, chloride traverses the epithelium through an extracellular pathway. As has been pointed out by Giebisch (1961) and others, the intracellular [Cl] is too high for chloride to be able to enter the tubule cells passively down the electrochemical gradient, so that any transcellular chloride transport necessitates the intervention of an active transport mechanism. Chloride may, of course, pass mainly through the junctional shunt pathways between the cells whose properties we have just been considering; but in this case, the ionic selectivity of the shunts becomes an important issue. For it can be argued that while anion-selective shunts would constitute a simple and efficient means for coupling together the fluxes of Na^+ and Cl^-, any appreciable cation permeability would merely result in wasteful leakage.

8. There is good evidence from studies of gastric secretion that Pump III is electrogenic, because the mucosal side is electrically negative. However, Pump IV must also be regarded as electrogenic, since the secretory potential can be reversed by simultaneously blocking the transport of Cl^- and stimulating that of H^+ (Rehm et al. 1963). Although in short-circuited rumen epithelium the net flux of chloride from rumen to blood is nearly as large as the net sodium flux (Ferreira et al. 1972), the transepithelial potential seems here to be sodium rather than chloride-dependent, and we have not yet discovered any way of unmasking the putative contribution from the chloride pump.

9. Although carbonic anhydrase inhibitors like Diamox (acetazolamide) seem to have a fairly specific blocking action on chloride transport, the precise role of carbonic anhydrase itself in anion transport remains a very open and intriguing question. It is now known (Carter & Parsons 1971) that there are two isoenzymes, one of which is catalytically much more active than the other; but their distribution cannot be correlated at all convincingly with the ion transporting capabilities of the various tissues in which they are found. One would also like to know something about the association between carbonic anhydrase and cell membranes.

10. There is a very active K^+ pumping system in *Cecropia* midgut (Harvey 1973, Zerahn 1973), which seems to have a lot in common with the K^+ pump in other insect tissues like the Malpighian tubule and rectal gland. It is a matter for speculation whether Pump V occurs in other animals. One vertebrate tissue in which the presence of an electrogenic K^+ pump has been postulated (Keynes 1969) is the stria vascularis of the mammalian cochlea. However, there is no doubt that Pump I also plays a part here, and as Johnstone & Sellick (1972) have explained, Pump III may be involved as well.

11. Can the membrane-bound, HCO_3^--stimulated, SCN^--inhibited ATPase found in gastric mucosa and pancreatic tissue (Kasbekar & Durbin 1965, Simon *et al.* 1972) be identified with Pump IV? It now seems that it can be distinguished from mitochondrial ATPase; but since mitochondria are quite likely to possess a proton pump, the objection that Kasbekar & Durbin's preparations might have been contaminated by mitochondrial ATPase has never appeared to me to be a very conclusive one.

12. There is no doubt that interruption of oxidative metabolism very rapidly brings Pump V to a standstill, and the obvious explanation would be that it involves a K^+-activated ATPase. However, unpublished attempts by Ellory & Wood in my laboratory to locate such an ATPase in a mitochondrion-free membrane preparation from *Cecropia* midgut gave uniformly negative results. Other nucleosides were not split either, nor did the energy source seem to be NADH or phosphoenolpyruvate; arginine phosphate was not tested. Berridge tells me that he has similarly failed to find any K^+-activated ATPase in insect salivary glands or Malpighian tubules. If the K^+ pump is coupled neither to the oxidation of pyridine nucleotide nor to a supply of energy-rich phosphate bonds, what alternative can be suggested? While I cannot produce a specific answer, there are two

unique structural features of these insect tissues that may well have functional significance. One is the remarkably close association between the surface membrane and the mitochondria, which are long thin bodies actually inserted into each of the microvilli at the apex of the cell (Berridge & Oschman 1969). The other is the presence in K^+-secreting insect tissues, but apparently nowhere else, of a coat of closely packed particulate subunits covering the cytoplasmic surface of the apical membranes (Gupta & Berridge 1966). Perhaps these participate in some kind of direct linkage between electron transport in the mitochondrion and potassium transport across the cell membrane.

REFERENCES

Berridge, M. J. & Oschman, J. L. (1969) A structural basis for fluid secretion by Malpighian tubules. *Tissue & Cell 1*, 247–272.

Carter, M. J. & Parsons, D. S. (1971) The isoenzymes of carbonic anhydrase: tissue, subcellular distribution and functional significance, with particular reference to the intestinal tract. *J. Physiol. (Lond.) 215*, 71–94.

Cross, S. B., Keynes, R. D. & Rybova, R. (1965) The coupling of sodium efflux and potassium influx in frog muscle. *J. Physiol. (Lond.) 181*, 865–880.

Ellory, C. J. & Wood, J. L. Unpublished findings.

Ferreira, H. G., Harrison, F. A., Keynes, R. D. & Zurich, L. (1972) Ion transport across an isolated preparation of sheep rumen epithelium. *J. Physiol. (Lond.) 222*, 77–93.

Giebisch, G. (1961) Measurements of electrical potential differences on single nephrons of the perfused *Necturus* kidney. *J. gen. Physiol. 44*, 659–678.

Gupta, B. L. & Berridge, M. J. (1966) A coat of repeating subunits on the cytoplasmic surface of the plasma membrane in the rectal papillae of the blowfly, *Calliphora erythrocephala* (Meig.), studied *in situ* by electron microscopy. *J. Cell Biol. 29*, 376–382.

Harvey, W. R. (1973) The route of cation transport across the Silkworm midgut. In Alfred Benzon Symposium V *Transport Mechanisms in Epithelia*, ed. Ussing, H. H. & Thorn, N. A., pp. 342–357. Munksgaard, Copenhagen.

Johnstone, B. M. & Sellick, P. M. (1972) The peripheral auditory apparatus. *Quart. Rev. Biophys. 5*, 1–57.

Kasbekar, D. K. & Durbin, R. P. (1965) An adenosine triphosphatase from frog gastric mucosa. *Biochim. biophys. Acta 105*, 472–482.

Keynes, R. D. (1963) Chloride in the squid giant axon. *J. Physiol. (Lond.) 169*, 690–705.

Keynes, R. D. (1969) From frog skin to sheep rumen: a survey of transport of salts and water across multicellular structures. *Quart. Rev. Biophys. 2*, 177–281.

Kirschner, L. B. (1973) Electrolyte transport across the body surface of fresh water fish and amphibia. In Alfred Benzon Symposium V *Transport Mechanisms in Epithelia*, ed. Ussing, H. H. & Thorn, N. A., pp. 447–458. Munksgaard, Copenhagen.

MacRobbie, E. A. C. (1970) The active transport of ions in plant cells. *Quart. Rev. Biophys. 3,* 251–294.

Rehm, W. S., Davis, T. L., Chandler, C., Gohmann, E. & Bashirelahi, A. (1963) Frog gastric mucosae bathed in chloride-free solutions. *Amer. J. Physiol. 204,* 233–242.

Schatzmann, H. J. (1953) Herzglykoside als Hemmstoffe für den aktiven Kalium und Natrium Transport durch die Erythrocytenmembran. *Helv. physiol. Acta 11,* 346–354.

Simon, B., Kinne, R. & Sachs, G. (1972) The presence of a HCO_3^--ATPase in pancreatic tissue. *Biochim. biophys. Acta 282,* 293–300.

Stevens, C. E. (1973) Transport across rumen epithelium. In Alfred Benzon Symposium V *Transport Mechanisms in Epithelia,* ed. Ussing, H. H. & Thorn, N. A., pp. 404–421. Munksgaard, Copenhagen.

Thomas, R. C. (1972) Electrogenic sodium pump in nerve and muscle cells. *Physiol. Rev. 52,* 563–594.

Zerahn, K. (1973) Properties of the cation pump in the *Cecropia* midgut. In Alfred Benzon Symposium V *Transport Mechanisms in Epithelia,* ed. Ussing, H. H. & Thorn, N. A., pp. 360–367. Munksgaard, Copenhagen.

Problems in Current Methods for Studying Transtubular Transport

Robert W. Berliner

A number of methods have been used for the study of transport in the renal tubules. The multiplicity of procedures reflect not only the progress of technical developments which have made feasible studies that would have been impossible earlier, but it also reflects the fact that although nearly all have their particular virtues for one or another purpose, all have limits in either applicability or reliability or both.

In the last 15 years there has been an enormous expansion of the study of function in individual segments of individual nephrons through the application of micro methods. Much of what you will hear in the remainder of this symposium will be based upon the use such procedures and 1 will devote most of my comments to the consideration of some of the variety of such techniques and attempt to point out some of particular virtues and possible limitations of each. Time will not permit consideration of more than a few points about several of the micropuncture techniques in use. For a more complete description and evaluation, I refer you to the review by Gottschalk & Lassiter which will appear in the Handbook of Physiology.

Although we currently look mainly to micropuncture, many of our conceptions of transport in the renal tubule have derived from application of the clearance technique to the intact kidney. Key to this method is the measurements of glomerular filtration, generally as the plasma clearance of inulin. The transport of other substances is then estimated as the difference between the amounts filtered and excreted. For substances for which there is a single transport process – such as the secretory process for organic anions or the reabsorption of sugars – the method has yielded quite reliable and interpretable quantitative results. Even for some of the more complicated

National Institutes of Health, Bethesda, Maryland 20014, USA.

»Transport Mechanisms in Epithelia«, Munksgaard, Copenhagen.

processes such as those involved in electrolyte transport, clearance methods have made possible a number of reasonably accurate inferences and in particular have provided leads that have subsequently been profitably explored with micropuncture procedures.

Obviously, however, there are a number of aspects of transport that cannot be evaluated from the sort of simple balance-sheet calculation that is available from clearance measurements. Specific localization cannot be made and, of course, electrical gradients, unidirectional fluxes, specific permeabilities, etc. cannot be estimated. Clearance procedures remain an important link between the study of single nephrons and the behavior of the kidney as a whole. However, in hopes of filling the many gaps in our knowledge of the behavior of the renal tubules and particularly with the goal of improving our understanding of the transport processes that constitute virtually the entire function of the renal tubules, many groups of investigators have turned to direct study of these structures through the application of microtechniques.

The technique of puncturing individual nephrons and collecting and analyzing the fluid was pioneered by Richards and his co-workers. For almost 20 years, during the 1920's and 30's Richards and his collaborators had the field virtually to themselves and were able to establish firmly not only that the glomerular fluid had the composition appropriate to an ultrafiltrate of plasma, but were able to determine the general pattern for the reabsorption of fluid and a number of solutes along the nephron (Richards 1938, Walker *et al.* 1941). When Richards and his collaborators left the pursuit of active work in this field, micropuncture was almost totally abandoned for nearly 15 years until it was revived largely through the efforts of Gottschalk. Unlike the situation in the earlier period, Gottschalk was not alone in the field for very long and a large number of laboratories have engaged in micropuncture work.

Unfortunately, the acquisition of reliable information has not increased in proportion to the number of workers in the field. A great many findings have turned out to be wrong or at least not reproducible on attempted repetition in another laboratory and sometimes have not been confirmed on repetition in the same laboratory. Some of the problems remain unexplained; others, it would appear, reflect inadequate regard for the difficulties of the techniques and some of their pitfalls. In all too many instances, in any case, one can find data to support a particular conclusion and other data to refute it. Some of these problems seem to be being gradually eliminated as a better appreciation of the methodological limitations is acquired.

In this connection it may be somewhat reassuring to note that even the unhurried Richards era was not without its missteps. For example, the first data on chloride concentrations published by the Richards group were grossly in error (Wearn & Richards 1925, Westfall *et al.* 1934). One of the only ventures of other laboratories into micropuncture work during the period ended in highly misleading and erroneous results; this laboratory gave up work in this field (White 1928).

On the whole, however, the work of the Ricnards laboratory has stood the test of time very well. One may assume that pressed neither by competitors in the field nor by the need to justify financial support, they had time to assure themselves of the validity of their results.

The essential feature of the original micropuncture procedure (which is now generally known as free-flow micropuncture) was the collection and analysis of fluid from identified sites within the nephron. For the overall qualitative description of the function of the tubule, or at least those segments accessible to puncture on the surface of the kidney, the procedure remains a highly useful and relatively reliable one, since it requires a minimum of manipulation and only a single puncture of each tubule.

The usual procedure is to inject a droplet of oil which is allowed to flow past the puncture site in order to block the tubule and assure the collection only of fluid reaching the site of puncture from higher in the tubule. Under normal conditions of pressure and flow, the dangers of error are relatively small,particularly when only small samples are collected for analysis and particularly if there is no need for quantitative collection of all the fluid reaching the puncture site. However, when intratubular pressure is increased, as in massive diuresis or when ureteral pressure is elevated, there is danger of retrograde flow from points distal to the puncture site. This is particularly a problem in the distal tubule, but we believe it has also been serious source of error in collections from the proximal tubule, even though the usual oil block has appeared adequate (Brenner *et al.* 1968).

Simple determination of the composition of fluid can reveal the fraction of filtered solute and water that has been reabsorbed between the glomerulus and the point of puncture. But to obtain an estimate of absolute rates of transport, it is essential to have a measure of the rate of glomerular filtration in the particular nephron punctured. This requires complete collection of the flow, preferably in such a fashion as not to change the rate of glomerular filtration. Whether or not the actual rate of glomerular filtration is markedly dependent upon the procedure used for the collection of fluid is subject to major differences of opinion.

Several studies have been reported to show that if the pressure proximal to the puncture site is monitored and maintained at its initial level, a lower value for the single nephron rate is obtained than that which is observed when the fluid collection is carried out without particular regard to maintaining the pressure in the tubule under study (Gertz et al. 1969, Schnermann et al. 1969). A peculiar feature of these studies has been the observation that the fraction of the filtered fluid reabsorbed between glomerulus and the puncture site has remained essentially unchanged despite the marked changes in apparent nephron filtration rate, some of these averaging close to 100%.

On the other hand, three independent studies on tubules, perfused by techniques that I will refer to a little later, have indicated that the absolute rate of reabsorption is independent of the rate at which fluid flows through the tubule (Burg & Orloff 1968, Morgan & Berliner 1969, Morel & Murayama 1970). The latter findings are, of course, directly in conflict with reabsorption of a constant fraction of the filtrate in the face of apparent changes in the rate of glomerular filtration, a finding that implies that reabsorption varies directly with flow.

Although it is difficult to specify the source of the error in these studies, the contradiction implied by apparently varying filtration rate and constant fractional reabsorption makes me incline strongly to putting my faith in several other studies in which the single nephron filtration rate was found to be largely unaffected by pressure changes at the site of puncture (Davidman et al. 1971, Daugharty & Brenner 1971). Also favoring this view is 1) the likelihood that the pressure changes at the puncture site are not transmitted back to the glomerulus, owing to the small hydrostatic pressure gradient available to maintain the patency of the lumen in the collapsible tubule and 2) the rather modest changes in net glomerular filtration pressure to be expected, even from rather marked percentage decreases in intratubular pressures.

In the so-called free flow micropuncture study, the experimenter is limited to the spontaneous rate of flow of fluid, and to a chemical composition that can be varied only to the extent that can be induced by administration of various substances to the animal. In order to be able to modify these

33*

experimental variables, rendering the flow and composition of the fluid
independent of the volume and composition of the glomerular filtrate, a
number of procedures have been devised that involve the instillation of fluid
of known composition into the lumen, either intermittently – so-called
stationary microperfusion – or continuously.

A wide variety of stationary microperfusion studies have been carried
out. Time will not permit a discussion of many of these. The instillation of
droplets of fluid between oil columns in the lumen and their subsequent
recovery for analysis has been particularly useful for determining the ability of
the epithelium of the tubule to maintain electrochemical potential gradients
between the lumen and the peritubular space (Giebisch *et al.* 1964). Droplets
containing only absorbable solutes are absorbed very rapidly from most
parts of the mammalian nephron. To allow time for the attainment of a
steady state of solute concentrations, it has therefore generally been
necessary to include some poorly reabsorbed solute, generally raffinose, to
impede reabsorption of the fluid droplet (Kashgarian *et al.* 1963). It is
claimed by some and denied by others that this procedure is not entirely
reliable because of possible damage to the tubule epithelium by the injection
of oil (Morel & Myrayama 1970).

On the other hand, the rapid rate at which fluid is reabsorbed from
droplets of saline, particularly in the proximal tubule, has suggested the
possibility of assessing the rate of solute absorption by following the loss
of volume from drops placed in the lumen between columns of oil. The
first attempt to take advantage of this phenomenon was made by Solomon
(1959) who injected Ringer's solution at a constant rate into a column of oil
placed in a rat proximal tubule. The length of the aqueous column when
this length became constant was a measure of the flux of isotonic fluid across
the tubule wall – the length of tubule required to reabsorb salt and water at
a rate equal to that of injection. The technique used was crude and the
results obtained of doubtful usefulness; furthermore the method had the
drawback that the region of injury at the site of puncture and fluid injection
might form a considerable part of the absorbing surface under study.
Nevertheless the procedure was an ingenious one and was the precursor of the
shrinking drop or split-oil droplet procedure introduced by Gertz (1963)
and subsequently widely adopted by many laboratories.

In the Gertz procedure, a column of castor oil is injected into the lumen
and then split by the injection of a droplet of saline. As the contents of the

droplet are reabsorbed, the length of the aqueous column shrinks and the ends of the oil columns approach each other. These events are recorded by photographing at intervals of several seconds and the length of the aqueous droplet is measured in the photographs. Since, if the cross-section of the lumen that contains the droplet is constant, both the volume of the droplet and the surface engaged in its reabsorption are proportional to its length, the length should decrease exponentially with a rate constant proportional to the reabsorptive activity per unit length. Gertz and his associates found the change in length to follow the predicted exponential course. Furthermore, they found that the reabsorptive rate did not seem to vary over the length of the proximal tubule (Gertz 1963, Gertz et al. 1965).

This seemingly simple procedure was widely and uncritically adopted since it seemed to offer a method for measuring reabsorptive activity of the tubule without the problems of measuring glomerular filtration and without any possible consequences that changing filtration rate might have. Thus it seemed to be ideal for evaluating the effect of various interventions on transport. Gradually, the pitfalls have come light, but only after a number of erroneous conclusions had been reached.

It should be noted that the method is based on the assumption that the rate of reabsorption of fluid is a measure of the rate of solute transport out of the droplet – an inference wholly justified only in the proximal tubule where isotonicity is assured by the high permeability to water. However, the procedure has also been applied in the distal tubule where the low water permeability leaves a question as to whether observed changes depend on changes in salt transport or in water movement.

More generally, a number of the assumptions underlying the procedure are not precisely correct in practice and are a source of errors under the best of conditions (Nakajima et al. 1970). I believe the greatest drawback is the extent to which the results obtained can be influenced by the unconscious prejudices of the experimenter. Of course, these subjective influences are in turn dependent upon departures of the real behavior of the shrinking drop from the ideal.

Sometimes droplets are hardly absorbed at all and are clearly not useful for obtaining rate measurements. The experimenter must decide when to begin photographing the process; if the reabsorption appears to be much slower than anticipated, he is not likely to begin taking the photographs. Later, in measuring the photographs, it is not always clear whether or not there are curves in the course of the tubule that are not in the plane of the photograph, nor is it certain that the radius of the tubule, always difficult to measure, remains constant throughout the reabsorptive process. Thus it has been found even in the most careful work that there are significant departures from pure exponential behavior and subjective decisions must be made in placing the line describing disappearance as a function of time.

Some of the errors attributable to bias can be minimized if the individual measuring the photographs and plotting and analyzing the data does not know the identity of the source. But there remain the possible biases in the manipulations themselves and in the decision to proceed with the photographs. The latter can be eliminated only if the experimenter does not know precisely what he is working with or the experimental condition of the animal, a situation that is sometimes difficult to achieve. If all these conditions are met, data can be obtained that are free of subjective bias but still subject to the inherent experimental errors and hence involving a rather large variability. This may still permit the detection of effects of relatively large magnitude, but it is clear that the usefulness of the method is limited.

Continuous *in vivo* microperfusion methods have the obvious disadvantage of greatly increased technical difficulty owing to the necessity of managing two separate micropipettes in the same tubule for relatively extended periods (Morgan & Berliner 1969, Morel & Murayama 1970, Wiederholt *et al.* 1967). They have the great advantage of making it possible to measure the effects of changing flow on the absolute and relative rates of reabsorption. The procedures are also very useful in making possible measurements of fluxes and permeabilities. However, it is clear that the methods may be subject to systematic sources of error. For example, it has been repeatedly observed that when reabsorption is plotted as a function of the perfused length of tubule, the line fails to intersect the origin, but rather suggests that reabsorption over the first part of the perfused segment is much lower than over the remainder (Morel & Murayama 1970, Wiederholt *et al.* 1967).

An additional source of error is failure to make the osmolality of the perfusion fluid match precisely that of the experimental animal so that part of the volume movement represents only osmotic equilibration rather than solute transport, this being particularly important since in many experiments only net volume movements have been measured (Morgan & Berliner 1969).

Both of these errors are minimized if the perfusion is carried out over an extended length of tubule and Morel & Marayama (1970) have found that they can be eliminated, at the expense of some loss of precision, if the measurements are made as the difference between those obtained at two sites of collection in the perfused segment.

A method that offers unique possibilities for the study of transport, and certainly the most elegant procedure available, is the method developed by Burg and his associates for perfusing single isolated segments of tubule *in vitro* (Burg *et al.* 1966). This method makes it possible to control not only the composition of the perfused fluid but to control the composition of the medium surrounding the tubule as well. Isotopes can be added to either one or both sides of the cells that line the tubule. Electrical measurements can be made without violating the integrity of the tubule epithelium by the usual puncturing electrodes. Perfusion rate can be varied at will over a wide range. The perfused segment of tubule is itself available at the termination of the experiment for either analysis or histologic study.

In addition, this method has the unique feature that it is possible to study segments of the nephron that are not accessible for direct study by any other procedure so far devised. It has thus been possible to study several segments whose characteristics have not previously been determined and in each case findings have revealed characteristics previously unsuspected, some more surprising than others.

The first segment subjected to examination was the cortical collecting tubule; in addition to the relatively precise determination of its permeability characteristics and their response to vasopressin, it was possible to demonstrate directly, for the first time, active transport of potassium into the lumen and the dependence of this transport on the reabsorption of sodium (Grantham *et al.* 1970).

The straight segment of the proximal tubule was found to have characteristics distinctly different from the convoluted portion, transporting less sodium and glucose and more para-aminohippurate (Burg & Orloff 1968, Tune *et al.* 1969). The most surprising finding, however, has been in the thick ascending limb of the loop of Henle where the reversed electrical potential difference, with the lumen positive to the bath, is clearly different from any other segment of the nephron (Burg, personal communication).

Although there had been reports of evidence of active chloride reabsorption in the distal tubule, most studies have failed to uncover such evidence. Certainly, no other segment has shown a chloride transport that dominates total solute absorption and this feature was completely unsuspected. Furthermore the profound inhibition of this chloride transport by furosemide (Burg, personal communication) finally offers an explanation for the discrepancy for the rather modest effects of this diuretic on sodium transport in segments accessible to micropuncture as compared with its extraordinary capacity to inhibit salt reabsorption in the kidney as a whole.

Since this talk is intended to deal with the problems of current methods, I should not leave the discussion of this technique of *in vitro* perfusion without mentioning a few minor drawbacks. Chief of these is the difficulty in dissecting out the individual tubule segments, a procedure that has so far proved impossible in any mammalian species except the rabbit. Although most features appear, from other types of study, to be similar from one mammalian species to another, there are enough differences so that it would be highly desirable to have a few interspecific comparisons of the results of *in vitro* perfusion. The permeability characteristics reported by Kokko (1970) for thin descending limbs of the loop of Henle of the rabbit are sufficiently different from those obtained by perfusion of the same segment on the surface of excised rat papillas *in vitro* (Morgan & Berliner 1968) that this difference might be a good comparison with which to start, particularly since the rabbit is a very poor former of concentrated urine and the rat an excellent one.

Another difficulty is the relatively short segment of tubule that can generally be studied. Particularly in the proximal tubule, this yields relatively low fractional reabsorption over the perfused segment and a relatively large experimental error that makes somewhat difficult the detection of the effects of experimental manipulations.

Finally, the separation of the tubule from its environment has both advantages and drawbacks. Interest has recently been directed to the possible role of the peritubular capillaries in regulating reabsorption in the proximal tubule. The isolated tubule may make possible a determination of whether or not the capillaries are involved. In fact, the question of whether the rate of reabsorption from isolated tubules is influenced by the concentration of protein in the surrounding bath is currently the subject of disagreement (Burg, personal communication). Should the answer be in the negative, as Burg and his associates believe, it is clear that the isolated tubule can not throw additional light on the mechanism involved.

REFERENCES

Brenner, B. M., Bennett, C. M. & Berliner, R. W. (1968) The relationship between glomerular filtration rate and sodium reabsorption by the proximal tubule of the rat nephron. *J. clin. Invest. 47;* 1358–1374.

Burg, M. B. Personal communication.

Burg, M. B., Grantham, J., Abramow, M. & Orloff, J. (1966) Preparation and study of fragments of single rabbit nephrons. *Amer. J. Physiol. 210;* 1293–1298.

Burg, M. B. & Orloff, J. (1968) Control of fluid absorption in the renal proximal tubule. *J. clin. Invest. 47,* 2016–2024.

Daugharty, T. M. & Brenner, B. M. (1971) Methodologic influences on measurements of proximal tubule function. *Fed Proc. 30,* 429.

Davidman, M., Lalone, R. C., Alexander, E. A. & Levinsky, N. (1971) Some micropuncture techniques in the rat. *Amer. J. Physiol. 221,* 1110–1114.

Gertz, K. H. (1963) Transtubuläre Natriumchloridflüsse und Permeabilität für Nichtelektrolyte im proximalen und distalen Konvolut der Rattenniere. *Pflügers Arch. ges. Physiol. 276,* 336–356.

Gertz, K. H., Braun-Schubert, G. & Brandis, M. (1969) Zur Methode der Messung der Filtrationsrate einzelner nahe der Nierenoberfläche gelegener Glomeruli. *Pflügers Arch. ges. Physiol. 310,* 109–115.

Gertz, K. H., Mangos, J. A., Braun, G. & Pagel, H. D. (1965) On the glomerular tubular balance in the rat kidney. *Pflügers Arch. ges. Physiol. 285,* 360–372.

Giebisch, G., Klose, R. M., Malnic, G., Sullivan, W. J. & Windhager, E. E. (1964) Sodium movement across single perfused proximal tubules of rat kidneys. *J. gen. Physiol. 47,* 1175–1194.

Gottschalk, C. W. & Lassiter, W. E. Micropuncture Methodology. In *Handbook of Physiology – Renal Physiology I.* (American Physiological Society. In press).

Grantham, J. J., Burg, M. B. & Orloff, J. (1970) The nature of transtubular Na and K transport in isolated rabbit renal collecting tubules. *J. clin. Invest. 49,* 1815–1826.

Kashgarian, M., Stockle, H., Gottschalk, C. W. & Ullrich, K. J. (1963) Transtubular electrochemical potentials of sodium and chloride in proximal and distal renal tubules of rats during antidiuresis and water diuresis (Diabetes insipidus). *Pflügers Arch. ges. Physiol, 277,* 89–106.

Kokko, J. P. (1970) Sodium chloride and water transport in the descending limb of Henle. *J. clin. Invest. 49, 1838–1846.*

Morel, F. & Murayama, Y. (1970) Simultaneous measurement of unidirectional and net sodium fluxes in microperfused rat proximal tubules. *Pflügers Arch. ges. Physiol. 320,* 1–23.

Morgan, T. & Berliner, R. W. (1968) Permeability of the loop of Henle, vasa recta, and collecting duct to water, urea, and sodium. *Amer. J. Physiol. 215,* 108–115.

Morgan, T. & Berliner, R. W. (1969) *In vivo* perfusion of proximal tubules of the rat: glomerulotubular balance. *Amer. J. Physiol. 217,* 992–997.

Nakajima, K., Clapp, J. R. & Robinson, R. R. (1970) Limitations of the shrinking-drop micropuncture technique. *Amer. J. Physiol. 219,* 345–357.

Richards, A. N. (1938) Processes of urine formation. The Croonian Lecture. *Proc. roy. Soc. B No. 844, 126,* 398–432.

Schnermann, J., Horster, M. & Levine, D. Z. (1969) The influence of sampling technique on the micropuncture determination of GFR and reabsorptive characteristics of single rat proximal tubules. *Pflügers Arch. ges. Physiol. 309*, 48–58.

Solomon, S. (1959) A method for investigating net water fluxes across individual proximal tubules. *Proc. Soc. exp. Biol. (N. Y.) 101*, 221–223.

Tune, B. M., Burg, M. B. & Patlak, C. S. (1969) Characteristics of p-aminohippurate transport in proximal renal tubules. *Amer. J. Physiol. 217*, 1057–1063.

Walker, A. M., Bott, P. A., Oliver, J. & MacDowell, M. C. (1941) The collection and analysis of fluid from single nephrons of the mammalian kidney. *Amer. J. Physiol. 134*, 562–595.

Wearn, J. T. & Richards, A. N. (1925) The concentration of chlorides in the glomerular urine of frogs. *J. biol. Chem. 66*, 247–273.

Westfall, B. B., Findley, T. & Richards, A. N. (1934) Quantitative studies of the composition of glomerular urine. XII. The concentration of chloride in glomerular urine of frogs and necturi. *J. biol. Chem. 107*, 661–672.

White, H. L. (1928) Observations on the intracapsular pressure and the molecular concentration of the renal capsular fluid in necturus. *Amer. J. Physiol. 85*, 191–206.

Wiederholt, M., Hierholzer, K., Windhager, E. E. & Giebisch, G. (1967) Microperfusion study of fluid reabsorption in proximal tubules of rat kidney. *Amer. J. Physiol. 213*, 809–813.

DISCUSSION

SCHMIDT-NIELSEN: I would like to ask: how good is the evidence that furosemide inhibits chloride transport specifically?

BERLINER: It is quite good. It knocks out the positive electrical potential that occurs in the segment and decreases the movement to a very low level.

GOTTSCHALK: Could you briefly summarize for us the available data relating to transport rates in tubules *in vivo* and in tubular fragments?

BERLINER: That's a little hard because there are relatively few data concerned with the over-all function in the rabbit. Burg says that, from his calculations (Burg & Orloff 1968) reabsorption in the proximal tubule is roughly what he would have expected it to have been in the intact rabbit.

GIEBISCH: I was wondering whether there is any information on sodium transport in the ascending limb. Is it only chloride or are both chloride and sodium ions actively reabsorbed?

BERLINER: The outward flux relative to the inward flux is too high to be accounted for entirely as a passive process. However, the difference from prediction is relatively small and the interpretation therefore somewhat uncertain.

ORLOFF: The P. D. calculated according to the Nernst relationship from the steady-state concentrations of Na^+ in lumen and bath (approximately 65 and 150 mM respectively), assuming no net flux, is equal to the observed P. D. under these circumstances, evidence for passive transport of Na^+. On the other hand the free flow flux ratio for Na^+ (with equal concentrations of NaCl in lumen and bath) is 0.65, whereas the ratio calculated from the observed P. D. is 0.79. Though the difference is statistically significant and could be interpreted as indicative of active Na^+ transport, in view of the uncertainties of the isotope permeability measurements, I prefer to reserve judgement (Burg & Green 1973).

Burg, M. B. & Orloff, J. (1968) Control of fluid absorption in the renal proximal tubule. *J. clin. Invest.* 47, 1016–2024.

Burg, M. B. & Green, N. (1973) Function of the thick ascending limb of Henle's loop. *Amer. J. Physiol.* 224, 659–668.

TOSTESON: You have to make assumptions about permeability to the cations and anions.

ORLOFF: I don't see why. Chloride is certainly actively transported. In the »Nernst« studies, flow down the tubule was stopped. The concentration of Na and Cl declined to a steady-state value, an assumption that the flux ratio was 1.0 at this point is reasonable; the mean value of E_{Na} so calculated was $+21.2$ mV, the observed $+24.8$.

BERLINER: I think I would like to summarize that the answer to the question is that we really don't know yet.

ULLRICH: May I shift the discussion to a point which non-kidney people may be interested in. This is the problem of applying the Donnan equilibrium or Terepka's filtration concept (Terepka et al. 1970) to the process of ultra-filtration in the glomerulus. In order to find out which one of these theories is applicable we measured the chloride concentration of ultrafiltrate obtained from Bowman's capsule and of venous serum.

The chloride content of a certain volume of serum determined by electro-metric titration was within the limits of the experimental error, the same as that of an equal volume of glomerular ultrafiltrate. The ratio $[Cl^-]$ in the ultrafiltrate: $[Cl^-]$ in the serum calculated by Donnan and plasma-water correction factors should be ~ 1.12 and by Terepka's concept about ~ 1.06. In our experiments however, they were 1.0. I would like to ask whether somebody in the audience has some experience in this field.

BERLINER: We did once do some ultrafiltration on artificial membranes and they gave the expected Donnan ratio, about 1.05 for the ultrafiltrate.

WHITTEMBURY: With respect to this chloride transport: how positive is the intraluminal potential?

BERLINER: The intraluminal potential starts out at about 57 mV positive when the fluid is isotonic and as the sodium and chloride concentration goes down, the potential goes up, because apparently the sodium movement is a shunt, keeping the potential low, and the potential goes up in the order of 25 mV.

Terepka, A. R., Chen Jr., P. S. & Toribara, T. Y. (1970) Ultrafiltration: a conceptual model and a study of sodium, potassium, chloride and water distribution in normal human sera. *Physiol. Chem. & Physics 2*, 59–78.

Microinjection Studies of Renal Tubular Permeability

Carl W. Gottschalk

INTRODUCTION

A variety of sophisticated micropuncture techniques which permit, at least in theory, quantitative description of permeability and transport properties have been employed in studies of renal tubule function. The results which I wish to present to you were obtained with the tubular microinjection technique developed in collaboration with Morel and are more qualitative in nature (Gottschalk *et al.* 1965). Despite this limitation, we have chosen to use this technique because of its technical simplicity. Technical simplicity should be associated with minimization of technical errors, as well as induction of fewer changes in the functional and structural properties of the structures under study.

With this technique a small volume of a solution containing radioactive inulin and one or more other isotopes is injected into some part of the nephron during free flow; the patterns of isotope recovery are determined in serial collections of urine from the injected and contralateral kidneys. The normal nephron is impermeable to inulin when studied by this method and errors of microinjection are revealed by excretion of inulin by the contralateral kidney. There is only one puncture site from which leakage may occur, and exposure of the tubular lumen to oil can be avoided. Comparison of the isotope ratio in the injectate and urine permit determination of relative volumes of distribution and detection of any loss from the tubule.

Using this method we have demonstrated what we believe to be a clearcut change in permeability of the rat renal tubule during increased intrarenal pressure (Lorentz *et al.* 1972). In addition to inulin the relatively large

Supported by a grant-in-aid from the American Heart Association, and by Grant HE-02334 from the National Institutes of Health.
Department of Medicine, University of North Carolina
Chapel Hill, N. C. 27514

»Transport Mechanisms in Epithelia«, Munksgaard, Copenhagen.

solutes creatinine, mannitol, sucrose and iothalamate were used as measures of tubular permeability. Under conditions of normal pressure the tubular epithelium is impermeable to all of these substances. Under conditions of increased intrarenal pressure the epithelium remains impermeable to inulin but becomes permeable to the smaller solutes. Bank *et al.* (1971) have also reported an increase in proximal tubular permeability during increased renal venous pressure.

METHODS

The left kidneys of Wistar rats anesthetized with sodium pentobarbital were prepared for micropuncture. To permit rapid serial urine collections, the animals were made diuretic by continuous infusion of 5 % mannitol and 0.85 % sodium chloride solution at 100 μl/min. A small volume (1–4.5 nl) of isotonic saline stained with nigrosin and containing inulin-3H and another isotope was injected into the tubule under microscopic observation. A microinjection was not considered satisfactory if there was visible leakage at the puncture site or retrograde flow toward the glomerulus.

Intratubular pressure, measured by the conventional Landis technique, was increased in some experiments by elevation of the polyethylene catheter placed in the ureter or by partial clamping of the renal vein. In other experiments intratubular pressure was decreased by partial constriction of the aorta with an adjustable metal clamp placed above the renal artery.

RESULTS AND DISCUSSION

Inulin: The tubular epithelium was impermeable to inulin under all experimental conditions. The recovery of inulin after early proximal injections at control and elevated intratubular pressures is shown in the upper part of Fig. 1. At control pressures recovery averaged 100.2 $\pm$ 2.4 %. Recovery averaged 99.9 $\pm$ 2.2 % when proximal intratubular pressure (ITP) was increased and 98.9 $\pm$ 4.4 % when proximal ITP was decreased.

Mannitol: The recovery of mannitol-14C injected simultaneously with inulin into early proximal tubules was complete under control conditions (99.8 $\pm$ 2.9 %). However, there was increasing loss of mannitol when the intratubular pressure was elevated. At a proximal ITP of 30 $\pm$ 2 mm Hg, 85 % of microinjected mannitol was recovered in the urine from

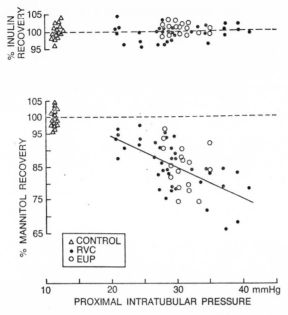

Fig. 1. Fractional inulin and mannitol recoveries as a function of proximal intratubular pressure after simultaneous early proximal injection during control conditions, renal venous constriction (RVC) and elevation of ureteral pressure (EUP). The broken horizontal lines represent 100 % recovery of injected substances. The regression line for mannitol recovery at elevated pressure was y = 112 − 0.93x. (Lorentz *et al.* 1972. Reprinted with permission from J. clin. Invest.)

microinjected kidneys. Approximately 60 % of the mannitol not recovered in urine from the experimental kidney was excreted by the contralateral kidney during the periods of urine collection, demonstrating that a significant fraction of the microinjected mannitol had in fact crossed the tubular epithelium.

To ascertain the location of mannitol loss from the tubular lumen, a series of microinjections was performed in early and late proximal and distal convolutions during renal venous constriction. Loss occurred after injection at all sites but was different at each of the three sites, increasing progressively after injection into more proximal parts of the tubule (Table I). These results demonstrate that loss occurred from all the major tubular segments, i. e. proximal convolution, loop of Henle and distal convolution and/or collecting ducts.

In addition to increasing the intratubular pressure, renal venous constric-

Table I. Percent Recovery of Mannitol after Microinjection at Various Tubular Locations under Control Conditions and during Renal Venous Constriction.

Injection site	ITP	Mannitol recovery	P*
Control	mmHg	%	
Early proximal, n = 18	10.5 ± 14.5	99.8 ± 2.9	
Renal venous constriction			
Early proximal, n = 19	28–32	83.1 ± 7.2	0.001
Late proximal, n = 16	28–32	90.6 ± 3.5	0.001
Distal, n = 18	22–28 ∓	96.2 ± 3.9	0.01

* Experimental versus control.
∓ Proximal ITP 28–32 mm Hg.

tion and elevation of ureteral pressure both prolong the transit time through the nephron. As it was conceivable that the measureable loss of mannitol from the tubule during elevated ITP was due to decreased tubular fluid velocity with increased contact time and not to a change in permeability *per se,* similar studies were done during partial aortic constriction. Aortic constriction prolongs transit time but *decreases* ITP. In Fig. 2 the recovery of mannitol is shown as a function of transit time under the three experimental conditions. Transit times were prolonged to a similar degree under all three experimental conditions but there was no measureable loss from the lumen during aortic constriction. As anticipated fractional loss from the tubule during increased ITP was a function of velocity of tubular flow.

Additional confirmation that measureable loss from the tubule is not simply a function of increased transit time, but represents a true change in renal tubular permeability, was obtained in animals made markedly diuretic by infusion of 2.5 % sodium chloride solution. When the diuresis was sufficient to produce intratubular pressures of 20 to 28 mm Hg, transtubular mannitol loss occurred and recovery averaged 86.7 ± 6.9 %, despite the fact that transit times were shorter than under control conditions.

In order to determine the time course of recovery of normal tubular permeability, paired microinjections were made at the same puncture site during and at varying intervals after release of elevated ureteral pressure. The results are shown in Fig. 3. Mannitol recovery was 86.4 ± 5.5 % at elevated pressure and 98.1 ± 3.3 % after release of pressure. A return to

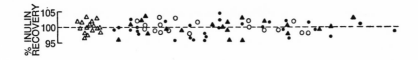

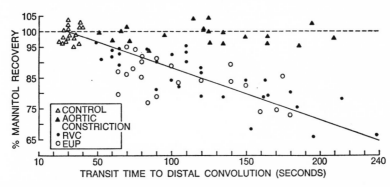

Fig. 2. Fractional inulin and mannitol recoveries after simultaneous early proximal microinjection as a function of transit time of the test solution to the distal convolution during control conditions, aortic constriction, renal venous constriction (RVC), and elevation of ureteral pressure (EUP). The broken horizontal lines represent 100 % recovery of the injected substances. The solid diagonal line, y = 105 – 0.15x, describes mannitol recovery during RVC and EUP. (Lorentz *et al.* 1972. Reprinted with permission from J. clin. Invest.)

normal permeability was evident at the earliest time, 5–6 min, that it was possible to obtain information after release of pressure. Similar studies were performed with sucrose-^{14}C, creatinine-^{14}C and ^{125}I sodium iothalamate. In each case recovery was less than 100 % during elevated ITP but returned towards complete recovery shortly after release of elevated ureteral pressure.

Comparison of the molecular weight and percent recovery of the various test substances following early proximal injection at ITP of 30 ± 2 mm Hg (Table II) reveals a direct relationship between molecular weight and recovery. An effect related to molecular size suggests diffusion through aqueous channels of increased diameter. These functional results provide, of course, no information about the location of such channels, but in view of the evidence indicating the presence of low resistance intercellular shunts (Frömter & Diamond 1972, Giebisch 1969, Whittembury & Rawlins 1971, Windhager *et al.* 1966) it is reasonable to postulate that the pathway involves the tight junctions and lateral intercellular spaces.

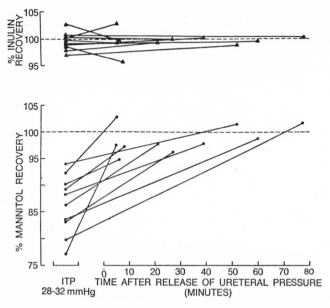

Fig. 3. Fractional inulin and mannitol recoveries after simultaneous proximal micro-injection during and after release of elevated ureteral pressure. Solid lines connect paired injections of the same puncture site. The broken horizontal lines represent 100 % recovery of the injected substances. (Lorentz *et al.* 1972. Reprinted with permission from J. clin. Invest.)

During elevation of ureteral pressure and saline diuresis dilatation of the tubular lumen is readily observable and one might presume this causes an increase in channel diameter. However during renal venous constriction with increased permeability there is no change or even a decrease in tubular diameter. Thus there is no basis on which to postulate that an assumed increase in channel size results from stretching.

An increased intrarenal pressure was a common factor associated with the change in tubular permeability but the precise loci at which pressure must be increased to produce this effect is unknown. Only proximal ITP was measured in these studies but the hydrostatic pressure in the peritubular vessels is also increased when ITP is increased. Presumably it is the gradient of pressure across some critical part of the epithelium, perhaps across the junctional complexes from tubular lumen to the lateral intercellular spaces, that produces the change in permeability.

Clearly and not unexpectedly it is not the gradient of pressure between

Table II. Comparison of Molecular Weight and Percent Recovery of Various Test Substances after Early Proximal Injection at ITP of 30 ± 2 mm Hg Produced by Elevation of Ureteral Catheter.

	Molecular weight	Recovery*
		%
Creatinine	113	73
Mannitol	182	85
Sucrose	342	89
Iothalamate	607	85
Inulin	5500	100

* Average recovery.

the tubular lumen and the peritubular capillaries, since the pressure gradient between these structures differs when produced by increased tubular or vascular pressure and loss of isotope was the same at a given intratubular pressure from either cause. Alternatively it is possible that a humoral substance produced in response to increased intrarenal pressure causes the presumed permeability change involving the tight junctions.

In order to determine the effect of various rates of microinjection on intratubular pressure, Dr. Romulo Colindres has made continuous recordings of proximal ITP with a servo-nulling device while microinjections were made in an adjacent upstream coil of the same proximal tubule. With injection rates of the magnitude (0.23 nl/sec on average) employed in the studies I have just described, increases of pressure of several mm Hg were usually observed in diuretic rats under control conditions and at elevated ITP. The pressure rapidly returned to baseline levels when the injection was terminated. With considerably higher rates of injection, pressure increases as large as 15 to 20 mm Hg were produced. Changes in pressure of this magnitude are sufficient to produce a readily measureable change in tubular permeability.

We have been fortunate in having the collaboration of Dr. Ruth Ellen Bulger in searching for a morphological basis for these changes in permeability. She has made electron micrographs of tubules fixed in several fashions at control and elevated pressures and has studied the distribution of colloidal lanthanum hydroxide microinjected into the tubular lumen under these circumstances. The studies to date must be considered preliminary, but they are highly suggestive of opening of the tight junctions under conditions

34*

of elevated intrarenal pressure. Colloidal lanthanum injected into a tubule of a nondiuretic rat with normal ITP is restricted to the lumen and does not gain access to the lateral intercellular spaces. This is in contrast to the findings during increased intratubular pressure where Dr. Bulger has demonstrated extensive amounts of lanthanum in the lateral intercellular spaces following intratubular microinjection. Under these conditions lanthanum hydroxide is apparently able to penetrate beyond the tight junctions and into the intercellular spaces.

In conclusion we believe that our results demonstrate a change in the permeability characteristics of the renal tubule during elevation of intrarenal pressure. The change appears to be rapidly reversible and to result from opening of the tight junctions such that solutes to which they are normally impermeable can pass through them. Although we have used as functional probes relatively large substances which are not normally present, we presume that the demonstrated changes in permeability may affect the transport of natural constituents of tubular fluid.

REFERENCES

Bank, N. W., Yarger, W. E. & Aynedjian, H. S. (1971) A microperfusion study of sucrose movement across the rat proximal tubule during renal vein constriction. *J. clin. Invest. 50,* 294.

Frömter, E. & Diamond, J. (1972) Route of passive ion permeation in epithelia. *Nature New Biol. 235,* 9–13.

Giebisch, G. (1969) Functional organization of proximal and distal tubular electrolyte transport. *Nephron 6,* 260.

Gottschalk, C. W., Morel, F. & Mylle, M. (1965) Tracer microinjection studies of renal tubular permeability. *Amer. J. Physiol. 209,* 173–178.

Lorentz, W. B., Lassiter, Jr., W. E. & Gottschalk, C. W. (1972) Renal tubular permeability during increased intrarenal pressure. *J. clin. Invest. 51,* 484–492.

Whittembury, G. & Rawlins, F. A. (1971) Evidence of a paracellular pathway for ion flow in the kidney proximal tubule: electromicroscopic demonstration of lanthanum precipitate in the tight junction. *Pflügers Arch. ges. Pysiol. 330,* 302–309.

Windhager, E. E., Boulpaep, E. L. & Giebisch, G. (1966) Electrophysiological studies on single nephrons. *Proc. 3rd Int. Congr. Nephrol. 1,* 35.

DISCUSSION

FRÖMTER: Have you also performed experiments with elevation of pressure in a single tubule. Such experiments could probably exclude unspecific effects from disturbances of blood flow etc.

GOTTSCHALK: Not yet. We can evaluate the pressure in single tubules by high rates of injection of test substances and we plan to do what you suggest.

HOSHIKO: Where do you think the barrier that is impermeable to inulin stands?

GOTTSCHALK: This obviously concerns us. We see what looks to be rather extensive permeation of these junctional complexes, as evidenced by accumulation of lanthanum in them, although we find no measurable loss of inulin. I suppose there are several possibilities. I don't know the size of the colloidal lanthanum particles. I suppose there is a variety of sizes, and some may be small. It probably doesn't require much lanthanum to show up in the electron micrographs. I suspect that there is also some loss of inulin which we don't pick up, because it is a small loss.

WHITTEMBURY: I would like to state that there is also some lanthanum going into the cells under the abnormal conditions.

GOTTSCHALK: Although it may have appeared this way on the projected lantern slides, the EM's themselves show no intracellular lanthanum.

GIEBISCH: I would just like to briefly mention that Weight & Mc Dougal (1972) in our laboratory have performed similar experiments in which prolonged periods of bilateral ureteral obstruction gave rise to a very profound diuresis upon release of the obstruction. Under those conditions there is even some inulin loss from the tubular lumen. Whether this is cell damage, we do not know but we consider it more likely that inulin passes between cells.

GOTTSCHALK: In some animals that have been treated with various nephrotoxins one can demonstrate a component of transtubular inulin leakage; I suppose that it is possible that at even higher pressures than we employed there might be some inulin leakage acutely.

Weight, F. F. & Mc Dougal, W. S. (1972) Defect in proximal and distal sodium transport in postobstructive diuresis. *Kidney Int.* 2, 304–317.

BERLINER: The collection of the urine from the other kidney is a much more sensitive way of detecting losses. I wonder if you have checked for inulin that way.

GOTTSCHALK: Yes, in all of these experiments we collected the urine from both kidneys. In none of them could we detect any measurable quantities of inulin in urine from the contralateral kidney. When there was mannitol loss, a large fraction of the mannitol which was not excreted by the microinjected kidney did appear in the urine from the other kidney. The mannitol was clearly going across the tubule and not being sequestered in or limited to the tubular cell.

SCHMIDT-NIELSEN: We have done considerable work with inulin and poly-ethyleneglycol and we find that tritiated inulin in the fish which we are working with is taken up by the renal tubular cells. It accumulates intra-cellularly to a greater degree than polyethyleneglycol.

Also, the tritiated inulin clearance is considerably lower, about 24 percent, than the polyethyleneglycol clearance. This varies with different species, but in most species we have worked with, the inulin clearance is always lower and inulin is taken up to a considerable degree in the cells. So I don't think the method of Dr. Berliner would be enough because inulin may leave the tubule by being taken up by the cell.

MEARES: If the permeability of the system is sensitive to one of the applied forces, the pressure in this case, this means the behaviour of the system should be non-linear. But I noticed your plot of total loss of mannitol against pressure was linear. This must have some particular significance. If the pres-sure driving force were a major contribution to the driving forces on the flux in addition to changing the properties in the system, one would not have expected a linear relationship there. This could imply that the driving force is simply a concentration gradient and that the pressure is not contributing to the driving of the mannitol or the other solutes, but merely to the change in the pore sizes. If the pores are large enough for mannitol to get through they must surely be large enough for water to go through. I would have thought that pressure would have made a considerable contribution to the water flux, because the concentration gradient of water must be very low.

ULLRICH: We have for many years been interested in the factors which influence the permeability across the proximal tubule. And we looked to see whether distension of the tubule, the intraluminal flow velocity, the concentration of calcium and the hydrostatic pressure influence the isotonic absorption and the permeability for small ions. The isotonic reabsorption is unchanged per tubular length in the proximal tubule when one increases the pressure from 20 to 40 and even 60 cm of water (Sato, unpublished results). The permeability of chloride, sodium, and urea is also unchanged when the intratubular hydrostatic pressure rises from 10–12 cm H_2O (i. e. microperfusion conditions) to about 30 cm H_2O (i. e. stopped flow microperfusion conditions). Our pressure changes lasted only for a few minutes but yours were for a much longer time. These longer lasting luminal pressures alone could change the permeability. Furthermore one has to consider secondary permeability changes due to disturbance of the cell metabolism. For instance, Dr. Lübbers and collaborators (Baumgärtl et al. 1972) using platinum O_2 microelectrodes, have shown that in some areas of the kidney in situ the oxygen pressure is very small, in the range of 1 to 2 Torr. I suggest that by clamping the ureter or the renal vein as done in your experiments one may get hypoxic areas where the permeability of the tubules may become augmented.

TOSTESON: Did you make electron micrographs of tubules only with increased ureteral pressure or did you also make them with increased renal venous pressure?

GOTTSCHALK: Tissue for electron microscopy was obtained from animals with increased ureteral pressure. Technically the pressure is much easier to regulate by a maneuver of the ureteral cannula than by construction of the renal vein.

OSCHMAN: Your studies as well as those of Dr. Ussing on the anomalous solvent drag in frog skin demonstrate reversible changes in the tight or limiting junctions. The freeze etch pictures of the limiting junction show a very complicated structure, and it will be important to study with freeze etch the changes that can occur here. The limiting junction is actually a complicated series of compartments as is shown in the accompanying freeze etch picture provided by Dr. L. Andrew Staehelin. I hope the physiologists will

Baumgärtl, H., Leichtweiss, H. P., Lübbers, D. W., Weiss, Ch. & Huland, H. (1972) The oxygen supply of the dog kidney: measurements of intrarenal pO_2. Microvasc. Res. 4, 247–257.

contemplate this picture and consider the functional implications of this compartmental arrangement.

FRÖMTER: I should perhaps mention that the so-called »tight« junction may have different aspects and that in proximal tubules it consists only of a single bar or ridge of approximately 200 Å thickness, in direction from lumen to interstititum, as can be inferred from the figures of Farquhar & Palade (1963).

Farquhar, M. & Palade, G. E. (1963) Junctional complexes in various epithelia. *J. Cell. Biol. 17*, 375–412.

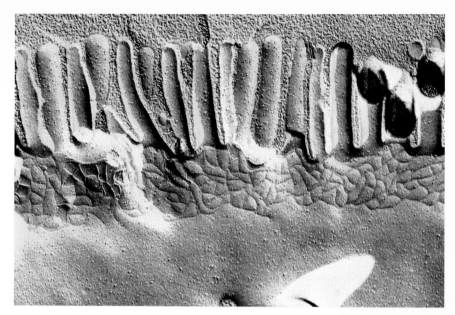

Freeze-cleaved tight junction (limiting junction) from the small intestine of a rat. The meshwork of interconnected ridges and furrows delineates the disposition of the junctional sealing elements which act as a series of seals in the intercellular pathway. The seals are more abundant in »tight« than in »leaky« epithelia (Claude & Goodenough 1972). Micrograph kindly provided by Dr. L. Andrew Staehelin, University of Colorado, Boulder. For further details, see Staehelin *et al.* (1969) and Chalcroft & Bullivant (1970).

Claude, P. & Goodenough, D. A. (1972) The ultrastructure of the zonula occludens in tight and leaky epithelia. *J. Cell. Biol. 55,* 46A.

Staehelin, L. A., Mukherjee, T. N. & Williams, A. W. (1969) Freeze-etch appearance of the tight junctions in the epithelium of small and large intestine of mice. *Protoplasma 67,* 165–184.

Chalcroft, J. P. & Bullivant, S. (1970) An interpretation of liver cell membrane and junction structure based on observation of freeze-fracture replicas of both sides of the fracture. *J. Cell. Biol. 47,* 49–60.

Coupled Ion Movement Across the Renal Tubule

Gerhard Giebisch and Gerhard Malnic

A variety of renal transport mechanisms are functionally interdependent. A large body of experimental evidence is consistent with the view that not only the tubular transport rates of strong electrolytes such as sodium, hydrogen and potassium ions are interdependent, but that solute and fluid movement are also coupled to each other. Thus, it is the active tubular transport of sodium ions which provides the major driving force for fluid reabsorption across the renal tubular epithelium.

In the following, some newer aspects of the relationship between tubular sodium and fluid reabsorption will be analyzed and some relevant aspects of these transport processes examined as they affect translocation of other ion species. In addition, the nature of coupling between proximal and distal tubular sodium, hydrogen and potassium transport will be explored.

PROXIMAL TUBULE

Occurrence of net sodium transport against an electrochemical gradient suggests its active nature (Giebisch & Windhager 1964). It is generally believed that such solute reabsorption generates an osmotic concentration difference at some site within the epithelium and induces passive water reabsorption from the lumen. In view of the isosmotic character of proximal tubular fluid transfer, it is thought that sodium chloride is pumped into a fluid pool of limited expandability *between* cells. Fluid moves from the lumen to the interspace compartment in response to an osmotic pressure gradient. In a terminal step the reabsorbate is moved from the interspaces

The Departments of Physiology, Yale University School of Medicine, New Haven, Connecticut, USA, and São Paulo University Medical School, São Paulo, Brazil.

»Transport Mechanisms in Epithelia«, Munksgaard, Copenhagen.

across the basal membrane and the capillary wall, by a force determined by the net balance between hydrostatic and oncotic pressure difference across these barriers (Giebisch 1969, Ussing & Windhager 1964, Windhager *et al.* 1969). Fig. 1. summarizes the salient points of the functional organization of proximal tubule cells.

The importance of a fluid compartment between proximal tubule cells is two- fold. It determines in a major way the passive permeability properties of the proximal tubular epithelium. In addition, changes in the properties of

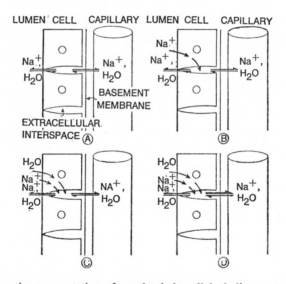

Fig. 1. Schematic representation of renal tubule cell including routes of sodium and water movement (Giebisch, 1972).

A: Tubule cells are joined luminally by tight junctions. They are very short in the proximal tubule. Located between tubule cells is an extracellular interspace. The tight junction, interspace and basement membrane constitute a separate, low-resistance extracellular pathway of solute and water movement.

B: Sodium is actively pumped from cells into the extracellular interspace. This renders their content hypertonic.

C: Water follows sodium passively along an osmotic gradient, probably also via the tight junction.

D: The intercellular compartment is hypertonic in the region close to the tubular lumen but becomes isosmotic due to water equilibration along the basal portion of the channel. Fluid and sodium movement out of the interspaces is achieved by slight elevation of the hydrostatic pressure in this compartment. A final transfer step is that of capillary uptake of sodium and water effected by the net balance of hydrostatic and oncotic pressure across the capillary wall.

tubular interspaces have emerged as a mechanism playing a major role in coupling peritubular events to net proximal sodium and fluid reabsorption.

The following lines of evidence support the view that proximal tubular interspaces represent a low resistance shunt path in parallel with the cell membrane resistance. (1) The transepithelial electrical resistance is unusually low-some 5 ohm cm² in the mammalian proximal tubule-and several orders of magnitude lower than that calculated on the assumption that the epithelium of the proximal tubule is lined with membranes having the resistance of proximal tubular cell membranes (Boulpaep 1971a, Frömter et al. 1971, Windhager et al. 1966).

(2) There is a marked difference between the ionic selectivity of individual cell membranes as compared to that of the proximal tubular epithelium as a whole. Measured by electrophysiological techniques (Boulpaep 1971a, Frömter et al. 1971, Boulpaep 1971b) individual tubular cell membranes show very high ionic selectivity (the peritubular membrane of single tubule cells is 25 × more permeable to potassium than to sodium ions), whereas the permeability of the tubular epithelium is such that the mobility ranking of cations and anions agrees with that predicted from ion mobilities in free solutions, with the exception that cation movement is generally faster than that of anions. It is safe to conclude that the permeability properties of the proximal tubule are determined to a large degree by fixed negative charges within a low resistance fluid path between tubule cells.

Several lines of evidence have clearly shown that change in the peritubular capillary fluid reabsorption, in particular changes in peritubular plasma protein concentration and the hydrostatic pressure in peritubular capillaries, affect proximal tubular sodium and fluid transfer (Windhager et al. 1969, Brenner et al. 1969).

Despite the relatively high permeability of the proximal tubule we do not believe that changes in the transepithelial oncotic pressure difference, induced by peritubular protein changes, provide a major driving force for reabsorptive fluid movement in the absence of net sodium transport (Green et al. 1972).

Fig. 2. summarizes relevant results. In continuous perfusion experiments of tubules and peritubular capillaries the effect on fluid reabsorption of changes in peritubular albumin concentration from 2.5 to 10 g% was tested. Whereas fluid reabsorption was significantly enhanced upon tubular perfusion with Ringer's solution only a minor effect on tubular fluid transport

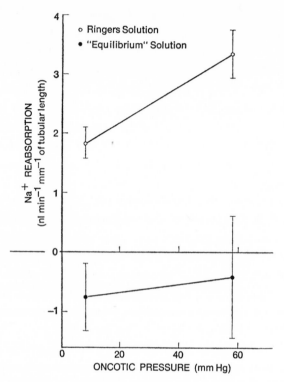

Fig. 2. Effect of changes in peritubular protein concentration on transepithelial fluid movement (Green *et al.* 1972).

was seen when the lumen was perfused with an equilibrium solution containing 110 mEq/L Na and raffinose. Essentially similar results were obtained when the lumen was perfused with a raffinose-free Ringer's solution containing 4mM cyanide. This supports the view of too low a tubular hydraulic conductivity to permit a substantial fraction of proximal tubular reabsorption to be driven by the transepithelial oncotic pressure gradient when peritubular protein changes are induced.

It appears that changes in capillary protein concentrations affect the sodium transport system by changing the conductance of the intercellular transport path. In volume-expanded animals, for example, hemodilution at the capillary side reduces the oncotic pressure (Brenner *et al.* 1971). At the same time an increase in peritubular hydrostatic pressure relative to that in the basal labyrinth may occur (Boulpaep 1972). Table I summarizes

Table I. Summary of effects of volume expansion on transepithelial conductance and permeability coefficients.

$$\frac{G_m \text{saline}}{G_m \text{control}} \text{ (from } R_{\text{input}}) = 2.80$$

$$\frac{G_m \text{saline}}{G_m \text{control}} \text{ (from } \lambda) = 3.18$$

$$\frac{P_{\text{NaCl}}^{\text{saline}}}{P_{\text{NaCl}}^{\text{control}}} = 3.21$$

$$\frac{P_{\text{raffinose}}^{\text{saline}}}{P_{\text{raffinose}}^{\text{control}}} = 6.17$$

Abbreviations: The superscripts indicate whether the value was obtained in control or in saline-loaded animals. G_m = transepithelial conductance, P_{NaCl} = transepithelial permeability coefficient for NaCl, $P_{\text{raffinose}}$ = transepithelial permeability coefficient for raffinose (Boulpaep 1972).

transepithelial conductance and permeability changes under such conditions (Boulpaep 1972). It is apparent that the total transepithelial conductance (i. e. ion permeability) increases drastically. Importantly, renal tubular cell resistance, not shown in the table, remains unchanged. Both the permeability of the proximal tubule to sodium chloride and to non-electrolytes such as raffinose is significantly augmented, a finding which underscores the relatively unspecific nature of the increased leakiness of the proximal tubule (Boulpaep 1972).

It has been proposed that a reduction of translocation of the reabsorbate from interstitial fluid into capillaries (low peritubular protein concentration, high peritubular hydrostatic pressure) leads to an enhanced rate of passive backflux of sodium from the interspace region into the tubular lumen (Windhager et al. 1969, Boulpaep 1972). An increased leak permeability of the proximal tubular epithelium to non-electrolytes has recently been also observed upon elevation of the renal venous or of the ureteral pressure (Lorentz et al. 1972, MacDougal & Wright, 1972.

Evidence is accumulating that it is the relative weakness of the tight junction at the luminal interface between neighboring tubule cells which renders it susceptible to pressure changes within the interspace region (Whittembury & Rawlins 1972, Bentzel *et al.* 1969). Furthermore, an increased back pressure due to slowing of reabsorbate removal and, as its consequence, an enhanced leakiness to sodium ions render proximal net sodium reabsorption less efficient (Windhager *et al.* 1969, Boulpaep 1972).

In view of the rather unspecific nature of this proximal permeability increase it is not surprising that diminished fluid and sodium transport are associated with a reduction of many other proximal tubular transport mechanisms. Reduction of proximal tubular sodium and fluid reabsorption, for instance during extracellular volume-expansion, leads to decreased reabsorption of potassium, chloride, calcium, phosphate and bicarbonate. It is of interest that during extracellular volume-expansion and reduction of proximal tubular sodium reabsorption even glucose reabsorption, at proximal tubular transport mechanism, is less efficient, as evidenced by the marked splay of the glucose titration curve (Robson *et al.* 1968). It is most reasonable to assume that most proximal tubular transfer mechanisms are rendered less efficient by increased proximal tubular permeability.

It has been generally accepted that hydrogen ions secreted along the proximal tubular epithelium are exchanged for sodium ions in the lumen (Malnic & Mello-Aires 1970, Rector 1971). Two lines of evidence indicate that the reabsorption of sodium and the secretion of hydrogen ions are only very loosely coupled.

Fig. 3 summarizes proximal tubular stopped-flow microperfusion studies in the rat in which the kinetics of acidification were studied when the tubular fluid either contained sodium or choline bicarbonate. It is apparent that in the absence of sodium in the luminal fluid the secretion of hydrogen continued in both proximal and distal tubules at near normal level. It is highly unlikely that choline ions could substitute for sodium if a tightly coupled carrier-mediated exchange mechanism were responsible for the process of hydrogen ion secretion. Rather it is likely that the mechanism of Na for H exchange is of unspecific nature and is the consequence of maintaining electroneutrality.

In the absence of an easily reabsorbable luminal counter ion such as sodium − (choline is reabsorbed less efficiently than sodium) − the secretion of hydrogen is probably accompanied by the inward movement of chloride along its electrochemical gradient (Malnic & Mello Aires 1970). The absence of net chloride secretion is not a cogent argument against such a mechanism

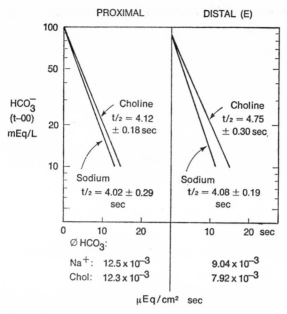

Fig. 3. Semilogarithmic plot of the approach of proximal and distal bicarbonate concentration to their steady-state values in tubular perfusion experiments in which either sodium or choline bicarbonate were used. Sodium or choline bicarbonate concentration: 100 mmoles/liter. Isotonicity maintained by raffinose (Malnic & Mello-Aires 1971).

under normal free-flow conditions since secreted chloride ions could be reabsorbed further downstream (Rector 1971).

However, it is also clear from other recent experiments that hydrogen ion secretion is also not tightly coupled to the secretory movement of chloride. In peritubular perfusion experiments with poorly permeant cyclamate anion replacing chloride, tubular hydrogen ion secretion continued at an only slightly reduced rate. These observations argue for the unspecific nature of coupling between hydrogen and other ion transport mechanisms. It is probably also true for the state of chronic hypokalemia which stimulates proximal tubular hydrogen ion secretion. It has been proposed that the electrical potential difference is elevated across the peritubular cell membrane in states of potassium deficiency, thereby accelerating passive peritubular bicarbonate extrusion from the cell and lowering intracellular pH (Rector 1971). Another unspecific coupling effect is that between chloride and hydrogen ion movement. It has been shown that the availability of chloride profoundly limits the tubule's ability to adjust its rate of hydrogen ion excretion (Kassirer *et al.* 1965). This effect is most likely due to the restraints on bicarbonate reabsorption (and hydrogen ion secretion) imposed by the necessity to maintain electroneutrality in the presence of distortions of the normal bicarbonate chloride concentration ratio in the glomerular filtrate.

DISTAL NEPHRON

Similar to the proximal tubule either active, tightly coupled carrier mediated ion transport (Berliner 1961) or ionic interactions of more unspecific nature could account for the well-established interrelationship between Na, K and H-ion movement at the distal tubular and collecting duct level (Giebisch

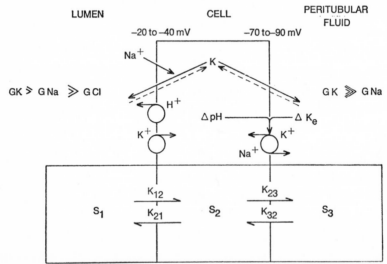

Fig. 4. Schematic presentation of some properties of a single distal tubule cell. Note different electrical polarization across the luminal and peritubular cell boundaries. Sodium entry across the luminal cell membrane is facilitated by a higher sodium permeability at this site than across the peritubular cell membrane. G_K, G_{Na} and G_{Cl} represent conductances. The coupling ratio of Na-K at the peritubular membrane is unknown. It is also unknown whether active K uptake across the luminal cell membrane is coupled to other ion transport.

Lower part: Schematic illustration of three-compartment system consisting of tubular lumen, the cell compartment and the peritubular fluid compartment. S_1, S_2, and S_3 denote amount of solute in individual compartments. k_{12}, k_{21}, k_{23}, and k_{32} are rate coefficients defining unidirectional solute movement across the luminal and peritubular cell membrane, respectively (Giebisch *et al.* 1971a, Wiederholt *et al.* 1971, Giebisch *et al.* 1971b).

et al. 1971a). We shall present evidence that many of the observed cationic exchange and coupling operations between lumen and cell are the result of complex ionic transfer processes across the luminal and peritubular cell membranes.

(1) At the luminal membrane, sodium reabsorption, low anion permeability (Malnic & Giebisch, 1972), generation of an electrical potential difference and *passive* movement of another cation species in the secretory direction account for a highly variable component of distal tubular potassium secretion (Giebisch 1971, Giebisch *et al.* 1971a).

Linkage between sodium reabsorption and potassium secretion is provided for by the electrical potential difference. In principle, the latter is generated by different rates of reabsorptive cation and anion movement. The magnitude of the transepithelial potential difference (lumen negative) along both the distal tubule and collecting ducts can be shown to depend critically upon the sodium concentration in the lumen (Grantham *et al.* 1970, Giebisch & Malnic 1966).

(2) Not less important than events on the luminal cell membrane are transport processes across the *peritubular* cell membrane (Giebisch 1971, Wiederholt *et al.* 1971, Giebisch *et al.* 1971b). At this cell site uptake of potassium is subject to modification by changes in the extracellular potassium concentration and by changes in pH. Evidence will be presented to indicate that alterations in peritubular potassium uptake determine the intracellular potassium concentration, the driving force for passive potassium egress from the cell into the lumen, and thus the net secretory transport rate of this ion.

Fig. 4 summarizes some relevant features of a distal tubule cell. It will serve to discuss the transport processes residing within this nephron segment as well as localize the site and mechanisms of some ionic coupling mechanisms. In particular, the role of potassium in tubular ion exchange mechanisms, especially its interaction with sodium and hydrogen ion movement, will be examined.

In the rat, net secretion of potassium along the distal tubule accounts for most of the excreted potassium (Giebisch 1971, Giebisch *et al.* 1971a, Malnic *et al.* 1964, Malnic *et al.* 1966a). Investigations dealing with the respective electrochemical potential differences across individual cell boundaries led to the definition of the transport elements incorporated in the cell schema.

It is proposed that potassium entry into the tubule lumen across the luminal cell membrane is passive and driven by the combined electrical and chemical potential difference. Changes in the activity of the peritubular potassium uptake mechanisms control the intracellular potassium concentration and thus, in part, the driving force acting on potassium as it moves across the luminal cell membrane.

Potassium entry across the luminal cell membrane is opposed by active potassium reabsorption. Evidence in support of such a mechanism are: observations of net K reabsorption from the lumen against an electrochemical potential difference in potassium depletion, the finding that the luminal K concentration is always lower than that to be expected from the transepithelial electrical potential difference (Malnic et al. 1966b), and the observation that the luminal K concentration increases subsequent to the administration of cardiac glycosides (Duarte et al. 1971).

It has long been recognized that a low rate of sodium excretion is associated with reduced urinary potassium excretion. What is the nature of this ionic interdependence? It has clearly been shown by micropuncture experiments done in low-sodium states that the amount of sodium in the distal tubule is adequate to permit sodium-potassium exchange at a rate far in excess of the observed one (Berliner 1961, Malnic et al. 1966a). It is along the collecting ducts that both sodium and potassium are avidly reabsorbed. Since a diminished distal tubular sodium load cannot be responsible for the curtailment of potassium excretion, it has been proposed that a major part of the interrelationship is indirect and due to changes in the electrical potential difference.

Reduction of the sodium concentration in the distal tubule (Malnic et al. 1966b) and within the collecting duct (Grantham et al. 1970) lowers the intratubular electrical potential difference (lumen negative). This event renders the potassium secretory mechanism less efficient, leading along the distal tubule to reduced tubular secretion and along the collecting duct to enchanced back leakage of potassium into the papillary interstitial space. In support of this view enhanced potassium reabsorption along the collecting ducts has been demonstrated during dietary sodium depletion by both indirect (Malnic et al. 1966a, Giebisch 1971) and direct (Diezi et al. 1970) methods.

In addition to changes in the electrical potential difference it is also possible that concentration changes along the collecting duct affect the rate of active secretion of potassium in exchange for sodium. Grantham and his associates have provided evidence for such a mechanism in the isolated collecting duct of the rabbit (Grantham et al. 1970). It differs from the situation in the distal tubule and could be an additional mechanism by which changes in the luminal sodium concentration affect potassium secretion. It is clear, however, that this mechanism could not account for the stimulation of potassium reabsorption which has been regularly observed in low-sodium states (Diezi et al. 1970, Malnic et al. 1966a).

The delivery of increased amounts of tubular fluid (and sodium) to the distal tubule is known to enhance tubular potassium secretion and urinary excretion (Malnic et al. 1966a). The previously mentioned observation that the normally reabsorbed amounts of sodium greatly exceed the secreted potassium-by at least an order of magnitude-excludes the possibility that an increased rate of Na delivery stimulates K secretion by specifically rendering a previously unsaturated transport mechanism more efficient. Rather, it appears that the distal tubular transport system of potassium is flow-limited and responds under many, but not all, conditions to the delivery of increased fluid loads rather unspecifically with an increased rate of potassium secretion.

The observation that potassium concentrations, in control free-flow samples from distal tubules, are quite similar to steady-state concentrations in stop-flow microperfusions (Malnic et al. 1964, Malnic et al. 1966a, Malnic et al. 1966b) and remain unchanged over a several-fold increment of distal tubular flow rates (Giebisch 1971, Wiederholt et al. 1970, Morgan & Berliner 1969), suggests that free-flow concentrations are always close to steady-state values. Obviously, if the distal tubular potassium concentration is independent of flow rate, the amount of potassium secreted at each point varies directly and proportionately with volume flow. Other factors remaining constant, net secretion into the distal tubule increases with volume flow past the secretory site. The kaliuresis frequently observed after administration of diuretics is probably a consequence of this phenomenon.

The greater efficacy of a distal sodium load as compared to a fluid load of nonelectrolytes, for instance mannitol (Malnic et al. 1966a) and urea (Khuri et al., unpublished observations) in promoting urinary potassium excretion is most likely due to the supportive role of sodium ions in maintaining intratubular negativity values at optimal levels.

It is also well established that the distal tubule is responsible for the wide range in potassium secretion rates which are associated with acid-base disturbances (Malnic & Giebisch 1972, Malnic et al. 1971). Alkalosis stimulates and acute acidosis depresses distal tubular potassium secretion. Since potassium transport does not appear to be carrier-mediated across the luminal cell membrane, it is virtually certain that potassium ions do not, as has been previously believed, compete for a carrier common to both K and H ions.

A direct comparison between the simultaneously measured rates of bicarbonate (H ion) and potassium transport during widely varying patterns of potassium excretion has also not supported a reciprocal relationship between the transport rates of these ions in acid-base disturbances (Malnic & Giebisch 1972, Malnic et al. 1971, Giebisch et al. 1971a). Compared to the normal acid-base status, the rate of distal tubular bicarbonate reabsorption and thus, by implication, that of hydrogen ion secretion has been found to be increased by enhancement of the delivery rate of bicarbonate ions to the distal tubule. This is a consequence of the fact that the latter is a highly unsaturated transport site with respect to both sodium and bicarbonate reabsorption (Malnic et al. 1972, Strieder et al. 1969). Accordingly, bicarbonate reabsorption increases during alkalosis at a time of augmented potassium secretion, which leads to a parallel increase of both potassium and hydrogen ion secretion.

It appears most reasonable to assign a key role to intracellular pH changes in the linkage between acid-base disturbances and potassium transport. It is our view that intracellular pH regulates the rate of uptake of potassium ions across the peritubular cell membrane, independent of whether such cellular pH changes are associated with an increase in hydrogen ion secretion rate or not (Giebisch et al. 1971a, Malnic et al. 1971).

Table II. Summary of kinetic parameters obtained in perfused rat distal tubules.

	Controls	Low K	High K	5 % NaHCO$_3$
S_2 10^{-8} mEq/mm tubule	12.6	6.5	22.2	24.5
$\emptyset_{32}$ 10^{-8} mEq/min. mm tubule	6.76	2.4	16.7	13.4

If the kinetics af transepithelial distal tubular potassium movement are studied in tracer flux studies (Wiederholt et al. 1971, Giebisch et al. 1971b), information on transport rate coefficients and on the size of the potassium transport pool can be obtained. The main conclusions which can be drawn from such studies are the following: (1) With stimulation of potassium secretion (either by alkalosis or by exogenous potassium loading), the potassium transport pool (S_2 in Fig. 4) increases. (2) This increase in the amount of potassium, partaking in the transport process, is due primarily to stimulation in potassium uptake across the peritubular cell border. (3) No evidence was obtained that K transport across the luminal cell boundary is affected in any other way than by changes in transmembrane potential or chemical gradients. Rather by primary changes in the transport properties of the luminal cell boundary, it appears that the driving force acting on passive potassium movement across the luminal cell membrane is regulated by changes in the intracellular K content. It is this latter variable which is subject to modulation by changes in *peritubular* potassium uptake.

In view of the evidence surveyed it appears doubtful whether carrier-mediated, tight coupling of strong electrolyte transport occurs at the luminal cell membrane of kidney tubule cells. The situation may be different at the peritubular membrane. At this cell site, the existence of a coupled Na-K exchange pump is supported by several lines of evidence. Important observations are the dependence of cellular sodium extrusion upon extracellular K concentration, and that of cellular potassium uptake upon cellular sodium concentrations. The effects of ouabain on Na and K movement and ATPase activity are also relevant (Giebisch *et al.* 1971a, Whittembury 1968, Whittembury & Proverbio 1970). Nevertheless, an additional and growing body of evidence indicates that added to such coupled ion exchange, shared by many cell systems, a second mode of Na extrusion exists in renal tubule cells. The latter is a directly electrogenic process, independent of K in the external media, particularly sensitive to the blocking effect of ethacrynic acid and relatively insentive to ouabain (Giebisch *et al.* 1971a, Whittembury 1968, Whittembury & Proverbio 1970). Both sodium transport processes participate in net tubular sodium reabsorption. Depending on the experimental condition, net sodium transport can be sustained by either of the two peritubular transport modes.

ACKNOWLEDGEMENT

Work of the authors was supported by the National Institutes of Health, the American Heart Association and the Fund. de Amparo a Pesquisa do Est. São Paulo.

REFERENCES

Bentzel, C. J., Parsa, B. & Hare, D. K. (1969) Osmotic flow across proximal tubules of Necturus. *Amer. J. Physiol. 217*, 570–580.

Berliner, R. W. (1961) Renal mechanism for potassium excretion. *Harvey Lect. 55*, 141–171.

Boulpaep, E. L. (1971a) Electrophysiological properties of the proximal tubule: Importance of cellular and intercellular transport pathways. In *Electrophysiology of epithelial cells*, ed. Giebisch, G., pp. 91–112, F. K. Schattauer Verlag, Stuttgart, New York.

Boulpaep, E. L. (1971b) Electrophysiology of proximal and distal tubules of the auto-perfused dog kidney. *Amer. J. Physiol. 221*, 1084–1096.

Boulpaep, E. L. (1972) Permeability changes of the proximal tubule of Necturus during saline loading. *Amer. J. Physiol. 222*, 517–531.

Brenner, B. M., Falchuk, K. H., Keimowitz, R. I. & Berliner, R. W. (1969) The relationship between peritubular capillary protein concentration and fluid reabsorption by the renal proximal tubule. *J. clin. Invest. 48*, 1519–1531.

Brenner, B. M., Troy, J. L. & Daugharty, T. M. (1971) On the mechanism of inhibition in fluid reabsorption by the renal proximal tubule of the volume-expanded rat. *J. clin. Invest. 50*, 1596–1602.

Diezi, J., Michoud, P., Aceves, J. & Giebisch, G. (1970) Micropuncture study of Na and K transport in collecting ducts of rats. *Fed. Proc. 29*, 271.

Duarte, Ch. G., Chomety, F. & Giebisch, G. (1971) Effect of amiloride, ouabain, and furosemide on distal tubular function in the rat. *Amer. J. Physiol. 221*, 632–640.

Frömter, E., Müller, C. W. & Wick, T. (1971) Permeability properties of the proximal tubular epithelium of the rat kidney studied with electrophysiological methods. In *Electrophysiology of Epithelial Cells*, ed. Giebisch, G., pp. 119–146, F. K. Schattauer Verlag, Stuttgart, New York.

Giebisch, G. (1969) Functional organization of proximal and distal tubular electrolyte transport. *Nephron 6*, 260–281.

Giebisch, G. (1971) Renal potassium excretion. In *The Kidney, Morphology, Biochemistry, Physiology*, ed. Rouiller, C. & Müller, A. F. Vol. 3, pp. 329–382. Academic Press, New York.

Giebisch, G. (1972) Coupled ion and fluid transport in the Kidney. *New Eng. J. Med. 287*, 913–919.

Giebisch, G., Boulpaep, E. L. & Whittembury, G. (1971a) Electrolyte transport in kidney tubule cells. *Proc. roy. Soc. B 262*, 175–196.

Giebisch, G., Curran, P. F., Mello-Aires, M. & Malnic, G. (1971b) Measurement of K-42 fluxes across single distal tubules of rat kidney. *Proc. Int. Union Physiol. Sci. 9*, 203.

Giebisch, G. & Malnic, G. (1966) Effect of ionic substitutions upon distal transtubular potential differences in rat kidney. *Amer. J. Physiol. 211*, 560–568.

Giebisch, G. & Windhager, E. E. (1964) Renal tubular transfer of sodium, chloride and potassium. *Amer. J. Med. 36*, 643–669.

Grantham, J. J., Burg, M. B. & Orloff, J. (1970) The nature of transtubular Na and K transport in isolated rabbit renal collecting tubules. *J. clin. Invest. 49*, 1815–1826.

Green, R., Windhager, E. E. & Giebisch, G. (1972) The effects of transepithelial oncotic pressure gradients on proximal tubular fluid movement in the rat. *Proc. 4th. Int. Congr. Nephrol.,* Mexico City. p. 145, 1972.

Kassirer, J. P., Berkman, P. M., Lawrenz, D. R. & Schwartz, W. B. (1965) The critical role of chloride in the correction of hypokalemic alkalosis in man. *Amer. J. Med. 38,* 172–189.

Khuri, R., Strieder, K., Wiederholt, M. & Giebisch, G. Unpublished obs.

Lorentz, W. B., Lassiter W. E. & Gottschalk, C. W. (1972) Renal tubular permeability during increased intrarenal pressure. *J. clin. Invest. 51,* 484–492.

Mac Dougal, S. & Wright F. (1972) *Kidney International* 2, 304–317.

Malnic, G. & Giebisch, G. (1972) Mechanism of renal hydrogen ion secretion. *Kidney International 1,* 280–296.

Malnic, G. & Giebisch, G. Some electrical properties of distal transtubular epithelium in the rat. *Amer. J. Physiol.* 223, 797–808, 1972.

Malnic, G., Klose, R. M. & Giebisch, G. (1964) Micropuncture study of renal potassium excretion in the rat. *Amer. J. Physiol. 206,* 674–686.

Malnic, G., Klose, R. M. & Giebisch, G. (1966a) Micropuncture study of distal tubular potassium and sodium transfer in rat kidney. *Amer. J. Physiol. 211,* 548–559.

Malnic, G., Klose, R. M. & Giebisch, G. (1966b) Microperfusion study of distal tubular potassium and sodium transfer in rat kidney. *Amer. J. Physiol. 211,* 548–559.

Malnic, G. & Mello-Aires, M. (1970) Microperfusion study of anion transfer in proximal tubules of rat kidney. *Amer. J. Physiol. 218,* 27–32.

Malnic, G. & Mello-Aires, M. (1971) Kinetic study of bicarbonate reabsorption in proximal tubule of the rat. *Amer. J. Physiol. 220,* 1759–1767.

Malnic, G., Mello-Aires, M. & Giebisch, G. (1971) Potassium transport across renal distal tubules during acid-base disturbances. *Amer. J. Physiol. 221,* 1192–1208.

Malnic, G., Mello-Aires, M. & Giebisch, G. (1972) Micropuncture study of renal tubular hydrogen ion transport during alteration of acid-base equilibrium in the rat. *Amer. J. Physiol. 222,* 147–158.

Morgan, T. & Berliner, R. W. (1969) A study by continuous microperfusion of water and electrolyte movements in the loop of Henle and distal tubule of the rat. *Nephron 6,* 388–405.

Rector, F. C. (1971) Renal secretion of hydrogen. In *The Kidney, Morphology, Biochemistry, Physiology,* ed. Rouiller, C. & Müller, A. F. Vol. 3, pp. 209–352. Academic Press, New York.

Robson, A. M., Shrivastava, P. L. & Bricker, N. S. (1968) The influence of saline loading on renal glucose reabsorption in the rat. *J. clin. Invest. 47,* 329–335.

Strieder, N., Khuri, R. & Giebisch, G. (1969) Recollection micropuncture study of distal tubular sodium reabsorption during graded extracellular volume expansion in the rat. *Amer. Soc. Nephrol.* 64 Abs.

Ussing, H. H. & Windhager, E. E. (1964) Nature of the shunt path and active sodium transport path through frog skin epithelium. *Acta physiol. scand. 61,* 484–504.

Whittembury, G. (1968) Na and water transport in kidney proximal tubular cells. *J. gen. Physiol 51,* 303s–314s.

Whittembury, G. & Proverbio, F. (1970) Two modes of Na extrusion in cells from guinea-pig kidney slices. *Pflügers Arch. ges. Physiol. 316,* 1–25.

Whittembury, G. & Rawlins, F. A. (1972) Evidence of a paracellular pathway for ion flow in the kidney proximal tubule: Electromicroscopic demonstration of lanthanum precipitate in the tight junction. *Pflügers Arch. ges. Physiol. 330,* 302–309.

Wiederholt, M., Sullivan, W. J., Giebisch, G., Curran, P. F. & Solomon, A. K. (1971) Potassium and sodium transport across single distal tubules of Amphiuma. *J. gen. Physiol. 57*, 495–525.

Windhager, E. E., Boulpaep, E. L. & Giebisch, G. (1966) Electrophysiological studies on single nephrons. *Proc. 3rd. Int. Congr. Nephrol. (Wash.) 1*, 35–47.

Windhager, E. E., Lewy, J. E. & Spitzer, A (1969) Intrarenal control of proximal tubular reabsorption of sodium and water. *Nephron 6*, 247–259.

DISCUSSION

WHITTEMBURY: You showed a slide of Malnic demonstrating an acidification at the same rate whether you have sodium in the lumen or choline. What is the rate of reabsorption of choline under those circumstances?

GIEBISCH: Compared to sodium chloride it is very slow.

ULLRICH: I think this is a very peculiar experiment. As far as I remember Dr. Malnic placed in the lumen a solution of 100 mEq bicarbonate sometimes with sodium and sometimes without sodium or with choline, therefore producing very large differences in the transtubular potential. I just think he has measured transient states and really to describe it as a transport of the buffer is very difficult, because we don't know which buffer component is really moving. It is very difficult from such experiments to infer that there is no coupling or to say that there is no tight coupling.

GIEBISCH: It appears to me if you have initially no sodium in the lumen and if the sodium were vitally important for rigid coupling between hydrogen ion secretion and sodium reabsorption that you would find a slowing of hydrogen ion secretion. It is true that with the high permeability, the lumen will not remain sodium-free; but the fact that you have a single exponential decrease in luminal bicarbonate concentration that doesn't show any change or any break with time would indicate that hydrogen ion secretion is not terribly sensitive to the intraluminal sodium concentration.

BERLINER: I would emphasize the point that the distal tubule has certain characteristics that are clearly different from some segments that are further down. In the distal tubule no matter how long you leave the sodium, there never is a sodium concentration below about 30–40 mEq/l, whereas we know we can get urines that have sodium concentrations of 1 mEq/l or even sometimes less. I think it has to remain an open question whether the characteristics of the sodium and potassium movements in the distal tubule are the same as those occurring in the collecting duct system.

GIEBISCH: However, there are also some additional interesting features of the collecting duct system in vivo. For instance, it is possible to demonstrate the presence of net potassium reabsorption along the collecting ducts under conditions of sodium and potassium depletion. This indicates the presence of more than a sodium-potassium exchange pump at this site. Apparently, under these conditions not only is potassium secretion suppressed but a component of active reabsorption unmasked. As far as I know this element of active potassium reabsorption has not yet been demonstrated in the isolated collecting tubule preparation.

FRÖMTER: I would like to come back to Dr. Malnic's experiment and ask first: what was the outside solution?

GIEBISCH. No peritubular perfusion under these conditions.

FRÖMTER: This means that you had also a considerable bicarbonate concentration difference from lumen to outside, which together with the small potential difference that we find under those conditions, must have driven bicarbonate ions out of the lumen passively. Do you correct for this or do you equate all bicarbonate flux as hydrogen ion secretion?

GIEBISCH: No, you cannot quantitatively equate it, one could have an element of passive bicarbonte loss, I'm quite aware of that.

FRÖMTER: Yes, and what is your idea about its quantitative importance.

GIEBISCH: The major element responsible for the apparent disappearance of bicarbonate from the tubular lumen is hydrogen ion secretion (see Malnic & Mello-Aires 1971). Not only does the process of bicarbonate loss from the lumen obey single exponential kinetics over the whole range of tubular concentrations but loss of bicarbonate is also dramatically slowed after administration of the carbonic anhydrase inhibitor Diamox. One would not expect this to occur if the loss of bicarbonate from the tubular lumen were predominantly determined by diffusion.

FRÖMTER: My data (Frömter et al. 1971) would indicate that the chloride versus the bicarbonate permeability is only 2:1 in rat proximal tubule.

TOSTESON: I don't understand the equilibrium fluid experiment with which you propose to rationalize the problem of whether there is so-called isotonic

Malnic, G. & Mello-Aires, M. (1971) Kinetic study of bicarbonate reabsorption in proximal tubule of the rat. *Amer. J. Physiol. 220,* 1759–1767.

Frömter, E., Müller, C. W. & Wick, T. (1971) Permeability properties of the proximal tubular epithelium of the rat kidney studied with electrophysiological methods. In: *Electrophysiology of Epithelial Cells,* ed. Giebisch, G., F. K. Schattauer Verlag. Stuttgart, New York, pp. 119–138.

reabsorption. I take it that the goal of the experiment is to distinguish between direct and indirect effects of lowering the peritubular protein concentration on reabsorption. What was the composition of the equilibrium fluid?

GIEBISCH: The "equilibrium" fluid contained 110 mEq/L sodium and enough raffinose to achieve isotonicity. The rationale of our approach was the following. The effect of changes in peritubular protein concentrations on proximal tubular sodium transport could be due to a direct osmotic effect of proteins or could be indirect by affecting the transfer of reabsorbate into the capillary circulation. A direct osmotic effect of the transepithelial protein concentration difference should manifest itself in the absence of active sodium transport whereas an effect of peritubular protein concentration differences on the transfer of reabsorbate from interspace region into the peritubular capillary compartment would critically depend on the presence of active proximal tubular sodium transport. Our results do not support the view that the transepithelial protein concentration difference acts as an important direct driving force across the proximal tubular epithelium since changing the peritubular protein concentration did result in only minor changes in the rate of proximal fluid transfer when net sodium transport was abolished by perfusion of the lumen with either an "equilibrium" fluid containing raffinose or with a Ringer's solution containing 4 mM cyanide.

TOSTESON: It seems to me that the result of this experiment depends upon the reflection coefficient for raffinose in the terminal bars.

GIEBISCH: This is correct but this complication would not apply to the experiments in which cyanide was used to suppress proximal tubular sodium transport.

BERLINER: Are you assuming that the raffinose can go through?

GIEBISCH: There is only a small loss of raffinose from the proximal tubule during perfusion.

ULLRICH: As I understand from your data you increased the albumin concentration from 2 to 10 percent and the oncotic pressure up to 60 mm of mercury. With the highest value an approximately 75 percent increase in sodium transport through the tubule was observed. With the normal oncotic pressure difference across the tubular wall, i. e. 25 mm of mercury the change in transport would be, if it is linearly related to the oncotic pressure, about 30 percent. We perfused the same individual tubule alternatively with ultrafiltrate, well balanced saline and serum at the outside and were unable to find any difference in the isotonic absorption surpassing the 10 percent margin. Therefore I doubt that even the smaller changes in oncotic pressure observed during volume expansion could account for the decrease in isotonic absorption.

GIEBISCH: Maybe Dr. Windhager would comment.

WINDHAGER: The Brenner data fall on the steep line between the two points which were shown by Dr. Giebisch. I shall discuss this during my presentation later during the symposium.

GIEBISCH: In saline diuresis you can reverse the inhibition of proximal tubular fluid transport by selectively perfusing the peritubular capillaries with a higher protein solution (Brenner *et al.* 1971).

ORLOFF: Dr. Ullrich, you say now that there is no difference in reabsorption with or without protein?

ULLRICH: Dr. Sato in our laboratory could repeat the shrinkage procedure in the same tubule up to 6–10 times. In these experiments, where the same tubule segment acted as its own control, the difference in isotonic absorption when serum or Krebs Ringer bicarbonate containing 5 mM glucose acted as the capillary perfusate was less than 10 percent.

BERLINER: This is one of the things people don't agree about because there are many observations on both sides. It should also be mentioned that the effect of protein has been found, both with free flow and with split-drop, in different laboratories; however there are data with the opposite implication.

The group in Dallas (Imai & Kokko 1972) have reported that when they use an ultrafiltrate of plasma as a medium they get a lower rate of reabsorption than they do when they use plasma on the outside of the perfused tubule. Dr. Burg has done this with protein added to the outside and he finds no effect. He finds that there are certain ultrafiltrates of plasma with which he gets very low absorption but adding protein to those ultrafiltrates does not

Imai, M. & Kokko, J. P. (1972) Effect of peritubular protein concentration on reabsorption of sodium and water in isolated perfused proximal tubules. *J. Clin. Invest.* 51, 314–325.

restore the reabsorption. On the other hand if he makes his ultrafiltrates with a different filter and I don't remember what the filters are that have been used for making those various ultrafiltrates, he can obtain ultrafiltrates that give normal rates of reabsorption. So these laboratories disagree even in that situation and that makes it a little difficult to interpret.

WINDHAGER: But I think it is correct to say in all free flow micropunctures in which peritubular concentration was measured and hydrostatic pressure was measured, at least in the acute situation all point in the direction that colloid osmotic pressure functions not as a driving force but is capable of regulating proximal tubular fluid reabsorption.

BERLINER: That is right. I should point out that with the isolated tubule there is a difference from the intact animal and a very important one in that there are no capillaries. This might be an effect produced across the capillary wall and thus absent in the isolated tubule but present in the intact animal.

TOSTESON: If the "equilibrium fluid" contains a solute like raffinose which presumably has a reflection coefficient of unity in the terminal bars, addition of protein to the peritubular fluid will produce reabsorption only to the extent necessary to make the increase in the osmolar raffinose concentration in the tubular fluid equal to the increase in osmolar protein concentration (including the Donnan effect of the protein on ion distribution) in the peritubular fluid.

BERLINER: The predicted changes might be very small.

TOSTESON: Perhaps the point can be made by a limiting case. If the proximal tubular fluid contains only a solution of raffinose isosmotic with the peritubular fluid (assuming that this solute does not penetrate proximal tubular epithelium) the extent of reabsorption produced by a given increase in the peritubular protein concentration will be limited to the extent required to make the tubular fluid once again isosmotic with the peritubular fluid. If, on the other hand, the tubular fluid contains a NaCl solution isosmotic with peritubular fluid (assuming that Na^+ and Cl^- both can penetrate the proximal tubular epithelium), the extent of reabsorption produced by a given increase in the peritubular protein concentration will continue even when tubular and peritubular fluids are isosmotic and will be limited only by the development of electrical potential and pressure differences across the peritubular epithelium necessary to bring the system to Donnan equilibrium. In the former case, reabsorption may be said to occur by osmosis. In the latter case, it may be said to occur by colloid osmosis. In the former case, the reabsorption is free of solute. In the latter case, the reabsorbate is a solution of NaCl isosmotic with tubular fluid.

GIEBISCH: The fact that the results are similar in perfusion experiments with raffinose ("equilibrium" solution) and without raffinose (Ringer's plus cyanide) argues strongly against proteins providing a major osmotic driving force for fluid movement.

LEYSSAC: On the regulatory function of these changes in oncotic pressure. These changes have been recorded all in the efferent arteriole but what should be regulating proximal tubular reabsorption is the average oncotic pressure in the capillaries. We know that the more there is filtered the more there is reabsorbed, so any effect will be very much minimized by the dilution from the reabsorbate. These changes of 25 mm mercury recorded in the efferent arteriole will be minimized or will be very much reduced in the mean oncotic pressure of the capillaries. They will reduce the force with one order of magnitude or so.

BERLINER: That's not true at least for the studies in our laboratory. We measured the peritubular protein concentration and the changes were observed there.

LEYSSAC: How close was your pipette to the star vessels and in which direction did you collect? Are you sure you did not collect efferent arteriolar blood?

BERLINER: Yes, because the concentration of protein is much lower in the peritubular than in the efferent arteriole. Brenner has done both. They are different.

LEYSSAC: And you find a relationship between proximal reabsorption rate and the protein concentration in the capillaries under spontaneous variation?

BERLINER: They are not spontaneous. The protein concentration change was produced by injection of albumin solution into the renal artery.

LEYSSAC: In these experiments you obtained a relationship only within the first 30 sec after injection, in which time the micropuncture technique does not permit a correct estimate of absolute rates of reabsorption. 5 min after the injection the apparent relationship had disappeared. From the point of view of regulatory mechanisms the major point is whether or not a correlation exists under more physiological steady state conditions; and whether the changes in peritubular oncotic pressure are large enough to account quantitatively for the changes in reabsorption rate, which may vary spontaneously by a factor of 2 or more.

BERLINER: Your statement that a correct estimate of absolute rates of reabsorption cannot be made in the first 30 seconds after injection is certainly debatable. However, we need not concern ourselves too much with whether or not this is correct since Barry Brenner has repeated these observations using continuous direct perfusion of the peritubular capillary circulation in which circumstance a steady state can be achieved and he has obtained the same result.

Coupling of the Transport Processes Across the Brush Border of the Proximal Renal Tubule

K. J. Ullrich, K. Sato & G. Rumrich with the technical assistance of S. Klöss

INTRODUCTION

The transport processes which take place across the brush border of the proximal tubule are: *reabsorption* of sodium, glucose and aminoacids from the lumen into the cell and *secretion* of hydrogen ions into the lumen. The question we were interested in was whether these transport processes across the brush border are coupled to each other. To answer this question we proceeded in 2 ways: 1. We attempted to drastically reduce one transport process and observed the effect on the others; 2. We attempted to modify the brush border chemically, either by applying large molecular size group reagents which do not penetrate the cell membranes, by fractional digestion of the brush border with protein- or glycoprotein-splitting enzymes or by applying specific brush border antibodies and observing how the different transport processes are affected. The data presented here indicate that sodium absorption is linked to the secretion of hydrogen ions and that the reabsorption of glucose and amino acids depends on sodium absorption.

METHODS

In the experiments to be reported, the sodium reabsorption from the proximal convolution of rats was measured by the shrinking droplet procedure (Gertz 1963, Ullrich *et al.* 1969). In this method the tubule is filled with coloured castor oil and a droplet of Ringer's solution is injected, which splits the oil column into two. During reabsorption of the injected Ringer solution, the oil columns approach each other, this process being photographed at 5 sec. intervals. From the velocity of approach and the luminal sodium concentration, the rate of isotonic sodium absorption per tubular length is calculated. One luminal (oil column) diameter for the meniscus correction was used

Max-Planck-Institut für Biophysik. 6 Frankfurt a. Main, Germany.

»Transport Mechanisms in Epithelia«, Munksgaard, Copenhagen.

(Györy 1971). Substances applied to the brush border were perfused through the lumen for periods of 30 sec. to 4 min. before the shrinking procedure was started. Usually the substances tested were also added to the shrinking droplet.

Glucose and histidine transport were evaluated by measuring the static head concentration (Kedem & Caplan 1965) which is the concentration difference between lumen and interstitium at zero net flux of water and solutes. To obtain this situation raffinose was added to the luminal perfusate at a concentration of 0 to 50 mM depending on the respective sodium transport rate. The droplet remained within the tubule for 45 sec. which is ample time for the static head concentration difference ($\triangle c$) for glucose (Loeschke et al. 1969) and histidine (Lingard et al. 1972) to be reached.

The concentration difference ($\triangle c$) could be evaluated by measuring the radioactivity of ^{14}C-glucose or ^{14}C-histidine in the capillary perfusate and the collected luminal fluid since the specific activities in the lumen and perfusate were the same. In the static head situation, the active outward transport (J_o) is balanced by the passive back flux or $J_o = P \cdot \triangle c$. When the permeability (P) remains constant, $\triangle c$ is a measure of the active outward transport.

Interrelation of Na$^+$ absorption and H$^+$ secretion

Fig. 1 demonstrates the influence of the ambient HCO_3^- concentration on the isotonic sodium absorption (Ullrich et al. 1971): The transtubular net sodium transport was greater than $450 \cdot 10^{-7}$ μMol/cm $\cdot$ sec when the bicarbonate concentration in the capillary perfusate was above 30 mMol/l. It fell to about $120 \cdot 10^{-7}$ μMol/cm $\cdot$ sec when the perfusate was bicarbonate-free. This means that the presence of bicarbonate is essential for normal isotonic absorption of sodium.

A further analysis (Table I), in which the bicarbonate buffer was replaced by other buffers, showed that the sulfonamide buffers glycodiazine and sulfamerazine and also butyrate can substitute for bicarbonate, but that propionate, acetate and α-aminoisobutyrate are much less effective. The common denominator of the buffers which could replace bicarbonate is that the undissociated buffer form is lipid soluble but the dissociated form is not. From these findings we are inclined to conclude that the H$^+$ secretion which leads to the reabsorption of the HCO_3^- buffer is an important factor which determines the isotonic reabsorption of sodium.

In an earlier study (Ullrich *et al.* 1973) we have shown that the active transport of Na⁺, with and without acetate in the capillary perfusion fluid, is only 39 and 35 % of the correspondent net sodium transport (Table II). The rest of the net sodium transport is due to diffusion and solvent drag. The corresponding active bicarbonate transport rates are 2.4 and $1.1 \cdot 10^{-8}$ mM cm^{-1} sec^{-1}. The ratio of active sodium to active HCO_3 transport is thus either 0.9 or 1.5. This means that the ratio of sodium ions entering the tubular cell to hydrogen ions secreted is either 1/1 or 3/2. Because of the theoretical uncertainties involved in calculating the active portion of the bicarbonate transport, these ratios are not accurate enough to allow evaluation of the degree of coupling. They nevertheless indicate that in the case of coupling the sodium transport must be seriously hampered when the H⁺ transport is abolished.

More recently, Aceves (1972) has shown that the electrical potential difference across the peritubular cell site depends largely on the sodium entrance from the lumen into the cell and that it is this process which is influenced by the ambient bicarbonate concentration. With the whole set of data mentioned, we suggest that the Na⁺ entrance into and the H⁺ secretion out of the proximal tubular cells are coupled processes. The problem which remains is whether this exchange is energetically driven by

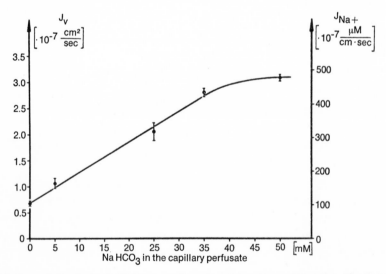

Fig 1. Relation of the isotonic fluid absorption J_V and sodium absorption (J_{Na^+}) to the bicarbonate concentration in the capillary perfusate, (Ullrich *et al.* 1971).

Table I. Effect of different buffers on the isotonic reabsorption in the proximal tubule
n = number of measurements: number of animals in brackets (Ullrich *et al.* 1971).

30 mM buffer (Na⁻salt)	pK for 20°C	Reabsorpion rate: J_V ($\cdot 10^7$-cm³ cm⁻¹ sec⁻¹)	
Bicarbonate	6.3 (38° C)	2.75 ± 0.07	n = 82 (15)
Glycodiazine	5.7	2.74 ± 0.07	n = 72 (9)
Sulfamerazine	6.7	2.74 ± 0.07	n = 38 (3)
Butyrate	4.8	2.48 ± 0.16	n = 34 (2)
Propionate	4.9	2.03 ± 0.11	n = 23 (2)
Acetate	4.8	1.85 ± 0.08	n = 32 (3)
α-Aminoisobutyrate	2.6 (pK₁), 4.4 (pK₂)	1.10 ± 0.07	n = 14 (2)

Table II. Na^+ and HCO_3^- across the proximal convolution (10^{-8}mM cm⁻¹ sec⁻¹) (Ullrich *et al.* 1973).

	active		net	
	with acetate	without acetate	with acetate	without acetate
Na^+ transport	2.2	1.7	5.65	4.85
HCO_3^- transport	2.4	1.1	0.92	0.92
ratio	0.92	1.54		

ATP or whether the downward running Na^+ influx provides enough energy to drive the H^+ ions in the opposite direction.

In this respect, the enzyme patterns described by Kinne *et al.* (1971) are of some interest. In their brush border preparation from kidney cortex they found a Mg^{++}-ATPase and low activities of a HCO_3^- stimulated ATPase, but no Na^+, K^+-ATPase (Kinne, personal communication). Since the HCO_3^- stimulated ATPase is assumed to be involved in the H^+-HCO_3^- transport in the stomach and pancreas (Kasbekar & Durbin 1965, Sachs *et al.* 1965, Simon *et al.* 1972) one may speculate that at least part of the H^+ secretion may be coupled energetically to ATP hydrolysis.

Interdependence of glucose-, aminoacid- and sodium absorption

Are the other brush border transport processes affected, when sodium transport is reduced? As shown in Fig. 2, in a bicarbonate free system, the

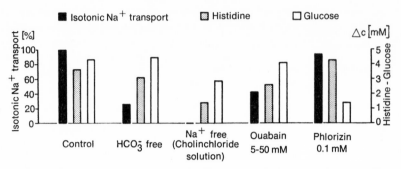

Fig. 2. Effect of omission of HCO₃⁻ og Na⁺ and the addition of Ouabain or phlorizin to the capillary and luminal perfusates on the isotonic Na⁺ absorption and the static head concentration difference ($\triangle$c) of histidine and glucose.

net transport of sodium is reduced by 75 % as compared to the control values, but the $\triangle$c of histidine is reduced by only 16 % and the $\triangle$c of glucose is unaffected. When all the perfusate sodium is replaced by choline chloride, the reduction of $\triangle$c is 61 % for histidine and 36 % for glucose.* It is probable, however, that the active outward transport of sodium was not zero, since a Na⁺ determination of the luminal stop flow microperfusion fluid revealed a mean concentration of 12 mEq/l with a range of 7 to 29 mEq/l.

Another set of experiments which could give us information about the mutual dependence of the brush border transport processes is the application of ouabain. After perfusion with ouabain saturated Ringer's solution $(2.5 \cdot 10^{-2}M)$, sodium transport is reduced by 58 %, the $\triangle$c histidine by 30 and the $\triangle$c glucose by 7 %. If it is assumed that the permeability of the tubular wall for glucose and histidine does not change, one may suggest from these data that histidine as well as glucose transport depends on the net sodium transport. The fact that with the remaining Na⁺ concentration and sodium transport the aminoacid transport is lower than the glucose transport may indicate that the affinity of Na⁺ for the glucose carrier is higher than for the histidine carrier.

The reverse test, the specific inhibition of glucose transport with $10^{-4}M$ phlorizin shows no inhibition of the net sodium and no decrease in the $\triangle$c of histidine. This is not surprising since under our experimental conditions

---

* When the perfusate contained 125 mM cholinechloride and 25 mM cholinebicarbonate the $\triangle$c glucose was reduced by 67 %.

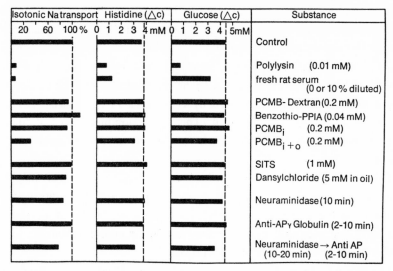

Fig. 3. Effect of basic polypeptides, SH-reagents, NH₂-reagents, neuraminidase and anti-aminopeptidase-γ-Globulin on the isotonic Na⁺-absorption and the static head concentration difference ($\triangle$c) of histidine and glucose.

the active glucose transport was only about 8 % of the active sodium transport, so that the inhibition of glucose transport should not affect Na⁺ transport even if both transport processes are tightly linked. On the other hand, the active glucose and active histidine transport are of the same order of magnitude. Therefore, the result that the specific inhibition of glucose transport had no effect on histidine transport indicates that both transport processes do not depend directly on each other.

The effect of chemical modifications of the brush border on the transport processes

Polylysine as well as other *positively charged macromolecules* such as polyornithine, polyethylenimine, histone, protaminsulfate and polymyxin B drastically inhibit the proximal isotonic sodium absorption when applied to the luminal surface (Sato & Ullrich 1972). This effect could be reversed by acidic macromolecules such as heparin, polyglutamic acid and polyaspartic acid. As shown in Fig. 3, simultaneously with the sodium absorption, the $\triangle$c of histidine and glucose are also strongly reduced by polylysine.

During experiments with antisera, we have observed that fresh serum of all mammals so far tested, contains a factor which, when applied to the tubular lumen inhibits isotonic Na^+ absorption as strongly as polylysine (Sato et al. 1972). This effect is also prevented by heparin. The factor is storage and heat labile, requires Ca^{++} ions for its action and can be destroyed by neuraminidase and precipitated by concanavallin A. This suggests that the serum factor is a glycoprotein. This factor inhibits both histidine and glucose transport, although the latter is affected to a much smaller extent than with polylysine.

The mechanism of action of the basic macromolecules is open to speculation. By analogy to the mode of action of polylysine on erythrocytes, it seems probable that they reversibly change the surface coat or the lipid bilayer of the microvilli either electrostatically or conformationally, thereby inhibiting all brush border transport processes.

The second group of substances tested were *SH reagents* (Ullrich et al. 1971) (Fig. 3): The mercurial compound PCMB bound to dextran and iodacetate bound to polyproline with the fluorescent marker benzothioxanthene attached to the molecule. These two macromolecular substances do not affect brush border transport processes although they inhibit pancreatic secretion when added to the vascular perfusate of the isolated gland (Wizemann et al. 1973). From this one may conclude that SH groups which may be important in the brush border transport processes could not be reached by large molecular SH reagents. As, however, the next line in the Fig. 3 shows, PCMB alone is also uneffective when added to the luminal perfutase. This is surprising since Bode et al. (1970) have shown that PCMB inhibits the binding of phlorizin to the isolated brush border microvilli. This indicates that PCMB is bound *in vitro* to SH groups at or near the phlorizin sensitive glucose binding site. As, however, is shown here, PCMB given into the lumen apparently does not inhibit the glucose transport. If, however, PCMB is applied to the lumen and the capillaries, the net sodium transport is inhibited by 68 %, histidine transport by 17 % and glucose transport by 15%. In this case it is not clear whether the Na^+, K^+-transport-ATPase located at the contraluminal cell site is primarily affected or if the result is due to effects on metabolic energy production.

To answer this question we perfused the whole kidney with artifical solutions and determined the ATP content of cortical tissue using the conventional freeze stop procedure (Bücher 1947). In an exposed kidney with its normal blood supply, an ATP content of 2.67 ± 0.12 SE μmol/g wet weight was found. After 1 min perfusion with Krebs-Ringer's (-glucose) solution, the ATP content had dropped to 1.42 ± 0.08 SE μmol/g. If the latter solution contained $2 \cdot 10^{-4}$ PCMB, an ATP content 1.39 ± 0.06 SE μmol/g was found after 1 min. and 1.43 ± 0.17 SE μmol/g after 3½ min. perfusion. Although the artificial perfusion *per se* caused a considerable drop in the ATP content of the cortical tissue, the PCMB concentration used in our flux experiments caused no further drop. From this one may suggest that PCMB acted on the plasma membranes rather than on the mitochondrial ATP-production.

The third substances listed in Fig. 3 were *NH₂ reagents*. Stilbene isothio-cyanate sulfonic acid (SITS) which does not enter the cells, and which has a strong anion flux controlling action on erythrocytes (Knauf & Rothstein 1971) had no effect on the brush border transport processes. Dansyl chloride, a fluorescent amino group reagent, was dissolved in mineral oil and applied into the lumen, with the result that the brush border soon showed a very strong fluorescence, indicating that dansyl-chloride was incorporated into or absorbed on the membranes. We do not know, however, how much of the reagent is dissolved in the membranes and how much reacts with NH_2 groups of membrane proteins and phospholipids. Dansyl-chloride did not change any brush border transport processes.

By *fractional digestion* of the isolated brush border and successive enzyme determinations and phlorizin binding studies, Pockrandt-Hemstedt *et al.* (1972) and Thomas & Kinne (1972) constructed the following model of the microvilli membranes. The surface is covered by a glycoprotein coat which could be partly removed by neuraminidase and completely removed by trypsin. Below this outer coat are particles containing aminopeptidase, the function of which is unknown. The aminopeptidase particles could be removed by papain. The remaining microvilli-core contains alkaline phosphatase activity and the phlorizin-sensitive glucose binding sites. This core could be dissolved by sodium dodecylsulfate (Thomas & Kinne 1972).

To gain information about the function of different brush border con-stituents we tested first the effect of neuraminidase. The neurominidase had no effect on the brush border transport processes (Fig. 3). From other cells it is also known, that the neuraminidase did not influence transcellular trans-port, nor the viability of the cells, although it removed a considerable part of the surface coat (Heath 1971). Trypsin and especially papain (1 mg/ml) destroyed the tubule when applied to the lumen for longer than 3–4 min. The brush border and then the whole tubule changed their colour to white which was followed by leakage and bleeding. However, within the first 3 min. before visible cell damage occured, isotonic fluid absorption was unaffected. Thus our attempt to eliminate the different brush border transport processes selectively by fraction digestion has been unsuccessful so far.

Finally we attempted to change the tubular transport processes with *antibodies* against brush border enzymes. We prepared rabbit γ-globulin with a strong antiaminopeptidase activity. The γ-globulin inhibited the amino-peptidase activity of isolated brush border membranes completely in vitro, but it had no effect on *in vivo* brush border transport processes. From this we may conclude that the brush border aminopeptidase has either no transport function or that the antibodies were prevented from reaching the aminopeptidase by the surface glycoprotein coat. In experiments where neuraminidase and Anti-LAP were applied together, the isotonic absorption was reduced by 23 %, $\triangle c$ histidine by 14 % and $\triangle c$ glucose by 18 %. This reduction is rather small and it remains questionable whether it could be attributed to a specific effect of these substances on the brush border or to an unspecific effect caused by the long contact time of these proteins with the brush border membrane.

SUMMARY AND CONCLUSIONS

Using the shrinking droplet and stop-flow microperfusion technique with simultaneous capillary perfusion it was observed that the net sodium transport across the proximal tubule of the rat kidney increased with increasing concentration of bicarbonate or related buffers in the capillary perfusate. The entrance of sodium from the lumen into the cell, which is the first step in the active transtubular sodium transport seems to be linked with the secretion of H^+ ions. These react in the lumen with the buffer base (buffer anion) to form undissociated acid which easily leaves the tubular lumen.

Furthermore, it was observed that the static head concentration difference for glucose and amino acids (histidine) decreased when sodium was absent from the capillary and luminal perfusates or when the active transport of sodium was inhibited. Assuming that the permeability of the tubular wall for glucose and histidine has not changed, the decrease of the static head concentration indicates a decreased rate of active transport for these substances. Thus the results indicate that both histidine and glucose absorption depend on the transport of sodium from the lumen into the cell supposedly in coupled Na^+-histidine and Na^+-glucose complex. The latter is in accordance with the findings of Frasch *et al.* (1970) who showed that the glucose binding to the isolated brush border is Na^+ dependent with an apparent K_m of 14 mM. Furthermore Frömter (1972) and Kokko (1972) observed that glucose as well as amino acids augment the electrical potential difference across the proximal tubular wall (lumen negative). The same phenomenon was seen by White and Armstrong on the bullfrog small intestine and may be explained by a coupled Na^+-glucose and Na^+ aminoacid transport across the brush border which is equivalent to an increased Na^+/K^+ conductance ratio across it (Armstrong & White 1971).

Specific site group reagents, fractional digestion and membrane antibodies were used in an attempt to eliminate the different brush border transport processes separately. It was found that if the transport of sodium was diminished, the $\triangle c$ values of aminoacid and glucose were also diminished, although to different degrees. An isolated effect on the aminoacid and/or the glucose transport mechanism could not be evoked.

REFERENCES

Armstrong, W. McD. & White, J. F. (1971) Electrical events during solute transfer in epithelial cells of small intestine In *Electrophysiology of Epithelial Cells,* ed. Giebisch, G., pp. 285–312. Schattauer Verlag, Stuttgart, New York.

Bode, F., Baumann, K, Frasch, W. & Kinne, R. (1970) Die Bindung von Phlorrhizin an die Bürstensaumfraktion der Rattenniere. *Pflügers Arch., ges. Physiol. 315,* 53–65.

Bücher, Th. (1947) Ein phosphatübertragendes Gärungsferment. *Biochim. biophys. Acta (Amst.)* 1, 292.

Frasch, W., Frohnert, P. P., Bode, F., Baumann, K. & Kinne, R. (1970) Competitive inhibition of phlorizin binding by D-glucose and the influence of sodium: a study on isolated brush border membrane of rat kidney. *Pflügers Arch., ges. Physiol. 320,* 265–284.

Frömter, E., K. Lüer (1973) Free flow potential profile along rat proximal tubule. Pflügers Arch. ges. Physiol. In press.

Gertz, K. H. (1963) Transtubuläre Natriumchloridflüsse und Permeabilität für Nichtelektrolyte im proximalen und distalen Tubulus der Rattenniere. *Pflügers Arch. ges. Physiol. 276,* 336–356.

570 K. J. ULLRICH, K. SATO & G. RUMRICH

Györy, A. Z. (1971) Reexamination of the split oil droplet method as applied to kidney tubules. *Pflügers Arch. ges. Physiol. 324*, 328–343.

Hetth, E. C. (1971) Complex polysaccharides. *Ann. Rev. Biochem. 40*, 29–56.

Kasbekar, D. K. & Durbin, R. P. (1965) An adenosine triphosphatase from frog gastric mucosa. *Biochim. biophys. Acta (Amst.) 105*, 472–482.

Kedem, O. & Caplan, S. R. (1965) Degree of coupling and its relation to efficiency of energy conversion. *Trans. Faraday Soc. 61*, 1897–1911.

Kinne, R., Schmitz, J.-E. & Kinne-Saffran, E. (1971) The localization of the Na^+-K^+ ATPase in the cells of rat kidney cortex. *Pflügers Arch. ges. Physiol. 329*, 191–206.

Knauf, P. A. & Rothstein, A. (1971) Chemical modifications of membranes. I. Effect of sulfhydryl and amino reactive reagents on anion and cation permeability of the human red blood cell. *J. gen. Physiol. 58*, 190–210.

Kokko, J. P. (1972) Proximal tubule potential difference: Dependence on glucose, HCO_3 and amino acids. *Clin. Res. 20*, Abstr. p. 598.

Lingard, J., Rumrich, G. & Young, J. A. (1972) Amino acid reabsorption in various sections of the proximal convolution of the rat nephron. *Proc. Austral. Physiol. Pharmacol. Soc. 3*, 106–107

Loeschke, K., Baumann, K., Renschler, H. & Ullrich, K. J. (1969) Differenzierung zwischen aktiver und passiver Komponente des D-Glukosetransportes am proximalen Konvolut der Rattenniere. *Pflügers Arch. ges. Physiol. 305*, 118–138.

Mendoza, V. and J. Aceves (1972) Effect of Na and HCO_3 on the peritubular potential of rat proximal tubule, V. Int. Congress of Nephrology Mexico City, Abstr. p. 150.

Pockrandt-Hemstedt, H., Schmitz, J.-E., Kinne-Saffran, E. & Kinne, R. (1972) Morphologische und biochemische Untersuchungen über die Oberflächenstruktur der Bürstensaummembran der Rattenniere. *Pflügers Arch. ges. Physiol. 333*, 297–313.

Sachs, G., Mitch, W. E. & Hirschowitz, B. I. (1965) Frog gastric mucosal ATPase. *Proc. Soc. Exp. Biol. (N.Y.) 119*, 1023–1027.

Sato, K. & Ullrich, K. J. (1972) Effect of acid and basic polyelectrolytes on the rat proximal tubular isotonic reabsorption. *Pflügers Arch. ges. Physiol. 332*, Abs. 60.

Sato, K., Kinne-Saffran, E., Thomas, L. & Ullrich, K. J. (1972) Serum factor which inhibits isotonic reabsorption by the proximal convolution of the rat kidney. *Pflügers Arch. ges. Physiol. 335*, Abs. 79.

Simon, B., Kinne, R. & Sachs, G. (1972) The presence of a HCO_3^- ATPase in pancreatic tissue. *Biochim. biophys. Acta (Amst.) 282*, 293–300.

Thomas, L. & Kinne, R. (1972) Studies on the arrangement of aminopeptidase and alkaline phosphatase in the microvilli of isolated brush border of rat kidney. *Biochim. biophys. Acta (Amst.) 255*, 114–125.

Ullrich, K. J., Frömter, E. & Baumann, K. (1969) Micropuncture and microanalysis in kidney physiology. In *Laboratory Techniques in Membrane Biophysics*. ed. Passow, H. & Stämpfli, R., pp. 106–129. Springer Verlag, Berlin, Heidelberg, New York.

Ullrich, K. J., Radtke, H. W., Rumrich, G. & Klöss, S. (1971) The role of bicarbonate and other buffers on isotonic fluid absorption in the proximal convolution of the rat kidney. *Pflügers Arch., ges. Physiol. 330*, 149–161.

Ullrich, K. J., Fasold, H., Salzer, M., Sato, K., Simon, B. & de Vries, J. X. (1972) Effect of SH, NH_2 and COOH site group reagents on the isotonic absorption from the proximal convolution of the rat kidney. *Abs. V. Int. Congr. Nephrology, Mexico*, p. 148.

Ullrich, K. J., Frömter, E., Sauer, F., Baldamus, C. A., Radtke, H. W. & Rumrich, G. (1973) Transport parameters of the proximal tubule of the rat kidney for sodium, chloride and the bicarbonate buffer. *Pflügers Arch. ges. Physiol.* In press.

Wizemann, V., Schulz, I. & Simon, B. (1973) SH groups on the surface of pancreas cells involved in secretin stimulation and glucose uptake. *Biochim. biophys. Acta (Amst.).* In press.

DISCUSSION

WHITTEMBURY: Was there time enough for the SH blocking agents to act?

ULLRICH: Yes, even if the fluid is reabsorbed, SH-blocking agent remains at the internal surface of the tubule. Furthermore we could repeat the procedure again and again at the same place and still observe no inhibition as compared with the pancreas where the large SH-reagents inhibit fluid secretion instantaneously (Wizemann *et al.* BBA, in preparation).

GIEBISCH: Dr. Ullrich, how relevant do you think is the very dramatic effect of changing peritubular bicarbonate concentrations on urinary sodium excretion? If one renders an animal acidotic by lowering its bicarbonate concentration to some 10 mEq/L one would expect a fairly extensive sodium diuresis from your data. Actually, the diuresis and the sodium loss are not very dramatic in states of metabolic acidosis. Of course one could argue that the sodium lost proximally is picked up by more distally located nephron segments.

ULLRICH: Even that strong proximal inhibitions could be completely counterbalanced by the transport processes in the distal tubule.

GIEBISCH: Have you done any studies in the whole animal rendering it acidotic to the same degre of lowering bicarbonate and then done micropuncture studies?

ULLRICH: No, we have not done those experiments.

GIEBISCH: I might make one final comment. We have perfused *Necturus* kidneys with solutions of various bicarbonate concentrations, keeping pH constant. We were unable to find any inhibition of fluid transport with low bicarbonate concentrations. We have checked this point in split-drop and free flow micropuncture studies. This may be a real species difference.

LEAF: May I ask Dr. Ullrich a question about the result he showed this morning with 5–50 mM ouabain. He finds there is still 50 percent of sodium reabsorption present in his preparation. Epstein found in the perfused rat kidney with the concentration of ouabain comparable (which is some four times that which will inhibit all the sodium potassium dependent ATPase in a homogenate) there still is 50 percent of the sodium reabsorption which cannot be inhibited by ouabain. I want to ask you is it correct that in the proximal tubule ouabain will inhibit all the sodium reabsorption? Were you using proximal tubule in your experiments. Was it rat?

ULLRICH: Yes, it is from rat. I think it is very astonishing. If one adds ouabain to a kidney Na$^+$K$^+$ ATPase preparation one can inhibit it half-maximal with $2 \cdot 10^{-4}$M (Györy et al. 1972). But it is not possible to obtain such an effect with this dose on the isotonic absorption of the intact tubule. Because it is known in some systems that the time before a complete effect of ouabain is seen may be as long as 30 minutes, we left the tubules in contact with ouabain for a considerable period. But we could not reach a higher inhibition than 60 percent.

BERLINER: Did you use Scillaren?

ULLRICH: Yes.

BERLINER: And it didn't do anymore.

ULLRICH: No.

LEAF: Do you think that in this preparation the cardiac glycosides may not totally have inhibited all the sodium potassium ATPase?

ULLRICH: I think we have not.

ULLRICH: Whether these are species differences amongst mammalian tubules is unfortunately not known. Dr. Maude (1971) who observed *in vitro* the same bicarbonate effect that we did, also worked with rat tubules.

EDELMAN: I wanted to ask what the average molecular weight of polylysine was and what concentrations you used.

ULLRICH: 10^{-5}M and we use several molecular sizes, the largest was 100,000, one was 30,000 and we went down to 5,000.

EDELMAN: I just wanted to mention that Mamelak et al. (1969) reported studies with polylysine on toad bladder and were able to show by radioauto-

Györy, A. Z., Brendel, U. & Kinner, R. (1972) Effect of cardiac glycosides and sodium ethacrynate on transepithelial sodium transport *in vivo* micropuncture experiments and isolated plasma membrane Na-K ATPase *in vitro* of the rat. *Pflügers Arch, ges. Physiol. 335,* 287–296.

Maude, D. L. (1971) Effect of bicarbonate, carbonic anhydrase and acetazolamide on proximal tubular NaCl transport *in vitro*. *Fed. Proc. 30,* 607 abstr.

Mamelak, M., Wissig, S., Bogoroch, R. & Edelman, I. S. (1969) Physiological and morphological effects of poly-L-lysine on the toad bladder. *J. Membrane Biol. 1,* 144–176.

graphs that the polylysine was specifically on the apical surface. From correlative morphological studies we concluded that inhibition of sodium transport was proportional to the number of cells that were damaged. There appeared to be a cooperative effect in which cells would either remain intact or would lyse and the short circuit current was reduced approximately in proprotion to the number of lysed cells.

ULLRICH: Your data discouraged us from testing polylysine on the kidney tubule – but Dr. Sato finally tried it and he could reverse the inhibitory effect of polylysine completely if he gave heparin.

EDELMAN: The reversal may have to do with a dual effect: there may be a fairly rapid replacement of damaged cells and the undamaged cells may show a compensatory increase in sodium transport. The reversal itself I think is not sufficient evidence to rule out the possibility that lowering all three parameters of transport (glucose, histidine, sodium) is a non-specific effect as a result of changes in permeability properties of the apical membrane.

NIELSEN: I want to ask Dr. Hogben and Dr. Zadunaisky what the ATP concentration is in the cell when they block their chloride pump with ouabain. The reason why I ask this question is that in kidney cortex slices of pH 8.2 the ATP concentration goes down when you add ouabain (Rorive *et al.* 1972).

HOGBEN: We have not explored this.

ZADUNAISKY: I don't know, I haven't measured that.

TOSTESON: That is a very interesting observation. What is the substrate in the slides under these conditions?

NIELSEN: Acetate.

TOSTESON: I am sure you know that in most systems the ouabain has just the opposite effect.

NIELSEN: Yes.

Rorive, G., Nielsen, R. & Kleinzeller, A. (1972) Effect of pH on the water and electrolyte content of renal cells. *Biochim. Biophys. Acta 266,* 376–396.

TOSTESON: How is it without ouabain at 8.2?

NIELSEN: Then it is perfect. Normal as with the other pH's.

LEAF: Do you have evidence that the tissue has not been killed. Is the tissue still alive?

NIELSEN: The control experiment is of course that the kidney cortex slices incubated at pH 8.2 have normal ATP levels, which they have.

LEAF: Without ouabain?

NIELSEN: Yes.

ULLRICH: We also did electronmicroscopic studies and were surprised to see no change in the tubular structure. Dr. Baumann in our group was also unable to observe any change in the permeability of the proximal tubule for ^{14}C-mannitol when it was perfused for up to 4 minutes with 10^{-5}M polylysine. With longer perfusion periods, however, an increase in the permeability took place. Furthermore Dr. Frömter made some measurements of the electrical potential differences in polylysine perfused tubules. I think he may comment himself on that.

FRÖMTER: When we perfuse the tubular lumen with 30 mmolar NaCl and the peritubular space with 150 mmolar NaCl, we observe a membrane diffusion potential of about 9 mV, lumen positive, which does not change after luminal application of polylysine. This is strong evidence that the passive permeability properties of the tubular wall, which reflect essentially the properties of the paracellular shunt path, have not changed. On the other hand, however, polylysine eliminated the active transport potential of 1 or 2 mV, lumen negative, which we observe during luminal perfusion with ultrafiltrate within 20 seconds indicating that it interrupted active sodium transport which is located inside the cells probably by blocking sodium entry through the brush-border membrane.

LEAF: But is it consuming oxygen or showing any other evidence of metabolic processes?

NIELSEN: Yes, it is consuming oxygen.

USSING: After the treatment with ouabain, does it still consume oxygen?

NIELSEN: Yes, but it decreased.

EDELMAN: How long was the ouabain in the system?

NIELSEN: It takes about an hour before it is completely down.

EDELMAN: In the presence of ouabain many tissues swell, and under these circumstances there may be an effect on distribution of intracellular calcium between organelles and cytoplasm. If the free intracellular calcium content is increased, oxidative phosphorylation may be impaired (*i. e.,* uncoupled).

Paracellular Pathway in Kidney Tubules: Electrophysiological and Morphological Evidence

G. Whittembury, F. A. Rawlins and E. L. Boulpaep.

The kidney proximal tubule reabsorbs considerable amount of ions and water. This process is envisaged as mainly transcellular (Giebisch 1960, Whittembury 1960). However, as schematized in Fig. 1, individual cells of the tubular epithelium are separated laterally from each other by interspaces which terminate in a junctional complex at the luminal end, called "zona occludens" or "tight junction". Electrophysiological studies support the view that there exists an important extracellular route for transport of salt and water

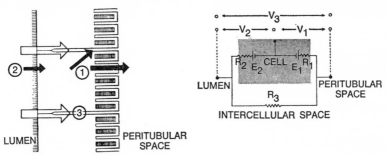

Fig. 1. A diagram of the proximal tubular wall is shown at the left. Arrows 1, 2 and 3 indicate transit through the peritubular cell membrane, luminal cell membrane and intercellular space. They correspond to resistances R_1, R_2 and R_3, respectively. To the right is shown a simplified equivalent circuit for the proximal tubule cell. $V_1 =$ peritubular membrane potential, $E_1 =$ peritubular e.m.f., $R_1 =$ peritubular resistance, $V_2 =$ luminal membrane potential, $E_2 =$ luminal e.m.f., $R_2 =$ luminal resistance, $V_3 =$ transepithelial potential, $R_3 =$ shunt or paracellular resistance (modified from Whittembury & Rawlins, 1971 and Boulpaep 1967).

Centro de Biofísica y Bioquímica, Instituto Venezolano de Investigaciones Científicas (IVIC), P. O. Box 1827, Caracas, Venezuela and Department of Physiology, Yale University, School of Medicine, 333 Cedar Street, New Haven, Conn., 06510, USA. (E.L.B.).

»Transport Mechanisms in Epithelia«, Munksgaard, Copenhagen.

across the proximal tubular epithelium (Windhager *et al.* 1967). This paper summarizes electrical and morphological evidence that locates the paracellular pathway at the level of the intercellular channel made up by the tight junction and the subsequent intercellular space. This paracellular path may play an important role in the overall passive permeation of ions across the renal epithelium.

Suggestive evidence regarding a paracellular pathway came out of studies on *Necturus* kidneys indicating that the water permeability coefficient for the whole epithelium (Whittembury *et al.* 1959), is about one order of magnitude larger than that of the cellular membranes (Whittembury 1967).

DIRECT ELECTROPHYSIOLOGICAL EVIDENCE FOR A MAJOR
PARACELLULAR PATHWAY

1) Evidence for the existence of a shunt pathway bypassing the cells came from observations of electrical interaction between the peritubular and luminal cell membrane. Changes in electrical membrane potential across the peritubular cell membrane were elicited by raising the peritubular fluid K concentration to alter the chemical concentration gradient across that cell membrane. In the absence of a modification of the intracellular composition and a diffusion potential occurring across a paracellular path, the contralateral luminal cell membrane potential should remain unchanged during such substitutions. However, (Boulpaep 1967), the luminal cell membrane underwent a depolarization nearly identical to that of the peritubular border, although the chemical concentration gradient across the luminal membrane (from cell to lumen) had remained unaltered.

Furthermore, the membrane potential of the luminal membrane could not be accounted for sufficiently in terms of the existing concentration gradients and of the estimated ionic permeabilities, which suggested the simplified equivalent circuit of Fig. 1 (Boulpaep, 1967).

Resistor R_3 is a shunt in parallel with the tubular cells, it allows current flow from the peritubular membrane to the luminal membrane and the generation of a voltage drop across the luminal membrane V_2. At the luminal border an e.m.f. E_2 smaller than E_1 might exist, but V_2 is mainly the consequence of a resistive voltage drop across R_2. The magnitude of V_2 is chiefly determined by the electromotive force E_1 and the shunt resistor R_3.

Three comments concerning this model are relevant here. a) In physiologi-

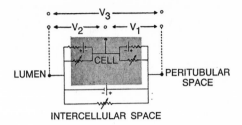

Fig. 1a. Unsimplified equivalent circuit of the proximal tubule cell drawn to emphasize that resistances 1, 2 and 3 of Fig. 1 may vary under given conditions.

cal conditions no appreciable ionic concentration gradients exist across the proximal epithelium. Consequently, an e.m.f. E_3 in the shunt pathway has not been included in Fig. 1, but it should be added whenever a transepithelial salt or ion gradient is created (Fig. 1a). b) Ionic substitutions across the tubular wall to test the ion selectivity of the paracellular pathway have to alter the ionic composition at the level of at least one single cell membrane. The resulting membrane potential changes of V_3 and V_1 (or V_2) cannot be easily translated in terms of changes of either E_3 and E_1 (or E_2) respectively, without a quantitative estimate of the individual ion conductances R_1, R_2, R_3. Therefore evaluation of such potential changes in terms of relative ion permeability of either the shunt pathway or the individual cell membranes is subject to errors without a full knowledge of all parameters of the circuit. c) V_2 will have a cell-inside-negative value if the value of E_1 (cell-inside negative) is large enough and that of R_3 is low enough, even if the value of E_2 were 0 or ÷ (cell-inside).

2) Evidence for the existence of electrical coupling between peritubular and luminal cell membranes was also obtained by passing electrical current across either R_1 or R_2 of Fig. 1. If no shunt existed ($R_3 = \infty$) the current flow across R_1 would result in a voltage deflection $\triangle V_1$, with no (or a minimal) deflection visible across R_2 i.e. $\triangle V_2 \to 0$, whatever the absolute values of R_1 and R_2 might be. On the other hand, if the shunt is large ($R_3 = 0$) the voltage deflections V_1 when current is passed across R_1 would not be larger than those across R_2, i.e. $\triangle V_1 = \triangle V_2$. It was found that within the accuracy of the method, the simultaneous deflections $\triangle V_1$ and $\triangle V_2$ were not statistically different, when current was passed across one of the cell membranes only. Thus the shunt resistor, R_3, must be smaller than the individual cell

37*

membrane resistance by at least an order of magnitude which is equivalent to the error in the measurement of $\triangle V_1$ and $\triangle V_2$ (Boulpaep 1967).

3) The overall transepithelial resistance can be estimated from application of cable analysis, treating the entire epithelial wall as a single layer and the tubular lumen its core. In *Necturus* proximal tubule values of about 70 ohm. cm² have been found for the specific transverse resistance (Boulpaep 1972), a value much lower than the resistance of most naturally occurring membranes. In order to compare the overall resistance with that of the individual membranes of the same preparation a double-core analysis has been performed (Windhager *et al.* 1967). Such approach indicates for the sum of the two cell membranes in series a value of about 7,900 ohm. cm², two orders of magnitude larger than the transepithelial value. It is concluded that more than 99 pct. of the transepithelial conductance of the *Necturus* proximal tubule is constituted by a paracellular shunt path.

INDIRECT ELECTROPHYSIOLOGICAL EVIDENCE FOR A MAJOR
PARACELLULAR PATHWAY

1) Cell membrane ionic permeability as tested by electrophysiological means indicates usually a rather wide range of ion selectivity. Cation permeability coefficients of the peritubular membrane of *Necturus* proximal tubule cover a range of about 1 to 25 (Giebisch 1961, Whittembury *et al.* 1961, Boulpaep 1967, Boulpaep 1971). Contrary to the single cell membrane selectivity, the transtubular range of cation selectivity, is smaller in the *Necturus* (Whittembury & Windhager 1961); in the dog proximal tubule it is about 1:2 (Boulpaep & Seely 1971), and about 1:6 in the rat proximal tubule (Frömter *et al.* 1971). Such discrepancy suggests that the main transepithelial permeation route is a channel less selective than the single cell membrane permeability would predict.

2) Relative transepithelial ionic permeabilities can also be compared with the sequence found for the same ions in individual cell membranes. In the dog proximal tubule (Boulpaep & Seely 1971) and in the rat proximal tubule (Frömter *et al.* 1971) the ranking of ion selectivity agrees fairly well with that predicted from mobilities in free solution within the separate cation and anion series (Boulpaep & Seely 1971). The sequence of relative ion permeabilities of the peritubular membrane of *Necturus* proximal cells (Whittembury *et al.* 1961, Boulpaep 1967), is completely at variance with

the transepithelial pattern (Whittembury & Windhager 1961). Therefore transepithelial ion permeation seems to occur across a single membrane barrier of a different ion selectivity, and this barrier could most likely be constituted by the tight junction.

3) Ionic transference numbers can be estimated from methods that either assume the existence of a single diffusion barrier or from approaches that do not require such postulate. Transference numbers of chloride ions have been estimated from potential changes subsequent to the application of transepithelial salt gradients, treating the epithelium as a single membrane. Direct determination of the partial chloride conductance yields an identical transference number. The later determination does not postulate the existence of a single membrane (Boulpaep & Seely 1971). The data reveal that both determinations were assessing the same extracellular diffusion path.

4) The electrical transtubular streaming current and the ion selectivity underlying it, is a function of the direction of the imposed osmotic flow in the proximal tubule (Boulpaep & Seely 1971). Independent measurements of transtubular conductance during identical osmotic flows of either direction shows that the overall transepithelial resistance correlates similarly with the direction of the osmotic water flow. Again, the second approach does not require the existence of only one barrier (Boulpaep & Seely 1971). Both measurements were thus probably testing the same paracellular pathway.

5) Instantaneous current-voltage relationships, in *Necturus* proximal tubule measured across the whole epithelium are quite linear (Boulpaep 1971). However, peritubular membranes of the same preparation show the phenomenon of anomalous rectification (Boulpaep 1966). The discrepancy in behaviour can be explained if an important bypass of the cell membranes exists which exhibits a linear conductance characteristic.

MAGNITUDE OF SHUNT IN AMPHIBIAN AND MAMMALIAN PROXIMAL TUBULES

Both direct and indirect lines of evidence establish the idea that the observed transpithelial permselectivity pattern resides actually in a single diffusion barrier, presumably the zonula occludens, rather than the two cell membranes in series. Such a barrier has a low electrical resistance and constitutes an important current path by which electrical interactions between the two membranes can occur.

582   G. WHITTEMBURY, F. A. RAWLINS AND E. L. BOULPAEP

Within the proximal tubule there are important species differences.

a) Amphibian tubules have a transepithelial resistance of about 70 ohm. cm² (Boulpaep 1972) as opposed to 4.9–5.7 ohm. cm² in the rat (Hegel *et al.* 1967), 5.6 ohm. cm² in the dog (Boulpaep & Seely 1971), and 6.8 ohm. cm² in the rabbit (Lutz & Burg 1972). b) This difference in resistances agrees with a higher transtubular potential difference in the *Necturus* proximal tubule (c. f. Boulpaep 1971).

c) Although in all species, essentially small chemical gradients prevail across the proximal tubule, some marked difference in ionic selectivity should be mentioned. Cation permeability dominates anion permeability in the mammalian proximal tubule, e.g. $P_{Na} > P_{Cl}$ (Boulpaep & Seely 1971, Frömter *et al.* 1971). In the *Necturus* proximal tubule the ratio of P_{Na} : P_{Cl} does not differ appreciably from that in free solution (Boulpaep 1967). Thus a less selective, more hydrated tight junction should exist in the amphibian as compared to the mammalian kidney, despite the higher resistance values.

d) Finally, the morphological counterpart of the transepithelial resistor R_3 of figure 1 might not be the same in the amphibian and the mammalian. As pointed out earlier (Boulpaep 1971, Boulpaep 1972) proximal tubule transepithelial resistance should be given in the *Necturus* (with tall cells) mainly by the long intercellular channels (beyond the zonula occludens) and in the mammalian (with small cells) mainly by the zonula occludens. Similar conclusions may be obtained from hydraulic resistance figures (Whittembury 1967).

CHANGES IN THE MAGNITUDE OF THE SHUNT

The most important physiological aspect of the paracellular pathway is its possible role in the regulation of net Na transport. Indeed the paracellular channel provides an appreciable leakage path, shunting the mechanism responsible for the net transfer of Na⁺ against its electrochemical potential gradient. The extent of passive backleak of Na⁺ through these channels will obviously affect the efficiency of the active Na transport.

Boulpaep has studied in the amphibian the condition of saline loading known to be associated with changes in net Na reabsorption to test whether changes in the magnitude of the shunt could be detected. The results are

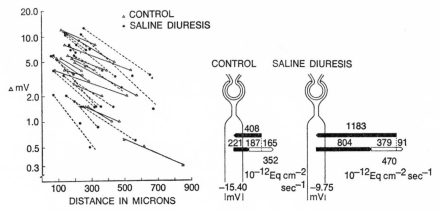

Fig. 2. Results of cable analysis are shown at the left. Transepithelial voltage attenuation of Necturus proximal tubule plotted against tubular distance for current application transepithelially at distance O. The dots and interrupted lines give the attenuation across the renal epithelium in saline diuresis, as opposed to the triangles and continuous lines for the control conditions· The average slope is steeper in saline diuresis than in control (Boulpaep 1972). At the right are shown estimates to the unidirectional fluxes in Necturus proximal tubule in control and saline diuresis. The dark arrows indicate the magnitude of the unidirectional passive fluxes, calculated on the basis of the electrochemical gradient for Na and the permeability coefficient. The empty arrow indicates the active flux component, being the sum of the net passive flux and the actually observed net flux. Net observed fluxes were 165 in control and 91 in saline diuresis (10^{-12} Eq. cm^{-2}. sec.$^{-1}$) (Boulpaep 1972). Transtubular potentials are also indicated.

summarized in Fig. 2 which shows the voltage attenuation curves during cable analysis under control and volume expansion conditions.

From the slopes of the lines (indicative of the length constant), the specific resistance of the core of the tubule and its geometry, it is possible to compute the transepithelial resistance of 70 ohm. cm^2 in control conditions and 22 ohm. cm^2 in saline diuresis. The cellular resistances did not change at the same time. It was concluded that the paracellular shunt increased during saline loading (Boulpaep 1972). The transtubular permeability for NaCl was increased by a factor of 3.2 and the permeability for large non-electrolytes was found to rise drastically after saline loading (Boulpaep 1972).

On the basis of determinations of net transport rate, permeability coefficient, and electrochemical gradient for Na, the data, shown at the right hand side of Fig. 2, were calculated. Passive unidirectional fluxes for Na increased (dark arrows) when the conductance of the paracellular pathway was enhanced. The active transport component (empty arrow) (estimated indirectly from actual net transport rate and predicted net passive Na flux) did not decrease during volume expansion. Therefore, the marked depression of net reabsorption should be mainly due to a change in the Na backflux, i.e. paracellular conductance.

Decreases in the oncotic pressure of the peritubular perfusate leads to a similar change in paracellular resistance and passive Na permeability as during volume expansion (Grandchamp & Boulpaep 1972). The available evidence thus suggests that the degrees of effectiveness of the sodium transport system can be controlled by the passive ion permeation through a paracellular channel, presumably the tight junction.

MORPHOLOGICAL EVIDENCE – INTERCELLULAR PASSAGE OF LANTHANUM

Whittembury & Rawlins (1971) have localized the paracellular pathway using lanthanum as an extracellular marker for electronmicroscopy in perfused amphibian kidneys. Lanthanum nitrate is soluble in water at pH lower than 7.5 while lanthanum salts of anion like sulfate, phosphate, bicarbonate and oxalate are insoluble in water and in acetone. In the amphibian kidney a solution perfused via the aorta filters through the glomerulus and bathes the tubular lumen. On the other hand, a solution perfused via the portal vein bathes mainly the peritubular spaces. Therefore, the kidney was perfused through the aorta with a solution containing soluble lanthanum (i.e. lanthanum nitrate) and through the portal vein with a solution containing precipitating anions. This procedure should result in the precipitation of insoluble lanthanum salt at the sites where the two solutions met. The electron dense precipitate was localized later with the electron microscope with the conventional techniques.

First kidneys of the toad *Bufo marinus* were perfused (Whittembury & Rawlins 1971). It was possible to conclude that lanthanum and the precipitating anions crossed the tight junctions and the intercellular spaces so that anions and cation could meet and precipitate there, because (a) lanthanum was found filling about 90 % of the tight junctions examined and 100 % of the intercellular spaces of the proximal tubules. (b) The concentration of the intercellular precipitate was higher towards the tight junction than towards the basement membrane. (c) Some lanthanum was also seen precipitated in

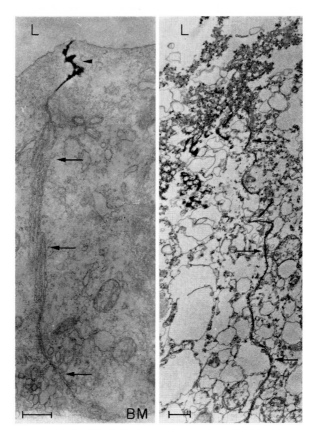

Fig. 3. Left. Electron micrograph of a Necturus kidney proximal tubule. The kidney was perfused with isotonic solution (control). A solution containing soluble lanthanum precipitating solution bathed peritubular space. Lanthanum is seen precipitated in the intercellular space only in that region close to the lumen (arrow head). (No lanthanum precipitated within the cell). The rest of the intercellular space (arrows) remain empty. Note the tortuosity of the intercellular space. L, lumen, BM, basement membrane. 16590X. In this and following E. M. the length of the bar denotes 0.5 micron. To the right is shown a Necturus proximal tubule of a kidney perfused with a solution containing in addition 100 mM urea to induce osmotic diuresis. Lanthanum is seen precipitated all along the intercellular space (arrows), in the lumen (L) and also intracellularly. Urea induce a vacuolation of the cytoplasm and diminution of the intercellular space tortuosity (11350X).

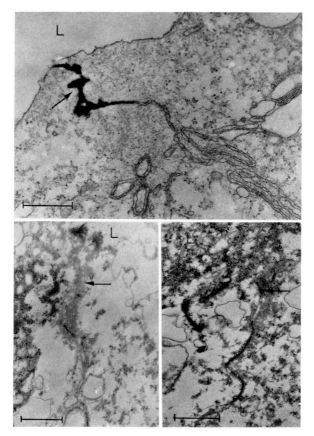

Fig. 4. Top. High magnification of the control proximal tubule showing concentration of lanthanum in the intercellular space towards the lumen (L) (34160X). High magnification of two urea treated proximal tubules is shown at the bottom (21960X and 24400X). The arrows show the presence of lanthanum at the level of the tight junctions. Lanthanum is also precipitated within the cell and in the lumen.

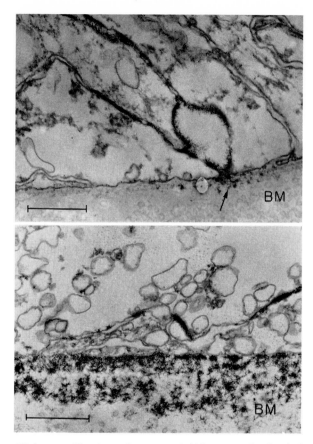

Fig. 5. Top. High magnification of a control kidney proximal tubule. Lanthanum is precipitated in the intercellular space in the region where it opens into the basement membrane (arrow) (32940X). High magnification of a urea treated kidney proximal tubule is shown at the bottom. Note the greater concentration of lanthanum at the basement membrane in comparison with the figure shown at the top (34160X).

the basement membrane. (d) Occassionally a few vacuoles containing only small quantities of lanthanum were observed within the cell.

Whittembury & Rawlins were able to rule out the following alternative possibilities. (1) That lanthanum could have precipitated in the intercellular spaces by entering there backwards from the peritubular capillaries (due to a mixture of aortic and portal circulations) and not by crossing the tight junctions. (2) That lanthanum crossed the cell and not the tight junction before precipitating in the intercellular space. (3) That perfusion of the kidney under the conditions used did not open the tight junctions abnormally. This last conclusion should be taken with caution since a perfused kidney may be considered as very near (but not at) a physiological condition. Consequently, a perfused kidney might be in a position intermediate between one *in vivo,* working normally, and one subjected to saline loading, which as we have seen above increases the magnitude of the paracellular shunt.

VARIATIONS IN THE MORPHOLOGY OF THE PARACELLULAR PATHWAY

Whittembury *et al.* (1972) have recently performed experiments in *Necturus* kidneys, to be able to compare two situations which should lead to variation in the morphology of the paracellular shunt. They used kidneys perfused with isosmotic fluids (controls) and others in which 100 mmole urea/liter were added to the aortic perfusate (containing the soluble lanthanum) to induce osmotic difference across the tubular wall (lumen hyperosmotic). Some representative electron micrographs are shown in Figs. 3, 4 and 5*. The main findings may be summarized as follows.

(1) In the control proximal tubules lanthanum crossed the tight junctions and precipitated mainly in the tight junctions and in the intercellular space near it. Much less precipitate was observed near the basement membrane. With urea more lanthanum was precipitated at the basement membrane and peritubular space indicating facilitated paracellular passage of lanthanum.

(2) In the controls, intracellular lanthanum precipitate was not observed, but with urea, there was intracellular lanthanum. Urea induced vacuolization. It is improbable that vacuolization is a fixation artefact since control and urea treated kidneys were processed alike for electron microscopy.

(3) The tortuosity of the intercellular space was markedly reduced with urea. The ratio of length of the intercellular space to cell thickness was of 2

* See figure insert opposite page 584.

or higher in the controls, while this ratio was of 1.4 or lower in the urea treated proximal tubules. The present electron micrographs do not allow an accurate estimate of the width of the tight junctions and of the intercellular space. However, a reduced length of the intercellular pathway might be present during saline loading. If this were so it could explain in part the lower values of R_3 reported by Boulpaep (1972) during saline diuresis.

(4) The increased transcellular passage of lanthanum with urea indicates that resistances R_1 and R_2, representing those of the cell membranes – – and not only R_3 – – may vary during given experimental conditions. This phenomenon may be illustrated with the aid of Fig. 1a which represents an equivalent circuit considerably more complicated than that of Fig. 1. Variable resistors have been drawn to represent R_1, R_2 and R_3. Also a shunt e.m.f. is included, which is relevant for example when solutes that create a lumen to peritubular space ionic-gradient are used.

In short, the degree of effectiveness of the Na transport system can be controlled by the passive ion permeation through the paracellular channel. Therefore this paracellular pathway plays a considerable role in the control of Na transport, but variations in the characteristics of the cell membranes should also enter into the picture.

A comparison of the properties of other segments of the nephron, e.g. the distal tubule (Boulpaep & Seely 1971), makes it clear that the paracellular route is tighter in the distal tubule. This segment is indeed characterized by much higher transepithelial electrical and chemical gradients, a wider range in ion selectivity, and a higher overall transepithelial resistance.

The paracellular pathway may open in amphibian skins under given experimental conditions (Rawlins 1967, Rawlins et al. 1970, Ussing 1970, Martínez Palomo et al. 1971, Erlij & Martínez Palomo 1972). This indicates that the magnitude of the paracellular shunt varies from one tissue to another, and within the same tissue, according to the experimental situation. Its physiological role might vary accordingly.

In summary, at the present stage we can provide information concerning the existence of the paracellular pathway, its localization and its variability. We can raise some questions as to its possible significance in ion and water transport. Let us hope we can also find some answers as to the physiological role of the paracellular shunt.

REFERENCES

Boulpaep, E. L. (1966) Potassium and chloride conductance of the peritubular membrane of proximal tubule cells of *Necturus* kidney. *Abstr. 10th Ann. Meeting Biophys. Soc.,* 133.

Boulpaep, E. L. (1967) Ion permeability of the peritubular and luminal membrane of the renal tubular cell. In *Transport und Funktion intracellulärer Elektrolyte,* ed. Krück, F., Munchen. Urban & Schwarzenberg, pp. 98–107.

Boulpaep, E. L. (1971). Electrophysiological properties of the proximal tubule. Importance of cellular and intercellular transport pathways. In *Symposia medica Hoechst. Electrophysiology of Epithelial Cells,* Giebisch, G., ed. F. K. Schattauer Verlag, Stuttgart, New York. pp. 91–112.

Boulpaep, E. L. (1972) Permeability changes of the proximal tubule of *Necturus* during saline loading. *Amer. J. Physiol. 222,* 517–531.

Boulpaep, E. L. & Seely, J. F. (1971) Electrophysiology of proximal and distal tubules on the autoperfused dog kidney. *Amer. J. Physiol. 221,* 1084–1096.

Erlij, D. & Martínez-Palomo, A. (1972) Opening of tight junctions in frog skin by hypertonic urea solutions. *J. Membrane Biol. 9,* 229–240.

Frömter, E., Müller, C. W. & Wick, T. (1971) Permeability properties of the proximal tubular epithelium of the rat kidney studied with electrophysiological method. In *"Electrophysiology of Epithelial Cells",* ed. Giebisch, G., pp. 119–146. F. K. Schattauer Verlag, Stuttgart, New York.

Giebisch, G. (1960) Measurements of electrical potentials and ion fluxes on single renal tubules. *Circulation, 21,* 879–891.

Giebisch, G. (1961) Measurements of electrical potential differences on single nephrons of the perfused *Necturus* kidney. *J. gen. Physiol. 44,* 659–678.

Grandchamp, A. & Boulpaep, E. L. (1972) Effect of intrarenal pressures on the permeability of the intercellular shunt pathway and proximal Na reabsorption. *5th Internat. Cong. Nephrology,* Mexico.

Hegel, U., Frömter, E. & Wick, T. (1967) Der elektrische Wandwiderstand des proximalen Konvolutes der Rattenniere. *Pflügers Arch. ges. Physiol. 234,* 274–290.

Lutz, M. D. & Burg, B. B. (1972) A single microelectrode to measure resistance of the isolated perfused rabbit proximal convoluted tubule. *Fed Proc. 31,* Abs.

Martínez-Palomo, A., Erlij, D., & Bracho, H. (1971) Localization of permeability barriers in the frog skin epithelium. *J. Cell Biol. 50,* 277–287.

Rawlins, F. (1967) Acción de la hormona antidiurética sobre las resistencias eléctricas del epitelio aislado de la piel de *Bufo marinus.* Universidad Central de Venezuela, Facultad de Ciensias, *Tesis* de Licenciado en Biología.

Rawlins, F. A., Mateu, L., Fragachan, F. & Whittembury, G. (1970). Isolated toad skin epithelium: Transport characteristics. *Pflügers Arch. ges. Physiol. 316,* 64–80.

Ussing, H. H. (1970) Tracer studies and membrane structure. In *Capillary permeability,* ed. Crone, C., & Lassen, N. A., pp. 654–656. Munksgaard, Copenhagen Academic Press, New York.

Whittembury, G. (1960) Ion and water transport in the proximal tubules of the kidney of *Necturus maculosus. J. gen. Physiol. 43,* suppl. 5, 43–56.

Whittembury, G. (1967) Sobre los mecanismos de absorpción en el tubo proximal del riñón. *Acta cient. Venez. Supl. 3,* 71–83.

Whittembury, G., Oken, D. E., Windhager, E. E. & Solomon, A. K. (1959) Single proximal tubules of *Necturus* kidney. IV Dependence of H_2O movements on osmotic gradients. *Amer. J. Physiol. 197,* 1121–1127.

Whittembury, G. & Rawlins, F. A. (1971) Evidence of a paracellular pathway for ion flow in the kidney proximal tubule: Electronmicroscopic demonstration of lanthanum precipitate in the tight junction. *Pflügers Arch. ges Physiol. 330,* 302–309.

Whittembury, G., Rawlins, F. A., Pérez-González, M. & Boulpaep, E. L. (1972) Paracellular pathway for ion flow in kidney tubule. *Abstract. 5th Internat. Congress of Nephrology,* Mexico., 804.

Whittembury, G., Sugino, N. & Solomon, A. K. (1961) Ionic permeability and electrical potential differences in *Necturus* kidney cells. *J. gen. Physiol. 44,* 689–712.

Whittembury, G. & Windhager, E. E. (1961) Electrical potential difference measurements in perfused single proximal tubules of *Necturus* kidney. *J. gen. Physiol. 44,* 679–688.

Windhager, E. E., Boulpaep, E. L. & Giebisch, G. (1967) Electrophysiological studies on single nephrons. *Proc. 3rd. Internat. Congress of Nephrology Washington, D. C. 1966, 1,* 35–47. Karger, Basel, New York.

DISCUSSION

ZADUNAISKY: Did you include lanthanum nitrate in the fixation media?

WHITTEMBURY: No, it was not included. We perfused with lanthanum and then at some point glutaraldehyde was used to fix the tissue, and in that case you might say that the fixative contains some lanthanum, but only at that state. Afterwards we took the kidneys out and put them in the medium with the precipitating anions, but not with lanthanum.

ZADUNAISKY: I wonder which action lanthanum in itself has. In doing some similar type of experiment in frog skin I found that it can replace calcium very easily.

WHITTEMBURY: If you use La in the outside you increase sodium transport across toad skin, at least.

LOEWENSTEIN: Did you compare the pictures in which you do not perfuse with La, but add the La later to your sections?

WHITTEMBURY: No, we never did that. We always had La in the perfusion.

LOEWENSTEIN: That would be very interesting simply for the following reason: La added to the outside can uncouple epithelia. That shouldn't worry you too much, becauce uncoupling doesn't necessarily mean that the cells come apart. In many cases, as you know they simply do not come apart. But that would be one way to assure yourself that La *per se* does not cause separation.

GIEBISCH: I wonder whether you had a chance to go into a problem that somebody brought up years ago. Dr. Philippa Claude in a thesis did an electronmicroscopic study of *Necturus* proximal tubule. In a number of instances she noticed the absence of any terminal bars between proximal tubule cells.

WHITTEMBURY: I cannot answer you directly. In the *Necturus* one does not observe the tight junctions with all the structures that you observe in others. However this might be due in part to unsatisfactory tissue preservation at the ultrastructural level becauce of the big size of the cells. Also cell size prevents you from sampling a large number of junctions.

OSCHMAN: There are at least 2 transporting epithelia that have been shown to be entirely lacking in zonula occludens or limiting junctions. These are the rectal gland (van Lennep 1968) and the rumen (Henrikson 1970)

SKADHAUGE: Following up Dr. Giebisch's question and Dr. Oschman's comment: When you are concerned with the concentration difference of solutes across the limiting area, its length is very important.

WHITTEMBURY: As you all know what limits diffusion is the reciprocal of the ratio of the area available for diffusion to the length of the diffusion pathway. I would like to stress that although the rest of the intercellular space is apparently wide (say 0.2 μ) as compared to the tight junction (say 10–20 Å), the length of the intercellular space (say 20–30 μ) is such that the intercellular space is at least as resistive to diffusion as the corresponding tight junction (length about 3000 Å). If tne tight junction were of 100 Å in width (instead of 10 Å) the corresponding intercellular space would become the main resistance to diffusion. To answer your question directly, I think $1/3$ or $1/2$ μ would be the length of the tight junction in the proximal tubule of *Necturus*.

SKADHAUGE: The real barrier with the very small channel is probably much shorter than $1/2$ μ.

WHITTEMBURY: It could well be.

SKADHAUGE: Would you speculate: 200 Å?

WHITTEMBURY: I just don't know. There is some little debate that we might pick up in the general discusion, if you want, about the morphology of the limiting junction. I don't think it is just a single compartment, even if you see it straight. There must be communication between cells, as Dr. Loewenstein showed us. Substances will have to go like a snake in a tortuous pathway in the perpendicular (lumen to peritubular) direction. And then you come to the intercellular space and you will have another compartment in series, anyway.

van Lennep, E. W. (1968) Electron microscopic histochemical studies on salt-secreting glands in elasmobranchs and marine catfish. *J. Ultrastruct. Res. 25*, 94–108.

Henrikson, R. C. (1970) Ultrastructure of ovine ruminal epithelium and localization of sodium in the tissue. *J. Ultrastruct. Res. 30*, 385–401.

GENERAL DISCUSSION

BERLINER: Again going back to Dr. Burg's data on the thick ascending limb, he finds very interestingly that the diuretic mersalyl has a very drastic effect in knocking down the transport of chloride in the ascending limb. He followed up an old observation of Miller & Farah (1962). They found in the intact animal that if you gave PCMB, you reverse the diuretic effect of the diuretic material. This same effect Dr. Burg has found occurs in the perfused ascending limb. If you add PCMB to the perfusing fluid in the lumen you immediately reverse the depression caused by mersalyl. Furthermore, if you then wash out the tubule and put in more mersalyl you can no longer get the diuretic effect. If the PCMB is there, it sticks and it blocks any further effect of mersalyl. These effects are not observed on the outside of the epithelia. On the outside if you add enough mercurial you knock out everything and the tubule falls apart. But you don't get this specific sort of inhibition that you observe in the lumen. So there is something there that reacts with sulfhydryl reagents. And it seems to be related, at least at the moment, with the chloride transport.

ULLRICH: It is known that the different sulfhydryl groups in the protein have different affinities for the different sulfhydryl reagents. By using sulfhydryl reagents one never knows whether only the conformation of the enzyme protein in question has changed or whether the specific substrate binding site has been attacked. If the structure of a protein is altered by the addition of a sulfhydryl reagent then cryptic site groups could undergo a conformational change such that they would react with reagent towards which they were unreactive in the native state. So one could change, by attacking one sulfhydryl group, the reactivity of the others. This may be the explanation for part of the experiments you mentioned. I recall our finding that PCMB in a short lasting exposure in vivo did not inhibit the transtubular glucose transport, but a longer exposure of the isolated brush border in vitro inhibits the glucose binding. So I also think there is the possibility that the glucose carrier in situ may be in a conformation such that the chloromercury benzonate which we added could not react with it, whereas *in vitro* it can.

Miller, T. B. & Farah, A. E. (1962) Inhibition of mercurial diuresis by nondiuretic mercurials. *J. Pharmacol. exp. Ther. 135,* 102–111.

CURRAN: I was really quite surprised to see that sulfhydryl reagents had no effect on the amino acid transport in the kidney, because the transport system for neutral amino acids in the intestine is very sensitive to these reagents. I wonder if there is any possibility that your experimental method could occasionally give a false result because the permeability was altered in such a way that the steady state concentration did not change even though there was an inhibition of the transport system.

ULLRICH: Yes, I mentioned that this is a screening method which we applied. The method for measuring fluid reabsorption which I mentioned is composed of an outward transport which is supposed to be an active one and a passive back flux.

EDELMAN: The results obtained with the large molecular weight carriers that were completely negative may not be meaningful because of a mechanical problem. In the toad bladder, polylysine reacts with polyanionic sites on the surface. And yet polycationic ferritin (Danon, D., Unpublished results) does not bind to the apical surface probably because it becomes enmeshed in the external filamentous coat.

The problem with the PCMB and the iodoacetic acid which are bound to large molecular weight carriers may involve a failure to penetrate to the surface membrane. Given sufficient time to titrate enough groups with high enough PCMB concentrations you should be able to at least show an inhibiting effect on the transport.

ULLRICH: We have used quite a lot of SH-reagents, some with net negative and some with net positive charge. The common denominator of all of them was their molecular size. When it was large they did not inhibit the proximal transport processes. With the smaller molecules we always found that they inhibited after a delay when similarly applied from the tubular lumen. But we are not sure whether they just diffused through the terminal bars and reacted on the outside of the tubule at the contraluminal cell site. All the SH reagents reacted much faster when we applied them, in the same concentration, to the outer surface of the tubule. However, we are not quite sure what happens to the cell metabolism, i. e. if a small amount of the inhibitors enters the cells and inhibits the energy supplying metabolism.

CURRAN: We have some results, in the rabbit small intestine, that sound very similar to those Dr. Edelman just described. The addition of PCMBS to the mucosal bathing solution initially causes a rather sharp rise in short circuit current followed by rather marked falls some 15 minutes later.

SCHOFFENIELS: Dr. Ullrich, do you have any idea, as to the effect of the sodium free medium on the intracellular potassium concentration?

ULLRICH: No, we have not. As the tubules in our preparation remain in situ. We perfuse only an area of 500–600 μ, where the capillaries are perfused.

TOSTESON: I would like to make the point that "shunts" occur not only between cells but also across the membranes themselves. This fact makes the assignment of the locus of a "leak" or "shunt" pathway in complex epithelia particularly difficult. The point can be made in terms of the equivalent circuit for a biological membrane in the figure. The membrane may be considered to comprise a parallel array of a capacitance (C_m), and pathways for the leak or dissipative transport pump or active transport of ions. Each leak pathway is represented by a conductance (G^L) and an equilibrium potential (E) arranged in series. Pathways exist which are (G^L_p, E_p) and are not (G^L_l, E_l) pumped. Pump pathways are represented as current sources for each pumped ion. Clearly, the problem of the quantitative identification of each of these pathways is formidable enough even when the membrane can be studied in isolation with complete control of the solutions bathing the two surfaces. The existence of shunts between adjacent membranes and lack of control of one bathing solution (the cytoplasm) complicates an already complicated situation.

EDELMAN: Shouldn't there be a shunt pathway across the membrane?

TOSTESON: Something like that is a minimum.

EQUIVALENT CIRCUIT OF BIOLOGICAL MEMBRANE

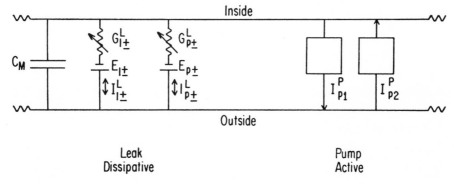

Inside

Outside

Leak
Dissipative

Pump
Active

LINDEMANN: To my knowledge there is also no compartmental analysis presently available that properly treats the intercellular compartment in addition to the cellular compartment.

FRÖMTER: Dr. Tosteson's point is essentially the point which I wanted to discuss in response to Dr. Giebisch's question yesterday. I want to add, however, that the situation is not in all respects as bad as it might appear on first view, since the error which you may expect depends on the relative magnitude of the individual resistors across which you do the potential measurements. The error will be small or even negligible, when you consider transepithelial measurements in extremely leaky epithelia like rat proximal tubule for example where the paracellular shunt conductance is perhaps at least 50 times larger than the cellular conductance. This is the basis which allows us to conclude from the salt gradient potential measurements under polylysine for example, that the permeability properties of the shunt path have not changed.

TOSTESON: If the shunt conductance is 100 times the membrane conductance, and if the shunt conductance is reduced to one-half without changes in ionic transference numbers, the change in potential will be small but the change in transport rate will be large.

FRÖMTER: In this special case you should not forget that polylysine is a positively charged substance of high charge density which must be expected first and above all to change the transference numbers.

HOSHIKO: I would think that the shunt conductance also should include an EMF term. Because depending upon the ionic gradients you may or may not have a finite ionic gradient there, too.

TOSTESON: Every resistance should be written as an array of parallel resistances, one for each charge carrier.

ULLRICH: May I come back to one other problem. This is the effect of the macromolecules or the proteins on the isotonic reabsorption when applied from the outside of the tubule. I just wonder whether the plasma proteins especially albumin which is used here, exerts, besides the colloid osmotic effect, other effects. It may, for instance, be absorbed to the membrane surface and influences the orientation of membrane proteins drastically

(Blasie & Worthington 1969). On the other hand, the albumin may act in the process of exchanging membranes lipids. To illustrate this I may mention experiments Dr. Hasselbach made recently (Fiehn & Hasselbach 1970). He observed that the uptake of calcium into isolated muscular endoplasmatic reticulum was unchanged when he added phospholipase A. But when he further added albumin the uptake of calcium broke down completely. His explanation is that albumin complexes and extracts the free fatty acids which remain after the phospholipase A splitting in the membrane. I suppose that the membranes in the kidney have quite a turnover of fatty acids and that albumin may be involved in it as a carrier.

ORLOFF: But the fatty acids are usually saturated.

ULLRICH: Yes, maybe.

ZADUNAISKY: In the experiments of Dr. Burg that were mentioned, has he tried the effect of ouabain on the active chloride transport?

BERLINER: It knocked it out.

Blasie, J. K. & Worthington, C. R. (1969) Planar liquid-like arrangement of photo-pigment molecules in frog retinal receptor disk membranes. *J. Mol. Biol. 39*, 417–439.
Fiehn, W. & Hasselbalch, W. (1970) The effect of phospholipase A on the calcium transport and the role of unsaturated fatty acids in ATPase activity of Sarcoplasmic vesicles. *Europ. J. Biochem. 13*, 510–518.

Peritubular control of proximal tubular fluid reabsorption

E. E. Windhager

There are currently two major lines of thought which dominate discussions of the hypothetical mechanisms involved in the regulation of proximal tubular salt reabsorption of mammalian kidneys. One concept emphasizes the role of factors acting upon the luminal side of the epithelial wall. It assumes that the glomerular load of salt and water determines primarily and directly the rate of epithelial sodium reabsorption. This hypothesis is supported by numerous studies in which a proportionality bewteen glomerular filtration rate and tubular transport was found. It finds further support in the fact that tubular sodium reabsorption is quantitatively related to the intratubular concentration of sodium ions. This finding alone leaves little doubt that, as a general principle, alterations in the composition and possibly also in quantity of tubular contents can, indeed, influence the rate of proximal tubular salt reabsorption.

The second concept assumes that changes in the peritubular environment can also control epithelial sodium transport. The factors that have been proposed to affect epithelial transport from the vascular side of the tubular wall are humoral, chemical and physical. At present, direct evidence for the existence of a hormone capable of altering sodium reabsorption at the proximal tubular level is still missing. The situation is somewhat more encouraging with respect to chemical factors. Support for this idea comes from several laboratories. Ullrich and his collaborators in Frankfurt (1971) have shown proximal tubular sodium reabsorption to increase in response to a rise of bicarbonate concentration in the peritubular fluid environment. Brandis and Keyes (Brandis *et al.* 1972) in our laboratory have recently obtained evidence that relatively small increments in potassium concentration in peritubular fluid significantly depress proximal tubular fluid reabsorption.

Dept. of Physiology, Cornell Univ. Med. Coll., New York, N.Y. 10021.

»Transport Mechanisms in Epithelia«, Munksgaard, Copenhagen.

Finally, a last factor, which today is generally recognized in most discussions of the regulation of renal salt transport, is the colloid-osmotic and hydrostatic pressure of postglomerular blood.

The basic thought behind this concept is that a radical reduction in capillary uptake must ultimately reduce epithelial reabsorption if infinite accumulation of interstitial fluid is to be avoided. Historically, the concept of peritubular control preceded by far the loading hypothesis. It was Ludwig (1861) who proposed that the peritubular protein concentration is an important factor for fluid absorption by renal tubules. In the post-Cushny era, particularly under the impact of Homer Smith's writings (1937), the pendulum swung in the direction of a primary and exclusive effect of glomerular load. There were a few dissenters, most notable among them Rehberg (1926), who postulated that filtration rate is changed secondarily in response to primary alterations in epithelial transport, but they were definitely in the minority.

The first experimental results about a transport-augmenting action of increased colloid-osmotic pressure were obtained in 1955 by Vogel et al. who worked with doubly-perfused amphibian kidneys. During the following years a number of papers appeared all of which claimed that the colloid-osmotic pressure of postglomerular blood is responsible for proximal tubular reabsorption. Bresler (1956) came to this conclusion on the basis of saline loading experiments, Barger (1956) on the basis of observations in experimental heart failure, Vander et al. (1958) on the basis of the then newly developed stop flow analysis, and Vereerstraeten & Toussaint (1965) from observations during renal arterial infusions of protein (Vereerstraeten et al. 1966).

There was, however, considerable reluctance among renal physiologists to accept this notion. One of the main reasons was the unequivocal demonstrations of active sodium transport (Windhager et al. 1959) and the very low degree of hydraulic conductivity of the proximal tubular wall of *Necturus* kidneys (Whittembury et al. 1959). In fairness, it must be stated that errors in judgment were probably made on both sides of the fence. The one group never came out clearly with the notion that capillary factors might *influence* rather than *be* the driving force for transport. The other, more or less, neglected experimental observations which suggested the possibility of a peritubular control mechanism.

The first investigator who bridged the gap between the two groups was

Earley. He proposed on the basis of clearance experiments that capillary fluid uptake can in some unknown fashion influence net reabsorption of fluid by proximal tubules (Earley et al. 1966). Lewy and I (1968) came to similar conclusions in a micropuncture study of proximal reabsorption during partial occlusion of the renal vein. But to us, even more important than the results of venous clamping, was the chance observation that in control kidneys a direct relationship existed between the filtration fraction of the kidney and the reabsorptive half-times of proximal tubular segments. It was this observation which led us to propose an hypothesis for glomerulo-tubular balance.

Our argument was as follows: both filtered load and peritubular oncotic pressure are altered by changes in the effective resistance of the efferent arterioles of the glomeruli. An increase in GFR is accompanied by increased peritubular oncotic pressure and secondary enhancement of reabsorption. Glomerulo-tubular adjustment in reabsorption would be achieved by the simultaneous and proportionate effects on load – that is GFR – and oncotic or hydrostatic pressure on the vascular side of the epithelium.

This interpretation was just one of several alternatives. A method had to be designed which permits one to measure tubular reabsorption under conditions of selective changes in the peritubular fluid environment. For this purpose an *in vivo* microperfusion of postglomerular capillaries was chosen, a technique which had been simultaneously and independently developed by Rumrich & Ullrich (1968). Two methods of estimating tubular transport can then be used: free-flow recollections or split-droplet experiments. The advantage of this rather cumbersome effort is that any effects of changes in local extracellular environment on tubular transport can obviously not be explained by humoral factors. Not even those which might be produced locally in the juxtaglomerular apparatus of the punctured nephron, because capillary flow is forced from the surface down into the depth of the tissue. This is worth mentioning because in all previous experiments on glomerulo-tubular balance, the action of locally produced intrarenal hormones could never be totally excluded.

We used this technique to test whether changes in colloid osmotic pressure of the capillary perfusion fluid induced by different concentrations of dextran lead to changes in proximal reabsorption. Some of our results, which were obtained by Adrian Spitzer (Spitzer & Windhager 1970) are shown

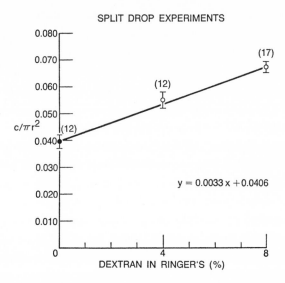

Fig. 1. Reabsorptive capacity of proximal tubular epithelium ($c/\pi r^2$) plotted as a function of dextran concentration in peritubular capillary perfusion fluid (Spitzer & Windhager 1970).

in Fig. 1. The reabsorptive capacity of proximal tubules ($c/\pi r^2$) is plotted as a function of dextran concentration in the peritubular perfusion fluid. An increase in dextran concentration was indeed accompanied by enhanced rates of proximal tubular reabsorption. The average reabsorptive capacity during perfusion of capillaries with colloid-free solutions was approximately one-half of the transport rate found during normal blood perfusion or when 8 g% dextran had been used. This observation deserves some attention, because it clearly shows that factors other than the colloid-osmotic pressure must play a role as a driving force in the reabsorption of fluid across the tubular wall.

Nevertheless, the observed relationship between dextran concentration in capillaries and tubular reabsorptive capacity demonstrated that a major portion of proximal tubular reabsorption of fluid can be influenced by changes in capillary oncotic pressure. On the basis of these experiments Spitzer and I concluded that the absorptive rate changes by 3 picoliters per sec. per mm length of tubule for a change of 10 mm Hg in colloid osmotic pressure of the infused perfusion fluid.

These experiments were repeated by Ullrich et al. (1971). They found a significant influence of dextran on proximal tubular reabsorption, but it was about half as effective as in our study. In other experiments also using peritubular capillary perfusion and split-droplet measurements, Ullrich (1970) and Lowitz et al. (1969) found no difference in reabsorption whether 6 g% albumin was present or absent in the perfusate. However, reabsorption *was* less than normal at 6 g% albumin and the conclusions rested solely on split-droplet experiments and not on free-flow recollections.

On the positive side of the balance sheet was the study of Koch et al. (1968) during acute hypertension, but really the strongest evidence in support of the importance of peritubular oncotic and hydrostatic pressure has come from Brenner et al. (1969). They have demonstrated that reabsorption in proximal tubules can vary directly with peritubular protein concentration when the latter was rapidly altered by bolus infusions of saline, isoncotic or hyperoncotic solutions. Subsequently, Falchuk et al. (1971) investigated glomerulotubular balance during partial constriction of the renal vein, the renal artery, carotid occlusion with vagotomy and the infusion of 6.0 or 15.0 g% albumin. In all of these conditions, changes in capillary colloid osmotic pressure and hydrostatic pressure were in the direction required by the hypothesis that capillary uptake regulates reabsorption in the proximal tubule.

Brenner and Troy (1971) then proceeded with a capillary perfusion study in which peritubular oncotic pressure was altered by changing the albumin concentration of the perfusion fluid. In agreement with Spitzer's findings, they found that absolute and fractional proximal reabsorption declined significantly during perfusion with colloid-free Ringer's solution and remained unchanged from control during perfusion with Ringer's solution to which albumin was added to make final concentrations of 9.0 to 10.0 g%. In the same rats, microperfusion with strongly hyperoncotic solutions containing 15.0 g% albumin, a significant increase in proximal reabsorption was noted. All these changes in reabsorption occurred in the absence of significant changes in glomerular filtration rate.

Considering the evidence that has accumulated over the past years, I am personally quite convinced that peritubular oncotic pressure can influence the rate of proximal reabsorption at least in acute situations. The question now becomes a quantitative one. Spitzer's data would suggest that about 1/3rd of the change in tubular reabsorption, which occurs in response to a drop in GFR to 50% of control, can be accounted for by changes in protein concentration. But this is a rough approximation and most likely under-estimates the quantitative significance. Our perfusion fluid contained no protein, only dextran, and it is known that absence of protein leads to increased leakiness of endothelial structures. Our calculation of the oncotic pressure may have resulted in overestimates if the reflection coefficient of dextran was abnormally low due to the absence of protein, and by distending the perfused capillaries whereby the latter may have lost their barrier function to colloids. Brenner's new results on this matter indicate that the bulk of the inhibition in absolute tubular reabsorption, in response to volume expansion with colloid-free solutions, is causally mediated by the accompanying decline in postglomerular protein concentration (Brenner *et al.* In press).

The second problem arising in connection with the oncotic pressure effect concerns of course the mechanism involved. Several observations make it likely that the compartments interposed between transporting cell membrane and basement membrane of the endothelium play an important role in this control mechanism. Some time ago, it was found that intercellular routes for transepithelial ion movement exist in the frog skin and it was proposed by Ussing & Windhager (1964) that sodium is actively transported into these interspaces. Subsequent studies (Windhager *et al.* 1967) have shown that in proximal tubules of *Necturus,* a significant fraction of ion transfer occurs via extracellular shunts in parallel to the cellular path. It is logical to assign to the intercellular spaces the same importance for osmotically induced water transfer, which the "middle compartment" plays in the model of Curran & MacIntosh (1962).

According to this concept, sodium ions which have entered the cellular transport pool are actively transported into the spaces between basal lamina and adjacent cell membranes. Water follows passively across the nearly semi-permeable cell membrane, thus elevating the hydrostatic pressure in the interspaces. A small hydrostatic pressure gradient leads to filtration across the tubular basement membrane. Oncotic forces may facilitate this process. The reabsorbate is finally moved from the interstitium into the capillary lumen by the net force of hydrostatic and oncotic pressure gradients across the capillary wall.

Lewy & Windhager (1968) considered a number of possibilities by which changes in peritubular capillary uptake might affect the operation of such a model. First, a reduction in capillary absorption could lead to volume expansion or a pressure increase in the intercellular spaces and the interstitium. This might result in widening of intercellular channels, probably at the tight junctions or increased cell membrane permeability. If sodium pumping continues normally, a greater backflux would still result in diminished net reabsorption. Diffusion of sodium might be importantly involved in the movement of sodium ions from the transporting cell membranes to the capillary wall. Volume expansion of the unstirred layer between these two sites, by a reduction in capillary fluid uptake, would increase the diffusional pathway and thus reduce the rate of diffusion. Sustained active sodium transport would lead to sodium concentrations higher than normal within the unmixed layer immediately adjacent to the transporting cell membrane. Back-diffusion into the cell and into the lumen across the tight junction would thus be increased, leading to decreased net movement of sodium and water.

Some of these ideas have since received experimental verification. Thus, Boulpaep (1972) found that intercellular electrical conductivity is markedly increased in *Necturus* undergoing saline diuresis, a condition in which peritubular oncotic pressure is reduced. Analyses of split-droplet experiments in *Necturus* have led Boulpaep to conclude that the permeability of proximal tubules is greatly enhanced, for non-electrolytes as well as for sodium chloride. Lorentz *et al.* (1972) found increased leakiness of the tubular wall to non-electrolytes under conditions of partial occlusion of the renal vein and during partial clamping of the ureter, situations in which peritubular oncotic pressure or hydrostatic pressure are known to change in such a direction as to decrease capillary fluid uptake (Falchuk *et al.* 1971). Similarly, Bank *et al.* (1971) found increased non-electrolyte permeability of the tubules in partial clamping of the renal vein. So it seems that backleakage is indeed increased under conditions of reduced capillary uptake. The work of Imai & Kokko (1972) on isolated proximal tubules leads to the same conclusion.

But there is one observation which does not fit this interpretation: Lorentz *et al.* (1972) found no sign of abnormal leakiness of tubules to non-electrolytes during partial clamping of the renal artery. We would have expected the opposite. Falchuk *et al.* (1971) found a significant decrease in postglomerular protein concentration during partial clamping of the renal artery. Hence, decreased capillary uptake should lead to the same chain of events that leads to increased leakiness during partial clamping of the renal vein, the ureter, which characterizes saline diuresis or mannitol diuresis. We cannot, therefore, completely exclude that renal ischemia has a specific influence on transport which is different from the oncotic effect.

On the other hand, some recent work by Asterita in our laboratory suggests that at least in *Necturus,* the thickness of the peritubular interstitial space is increased when the peritubular capillary blood flow is sluggish. This was an accidental finding in a study aimed primarily at obtaining information about possible changes in the distance between transporting cell membrane and capillary endothelium during saline diuresis. This work is still preliminary, but I would like to describe it here briefly because it deals with a part of our model for peritubular control which has generally been postulated, but not investigated directly.

It concerns the postulate that diminished capillary fluid uptake leads to fluid accumulation on the epithelial side of the capillary wall. If such interstitial edema does indeed occur, the time taken by depolarizing ions such as potassium, to travel from capillary endothelium to peritubular cell membrane should be prolonged. Assuming that fluid accumulation during reduced capillary uptake would involve interstitial spaces on all sides of the peritubular circumference of the epithelial cylinder, measurements of the relative distance between cell membrane and capsule of the kidney would serve the same purpose.

Peritubular membrane potential differences of proximal tubules of *Necturus* kidney were measured by conventional microelectrode techniques. The effect of sudden changes in K-concentration of SO_4 – Ringer's solution superfusing the kidney surface on the time course of changes in membrane potential was then recorded. In controls, the half-time of K-induced depolarization was 4.7 sec. In contrast, the half-time of depolarization in saline expanded animals was increased to 8.3 sec. A similar prolongation of the half-time of depolarization was observed in animals with impaired peritubular blood supply.

These results suggest a significant increase in thickness of the compartments interposed between peritubular membrane and capillary wall under conditions of reduced capillary absorption of fluid. This may have either of two consequences: an increase in hydrostatic pressure in this compartment leading to increased leakiness of the tight junctions or a preponderance of back diffusion of sodium into the lumen across the tight junction, rather than into the capillaries because of the increase in length of this latter pathway. In any event, the postulate of interstitial fluid accumulation as a necessary link in the chain of events between changes in capillary absorption and epithelial net transport apparently holds, at least in the two situations which we have studied so far.

Another question, which is of considerable importance for understanding the overall control mechanism of epithelial ion transport, concerns the site at which colloids exert their osmotic effect. Grantham *et al.* (1972) have recently found an increase in tubular reabsorption as a function of protein concentration in the external fluid surrounding isolated proximal tubules. It appears therefore that one of the critical barriers for protein is the basement lamina of the epithelium. Our own observations lend indirect support to this view.

In perfusion studies of peritubular capillaries, Windhager & Anagnostopoulos (1972) could detect significant fluid transfer across the endothelium when the capillary oncotic pressure was changed by using high-molecular weight dextran. On the average some 10^{-4} ml/ cm² per cm H_2O of colloid osmotic pressure was transferred when this colloid osmotic pressure was exerted by dextran. However, with the same technique and the same calculated oncotic pressure, but now exerted by albumin, the fluid transfer could not be detected. This should not be taken as evidence that albumin exerts no oncotic effect. We know from the perfusion studies of Brenner & Troy (1971) and Brenner *et al.* (In press) that it is at least as effective as dextran in altering fluid reabsorption out of the tubular lumen. The only logical conclusion is that albumin leaks somewhat across the capillary wall, sufficient to decrease the resultant fluid movement across this barrier to become undetectable with our methods of measurements. Its strong influence on epithelial reabsorption can then be explained only if we assume that Starling's principle applies not only to the capillary itself but also to epithelial basement membranes.

REFERENCES

Asterita, M. F., Unpublished Findings.
Baldamus, C. A. (1969) Sodium transport in the proximal tubules and collecting ducts during variation of the sodium concentration in the surrounding interstititium. *Pflügers Arch. ges Physiol. 310,* 354–368.
Bank, N., Yarger, W. E. & Aynedjian, H. S. (1971) A microperfusion study of sucrose movement across the rat proximal tubule during renal vein constriction. *J. clin. Invest. 50,* 294–302.
Barger, A. C. (1956) The pathogenesis of sodium retention in congestive heart failure. *Metabolism 5,* 480–489.

Brandis, M., Keyes, J. & Windhager, E. E. (1972) Potassium-induced inhibition of proximal tubular fluid reabsorption in rats. *Amer. J. Physiol. 222*, 421–427.

Brenner, B. M., Falchuk, K. H., Keimowitz, R. I. & Berliner, R. W. (1969) The relationship between peritubular capillary protein concentration and fluid reabsorption by the renal proximal tubule. *J. clin. Invest. 48*, 1519–1531.

Brenner, B. M. & Troy, J. L. (1971) Postglomerular vascular protein concentration: evidence for a causal role in governing fluid reabsorption and glomerulotubular balance by the renal proximal tubule. *J. clin. Invest. 50*, 336–349.

Brenner, B. M., Troy, J. L. & Daugharty, T. M. A causal relationship between postglomerular vascular protein concentration and fluid reabsorption by the renal proximal tubule of the volume-expanded rat. *J. clin. Invest.* In press.

Boulpaep, E. L. (1972) Permeability changes of the proximal tubule of *Necturus* during saline loading· *Amer. J. Physiol. 222*, 517–531.

Bresler, E. H. (1956) Problem of volume component of body fluid homeostasis. *Amer. J. med. Sci. 232*, 93–97.

Curran, P. F. & MacIntosh, J. R. (1962) A model system for biological water transport. *Nature (Lond.) 193*, 347–348.

Earley, L. E., Martino, J. A. & Friedler, R. M. (1966) Factors affecting sodium reabsorption by proximal tubule as determined during blockade in the rat kidney. *J. clin. Invest. 45*, 1668–1678.

Falchuk, K. H., Brenner, B. M., Tadokoro, M. & Berliner, R. W. (1971) Oncotic and hydrostatitic pressures in peritubular capillaries and fluid reabsorptoin by the proximal tubule. *Amer. J. Physiol. 220*, 1427–1433.

Grantham, J. J., Qualizza, P. B. & Welling, L. W. (1972) Influence of serum proteins on net fluid reabsorption of isolated proximal tubules. *Kidney Int. 2*, 66–75.

Imai, M. & Kokko, J. P. (1972) Effect of peritubular protein concentration on reabsorption of sodium and water in isolated perfused proximal tubules. *J. clin. Invest. 51*, 314–325.

Koch, K. M., Aynedjian, H. S. & Bank, N. (1969) Effect of acute hypertension on sodium reabsorption by the proximal tubule. *J. clin. Invest. 47*, 1696–1709.

Lewy, J. E. & Windhager, E. E. (1968) Peritubular control of proximal tubular fluid reabsorption in the rat kidney. *Amer. J. Physiol. 214*, 943–954.

Lorentz, W. B., Lassiter Jr., W. E. & Gottschalk, C. W. (1972) Renal tubular permeability during increased intrarenal pressure. *J. clin. Invest. 51*, 484–492.

Lowitz, H. D., Stumpe, K. O. & Ochwadt, B. (1969) Micropuncture study of the action of angiotensin–II on tubular sodium and water reabsorption in the rat. *Nephron 6*, 173–187.

Ludwig, C. (1861) *Lehrbuch der Physiologie des Menschen.* C. F. Winter'sche Verlagsbuchhandlung, Leipzig.

Rehberg, P. B. (1926) Studies on kidney function. I. The rate of filtration and reabsorption in the human kidney. *Biochem. J. 20*, 447–482.

Rumrich, G. & Ullrich, K. J. (1968) The minimum requirements for the maintenance of sodium chloride reabsorption in the proximal convolution of the mammalian kidney. *J. Physiol. (Lond.) 197*, 69–70.

Smith, H. W. (1937) *Physiology of the Kidney.* Oxford Univ. Press, London.

Spitzer, A. & Windhager, E. E. (1970) Effect of peritubular oncotic pressure changes on proximal tubular fluid reabsorption. *Amer. J. Physiol. 218*, 1188–1193.

Ullrich, K. J. (1970) Recent advances in nephrology, physiological aspects. *Proc. 4th*

Internat. Congr. Nephrol. Stockholm 1969, vol. 1, pp. 8–19, Karger, Basel/München/New York.

Ullrich, K. J., Radtke, H. W. & Rumrich, G. (1971) The role of bicarbonate and other buffers on isotonic fluid absorption in the proximal convolution of the rat kidney. *Pflügers Arch. qes Physiol. 330,* 149–161.

Ussing, H· H. & Windhager, E. E. (1964) Nature of shunt path and active sodium transport path through frog skin epithelium. *Acta physiol. scand. 61,* 484–504.

Vander, A. J., Maltin, R. L., Wilde, W. S. & Sullivan, L. P. (1958) Re-examination of salt and water retention in congestive heart failure. *Amer. J. Med. 25,* 497–502.

Vereerstraeten, P., de Myttenaere, M. & Lambert, P. P. (1966) Reduction de la natriurese par la perfusion de proteines dans l'artere renale du chien. *Nephron 3,* 103–108.

Vereerstraeten, P. & Toussaint C. (1965) Reduction de la natriurese par la perfusion d'albumine dans la veine porte renale du coq. *Nephron 2,* 355–402.

Vogel, G., Heym, E. & Andersohn, K. (1955) Versuche zur Bedeutung Kolloid-osmotischer Druckdifferenzen für einen passiven Transport Mechanismus in den Nierenkanälchen. *Z. ges. exp. Med. 126,* 485–489.

Whittembury, G., Oken, D. E., Windhager, E. E. & Solomon, A. K. (1959) Single proximal tubules of *Necturus* kidney. IV. Dependence of H_2O movement on osmotic gradients. *Amer. J. Physiol. 1972,* 1121–1127.

Windhager, E. E. & Anagnostopoulos, T. Hydraulic permeability of peritubular capillaries. Symposium on Renal Handling of Sodium. Brestenberg, Switzerland, 1971. Ed. H. Wirz. Karger Publ. Co., 1972, p. 73–74.

Windhager, E. E., Boulpaep, E. L. & Giebisch, G. (1967) Electrophysiological studies on single nephrons. *Proc. 3rd Int. Congr. Nephrol.,* Wash. 1966, vol. 1, pp. 35–47. Karger, Basel, New York.

Windhager, E. E., Whittembury, G., Oken, D. E., Schatzmann, H. J. & Solomon, A. K. (1959) Single proximal tubules of the *Necturus* kidney. III. Dependence of H_2O movement on NaCl concentration. *Amer. J. Physiol. 197,* 313–318.

DISCUSSION

WHITTEMBURY: Which molecular size dextran did you use?

WINDHAGER: Large molecular, dextran 110, and I think the batch we used was about 85,000.

Renal Transport of Urea in Elasmobranchs

B. Schmidt-Nielsen

INTRODUCTION

Urea moves across renal tubules by secretion into the tubules and by passive and active transport out of the tubules (see Table I). The passive movement shows specificity in toad bladders (Leaf 1970) and in the mammalian collecting duct (Truniger & Schmidt-Nielsen 1964, Danielson & Schmidt-Nielsen 1972). The secretory process has been well established in the renal tubules of the frogs *Rana catesbeiana* and *Rana clamitans* (Marshall & Crane 1924, Schmidt-Nielsen & Forster 1954). The secretion process shows saturation kinetics (Forster 1954), and the amount secreted is proportional to the glomerular filtration rate (GFR) (Long 1970). Urea secretion is found in other species of amphibians (Carlisky 1970), in reptiles (Schmidt-Nielsen & Schmidt 1972) and in birds (Stewart *et al.* 1969). However, in most of these instances, back diffusion plays an important role, and urea secretion is only seen during extreme osmotic diuresis.

UREA REABSORPTION AGAINST A CONCENTRATION GRADIENT

The mechanism of urea reabsorption against a concentration gradient is more elusive and difficult to characterize. Transport of this type is found in the elasmobranch renal tubules and in mammalian collecting ducts (Smith 1936, Truniger & Schmidt-Nielsen 1964, Ullrich *et al.* 1967, Danielson *et al.* 1970). Baglioni (1906) first showed that the urea concentration of the urine of elasmobranchs rarely rose above one-third of that of the blood. Smith (1936) pointed out that the elasmobranchs have numerous active glomeruli and excrete filterable molecules such as inulin, xylose and creatinine. Therefore, there is no reason to believe that urea is not filtered in the same concentration as in the blood and subsequently removed from the tubule against

Mount Desert Island Biological Laboratory. Salisbury Cove, Maine 04672 U. S. A.

»Transport Mechanisms in Epithelia«, Munksgaard, Copenhagen.

Table I. Urea transport.

	Passive	Uphill Serosal → mucosal	Uphill Mucosal → Serosal
Tissue	Renal tubules Toad bladders Frog skin Mammalian coll. duct	Amphibia renal tubules Reptile renal tubules Bird renal tubules	Elasmobranch renal tubules Mammalian coll. duct Toad skin
Selectivity	Amide or amide like compounds	Urea and thiourea	Amide or amide like compounds
Saturation	0	K_m: 0.7 mM V_{max}: 300 μM/Kg. hr.	0
Solvent drag	$\sigma = 0.8$	GFR	Na – linked
Inhibitor	Phloretin	DNP Probenecid	Ethacrynic acid Furosomide Epinephrine
Competition	0	Thiourea PAH	0

Passive or downhill movement of urea takes place across the epithelia of renal tubules (Ullrich et al. 1970), toad bladder (Leaf & Hays 1962), frog skin (Ussing 1963), and mammalian collecting duct (Gardner & Maffly 1964). In toad bladder and mammalian collecting duct, the permeability is greatly increased by ADH. A strong selectivity of the membrane for amide-like compounds is observed (Leaf 1970, Truniger & Schmidt-Nielsen 1964, Danielson & Schmidt-Nielsen 1972). The reflection coefficient for urea has been measured in a number of these membranes and has been found to be close to 0.8 (Ullrich et al. 1970, Leaf & Hays 1962, Whittembury 1962). In the toad bladder, phloretin has been found to inhibit urea movement without interfering with water movement (Hays, personal communication). It is not possible to show saturation or competition.

Uphill movement, serosal to mucosal side, is observed is renal tubules of amphibians, reptiles, and birds. In the frog, thiourea is also secreted into the tubules (Schmidt-Nielsen & Shrauger 1963), but amide or amide-like compounds are not. The transport of urea takes place across the basal membrane of the cell and the cells have higher urea concentrations than the plasma. The transport system exhibits saturation kinetics (Forster 1954) with a K_m of about 0.7 μM/kg. hr., but V_{max} varies in direct proportion to GFR (Long 1970). Apparently, urea diffuses passively across the apical border into the tubular lumen. The urea transport and cellular urea accumulation is effectively inhibited by dinitrophenol (DNP) and by probenecid (Forster 1954). Competition for transport is seen with thiourea (Schmidt-Nielsen 1970) and PAH (Forster 1954).

Uphill transport, mucosal to serosal side, is seen in elasmobranch renal tubules, across toad skin but not frog skin (Ussing & Johansen 1969), in the mammalian collecting duct in animals fed a low protein diet (Truniger & Schmidt-Nielsen 1964, Ullrich et al. 1967). In the elasmobranch tubule and mammalian collecting duct, a selectivity for urea, acetamide and methyl urea is seen (Schmidt-Nielsen 1970). Saturation of the transport system has not been observed. In the dogfish, urea transport appears to be linked to sodium transport and is partially inhibited by saluretics and epinephrine.

a chemical gradient. Kempton (1953) working on the smooth dogfish, *Mustelus canis* found that there is no indication of saturation of the urea transport mechanism. When plasma urea concentrations were more than doubled, the rate of reabsorption increased proportionally.

Our micropuncture studies on *Squalus acanthias* confirmed the notion that urea is freely filtered and showed that the urea concentration along the proximal tubule is almost identical to that of the blood (Schmidt-Nielsen *et al.* 1966). Because early collecting duct fluid has a urea concentration and osmolality lower than that of the systemic blood, urea reabsorption must occur in the distal tubule, which is not accessible to micropuncture.

Sodium Linked Transport In Squalus acanthias

My colleagues and I (Schmidt-Nielsen *et al.* 1972) studied renal function in the dogfish *Squalus acanthias* to determine the effects of body fluid expansion and exposure to dilute sea water. The most interesting finding was an extremely constant relationship between tubular urea and sodium reabsorption (Fig. 1). The ratio between moles of urea and moles of sodium reabsorbed was 1.60. This ratio was unaffected by the amounts reabsorbed. It was constant over the range from 50 to 7,000 μM urea reabsorbed per kg body weight per hour.

The question then arises: is the reabsorption ratio related to the concentration ratio to which the apical membrane of the tubular cells are exposed? The answer is no. In the plasma, and thus in the glomerular filtrate, the ratio between urea and sodium concentrations was 1.39 in sea water acclimated fish and 1.29 in 75 % sea water acclimated fish (Table II). In the final urine the average ratio was 0.246 in sea water and 0.81 in 75 % sea water acclimated fish. In spite of these differences, the reabsorption ratio was unaffected by acclimatization to sea water or dilute sea water.

If the transport of urea across the apical membrane were due to solvent drag, we would expect the reabsorption ratio to vary with the concentration of urea in the tubules and with the amount of water accompanying sodium

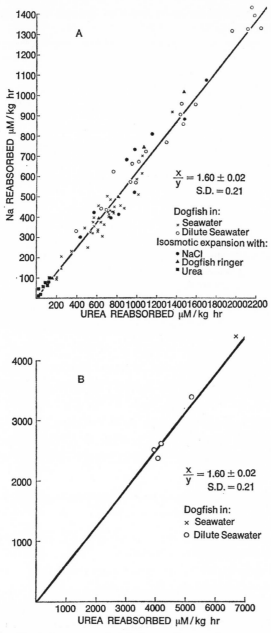

Fig. 1. Relationship between renal tubular sodium and urea reabsorption in the dogfish, Squalus acanthias. (Schmidt-Nielsen *et al.*, 1972). Reabsorption rates given in μMoles per kg body weight per hour. The rates too high to be plotted on the scale used in "A" are plotted in "B".

Table II. Plasma concentrations, water excretion and Na reabsorption in *Squalus acanthias* in sea water and in 75% sea water

	Sea water	75 % Sea water
P_{osm},	981.7 ± 3.9 (67)	861.7 ± 2.8 (81)
	S.D. 31.8	S.D. 25.3
P_{Na}/P_{osm}.	0.266 ± 0.0013 (53)	0.273±0.0015 (61)
	S.D. 0.010	S.D. 0.011
P_{urea}/P_{Na}	1·39±0.014 (68)	1.29 ± 0.014 (64)
	S. D. 0.12	S.D. 0.11
C_{H_2O}/GFR %	7.49 ± 0.81 (34)	20.06 ± 0.97 (19)
	S.D. 4.74	S. D. 4.24
$Q_{Na\ reab.}/GFR.P_{Na}$ %	75.92 ± 1.32 (34)	68.67 ± 1.15 (18)
	S.D. 7.68	S.D. 4.88

Mean value ± S.E.
P<0.001 for all pairs of values.
Figures in parantheses indicate numbers of samples.
(Based on data from Schmidt-Nielsen *et al.* 1972).

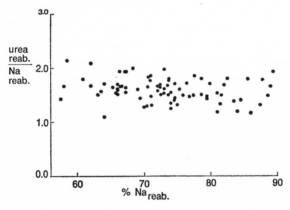

Fig. 2. The ratio of urea to sodium reabsorption versus fractional sodium reabsorption given as per cent of filtered sodium reabsorbed.

movement. From Fig. 2, it can be seen that the reabsorption ratio is independent of fractional sodium reabsorption (% of filtered sodium reabsorbed). The reabsorption ratio is also independent of the free water clearance (C_{H2O}/GFR) (Fig. 3).

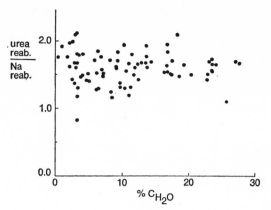

Fig. 3. The ratio of urea to sodium reabsorption versus free water clearance (C_{H_2O}), in per cent of glomerular filtration rate. (Data replotted from Schmidt-Nielsen *et al.* 1972)

Effect of Inhibition of Sodium Transport

From the preceding results it is clear that the major part of the urea reabsorption by the shark tubules occurs in the ratio indicated. What happens if sodium transport across the tubular wall is substantially reduced? During the past two summers, Myers and Murdaugh, Jr. have studied the effects of the saluretics furosomide and ethacrynic acid and of epinephrine upon excretion in *Squalus acanthias* (Myers *et al.* 1971) They have kindly permitted me to use and recalculate their data. The saluretics cause a substantial decrease in the fractional sodium reabsorption. The average fractional sodium reabsorption varied from 0.77 in the control periods to 0.27 in periods when the sharks had received furosomide (Table III).

Table III. Effect of Saluretics and Epinephrine on Na and urea Reabsorption in the Renal tubules of *Squalus acanthias*

	% Na reabsorbed	μM urea reabsorbed/ μM Na reabsorbed
Control (50)	77 ± 0.07	1.43 ± 0.03
Furosomide (20)	27 ± 0.03	3.44 ± 0.38
Ethacrynic acid (17)	40 ± 0.04	1.86 ± 0.18
Epinephrine (24)	71 ± 0.02	1.33 ± 0.04

Calculated from data from Myers *et al.* (1971) Mean values ± S.E. Figures in parantheses indicate numbers of samples.

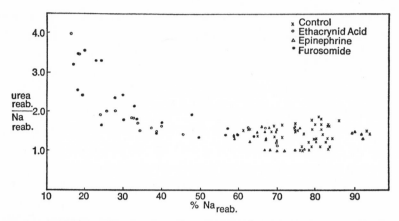

Fig. 4. Inhibition of sodium reabsorption with furosomide, ethacrynic acid and epinephrine. Plotted from data kindly lent me by Myers and Murdaugh (Myers *et al.* 1971). The effect of epinephrine was studied because it had been shown by Forster *et al.* (1972) that epinephrine caused a fifteen fold increase in urea excretion in the dogfish.

When the ratio of urea to sodium reabsorption is plotted against per cent sodium reabsorption (Fig. 4), it is seen that the ratio is independent of percent sodium reabsorbed over a range from 25 % to 94 %. When percent sodium reabsorption falls below 25 %, the urea reabsorption rises sharply relative to the sodium reabsorption, indicating that although the major fraction of the urea reabsorption appears to be linked to sodium reabsorption, a portion is not linked. One may assume that this unlinked urea reabsorption also takes place against a concentration gradient since in all the urine samples, including those where sodium rebsorption was less than 25 % of the amount filtered, the urine urea concentrations were lower than the plasma urea concentrations by at least 100 mM. In one urine sample, where the sodium reabsorption was only 5.8 %, the ratio between urea and sodium reabsorption was 12.3 (not shown in Fig. 4). Even in this sample, the urine urea concentration was more than 100 mM lower than the plasma urea concentration.

DISCUSSION

Two tentative models are presented here to explain the findings. One model involves a passive movement of urea out of the tubule, the other model requires active transport of urea into the tubular cells.

The relationship of the proximal tubule to the terminal segment of the tubule in *Squalus acanthias* has been studied in some detail by Thurau & Acquisto (1969) and in the skate *Raja erinacea* by Deetjen *et al.* (1970). The terminal segment of each nephron, which is the segment in which urea reabsorption most likely takes place (Schmidt-Nielsen *et al.* 1966, Thurau & Acquisto 1969), is intimately surrounded by loops of proximal segments (Fig. 5). Thurau (1970) has proposed that the proximal tubule creates an environment of low urea concentration around the terminal segment, through reabsorption of sodium and water. Micropuncture studies indicated that the terminal segment has a low water permeability (Schmidt-Nielsen *et al.* 1966, Thurau & Acquisto 1969). Thurau (1970) and Boylan (1972) suggested that urea moves passively out of the terminal segment of the nephron into the surrounding space.

This model might explain the relationship of urea reabsorption to sodium reabsorption that we have found, because urea transport in this model would be secondary to sodium transport. It does not readily explain the urea transport which takes place when fractional sodium reabsorption is

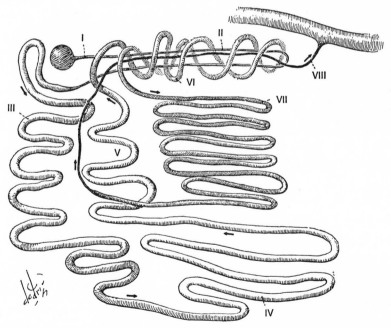

Fig. 5. Schematic drawing of a Skate nephron (Deetjen *et al.* 1970). The nephron was reconstructed from surface microscopy and microdissected segments.

drastically reduced by saluretics. The finding that the reabsorption ratio is independent of the free water clearance can be explained on the basis that the passive reabsorption of water, which regulates the osmotic concentration of the final urine, takes place in the collecting duct or in the ureter. The similarity between the passive urea transport system across the toad bladder epithelium and the uphill transport across epithelium in the elasmobranch tubules is in agreement with this model (Table I). The two systems show the same specificity for amide or amide-like compounds and neither of the systems show saturation kinetics. Against the model is the argument that the elasmobranch tubule is not unique in its peculiar arrangement of the tubule since some invertebrates show similar arrangements of the excretory tubules (Graszynski 1963, Boroffka *et al.* 1970).

Another possible model for the uphill transport of urea in the elasmobranch tubule and in the mammalian collecting duct is that urea moves across the apical membrane into the cell by active transport. This transport mechanism maintains a constant ratio between sodium and urea concentration in the cell. Urea then moves from the cell into the intracellular space following sodium and water movement. This passive part of the urea transport involves a carrier with a specificity for amide or amide-like compounds, similar to that found in the toad bladder (Table I).

The secretion of urea into the cell would then be similar to the secretion of urea into the tubular cells of the frog, in that it involves an uphill movement into the cells. It would differ from the mechanism in the frog renal tubules in that the active movement takes place across the apical rather than across the basal plasma membrane. It would also differ from the transport mechanism in the renal tubules of the frog in that the amount transported is at least 60 to 70 times greater than in the frog. Calculating from clearance data in *Rana clamitans* (Schmidt-Nielsen & Forster 1954) the maximal amount of urea secreted by the frog tubules with the maximal GFR is about 1,000 μM/kg. hr. If an average frog weighs 150 grams, then the amount secreted per frog is 150 μM/hr. The number of glomeruli in the two kidneys of a bullfrog is approximately 5,000, thus, the rate of secretion is 0.03 μM/nephron hr. In the dogfish kidney, the maximal rate of reabsorption was 6,000 μM/kg body weight/hr. With an average body weight of 3 kg. and the number of glomeruli in the two kidneys being 8,800, the amount reabsorbed is 2 μM/nephron/hr. This is almost two magnitudes greater than the secretion rate in the frog renal tubules.

In addition to the rate per nephron, we must consider the fact that in *S. acanthias,* the distal segment, in which the urea reabsorption takes place, is only 0.05–0.15 mm long, while the length of the entire nephron averages 3.3 cm (Ghouse *et al.* 1968). Thus, only a very small part of the nephron is involved in the urea transport. In the frog, on the other hand, the active secretion takes place along the entire proximal and distal tubule. Therefore, unless the transport into the cells in the elasmobranchs is also in some way driven by the sodium pump, it is difficult to imagine a urea pump with such a high capacity.

Presently, we are left with no firm evidence in support of either of the two models.

REFERENCES

Baglioni, S. (1906) Einige Daten zur Kenntnis der quantitativen Zusammensetzung verschiedener Körperflussigkeiten von Seetieren. *Beitr. chem. Physiol. Path. 9,* 50–66.

Boroffka, I., Altner, H. & Haupt, J. (1970) Funktion und Ultrastruktur des Nephridiums von *Hirudo medicinalis.* I. Ort und Mechanismus der Primärharnbildung. *Z. vergl. Physiol. 66,* 421–438.

Boylan, J. W. (1972) A model for passive urea reabsorption in the elasmobranch kidney. *Comp. Biochem. Physiol. 42,* 27–30.

Carlisky, N. J. (1970) Urea excretion and arginase in anuran kidney. In *Urea and the Kidney,* ed. Schmidt-Nielsen, B. & Kerr, D. N. S., pp. 263–271. Excerpta Medica Press, Amsterdam.

Danielson, R. A., Schmidt-Nielsen, B. & Hohberger, C. (1970) Micropuncture study of the regulation of urea excretion by the collecting ducts in rats on high and low protein diets. In *Urea and the Kidney,* ed. Schmidt-Nielsen, B. & Kerr, D. N. S., pp. 375–384. Excerpta Medica Press, Amsterdam.

Danielson, R. A. & Schmidt-Nielsen. B. (1972) Recirculation of urea analogues from collecting ducts of high and low protein fed rats. *Amer. J. Physiol. 223,* 130–137.

Deetjen, P. & Antkowiak, D. (1970) The nephron of the skate, *Raja erinacea. Bull. Mt. Desert Is. Biol. Lab. 10,* 5–7.

Forster, R. P. (1954) Active cellular transport of urea by frog renal tubules. *Amer. J. Physiol. 179,* 372–377.

Forster, R. P. & Berglund, F. (1957) Contrasting inhibitory effects of probenecid on the renal tubular excretion of P-aminohippurate and on the active reabsorption of urea in the dogfish, *Squalus acanthias. J. cell. comp. Physiol. 49,* 281–285.

Forster, R. P., Goldstein, L. & Rosen, J. K. (1972) Intrarenal control of urea reabsorption by renal tubules of the Marine elasmobranch *Squalus achanthias. Comp. Biochem. Physiol. 42A,* 3–12.

Gardner, K. D. & Maffly, R. H. (1964) An *in vitro* demonstration of increased collecting tubular permeability to urea in the presence of vasopressin. *J. clin. Invest. 43,* 1968–1975.

Ghouse, M. A·, Parsa, B., Boylan, J. W. & Brennan, J. C. (1968) The anatomy,

618   B. SCHMIDT-NIELSEN

micro-anatomy, and ultrastructure of the kidney of the dogfish, *Squalus acanthias*. *Bull. Mt. Desert Is. Biol. Lab. 8*, 22–29.

Graszynski, K. (1963) Feinstruktur des Nephridiakanals von *Lumbricus terrestris*. *Zool. Beitrage 8*, 189–296.

Hays, R. M. Personal communication.

Kempton, R. T. (1953) Studies on the elasmobranch kidney. II. Reabsorption of urea by the smooth dogfish, *Mustelus canis. Biol. Bull. 104*, 45–56.

Leaf, A. (1970) Transport of urea across a living membrane. In *Urea and the Kidney*, ed. Schmidt-Nielsen, B. & Kerr, D. N. S., pp. 83–88. Excerpta Medica Press, Amsterdam.

Leaf, A. & Hays, R. M. (1962) Permeability of the isolated toad bladder to solutes and its modification by vasopressin. *J. gen. Physiol. 45*, 921–932.

Long, W. S. (1973) Renal handling of urea in *Rana catesbiana. Am. J. Physiol. 224*, 482–490.

Marshall, E. K. & Crane, M. M. (1924) The secretory function of the renal tubules. *Amer. J. Physiol. 70*, 465–488.

Myers, J. D., Murdaugh, Jr., H. V., Davis, B. B., Blumentals, A., Eichenholz, A., Ragn, M. V. & Murdaugh, E. W. (1971) Effects of diuretic drugs on renal function in *Squalus acanthias. Bull Mt. Desert Is. Biol. Lab. 11*, 71–72.

Schmidt-Nielsen, B. (1970) Urea analogues and tubular transport competition. In *Urea and the Kidney*, ed. Schmidt-Nielsen, B. & Kerr, D. N. S., pp. 252–262. Excerpta Medica Press, Amsterdam.

Schmidt-Nielsen, B. & Forster, R. P. (1954) The effect of dehydration and low temperature on renal function in the bullfrog. *J. cell. comp. Physiol. 44*, 233–246.

Schmidt-Nielsen, B. & Rabinowitz, L. (1964) Methylurea and acetamide: active reabsorption by elasmobranch renal tubules. *Science 146*, 1587–1588.

Schmidt-Nielsen, B. & Schmidt, D. (1973) Renal function of *Sphenodon punctatum. Comp. Biochem. Physiol. 44A*, 121–129.

Schmidt-Nielsen, B. & Shrauger, C. R. (1963) Handling of urea and related compounds by the renal tubules of the frog. *Amer. J. Physiol. 205*, 483–488.

Schmidt-Nielsen, B., Truniger, B. & Rabinowitz, L. (1972) Sodium linked urea transport by the renal tubules of the spiny dogfish, *Squalus acanthias. Comp. Biochem. Physiol 42A*, 13–25.

Schmidt-Nielsen, B., Ullrich, K., Rumrich, G. & Long, W. S. (1966) Micropuncture study of urea movements across the renal tubules of *Squalus acanthias. Bull. Mt. Desert Is. Biol. Lab. 6*, 35.

Smith, H. W. (1936) The retention and physiological role of urea in the elasmobranchii. *Biol. Rev. 11*, 49–82.

Stewart, D. J., Holmes, W. N. & Fletcher, G. (1969) The renal excretion of nitrogenous compounds by the duck *Anas platyrhynchos. J. exp. Biol. 50*, 527–539.

Thurau, K. (1970) Regulation of salt excretion. Address to the 4th Annual Meeting, *Amer. Soc. Nephrol.*, Washington, D. C.

Thurau, K. & Acquisto, P. (1969) Localization of the diluting segment in the dogfish nephron: a micropuncture study. *Bull. Mt. Desert Is. Biol. Lab. 9*, 60–63.

Truniger, B. & Schmidt-Nielsen, B. (1964) Intrarenal distribution of urea and related compounds. – – effects of nitrogen intake. *Amer. J. Physiol. 207*, 971–978.

Ullrich, K. J., Rumrich, G. & Baldamus, C. A. (1970) Mode of urea transport across the mammalian nephron. In *Urea and the Kidney*. ed. Schmidt-Nielsen, B. & Kerr, D. N. S., pp. 175–185. Excerpta Medica Press, Amsterdam.

Ullrich, K. J., Rumrich, G. & Schmidt-Nielsen, B. (1967) Urea transport in the collecting duct of rats in normal and low protein diet. *Pflügers Arch. ges. Physiol.* 295, 147–156.

Ussing, H. H. (1963) Passage of materials across biological membranes. *Proc. First Int. Pharm. Meet. 4,* 15–21.

Ussing, H. H. & Johansen, B. (1969) Anomalous transport os sucrose and urea in toad skin. *Nephron 6,* 317–328.

Whittembury, G. (1962) Action of antidiuretic hormone on the equivalent pore radius at both surfaces os the epithelium of isolated toad skin. *J. gen. Physiol. 46,* 117–130.

DISCUSSION

USSING: It is just a very minor point with respect to the ADH effect on permeability to different substances. Dr. Andersen and I showed several years ago that the thio-urea permeability is increased in the toad skin, although it is not increased in the toad bladder. And, likewise, I mentioned in my introductory talk that in toad skin even glycerol permeability is increased. So I have the vague idea that what the antidiuretic hormone does is really to produce a rather general admittance to whatever specific mechanisms a given cell possesses. That is not in disagreement, I guess, with what you said.

SCHMIDT-NIELSEN: No, it is full agreement with what I said.

BERLINER: Is it not true that the Thurau theory here requires that this membrane be highly permeable to urea but not to water which is a little difficult.

SCHMIDT-NIELSEN: Yes, that is a little difficult, and the free water clearance varies. But of course that could be at another site. But I would think that micropuncture studies have not resolved that problem.

SKADHAUGE: One thinks, of course, whether it could be entrainment of urea in a solute linked water flow, presumably caused by the sodium absorption. And this would do, were it not for the fact that the urea concentration gets so low in the urine. Should it be something like the Ussing anomalous solvent drag, the flux ratio for urea should be just opposite proportional to the concentration ratio in equilibrium. Since these animals in fact end up with a urine urea concentration only $1/10$ of the plasma concentration, this would require a flux ratio of 10, which would need a high linear velocity of the solute linked water flow which would require much longer cells than those present. Although attractive, this model doesn't seem to work quantitatively.

SCHMIDT-NIELSEN: It works on this level, only when the concentration of the cell is maintained high.

SKADHAUGE: Yes, if you also allow an active, uphill, urea transport into the cells, then it could work.

WHITTEMBURY: Would there be any possibility that urea goes between cells. Also do you know the cellular concentration of urea?

SCHMIDT-NIELSEN: No. I only know an average of the kidneys. We cannot get this part of the tubule out for analysis. Maybe we should try. I think it would be difficult to tease it apart, before you had lost the functioning urea concentration.

ZADUNAISKY: What is the situation for synthesis of urea in the amphibian kidney?

SCHMIDT-NIELSEN: There is very little synthesis. We have shown previously with C_{14} urea that the secretion and the intracellular concentration cannot be accounted for by synthesis at all.

ZADUNAISKY: In the elasmobranchs?

SCHMIDT-NIELSEN: I don't know.

ZADUNAISKY: But it is not enough in the amphibian.

SCHMIDT-NIELSEN: No, not at all. You can induce urea synthesis in the frog by giving arginine, and if you then block the urea transport with 2,4 dinitrophenol you get an apparent secretion of urea which has nothing to do with the secretory mechanism. Normally the plasma arginine concentration is too low to account for more than a few per cent (Schmidt-Nielsen & Shrauger 1963).

Schmidt-Nielsen, B. & Shrauger, C. R. (1963) Handling of urea and related compounds by the renal tubules of the frog. *Amer. J. Physiol. 205*, 483–488.

Alphabetic Author Index

Subject Index

Subject Index

For entries with multiple page references, the main reference is indicated by the use of italic-face number.